Machines That Learn

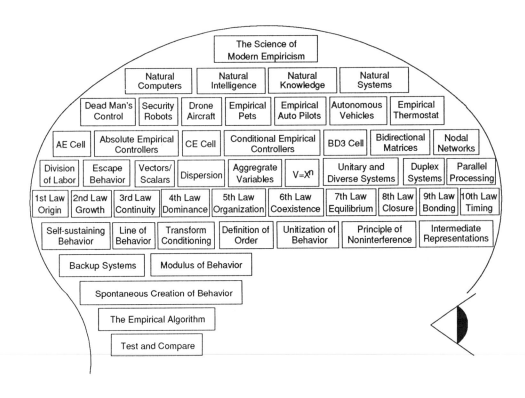

Machines That Learn

Based on the Principles of Empirical Control

Robert Alan Brown

New York   Oxford
OXFORD UNIVERSITY PRESS
1994

Oxford University Press

Oxford   New York   Toronto
Delhi   Bombay   Calcutta   Madras   Karachi
Kuala Lumpur   Singapore   Hong Kong   Tokyo
Nairobi   Dar es Salaam   Cape Town
Melbourne   Auckland   Madrid

and associated companies in
Berlin   Ibadan

Library of Congress Cataloging-in-Publication Data
Brown, Robert Alan, 1938-
Machines that learn: based on the principles of empirical control/
Robert Alan Brown
p. cm. Includes index.
ISBN 0-19-506966-8
1. Machine learning. 2. Artificial intelligence.
3. Neural computers—Design and construction.
I. Title. Q325.5.B76 1993 006.3'1—dc20 93-23985

Printed from a camera ready manuscript
supplied by the author.

2 4 6 8 9 7 5 3 1

Printed in the United States of America
on acid-free paper

In memory of my father, Harold James Brown,
for whom engineering was a way of life.

Preface

The purpose of this book is to describe the design of a natural computer that learns and acts like the animal brain. The subject is approached from an engineering point of view. That is to say it answers the question: How does one go about building a device that learns to create and execute useful behavior by itself? The book attempts to answer this question in concrete and measurable engineering terms.

While designing a robot or any other automatic machine, the development of the control portion of the system is usually the last phase of the project. This is because it is easier to design a control system when the designer knows exactly what to control. The author is faced with the opposite situation. He is faced with designing a control system that must deal with many different tasks. Without a specific application, it was necessary to establish concepts and principles that support many different applications.

Writing this book forced the author to organize his ideas into a logical sequence or string of thoughts so that the correct new idea appears as the reader pulls the string. The author also tried to establish and hold to a central concept (machines that learn) and the outer limits of the study (living systems).

The goals of the study of empirical control are presented in Part I in terms of a machine that creates and maintains successful behavior through an empirical process of doing more of what it can do and less of what it cannot do. The elements of machine behavior and control behavior are presented to introduce the concept of a controlled machine.

The designs of absolute and conditional predetermined electromechanical controlled machines are presented in Part II along with a detailed definition and explanation of the behavior of controlled machines in terms of value, variable, state, line of behavior, variety, and vector and scalar quantities. The designs of multivariable duplex units are presented. Digitized units are shown to be able to deal effectively with multiple variables requiring a high resolution.

The designs of absolute and conditional empirical controllers are presented in Part III. These control units gather information from their sensors and actuators about what actions can and cannot be carried out in given sensed conditions. They use this information to produce doable behavior. This behavior is likely to endure because it reflects a knowledge of the restrictions imposed by its environment and its operators. The design of a duplex empirical controller is presented having a bidirectional output matrix that sends a signal to the input matrix representing value of the intermediate variable that is most likely to produce the actual output state.

Different networks are presented in Part IV that allow empirical matrices to organize themselves by attempting to resolve questions of competition, alliance, dominance, and separation through the empirical process to form compatible systems that can deal effectively with multivariable, diverse, and unitary environmental tasks. These units can be connected directly in series by virtue of the design of a universal bidirectional duplex controller.

Applications for empirical systems are explained in Part V, starting from the simplest and moving to the most complex systems. The behavior of these systems is based upon the ten laws of empirical behavior, which deal with the origin, change, continuity, force, organization, coexistence, equilibrium, closure, bonding, and timing of behavior units. Information and order are defined and are related to empirical machines. Questions about the animal brain also are raised as part of this study of empirical control and the new science of modern empiricism.

Some readers may want to concentrate their reading in one part or another according to their interest. However, the book is "built" by developing each idea from previous ideas, so the author recommends the reader start at the beginning and work through each part. The book's basic building blocks are the indexed terms, which are placed in *italics* close to where they first appear. A summary of the main ideas in each section is presented at the beginning of each section to help "speed read" the book. Illustrations such as drawings of machines, flow diagrams, and circuit designs are used whenever a picture contributes to the clarity of an idea. Each figure contains a caption that defines the object in the illustration, and both are placed within paragraphs that provide more information about the object in the figure. At each stage, the author tries to show the exact design and operation of a working system — what it can do and what its limitations are. Each section deals with the problems raised in the preceding one. Examples of useful applications are given at each stage and are summarized in the last part of the book covering the behavior of empirical machines.

This book attempts to show the similarity between living and nonliving systems. Though the book has been prepared from an engineering point of view, it draws upon many subjects in the behavioral and social sciences. It attempts to build a strong case for the existence of behavioral bonding through the development of cooperative behavior in groups of empirical machines. A comparison is made between large human systems such as an economy and networks of empirical controllers.

The empirical controller uses feedback, but not like a typical regulator. An empirical controller attempts to produce action. It measures and remembers what it actually does. Its goal, although it is not aware of it, is to learn what it can do within given conditions. It keeps changing until it finds behavior that it can carry out successfully. This empirical algorithm leads to many unexpected results.

This book is written for those interested in living beings and the design of machines that perform better the longer they work in contrast to most machines that tend to wear out and break down with use. These machines are made up of simple components like switches and relays. Detailed circuits allow the reader to trace the path of electrons around a completely functioning natural computer. These circuits define the logical operations in this computer. The hardware used to describe this machine is not intended to be state-of-the-art electronics. However, the circuit designs can be readily transformed into integrated circuits (ICs) by those skilled in the art.

The theoretical and philosophical discussions are also intended to form the basis of the study of a modern empiricism, which attempts to contribute to an understanding of the design and operation of all natural systems that learn from experience, including the animal brain and other empirical systems in the study of sociology, linguistics, biology, and economics.

Mattapoisett, Mass.   R.A.B.
May 1993

Contents

Contents

Contents

Contents

Contents

PART I
GOALS IN THE STUDY OF
EMPIRICAL CONTROL

Part I sets the goals of the study of empirical control, explains new applications using empirical control systems, and establishes the definitions of machine behavior, control behavior, and empirical behavior.

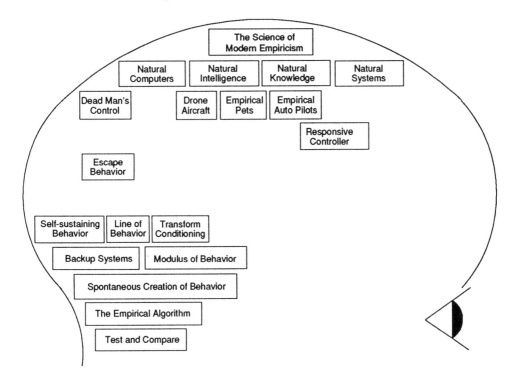

Chapter 1
Opportunities in Empirical Control

An explorer or navigator must start at a known position to move with assurance into the unknown. We also will start a study of Machines That Learn at a point that is familiar to us, and then proceed into a world of new and useful ideas about control systems that learn and act like living beings.

Successful behavior and existence

We are all engaged in the search for self-sustaining behavior. When we see how autonomous empirical machines establish self-sustaining behavior, we may learn something about our own search for successful behavior.

Control behavior

We sense conditions in our environment every moment we are awake, and we produce actions that allow us to meet our needs and goals under these conditions. For example, when we drive a car, we sense where we are and then steer our car in the direction we wish to go from that point. This process is represented by Figure 1.1.

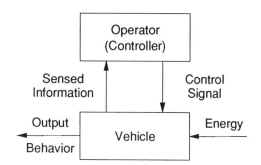

Figure 1.1. An *automobile motion control system* consists of a driver, the automobile, and a signal channel to and from the driver and the automobile.

The driver senses the relation of the car to the road and acts upon the steering wheel and pedals to keep the car moving toward the driver's destination. The set of actions produced by the driver for each sensed condition is defined as *control behavior*.

Useful behavior

Pilots, helmsmen, and truck drivers earn money through their control behavior. The more useful this behavior is to people, the more money people are willing to pay for the control behavior. This behavior can be useful because it helps people meet real needs. The purpose of this book is to describe machines that can learn to produce useful control behavior.

The problem

A truck driver falls asleep behind the wheel of his eighteen wheeler and careens into a ditch, or worse, swerves into a lane of oncoming traffic. The control signal from the remote pilot to a drone aircraft is cut off or intentionally jammed by radio interference. It banks and crashes. An individual is cut off from the people with whom he or she has shared many experiences and responsibilities. This individual now feels isolated and lonely. These are real problems that occur every day. What can be done to help these people deal with interruptions in the usual flow of usefull control information?

Interferences in behavior

In these examples, something interferes with a person's normal behavior. These interferences (Fig. 1.2) may not allow people to complete their tasks.

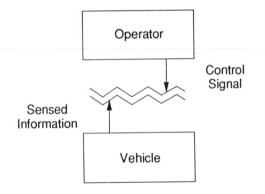

Figure 1.2. In an *interrupted vehicle control system*, an interference in the control signals to and from the operator will disrupt the flow of desired results produced by the control system.

Running off the road does not allow the driver to complete a delivery. Crashing does not allow the pilot of the drone airplane to carry out a mission successfully. A person may find it difficult to live alone.

The solution

One solution to these problems is to have a second person help each individual complete each task. For example, a co-driver can be placed in the truck who has access to the same set of controls as the driver. If the driver falls asleep, the co-driver can keep the truck on the road. A pilot can be placed in the remote-controlled aircraft to take over the flight of the aircraft if the control link is broken. A person can be hired to be a companion to the individual who is alone. The presence of the companion restores some social behavior.

Cooperator

In the examples given, the *cooperator* may supplement the behavior of the operator, and provoke the operator into action. In general, the operator and cooperator must want the same results (Fig. 1.3).

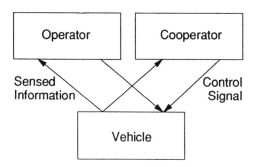

Figure 1.3. In a *cooperative system*, a cooperator can act alongside the primary operator if both produce the same kind of behavior.

It may be difficult to find a person to do these tasks at a reasonable cost. It may be too dangerous to place a human being in some situations. It would be useful to have a device that acts like a person and produces the same kind of behavior as the human operator when the need arises.

Cooperating controller

A *cooperating controller* can be put in the cab of a truck to monitor the behavior of the driver and the truck and learn what the driver does under different driving conditions (Fig. 1.4). It can then produce that behavior if the driver withdraws from the task.

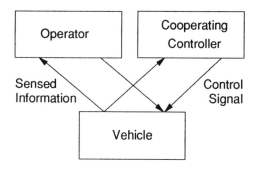

Figure 1.4. A *cooperating controller* may produce control behavior that the operator is unable to supply.

This type of cooperating controller also can be put in the remote controlled aircraft. It can monitor the instrument in the craft and the control signals to the craft. It can produce the same kind of behavior provided by the remote pilot if the remote signal is interrupted. This type of controller also can be used to operate a small personal robot. It can be taught to retrieve personal items that an individual may need, such as a radio, telephone, reading and writing materials, and grooming items. The robot can be taught to come when summoned and to greeting visitors. It may help an individual feel that he or she is no longer in an empty and unresponsive environment.

Backup controller

The truck driver may teach the cooperating controller an "emergency stop" procedure. If the driver abandons the driving task by falling asleep, the *backup controller* (Fig. 1.5) can assume control and stop the truck.

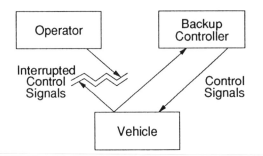

Figure 1.5. A *backup controller* can keep the system running even if the control signals from the operator are interrupted and/or the operator is removed from the task.

If the control signals from the pilot of the remote vehicles are cut off, the backup controller in the craft can take over the control function and complete the task of bringing the vehicle back to home base. The personal robot can answer the door, take messages, and act as a guard when other people are not present. Standing in their place, a backup controller can respond quickly and appropriately when people are not available. These systems must know what to do. They must be ready to act. They must not be costly, and they must not be difficult to teach or program.

Successful behavior

A specific kind of behavior is needed from the backup controllers to help these individuals complete tasks successfully. The controller must know how to deal with each interference in behavior. At any given moment, the controller must know what successful behavior is required to keep the system going. Thus, any action or sequence of actions that results in the completion of activities initiated by a system is considered *successful behavior*. Our main concern will be the study of successful behavior — what it is, how it can be learned, and how it can be produced by inanimate controllers. This successful behavior can come from any source. It can be supplied by a friend, a copilot, or a computer. It can act in place of the behavior that an operator, or can have an identity of its own. We will assume that some combination of useful behavior will allow tasks to be completed successfully. Creating this successful behavior is the major concern of all organizations, individuals, and machine systems.

Physical and behavioral existence

In some cases, the failure of a system to develop successful behavior leads to the end of the *physical existence* of the system. The truck driver may be killed because the driver fell asleep behind the wheel. The truck may be destroyed. The remote controlled vehicles may be lost or destroyed. However, interferences in behavior often result in the termination of a task — not the termination of the physical existence of the controlled system.

In most cases, the failure to develop successful behavior does not terminate the physical existence of a system. The system may continue to exist, but a particular pattern of behavior may end if that system fails to complete a task. For instance, a truck driver may lose his or her job if his or her license is revoked by having a serious accident. However, the individual will continue to exist physically though he or she may have to do something else for a living. In this case, the driving behavior and the income produced by the driving behavior are lost because the driver failed to complete his or her task successfully.

Usually, the most valuable part of a system is what it does. For instance, a truck can produce revenue only when it is driven successfully. If for some reason the truck cannot be driven, the revenue-producing value of the truck and its cargo is lost until it can be driven again. For that reason, we will be concerned primarily with the *continued existence of particular patterns of behavior* rather than the survival or loss of the physical part of a system. Thus, we will focus upon the *continued behavioral existence* of machines and living beings primarily rather than their continued physical existence.

Self-sustaining behavior

The behavior of a system is more likely to continue if it is has a set of actions that can be carried out successfully for a wide range of possible interferences. Backup controllers can strengthen and preserve the operation of a system by helping it operate over a wider range of interferences. Since they allow a system to operate without the infusion of new behavior from an outside source, they contribute to the *self-sustaining behavior* of the system.

What device can perform as if it were the copilot of an airplane? How can a device figure out how a pilot is flying and then take over the task of flying? How can a robot controller produce behavior that will draw another person into an interactive relationship? How is it designed? How does it work? What can it do? What is the relationship between human beings and this device? Can we learn anything about our own search for self-sustaining behavior by studying these devices? The purpose of this book is to describe a machine that learns from experience and helps people achieve successful behavior.

Natural systems

Many systems in nature develop self-sustaining behavior. They do so without a complex predetermined master plan. Instead, they conduct simple tests upon the environment under actual conditions and measure the results. The results are used to mold the future behavior of these systems. In many cases, the behavior of these natural systems can be more effective than behavior that is established beforehand. What are these natural systems and how do they establish and maintain their existence?

Biological evolution

Biological evolution is often cited as an example of a *natural system* that develops seemingly desirable characteristics without an outside source of intelligence. The process of evolution is sometimes called "the survival of the fittest." Organic species living in the wild are responsible for obtaining their energy and other nutritional resources, maintaining their health, protecting themselves from injury, preserving a secure environment, and propagating their species. They may have to fight for their lives at times, and must compete with other individuals for physical survival.

Evolutionary changes occur in species over many generations. These changes are based upon the following rule: Those inheritable characteristics that contribute to an increase in the reproduction rate of the set of individuals become more widespread than those characteristics that do not contribute to or diminish the reproduction rate. The result of this process is that the characteristics of species change, species disappear, and new species appear that seem to meet the need to survive in their particular environment more effectively.

Many other natural systems evolve, grow, or develop by themselves without a conscious effort by someone or something to make them the way they are. Examples of these natural systems are: economic and some business systems, language, a unique combination of organisms in a region known as an ecosystem, animal groups, human culture, language, and the animal brain.

Originative systems

The remarkable feature of evolution and other natural systems is that they come up with unique designs and ways of behaving. They do not depend upon someone or something to make them what they are. Nor is there a predetermined pattern, given set of alternatives, or models established for them to follow. These natural systems establish their physical and behavioral characteristics by a *trial and error process*. They try a possible solution to a problem and then see if it works. If it works, they use it again. If it fails to work, they discard that solution and try something else. Thus, all natural systems "try something out", as shown in the following examples:

- The brain produces ideas and behavior, and then determines what works and what does not.
- In our economic system, businesses with specific charters and goals attempt to operate with a profit. Some succeed and some fail based upon the effectiveness of their way of doing business.
- Individuals make financial investments based upon their "investment philosophy." Some investments increase the assets of the individual while others decrease the assets of the individual.
- Social groups try out or commit themselves to specific laws and codes of conduct. The more these values are supported by the members of the group, the more influence these values are likely to have and the more likely they are to be preserved.
- Biology commits itself to specific species. The population of some species increases, while the population of other species decreases. Some species may become extinct and lose the ability to influence evolution.

A key element of these natural systems is that they originate their trials. Each system decides what the trial will be and when it will be carried out. These *originative trials* may be selected at random. Subsequent trials may be chosen according to the results of earlier trials. In all natural systems, *the selection of future trials is based upon the degree of success of previous trials*. This is exactly the kind of system needed to create self-sustaining behavior.

Trial and incorporation of success

Instead of "trial and error," a more appropriate term for this process of evolution is *trial and incorporation of success,* because these natural systems learn from their successes and disregard their failures, as shown in the following examples:

- The genetic makeup of a successful species is determined by those organisms that can survive and reproduce. Extinct species can no longer contribute to the process of evolution. However, the surviving species provide the gene pool for future populations and provide the gene pool from which new species evolve.
- The marketplace is made up of successful businesses. They form the basis of future changes in the marketplace. Failed businesses quickly disappear from the marketplace and have little or no influence upon future business activity.
- A failed investor can no longer invest, so cannot influence the investment market through his or her sales or purchases.
- A living society can decide to change its institutions, whereas the institutions of an extinct society remain forever unchanged.

In the same way, successful individuals tend to build new behavior upon their successful behavior and attempt to leave their unsuccessful behavior behind.

The prerogative of nature

The ultimate arbiter of the trials of natural systems is Nature herself. The present view of most people involved in science is that all systems must obey the principles and laws of nature, whether they are living or artificial, mechanical or electrical, chemical or computational. Examples of the inherent relations in nature are shown in Table 1.1.

Table 1.1. Inherent relations in nature

- The definition of matter, which says that two or more objects cannot occupy the same space at the same time.
- A corollary that says that an object cannot be in two or more different places at the same time.
- Another corollary that says that an organism cannot carry out two different tasks at the same time if each task has to be done at a different location.
- Another corollary that says that an object cannot move in two different directions at the same time.
- The Mechanistic Assumption that says the following: If something happens in a given way under given circumstances at one time, it will happen the same way at another time under the same circumstances.
- The law of acceleration, which states that there cannot be instantaneous changes in the velocity of objects that have mass.
- The law of the equilibrium of forces that states that at every place where a force acts upon a body, there is an equal force in the opposite direction acting upon that body.
- A principle involving change that says that all objects and living beings are either in a process of coming together, falling apart, or staying the same. If they are coming together as fast as they are falling apart, they may be changing into something else.
- The definition of a variable that says that a variable can have only one value at a given time.
- An assumption about time that says that time cannot stand still or go backward.
- A characteristic of information that says that two or more pieces of information cannot occupy the same space at the same time like matter, but the velocity of information can change instantaneously. Thus information is like matter but is intrinsically massless.

The list can go on for as long as one wishes to think about science. We will accept the predominance of nature's laws over all living and nonliving systems. We will assume that the laws of nature will determine the outcome of the trials of natural systems. We will show that these *natural relations* mold the behavior of *natural empirical systems* and determine how they grow and develop. We will consider what happens when people determine the outcome of trials of *synthetic empirical systems* and see how people can influence the actions of these systems to create behavior they consider to be desirable.

Natural relations

Most animals do not know or are not aware of the laws of nature. However, they still function effectively in their environment. What do they need to know to carry out successful self-sustaining behavior and maintain their physical existence? Natural systems must know something about the specific details in their environment that are essential to their success. For instance:

- To succeed, a *business* must know what specific goods or services will sell and what will not sell.
- The *investor* must know the difference of specific sound investments and poor investments with little or no return.
- The specific physical properties of a *species* must reflect what is possible and practical within the features of it surroundings and compatible with the laws of nature.

These natural systems must learn about the specific *natural relations* or *heuristics* that are necessary to their survival. This knowledge may be very incomplete and inaccurate from a scientific point of view. The business person may think that the color or shape of his or her product is what makes it sell when in fact there are other properties that appeal to the customer. If the pertinent properties are retained, the misconception may not harm sales.

Natural information

We may not be aware of the real underlying physical properties behind our behavior. For example, we can ride a bicycle without knowing the physics involved in the balancing process. We may forget the underlying principles involved in some behavior, yet still produce that behavior. For example, we may learn to hit a tennis ball with top spin so the ball will fly down into the court across the net, but may not think about top spin when we hit the ball later. We take many other relations for granted. For example, we may reach for a handrail when we come to stairs. We may *act as if we know* the underlying properties within our environment though we may not be consciously aware of them. If this body of *natural information* in an environment is discovered and then incorporated in our behavior, it may allow us to operate more successfully in that environment.

Natural knowledge

This *natural knowledge* is not what we learn in school or from books. It is not theory or mathematical relations. Natural knowledge is embodied in the set of seemingly disconnected bits of action (heuristics) that allow us to navigate successfully in our environment and carry out actions without interfering with other objects and other living beings. It is the knowledge of what can and cannot be done. It is the practical knowledge needed to keep things running, to complete tasks, to create successful behavior. This natural knowledge seldom violates the laws of nature, and successful behavior reflects a knowledge of the natural information in the environment.

Natural intelligence

A natural system or individual that behaves successfully, appears to understand the requirements and information in its natural environment, exhibits behavior that keeps it within that environment, and has acquired this ability through its own experience with its natural surroundings is said to possess *natural intelligence.* Natural intelligence differs from *artificial intelligence* in that artificial intelligence is created by an *outside* source such as an "expert" programmer, whereas natural intelligence is created by behavior that originates from *within* an individual system.

Thus, natural intelligence is the result of a system's attempt to *create its own* behavior by finding out what can and cannot be done within the constraints of a natural computer/environment system rather than having its behavior predetermined by a programmer. In some cases, we may not know what a natural computer should do to establish its existence. In many cases, a natural computer may be better able to establish successful behavior by itself rather than being programmed by an outside source. The differences between natural and artificial intelligence are shown in Table 1.2.

Table 1.2. Natural intelligence

- Natural intelligence may be created without any intelligent contribution from a human source.
- A natural system may develop a new approach to problems that is different from any human sources.
- A system with natural intelligence may operate on the basis of its own knowledge and misconceptions rather than the knowledge and misconceptions of a human source.

Thus, natural intelligence is the set of self-discovered and self-proven understandings about relations in the environment that results in the successful behavior of an individual, organism, mechanism, or another *nonliving being.*

Indicators of success

After a period of trial, error, and success, some natural systems may become stronger as they discover and carry out behavior that allows them to exist within the natural constraints of their surroundings. Examples of these *indicators of success* are as follows:

- Businesses that offer desirable goods and services that customers want and are willing to pay for end up with a *larger customer base* and more money to invest in their business.
- Individuals who make wise investments end up with *more money* to reinvest. Those who are wrong end up with less money to invest.
- As stated, the *population level* of a successful species may grow in number. This increases the number of members of that species that can be produced each year. This provides a higher resistance to the effects of death-causing factors such as disease and predation.

In most natural systems, success begets success. This potentially explosive chain reaction allows those experiments that succeed to prevail.

Oversuccess

If the indicators of success are excessive, they may lead to the failure of a system, as shown in the following examples:

- A business that makes too much money may become lax in its responsiveness to its customers while its high profit margins may attract competitors who may be more attentive to the needs of the market.
- An investor who makes a lot of money may not work as hard as before to secure good future investments.
- A successful species may overrun its food supply and/or environmental resources.

Oversuccess can be a great threat to any natural system. We are becoming more concerned about the ever-increasing population level of our species that threatens our own well being, the well being of other life, and the stability of the environment of our planet.

Continued existence of successful behavior

Successful behavior contributes to survival or *continued existence* of natural systems because it permits them to complete whatever behavior they have selected. To complete tasks, successful behavior must meet the three requirements in Table 1.3.

Table 1.3. Requirements for the continued existence of behavior

---

1. Successful behavior must reflect a knowledge of the natural relations in its environment.
2. Successful behavior must include the specific behavior needed to produce additional successful behavior.
3. Successful behavior must cause a system to stay within the surroundings in which this behavior is successful.

Some behavior that can be carried out leads to behavior that cannot be carried out. For example, a mobile robot can move toward a wall, but is stopped once when it arrives at a wall. Behavior that works well in one environment may not work well in another environment, and a species that adapts superbly to one environment may be decimated when placed in another environment. For example, fish have evolved to survive under water, but perish quickly when placed on land.

Self-perpetuating behavior

However, some behavior may increase the likelihood that it can continue to occur. For example, learning to swim successfully does two things: it may save our life someday, and we may choose to go swimming more often. Thus, learning to swim may sustain our *physical existence* and cause us to continue to produce that swimming behavior. So the behavior we learn that can be carried out becomes the source of our future behavior. It fortifies our physical and *behavioral existence*.

Thus, successful behavior may uphold and perpetuate itself. We sometimes view self-serving or *self-perpetuating behavior* as undesirable. Some of us admire individuals who are willing to throw themselves into new settings, and have disdain for those individuals who hold onto the old ways of doing things. However, the individual who learns behavior that can be used again may be more successful in carrying out behavior than the individual who must learn something new at every turn.

Spontaneous creation of self-sustaining behavior

The behavior of most living systems develops by a process of coming up with some kind of behavior and then continuing to do those things that work. This process can be described as the *spontaneous creation of self-sustaining behavior*. It is based upon the ability of an individual to originate behavior that allows the individual to stay within the boundaries of a known environment and contributes to the physical and behavioral existence of the individual. This process of spontaneously creating and using successful behavior is the main concern of the study of empirical control.

Natural computers

This spontaneous creation of self-sustaining behavior can be carried out by a computer-like device that keeps track of the success or failure of its experiments, and tries out new behavior based upon this record of natural information. What are the features of this natural computer? These natural systems contain *natural computers* that carry out the three functions in Table 1.4.

Table 1.4. The three functions of a natural computer

---

1. A natural computer makes its own selection about what actions it will initiate.
2. A natural computer keeps track of which experiments succeed or fail.
3. A natural computer uses this information to re-select its most successful trials in the future.

The natural computer in a mail order business may be a list of customers and their accounts. A business with a list of customers who buy often may earn more money than a business whose customers seldom buy. Each mailing is a trial that can be used to establish the degree of confidence the business has in each customer. The results of each trial can be used to determine which customers are sent catalogues in the future. Thus, each trial is used to influence future actions.

An investor may be considered a natural computer. Investors who make wise investments end up with more money. Their account balances are a measure of the success of their investments. Successful investors are more likely to invest again in their successful sources of income. A species may be considered a natural computer. Successful species beget more of their kind. The population of unsuccessful species grows smaller. The basis of the natural computer in biological evolution is the number of reproducing pairs. The number of reproducing pairs determines what animals will be produced in the future. Note that information as to the number of nonreproducing pairs is quickly lost.

Individuality

Most natural computers exhibit the unique properties of an individual, as shown in Table 1.5.

Table 1.5. Properties of an individual

- Individuals can initiate action.
- Individuals have an identity and existence of their own.
- Individuals flourish, wane, and sometimes disappear because of their capacity to change themselves according to the circumstances around them.
- Individuals act according to their internal decision-making process.
- Individuals usually can act alone without outside help or external controls.

Natural computers embody the elements of *individuality* in that they come into existence as a unique collection of parts, they can initiate action, they can be self-governing, and their continued existence is based upon how successfully they act.

Empirical systems

Much of our ability comes about through practice and experience. The value of experience is demonstrated by comparing our first attempt to ride a bicycle or play tennis to our performance after years of training and practice. Why is it necessary to experience events more than once to develop these skills? And is it possible for a machine to improve its performance as it continues to operate?

Empiricism

In philosophy, Empiricism is the theory that sensory experience is the only source of knowledge. The early Empiricists claimed that this knowledge is developed *a posteriori*, which means after the individual is created. They asserted that knowledge comes about because of the experience of a being within the physical world. This was a revolutionary idea put forth in the seventeenth century by Locke, Hume, and Kant in a period in which many believed that we were born with an *a priori* (built-in) religious and practical knowledge and the innate ability to reason like our built-in ability to see, hear, and move about.

Today we accept the idea that we learn through experience. We expect that we must go through a learning process, including drills and practice, to develop skills. The question that will be addressed is: How does one go about designing and building a machine that can develop useful behavior through experience *a posteriori* (after the machine is completely built and put into use)?

Empiricism is sometimes used in a negative sense when applied to those who rely solely upon practical experience without regard to scientific principles, as in "an empirical remedy" in medicine. However, empirical or natural knowledge is valuable to machines because it contains the natural information about what will and will not work within a given environment.

The empirical process

It is often difficult for someone to know beforehand what natural information is necessary for a given device to operate successfully. A more effective procedure may be to let a machine try to operate in a given environment and then see what problems it encounters through a process of *trial and incorporation of success*. So, we will describe how a machine can discover the natural information in its surroundings, by itself and put this knowledge to use to create successful behavior. This process of experimentation, observation, and incorporation of successful behavior is called the *empirical process*.

Another possible meaning of the term empirical process applies to the *scientific method* that refers to the process shown in Table 1.6.

Table 1.6. The scientific method of investigation

---

1. Create an hypothesis usually with an equation or theorem,
2. Test the hypothesis with experiments,
3. Revise the hypothesis based upon the empirically derived results.

This empirical process is closely related to the process used by an empirical machine except that the hypotheses of the empirical machines shown in this book are not generalized or summarized by equations or theorems. Instead, they are embodied in the data field of a memory matrix and expressed in the form of successful behavior, as shown in later chapters.

Self-generated behavior

One requirement of the backup controller described earlier is that it must learn by itself. The truck driver may not know how to program a device to take over in case the driver falls asleep. However, if the controller *initiates behavior* and measures its own success and failure, the driver does not have to do anything except to prohibit the controller from producing undesirable behavior. These decisions are based upon the driver's skill and experience as a driver, not as a programmer.

A pilot flying a remote-controlled drone aircraft may not know how to program a complex control system. An onboard empirical controller may have access to more of the flight instrumentation and may be in a better position to measure the results of the actions taken to control the aircraft than the remote pilot. The remote pilot only needs to control the remote aircraft to teach the onboard empirical controller how to fly a mission. If the onboard empirical controller can initiate behavior and the remote pilot can overrule the controller's behavior, the controller can compare the outcomes of different behavior and select that behavior that has the highest probability of being carried out without interference from the remote pilot.

Likewise, all *empirical systems* must "try something out" in the same way that natural systems initiate action. The backup controller for the truck driver must try to drive the truck in a given way under given circumstances. The drone controller must try to fly the aircraft in some manner. The pet robot must try to move around in a room and produce music or recite stories. This spontaneous or self-generated behavior can be called *originative behavior*.

Interference

Since empirical systems actively attempt to produce behavior, something may have to prohibit that behavior from being carried out, as shown in the following examples:

- The backup controller for the truck driver must try to drive the truck in a given way under given circumstances. But, if that behavior is not satisfactory, the driver must be able to *prohibit* that behavior.
- The drone controller must try to fly the aircraft in some manner. If that behavior is not acceptable, the drone pilot must be able to *prohibit* that behavior from being carried out.
- The pet robot must be able to move around in a room and attempt to service its master. The individual in the room with the robot must then be able to *restrict* that behavior that is not wanted at a given time.

Thus, the operator of an empirical system may have to prohibit or otherwise influence behavior produced by the empirical systems.

Dominance

An operator can interfere with the behavior of an empirical system only if the operator is in a position of *dominance*. Dominance is a fundamental property of natural and synthetic empirical systems, as shown in the following examples:

- Businesses will reach their goals only if they select marketing plans that reflect the wants and needs of the buying public. Thus, the behavior of the buyer *dominates* the market.
- Investment philosophies that do not interpret business cycles correctly will not be successful. Business cycles may *dominate* the investment markets.
- Social groups that commit themselves to unworkable behavior codes will cease to exist. Thus, the fundamental needs of human individuals may *dominate* sociology.
- Species that evolve in ways that do not work in a given environment will perish. Thus, the laws of nature *dominate* evolution.

In all cases, empirical systems produce some kind of action or commitment. Then the natural physical laws or other external factors such as human operators or other control systems determine if that action can be carried out. We will be concerned with the design of an empirical control system that attempts to produce its own behavior while something else determines if it can be carried out.

Operators in an empirical system

The role of an operator of an empirical systems is to try to *influence the behavior* of an empirical system in the direction of what is considered desirable. An operator may not have the ability, in some cases, to force an empirical system to produce specific behavior though the operator may be able to prohibit an empirical system from carrying out specific behavior. Our concern will be: How do we cause an empirical system to decide to do what we want it to do?

Influencing behavior

There are three ways of influencing the behavior of an empirical system as shown in Table 1.7.

Table 1.7. Methods of influencing the behavior of an empirical system

---

1. We may *cause* a particular action to occur, as in the case of the pilot of the drone aircraft who can manipulate the flight controls. The empirical system on board the aircraft must alter its behavior to correspond to the commands of the remote pilot.
2. We may *prohibit* undesirable actions from being carried out. When the empirical system attempts an unwanted maneuver, the dominant pilot must be able to prohibit those actions from being carried out.
3. We may *alter the environment* so as to increase the probability that the empirical system will carry out the behavior we want.

Our means of influencing the behavior of an empirical system will be to exercise a physical dominance over the system or to trick it into doing what we want.

Causing behavior

In some cases we can *cause* the controller to learn specific behavior. We may be able directly to cause the controller's actuators to assume the values we want. For example, as the truck driver steers the truck, the controller can be designed so that it is forced to act the same way as the driver. The only requirements are that the driver is stronger than the controller's steering actuator and the driver is persistent and consistent in the way he or she attempts to control the truck.

Restricting behavior

In some cases, we may not wish to prescribe the behavior of an empirical system. We may find that it is better to let the system generate its own behavior and only restrict or *prohibit* that behavior that we think is dangerous or undesirable.

In many applications, it may be easier for an operator to redirect the spontaneous behavior of an industrial robot into what is wanted than to learn an arcane robot programming language needed to boot up, program, and run the robot from scratch. This is also true of computer programming. It often takes less programming knowledge to edit an existing program than to write a program from scratch.

In many applications, it may be easier to create the desired control behavior using these *reactive techniques* than to program the controller directly. This is particularly true when it is difficult for an operator to know exactly how an empirical system must perform. This may be the case with a security robot that must traverse a difficult terrain. The operator may let the robot learn how to negotiate the terrain by itself and then impose additional restrictions to obtain the behavior he or she wants of the robot. Even in this case, the operator must have a physically stronger influence than the empirical system and the operator must act with persistence and consistency.

Controlling the environment

In some case, we may not be able directly to cause empirical systems to produce specific behavior, or be strong or diligent enough to restrict the behavior of an empirical system. Yet there still are ways to make an empirical system behave the way we want. We will assume that we can alter, manipulate, or restrict parts of the controller's environment. If we know what a controller will do under given circumstances, we can create those environmental circumstances that will elicit the desired behavior and eliminate those circumstances that cause undesirable behavior.

We also may set up situations in the controller's environment that increase the probability of the controller's discovering what we want it to do. This *indirect teaching* and learning process becomes more important as the empirical system becomes physically stronger. The study of how to control the environment of an empirical system to influence its behavior indirectly is a major part of our study, since these methods are similar to the techniques needed to teach living organisms, including humans (see Part V).

Problems with self-governing systems

The *self-governing* feature of empirical systems leads to many new problems. An empirical unit cannot be left alone for long periods of time because it may develop some undesirable or destructive behavior. It also may develop behavior that omits the desired behavior. It may wander away into an unfamiliar environment where its behavior is not applicable. New supervisory procedures have to be established to deal with these empirical systems.

Does this kind of activity sound familiar? People who deal with children are often confronted with this kind of self-governing behavior, which tends to follow its own course unless redirected from time to time by those in charge. The effects of self-governing behavior is seen in biological evolution, selective breeding, training empirical robots, and in many other human activities. For example, carefully bred animals will lose their special characteristics if allowed to mate at random with other breeds. So the breeder must dominate their mating behavior.

Autonomous behavior

Some empirical systems may act entirely on their own. This feature is a by-product of the requirement that empirical systems initiate behavior to find out what behavior will and will not work. The main feature of the empirical process is that it describes a method of learning, growth, and development in which empirical systems are expected to come up with their own knowledge and understanding of what is possible and impossible by initiating action and measuring how well it works.

A predetermined system can be designed to have much or all of its behavior predetermined and unalterable (see Part II). An empirical system can be designed so that much or all of its behavior is predetermined but subject to change according to experience (see Part III), or a system can be designed so that it contains no preconceived ideas about how it should behave and develops all of its behavior through experience.

Most practical empirical control systems contain some predetermined (instinctive) behavior, and the rest is established by experience to meet the needs of its operators. However, some people may be interested in determining the limits of *autonomous behavior*, which is initiated, discovered, and incorporated without the aid of human operators based upon the experience and the unique relation of an empirical system to its environment. An autonomous system must *select* behavior in its own way and may have to invent new ways of accomplishing tasks that may not have been thought of by the trainer, operator, or system designer.

Autonomous empirical systems

We seldom have reasons to design devices that have the freedom to determine all of their behavior. We usually have some predetermined idea of what they should do and how they should do it. But this is not the case with some *autonomous empirical systems*. Though we may have a use in mind for a given system, we may wish to create an autonomous empirical system that can establish its own behavior by itself based entirely upon its own experience with its environment.

We may not know what opportunities for behavior exist for a given empirical unit. Yet an *empirical controller* can determine what behavior can be produced most consistently, measure how well it is executed, and incorporate the behavior that can be carried out.

Another benefit of allowing an empirical system to develop its own behavior without outside intervention is that the system may organize itself according its own needs and abilities that a programmer may not recognize. An autonomous empirical system may be in a better position than we are to determine how to join with other units to produce behavior or to separate from other units to produce successful behavior.

Opportunities for behavior

We may not be able to control the outputs of an autonomous system directly. We may only be able to deal with its environment. We can still cause an autonomous unit to behave in some desired way by creating specific opportunities for behavior in this environment. For example, we can "teach" an autonomous empirical robot to follow a particular path through a house by opening doors along the path we want the robot to follow and locking the rest of the doors. The only *opportunities for behavior* that exist in that situation are based upon the pattern of open doors.

Operant conditioning

All animals may be considered autonomous beings. We cannot directly control the position of their limbs when given sensed conditions occur and expect that they will produce these limb positions when those sensed conditions occur again. However, an *operant conditioning process* can be used to influence the behavior of all animals including humans. An animal trainer may know of a specific condition that will evoke a given desired response. For example, a lion trainer knows the lion will open its mouth when a stick is placed upon its nose. If the lion trainer wants a lion to open its mouth upon command, the trainer places a stick on the lion's nose and gives a specific command. By repeating this training procedure, the lion may eventually open its mouth when that command is given. In a similar manner, an operator can teach an autonomous empirical unit to produce specific behavior.

Apprenticeship

An empirical unit and an operator work in the same environment in much the same way that a master craftsperson and apprentice work together in an *apprenticeship system*. The master and the environment may create certain opportunities for behavior for the empirical unit. The environment will prohibit the empirical unit from carrying out behavior that violates the laws of nature, and the operator can prohibit the empirical unit from carrying out behavior that is not wanted. Thus, the behavior of the apprentice will change until the apprentice produces the behavior that would have been produced by the master had the master been doing the job. At this juncture, the master can withdraw from the process with the assumption that the apprentice will perform in a satisfactory manner. In this study, this apprenticeship system is the basic approach used to influence the behavior of empirical units.

Empirical control

We have presented the need for a new kind of control system. It must help people or other control systems complete behavior. It must originate and maintain successful behavior within the requirements of an environment and specific operators. To do this, the controller must gather information from its environment and create actions that reflect a knowledge and understanding of the laws of nature, the structures, desires of other individuals and the organization of the world around it.

The purpose of this book is to describe a device that can discover and then incorporating the natural relations in its environment into its behavior without the need for predetermined programming from an outside source. This ability is very useful in backup controllers that are intended to aid people in their efforts to produce successful behavior. This ability is also very useful in situations where it is difficult or impossible to know beforehand what behavior is required. This application occurs in space exploration, where we do not have firsthand experience about the conditions that exist on other planets.

To get things done, to keep things running, to complete tasks, and to create successful behavior, these control systems must know certain things about the workings of their environment. A robot controller must know that its robot cannot walk through walls. The drone controller must know the relation between aileron settings and degree of banking. The submarine controller must know that it cannot swim through solid objects.

So what are the basic requirements for the design of a controller that will produce useful behavior? We have said that its behavior must be originative, self-governing, useful, and self-sustaining.

Requirements of a self-learning system

What are the fundamental requirements of a system that can create and carry out successful behavior? One would think that this complex behavior would require a complex process. This is not so. All of the objectives of empirical control systems can be met by a system that is designed to carry out just the three steps in Table 1.8.

Table 1.8. Three steps used by a self-learning system

---

1. Produce certain behavior under certain conditions.
2. Measure whether that behavior is carried out.
3. Produce the behavior that has the highest probability of being carried out successfully under those conditions.

This procedure is similar to the test and compare processes used in science. It is also the basis of the *empirical algorithm*, which is the fundamental process used by an empirical controller to produce and maintain successful behavior.

Empirical algorithm

The empirical algorithm acts according to the four general rules in Table 1.9.

Table 1.9. Rules of the Empirical Algorithm

---

1. The empirical controller must select an output (after a given time delay) having the highest confidence level, according to its memory, for a specific given input.
2. If the output selected can be carried out, the confidence level stored in its memory of that output for that input must be increased so that the probability of later selecting that output for that input is increased.
3. If the output selected for a specific given input cannot be carried out (because it is inhibited, restricted, or otherwise interfered with by the environment, an outside influence such as a trainer, or by some internal signals, actuators, or structure of its own), the confidence level stored in memory must be decreased so that the probability of the controller later selecting that output for that input is decreased.
4. If another output is carried out, the confidence level stored in its memory of that output for that input must be increased so the probability of later selecting that output for that input is increased.

These four simple steps create the growth, evolution, and success of all empirical systems.

Test and compare process

The unique feature of this algorithm is the *test and compare process* used by an empirical system to establish knowledge. It actively tries to do things — to produce behavior — and then measures whether it carried out these actions successfully. It uses this information to select actions in the future that have the highest probability of being carried out successfully in the past. Thus it *creates its own behavior* based upon what it can and cannot accomplish within the context of an environment or other constraints.

Because it initiates its own behavior, it is not dependent upon another source to produce behavior. For this reason, it may discover new features in its environment that are not apparent to others. It can continue to grow by itself. It will not just wait for someone to change it. If left alone, it will continue to find ways to exist in whatever environment it is placed. It will change by itself if conditions change in that environment.

Transform conditioning

A natural computer acquires behavior by remembering what action actually occurs after each input condition. In the terminology used in the behavioral sciences, inputs are called *operands* and outputs are called *transforms*. The learned and recorded data consist of the actual output or transform that is stored at the input or operand address. To emphasize that the actual output is recorded, this learning or data storage process is called *transform conditioning*.

The term *transform conditioning* is used to emphasize that the actual action produced, or "transform," is the behavior learned for given conditions. The term *operant conditioning*, used by B. F. Skinner, emphasizes the process of learning to respond to specific input conditions or stimuli. The results of both processes are basically the same in that specific actions are produced for specific sensed conditions. Both results are created by the same basic process inherent in the empirical algorithm that defines an *operand/transform conditioning* process.

A natural computer can be programmed by producing or otherwise defining an output for each input. But what happens if two or more outputs are defined for a given input? What output should the controller select for that input? The only logical choice for the natural computer is to select the output that has historically occurred most consistently with a given input.

Therefore, the natural computer must remember how often a given action occurs for given conditions in relation to what other actions occur for these conditions. When those conditions occur again, it must select that action that occurs most often under those conditions.

In autonomous empirical systems, most of the outputs will be produced by the controller itself rather than an outside agent like an operator. The environment will determined which outputs can actually be carried out. If the output is allowed to occur, it will be remembered. In empirical systems with a high degree of interaction with human operators, the operators may determine much of its behavior. In this case, the behavior of the empirical system will reflect the wants and needs of its operators.

Because of the empirical algorithm, the natural controller only learns behavior that can actually be carried out within the constraints of its own action generating system (actuators), its operators, and a specific environment. So, what is *actually done* drives the learning process.

Positive reinforcement

This learning process can be viewed as *positive reinforcement* in that an empirical controller learns the positive or real actions that take place and it discards the behavior that cannot happen. Punishment or *negative reinforcements* prohibit actions from being carried out. But that behavior is soon abandoned. The empirical behavior that remains and continues to be exercised by an empirical controller based upon the *empirical algorithm* is the behavior that can be carried out. An empirical controller remembers and tries to execute those actions that can occur with the least interference.

Praise and rewards in themselves have no bearing upon the learning process of an empirical controller unless they contribute to the completion of an action. We may help a person learn a task by encouraging him or her to complete the task. They learn from what actually happens to them when they complete the task, not necessarily from the encouragement.

An individual and/or an environment can be the trainer. Each can either permit or prohibit the action behavior of the empirical controller. In some cases each can encourage or even cause behavior. In any case, the empirical controller will be designed to produce behavior that it can carry out and abandon behavior it cannot carry out. After some number of repeated trials, the behavior of an empirical controller will reflect the opportunities for behavior in its environment and/or the intentions of its trainer.

Components of an empirical controller

An empirical controller is made up of the five components listed in Table 1.10.

Table 1.10. Five elements of an empirical system

---

1. Sensors to determine what conditions exist;
2. Actuators to produce changes in the physical environment;
3. A method of measuring what actions are actually carried out;
4. A natural computer (brain) to keep track of what actions can and cannot be carried out;
5. An environment that contains consistent natural information upon which the natural computer can operate.

All empirical systems must have these five elements, although the functions of the sensors and actuators may be combined in some cases (see sensor/sensor machines and actuator/actuator machines).

Environment

The controller can never develop behavior without being connected to an *active environment* (Fig. 1.6) that can provide new conditions in response to the actions of the controller and imposing restrictions on what actions can be carried out. The purpose of the controller is to produce behavior and measure the success of this behavior. An empirical controller cannot learn new behavior without an environment that imposes restrictions upon its behavior.

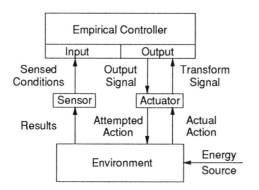

Figure 1.6. An *empirical control system* must include a responsive environment to function.

Survival of behavior

The *survival of the behavior* of an empirical control system is based upon its ability to produce actions that create the conditions for which it has actions that can be carried out. So it must discover and incorporate behavior that is allowed by the whole system *and* keeps the system within the field of behavior with which the controller is familiar. In this sense, the *behavioral existence* of the controller depends upon its ability to produce *doable behavior* and stay within the environment in which it has learned this doable behavior. So it must discover the behavior that can be carried out within a given environment and use some of this behavior to keep it in this environment. Thus, some of its behavior must be used to sustain the rest of its behavior. Without this *self-sustaining behavior* the empirical controller may not be able to use the successful behavior it may have developed.

An empirical thermostat

An empirical thermostat is an example of an empirical controller that can learn to produce useful self-sustaining behavior.

Design requirements

An empirical thermostat must have the following five components:

1. A sensor to measure temperature;
2. An actuator to turn the furnace on and off;
3. A transform signal to determine if the actions called for are carried out;
4. A method of storing information about the probability of success of each decision;
5. A consistent environment that completes a connection between the actuator and the sensor.

Diagram of an empirical thermostat

The empirical controller, shown in Figure 1.7, measures the temperature of the room and makes a decision to turn on the furnace, turn off the furnace, or leave it in its existing state based upon its memory. These decisions influence the temperature of the dwelling in some way.

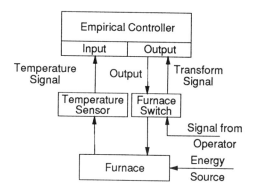

Figure 1.7. The *diagram of an empirical thermostat* shows how each component influences other components.

The operator also can turn the furnace on or off. If the operator produces an action that is different from that produced by the controller, an interference occurs. Information about the actual state of the furnace is sent back to the controller by the transform signal. An interference changes the state of the controller's memory so that it is less likely to make the decision that produced the interference, and is more likely to make the decision represented by the action of the operator.

Operation

An empirical thermostat works according to the steps shown in Table 1.11.

Table 1.11. Operation of an empirical thermostat

1. The controller senses temperature and makes a decision as to the position of the furnace switch. The choices are: turn the furnace on, leave it in its existing state, or turn it off.
2. If the person living in the dwelling is satisfied with the temperature, then he or she will not change the position of the furnace switch. The controller increases the confidence level of whatever decision it had made.
3. If the person living in the dwelling is *not* satisfied with the temperature, then he or she will naturally change the position of the furnace switch in a way that will satisfy his or her need. The controller will decrease the confidence of any decision that is overridden and increase the confidence of the actual decision made by the operator.
4. Eventually, the controller will "anticipate" the decisions of the operator and take the actions previously called for by the operator. The controller will in fact take over the control of the furnace and keep control if it produces the control behavior wanted by the operator. (This control behavior becomes identical with what the operator would have produced if the empirical thermostat were not in use.)

The empirical thermostat eventually learns the desired control behavior for each temperature condition encountered. Then the operator does not have to get up and change the furnace switch to maintain the desired temperature. This frees the operator to do other tasks.

This empirical thermostat can be trained automatically (without an individual being involved) under some circumstances. For example, the high/low limits on the furnace can be set to produce a given temperature. The only behavior that can occur without the high/low limits interfering is control behavior between these limits. Once the empirical thermostat establishes this behavior, the high/low limits will not come into play, and the empirical thermostat will in effect be in control of the furnace. This is an example of an *autonomous empirical controller.*

Features of empirical behavior

This application of an empirical thermostat shows the importance of dominance and transform condition in the development of useful, self-sustaining, empirical, originative, and autonomous behavior (Table 1.12).

Table 1.12. Features of empirical behavior

- The operator can *override* the output behavior of the controller (dominance).
- The controller learns from the actual output (transform conditioning).
- The learned behavior frees the operator (does useful behavior).
- The control behavior can continue unchanged once learned (self-sustaining behavior).
- The resultant behavior comes about through experience (empirical behavior). It is not predetermined.
- The control behavior originates within the controller (originative behavior).
- The control behavior can come into existence without outside human influence in the example of the system with preset high/low limit switches (autonomous control).

Obviously, a second controller with specific control behavior can teach the empirical controller in the same way the operator or a built-in high/low limit controller taught the empirical thermostat. This second controller can train another empirical controller if it can dominate the control function. This is an example of how two or more empirical controllers can learn to coexist much the same way an operator and a controller can learn to coexist (see Part IV).

Definition of empirical control

Thus, an empirical controller gathers and uses natural information in its surroundings to create successful behavior. It then behaves as if it understands its environment. The study of this process will be called *the study* of empirical control. It pertains to control systems that direct, regulate, or conduct behavior by means of a process that *compares* the action produced by the controller against the action prescribed by the controller within the constraints imposed by an environment or other controllers, and *selects* those actions that have the highest historical probability of being carried out.

The purpose of the empirical controller is to produce successful behavior. It does this by causing behavior that can be carried out and avoiding behavior that cannot be carried out. There are many new and interesting questions raised by systems that operate in this way. The purpose of the rest of this book is to uncover these questions and offer solutions that permit us to design and build an empirical controller that will create and maintain useful behavior.

Goals in the study of natural and empirical systems

The empirical process of observation and experiment has been introduced to show how information can be acquired, recorded, and used to produce useful behavior. This introduction has included examples of many living and nonliving natural empirical systems such as the brain, business and economic systems, social and ecological groups, and the process of biological evolution, all of which operate according to the same set of rules.

We are particularly interested in the design and operation of *synthetic empirical* systems. These include robot and autonomous vehicle controllers, empirical thermostats, and other types of controllers that are designed and built by people to carry out useful tasks.

The terms *natural system* and *empirical system* can be used interchangeably in many cases. However, the term *natural system* generally applies to systems found in fields such as economics, biology, or sociology that develop self-sustaining behavior in a natural setting without a conscious effort by anyone to create the system or to specify their behavior.

When the term *empirical system* is used, more emphasis is placed upon the creation or *learning* of specific behavior. Usually, these are systems that have been designed and built by human beings with some purpose in mind. Methods of teaching and learning useful behavior are the primary concerns of people involved with empirical systems. For example, *evolution* is a natural process, while *selective breeding* of plants and animals would be considered an empirical process.

Empirical robot controllers, empirical thermostats, empirical governors, regulators, auto-pilots, and autonomous vehicles are synthetic empirical systems because they are designed and built by people, because people attempt to determine their behavior and because people have some use in mind for their behavior. An economy is an example of a natural empirical system since people cannot build an economy in most cases, though they may wish to influence its behavior.

An understanding of natural systems will contribute to design of synthetic empirical systems that can produce useful behavior spontaneously. They will be electromechanical machines that can act as backup or standalone *nonliving individuals* capable of producing and sustaining their behavior.

To deal effectively with these control systems, it is desirable to develop a more concrete understanding of what is meant by a machine, behavior, and control.

Chapter 2
Machine Behavior

The usefulness of any machine resides in its behavior. This behavior can be successful if it can be carried out and can continue without outside assistance in spite of interferences. The purpose of this chapter is to understand the behavior of machines in concrete terms so that we may deal more effectively with the behavior of empirical machines.

A *machine* can be viewed as a responsive device that produces some action in response to the conditions it encounters. The *behavior of a machine* is defined by the sequence of action states it produces for every condition state, and the *modulus of behavior* of a machine is defined as the set of condition/action relations that the machine can produce. A machine can be *continuous* if its condition/action variables change smoothly over time, or a machine can be *discontinuous* if it operates in discrete steps like a digital computer. A machine can produce different *modes of behavior*. For example, a punch press attempts to *repeat* a sequence of states, a spring attempts to *seek* a selected state called its *rest position*, or a *toggle* mechanism attempts to *avoid* a *point of instability*. These modes of behavior can be built or programmed into the machine before it is used in a *predetermined* manner, or behavior can be established while the machine is in service according to the opportunities for action as each condition is encountered within the task environment in an *empirical* manner. A machine can *standalone* or work with other machines in a *network*.

Machines

A machine has been defined as a device that produces actions according to some set of conditions. Most machines modify force or motion in a specific manner according to sensed conditions. A *spring* may be considered a simple machine in that it creates a varying force in some relation to its length. Machines inevitably influence force or position variables in the environment in some specific way in relation to sensed conditions.

Variable

The *variable* is the fundamental entity of science. It is also the basic building block of the science of empirical control. A variable has two parts — its *specification* and its *value*. This can be represented by the thermometer shown in Figure 2.1.

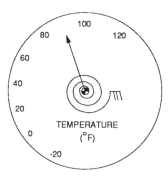

Figure 2.1. A *variable* is specified by some instrument or device and has only one unique value at a given time.

All variables are considered to have only one value at a time (single valued) and will be considered valid only if they can be represented the reading of an instrument or other device. A variable also must be defined by the specification of the instrument or device used to measure or represent it, like *Thermometer #1*, along with its reading or state such as (68^0F). The specification of the measuring instrument used to represent a variable allows other investigators to attempt to duplicate the measurements of that variable.

Machine

A machine is any device that senses conditions in its surroundings and produces some action in response to these conditions. The thermometer shown senses air temperature and positions its indicator arm in some relation to that temperature. Thus, the thermometer may be considered a simple machine. A relay (Fig. 2.2) also may be considered a machine because it senses the voltage at its input terminal and produces an output that relates to this voltage.

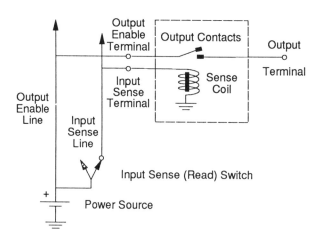

Figure 2.2. A *relay* (gate) is a simple machine that senses the condition of its input terminal and produces an output that relates to this sensed condition.

Sensor and actuator variables

Thus, there are two different types of variables with which a machine must deal — sensor and actuator variables. A *sensor variable* is considered a measure of the conditions surrounding or within a machine. Sensor variables, such as temperature or light level, usually do not influence other variables directly since they usually cannot produce a change in force or position without the aid of a machine such as a bimetal spring, heat engine, or a photocell.

However, an *actuator variable* such as a furnace burner or an air cylinder can produce some action that can influence other variables directly. A heater or air conditioner (actuator variables) can directly influence the temperature of a room (sensor variable), but the temperature of a room cannot influence the operation of a heater or air conditioner (unless the sensor and actuator variables are connected by another machine called a thermostat).

Of course, the temperature in a room can increase the rate at which food spoils. Photons can produce a small force. So whether a variable is a sensor or actuator variable is often determined by the definition of the machine system.

Machine with sensors and actuators

Most machines consist of a *sensor* or sensors that convert the sensor variable(s) in an environment into the input variable(s) of the machine, some means of converting the values of the input variables into values of the output variables of the machine based upon its modulus of behavior, and an *actuator* or actuators that convert the output variable(s) of the machine into value(s) of the actuator variables in the environment, as shown in Figure 2.3.

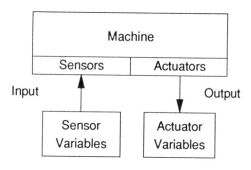

Figure 2.3. A *machine with sensors and actuators* can sense conditions in an environment and produce actions in the environment that may change those conditions.

For example, a spring can be viewed as a machine that has the intrinsic ability to sense its length and produce a force at it ends according to this length. The force produced by a spring is determined by its length and its modulus of behavior. In some cases, the sensor and actuator variables seem to be combined. For example, the bimetal spring in the thermometer senses temperature and produces an action. However, air temperature and indicator position are different variables related by the bimetal spring just as the force on a spring and its length are separate variables related by the stiffness of the spring.

Energy

Some form of *energy* may be required for a machine to produce actions. In the case of the thermometer, the energy involved in the movement of its indicator is produced by the heat energy in the air. In the example of the spring, a force must be applied to the spring to produce a displacement. The amount of energy required to move the spring is equal to the produce of its displacement and the average applied force. This energy must be supplied or the machine will not work.

Passive machine

A *passive machine* such as a thermometer (Fig. 2.4) does not supply the energy involved in the action of the machine. The heat energy in the air causes the bimetal spring in the thermometer to coil or uncoil. A spring, capacitor, resistor, pendulum, and coasting cart are examples of other passive machines.

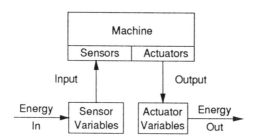

Figure 2.4. A *passive machine* does not increase the total amount of energy involved in its behavior.

Active machine

A relay is considered an *active machine* (Fig. 2.5) because it releases or inhibits the energy of its output enable power source. Amplifiers, servomotors, temperature control systems and heat engines are examples of active machines.

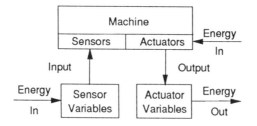

Figure 2.5. An *active machine* may increase the energy available to produce behavior.

Behavior

Behavior has been defined as a sequence of conditions and the resultant actions. The *modulus of behavior* defines this set of condition/action relations. Behavior can be *random*, such as the sequence of heads and tails encountered by repeatedly tossing a coin; behavior can be *state determined*, such as the swing of a pendulum; or behavior can be *probabilistic*, such as the ratio of heads to tails obtained after a sufficient number of coin tosses.

Modulus of behavior

The term *modulus* is often used to describe the physical behavior of materials in nature. For example, the *modulus of elasticity (E)* defines the ratio of the change in unit load (*f*) to the change in unit deformation (*e*) of a material according to the equation:

$$E = f/e. \qquad (2.1)$$

If a specimen is stretched in a stress/strain testing machine, the change in unit deformation *e* (the change in length divided by its total length) of the specimen may be considered the *input* and the resulting unit load *f* (the total force divided by the cross sectional area) on the specimen may be considered the *output*. The term *E* describes the relation between input and output, as shown in Figure 2.6. The relationship between the results of any operation on a system and the operation that caused these results can be called the *modulus of behavior* of that system.

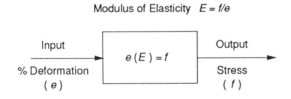

Modulus of Elasticity $E = f/e$

Input
% Deformation
(*e*)

$e (E) = f$

Output
Stress
(*f*)

Figure 2.6. The *modulus of elasticity (E)* specifies the elastic behavior of a material.

Spring constant

The modulus of behavior of a machine like a linear spring can be described by a single number called the *spring constant (k)* that defines how much force (*F*) is produced when a spring is deflected by a given distance (*x*), as shown in Figure 2.7. In the example shown in this figure, the value of the spring constant can be determined when

$$k = \text{force/deflection} = 10(\text{lbs/2in.}) = 5(\text{lbs/in.}).$$

This number defines how the spring will behave. A larger spring constant identifies a stiffer spring.

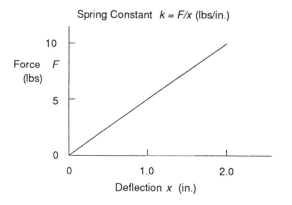

Figure 2.7. A *spring constant* defines the behavior of a linear spring.

Representation of the modulus of a spring

The relationship between the deflection of a spring and resultant force produced by the spring can be represented by Figure 2.8.

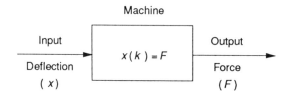

Figure 2.8. This representation of the *modulus of a spring* defines the relation between an input deflection and a resultant output force.

   The spring can be considered a machine that produces a specific output because of a given input.

Nonlinear spring

If the spring is nonlinear, as in the case of an elastic ball, the relation between force and deflection can be shown in Figure 2.9.

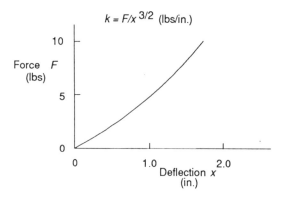

Figure 2.9. A *nonlinear spring* requires a more complex equation to define the relation between the input deflection and output force.

A solid rubber ball produces a greater restoring force for a given deflection when it is highly compressed than it does when it is only slightly compressed, as shown by the following equation:

$$k = \text{force/deflection}^{3/2}.$$

Its modulus of behavior (in parentheses) also can be written as follows:

$$x(kx^{1/2}) = F.$$

Modulus of a nonlinear spring

The input/output relation between a nonlinear spring also can be represented by Figure 2.10. The behavior of all springs can be defined by moduli like those shown in this figure, although they may be even more complex depending upon the nature of the springs. These moduli allow the output force to be determined for each input condition.

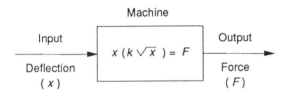

Figure 2.10. The *modulus of a nonlinear spring* cannot be represented by a single number.

The behavior of all of the machines described in this study must be defined by a modulus of behavior of some kind. The purpose of the study of empirical control is to describe the process of creating the modulus of behavior of an empirical controller that can learn to produce behavior that can be carried out successfully within a given environment.

Discontinuous machines

There are two basic types of machines — continuous and discontinuous machines. Sometimes the term *analog machine* is used to describe a continuous machine and the term *digital machine* is used to describe a discontinuous machine. The variables of a continuous machine, like the spring described here, can be represented by continuous electrical signals that are *analogous* to the values of those variables. The variables of a discontinuous machine are represented by numbers that are established by sampling its variables at discrete moments in time and converting the values of these variables into discrete (digital) numbers. This process is sometimes called *digitization*. However, I prefer to call it *descretizing* because I wish to use the term *digitizing* to describe a process of converting the values of a single variable into the values of a set of *aggregate variables* or of converting from one number system into another. I prefer to use the terms *continuous* and *discontinuous machines* rather than the terms *analog* and *digital machines* because I wish to reserve the term *analog* to describe a type of memory unit that can assume an almost infinite number of different values. The discontinuous machines described in this book are not digital in the sense of operating entirely with numbers.

Continuous physical model

Most machines are considered continuous. That is to say, the values of their position, velocity, and acceleration change in a smooth, continuous manner from one moment to the next. This allows these systems to be modeled and described by the equation shown in Figure 2.11.

$$m\,\ddot{x} + c\,\dot{x} + k\,x = F$$

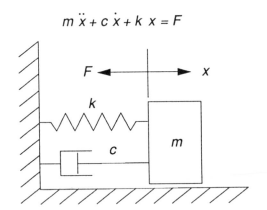

Figure 2.11. Most machines are described in terms of a *continuous physical model*.

The output of this machine can be calculated if enough input conditions are known. Science is richly endowed with equations for continuous systems. These equations make it possible to calculate many important quantities, such as the position of planets or the state of a complex control system.

Sampling continuous variables

However, the machines described in this work and the animal brain cannot compute behavior. However, they can produce specific output states for specific input states. The continuously changing values of their input and output variables can be sampled at specific moments in time, and these values can be recorded as a *list of input/output states*. The rate at which these variables are sampled and the resolution of the recorded values can be varied to provide whatever degree of similarity is wanted between the discontinuous lists and the continuous variables.

Modulus of behavior as a list

The behavior of most real machines such as automobiles, sewing machines, and clocks are usually defined in terms of the continuous model shown. However, the behavior of these machines also can be defined by a *list* (*U*) of input/output relations, as shown in Figure 2.12.

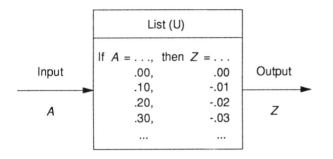

Figure 2.12. The behavior of a *discontinuous machine* can be defined by a list of "If (input), then (output)" statements.

The list of input/output relations defines the modulus of behavior (*U*) of a *discontinuous machine*, in contrast to the equations used to define the modulus of behavior (*M*) of a continuous machine. A discontinuous machine can be called a *responsive machine* because it responds to each input "if. . .," statement with an output "then. . .." statement.

Discontinuous state ontology

To use a list of input/output relations to define the behavior of a real machine, it is necessary to specify each input and output state and the time relations in which these input and output states occur, according the assumption of the *discontinuous state ontology (DS)*. *Ontology* is used here to mean the definition of the universe in which a machine operates. A *system state* is the unique set of input and/or output *variables* and the unique set of *values* of those variables that occur at a given moment in *time* and for a given *time duration* in the specific *region of variables* surrounding a machine.

Assumptions of the discontinuous state ontology

We will assume that we can *sample variables* at given moments in time. We also will assume the rate at which values change is determined by the physical characteristics of the environment and that we can adjust the sampling rate to capture the important characteristics of these variables. An attempt will be made to sample variables at the available stopped or at rest (*accessible*) values of the variables that make up the *system states*. We will assume that a variable can have only one value at a given time and that whatever conditions cause a variable to behave in a given way will cause the variable to behave in the same way when the same conditions occur again. For the sake of simplicity, we will assume that the set of values of some set of input variables determines each output state completely, and that the values of these input variables preceding a given input state have no bearing upon the outcome of that input/output relation. We also will assume that variations in input/output relations are due to the presence of *hidden variables*. Changes in these hidden variables create changes in the system states that make them appear to be *probabilistic* rather than *deterministic*.

Machine as a list

The modulus of behavior (U_1) of a discontinuous machine can be defined by the list of each output state that will occur for each input state according to the assumptions and definitions of the discontinuous state ontology (Fig. 2.13). Since the output states of a machine act in response to input states, a machine may be considered a *responsive device* having a modulus of behavior defined by (U_1) based upon the discontinuous state ontology.

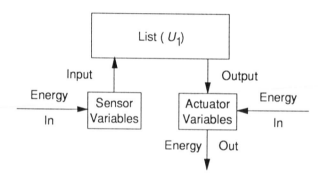

Figure 2.13. The *modulus of behavior of a machine* can be defined by a list (U_1) of relations between its input and output states.

    The *input variety* of a machine is equal to the total number of possible input states it can identify, and its *output variety* is equal to the total number of possible output states it can produce. The *potential machine variety* is equal to the input variety multiplied by the output variety, and is equal to the number of possible input/output relations. The *actual machine variety* is equal to the number of actual input/output relations produced by the machine.

Input/output transformation

The modulus of behavior of a discontinuous machine can be defined by a *list of transitions* showing each input and output state, and the probability, time delay, and time of action of each output state that is produced for each input state according to the discontinuous state ontology, as shown in the example of an *input/output transformation* in Table 2.1.

Table 2.1. Input/output transformation (*U*1)

Input state (Operand)	Time delay (t_d) (sec.)	Probability (p)	Output state (Transform)	Time period (t_p) (sec.)
1	>-(1)->	(1)	K	(.5)
2	>-(1)->	(1)	B	(1.5)
3	>-(2)->	(1)	C	(1.5)
4	>-(1)->	(1)	M	(.5)
5	>-(1)->	(1)	M	(.5)
14	>-(1)->	(.1)	M	(.5)
15	>-(1)->	(.75)	K	(.5)
.	.	.	.	.

An input/output transformation defines the output state (*transform*) of the machine for each input state (*operand*) likely to be encountered. (The term *operand* is used to designate the state that precedes the transform state. This avoids confusion in systems having two or more operating units in which the output of one is the input to another.) In a *state-determined transformation* in a discontinuous state system, the machine produces the output state (transform) called for by the transformation, with a *probability (p)* of 1, after some prescribed *time delay (t_d)*, and within a specific *time period (t_p)*, according to the input state (operand) encountered at some moment in time. Each input state may be encountered at specific time intervals based upon the *sampling rate* of a system clock. In a *probabilistic transformation*, the probability of a given transform occurring after a given operand may be less than 1.

Environment

The *environment* may be viewed as a device that transforms the actions of a machine into specific new input states to the machine according to some transformation (U_2), as shown in Figure 2.14.

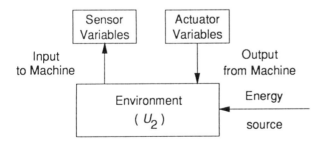

Figure 2.14. The *environment (U_2)* may be viewed as a device that supplies new input states to the machine.

If the environment is state determined, the environment always produces the same input state to the machine for given output states from the machine after some prescribed time delay determined by the operating characteristics of the environment. For example, the time delay in a low-gravity environment may be considered greater than the time delay in a high-gravity environment. Also, the time delay for an auto-pilot for a large ship would be much longer than the time delay for that of small boat. If the environment does not always produce the same input state to a machine for given output states from the machine, it is considered a *probabilistic environment.* Since the variables in the environment respond to the action of a machine, the environment also may be considered a state determined or probabilistic responsive machine.

Modulus of the environment

The modulus of behavior of the environment also can be defined by a transformation (U_2) such as the one shown in Table 2.2.

Table 2.2. Modulus of behavior of an environment (U2)

Output state (Operand)	Time delay (t_d) (sec.)	Probability (p)	Input state (Transform)	Time period (t_p) (sec.)
A	>-(1)->	(1)	13	(.5)
B	>-(1)->	(1)	3	(.5)
C	>-(2)->	(1)	1	(.5)
H	>-(3)->	(1)	2	(.2)
K	>-(3)->	(1)	4	(1)
M	>-(1)->	(.75)	15	(2)
M	>-(1)->	(.25)	14	(2)
.	.	.	.	.
.				

This list specifies the input states that the environment supplies to a machine for each output produced by the machine. The environment transformation (U_2) is a manifestation of the laws of nature and the physical structure of the environment and the machines connected to this environment. The environment may be probabilistic if it contains one or more hidden variables or state determined if each combination of output values of a machine connected to the environment produces a unique combination of input values to that machine.

Line of behavior

A *line of behavior* is a sequence of input and output states that a machine undergoes while producing behavior. If every line of behavior that can be produced by a machine is shown in a diagram, the result looks like a road map that shows all of the paths that a machine can take while producing behavior. In a state-determined line of behavior, there is only one road and no intersections. In a probabilistic line of behavior, there are many roads and intersections, and the probability of turning one way or another at each intersection is determined by the probability of each transformation.

Operating system

When a machine is connected to an environment or another machine, as shown in Figure 2.15, it may start to produce a sequence of states. If the machine output states produce new input states that lead to output states that can be carried out, the operation of the system will continue.

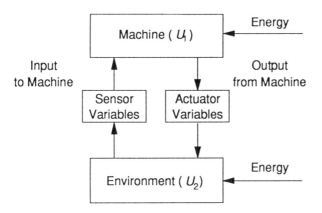

Figure 2.15. An *operating system* is created when at least two responsive devices such as a responsive machine and an environment are connected.

A second machine or a human operator can be used in place of the environment. Without the second responsive machine or an environment to supply new input states, a responsive machine will cease to operate.

State-determined line of behavior

A *line of behavior* is produced when a machine is connected to an environment that produces input states for which the machine has doable output states and when the system is released from an initial state, as shown in Figure 2.16.

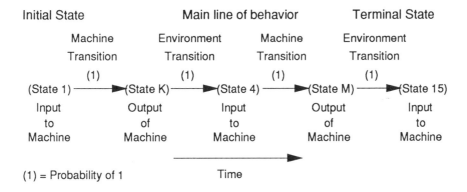

Figure 2.16. A *line of behavior* consists of a sequence of sensed input states and the output action states connected by the transitions produced by the machine and its environment.

In a *state-determined line of behavior,* there is only one output state for each input state, and the environment produces only one input state for each output state. The characteristic way in which a machine operates is defined by its *main line of behavior.* To reduce the difficulty in representing lines of behavior, the time variable will be omitted in most diagrams though the *time variable* is an intrinsic part of a line of behavior, as indicated by the *time direction arrow.*

Convergent line of behavior

A machine may produce the same output state for two or more different input states creating a *convergent line of behavior* as shown in Figure 2.17. The convergent line of behavior draws the system states to the main line of behavior.

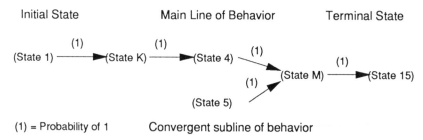

Figure 2.17. A *convergent line of behavior* occurs when one of two or more different system states can produce a single system state.

Finding our way back to the travel route we have chosen after getting lost is an example of a convergent line of behavior.

Probabilistic line of behavior

Some variables in a machine or its environment may not be identified and included in the definition of the system states. If the values of some of these *hidden variables* in the environment change over time, a different input state to the machine may occur at different times for a given output state. Some hidden variable within the machine may change causing a different output state to occur at different times for a given input state. Even if the machine is state determined, someone or something may not allow a given machine output state to occur for a given input state. These conditions create a *probabilistic line of behavior* having *bifurcations* in the line of behavior that lead to *divergent lines of behavior*, as shown in Figure 2.18.

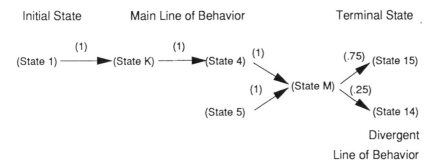

Figure 2.18. A *probabilistic line of behavior* is like a road map that shows all of the possible roads (transitions) and the probability of taking each possible road at each intersection (system state).

Straying from our travel route and getting lost is an example of a divergent line of behavior. If the behavior of the system is observed over a sufficient number of trials, the *historical probability* of each transition can be accurately determined. In general, after traveling the same route many times, we learn the desired turns at each intersection and lose our way less often. Our travel behavior becomes more state determined and less probabilistic after repeated attempts to go from one place to another.

Closed line of behavior

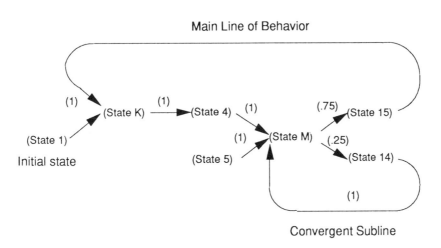

Figure 2.19. A *closed line of behavior* has convergent sublines of behavior for every divergent subline of behavior and eventually returns to a common state.

If a machine with a finite number of transitions is to operate for an extended period without changing its behavior or requiring assistance from an operator, it must repeat its line of behavior. This is represented by the *closed line of behavior* shown in Figure 2.19. Most machines used in the production of industrial goods operate in a repeating or closed line of behavior. The ability of a machine to maintain *closure* is a measure of its successful behavior or *reliability*. Some *automatic machines* use convergent sublines of behavior based upon *fault detection, fault correction,* and *fault recovery* to create a successful *ongoing line of behavior* despite potentially disruptive *faults*.

Line of behavior and closure

Many animals are endowed with very efficient tools for securing food. A giraffe has a long neck to reach leaves high in trees, and a tiger has sharp claws to help capture prey. However, both must still *act successfully* to use these tools effectively. The giraffe must attempt to eat the leaves. The tiger must chase prey. What makes their behavior successful?

Responsive system

We have shown that a responsive machine must be connected to an environment to operate. The environment can be a physical territory, another machine, an empirical controller, or a person. It can be any device that can generate inputs to the modulus of behavior of the controller. We will call this mating machine Unit B (Fig. 2.20).

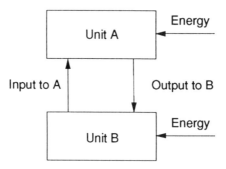

Figure 2.20. A *responsive system* must consist of at least two devices that can respond to and produce behavior.

As stated before, a responsive machine needs a constant supply of new input data to generate new behavior. The purpose of the second Unit B is to supply these input data. The output of Unit B becomes the input to the responsive machine. This information is then used by the responsive unit to decide what its next output is going to be. This output then becomes the input to Unit B, which in turn "causes" a new input to Unit A. Putting two responsive units together may cause a "chain of events" in which the output event of one unit becomes the input event of the other unit after certain time delays. This *sequence of events* becomes a line of behavior. Without Unit B, Unit A cannot generate new information to form new outputs. Without new inputs, a responsive controller will stop producing behavior.

Line of behavior in time

An active line of behavior appears as a changing state (a changing vector if each state is made up of two or more variables) or as a constant or null function when the system variable(s) remain inactive, as shown in Figure 2.21.

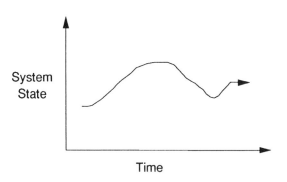

Figure 2.21. A line of behavior in time shows the sequence of states that occurs when two machines are connected.

The line of behavior changes whenever any component variable comprising the line of behavior is active, and the line of behavior remains constant only when all of its component variables are inactive.

Doable outputs

Each unit must produce *doable outputs* (actions that can occur as called for by the machine). If a machine cannot produce doable outputs for each given input, the operation of the system will stop, and the line of behavior will become inactive. This condition occurs, for example, when an automatic machine encounters a fault such as running out of parts or having a part jam in the machine so as to prohibit further action.

Familiar inputs

Since a responsive machine unit produces given outputs in response to given inputs, the inputs must be ones that produce doable outputs if the machine is to operate successfully. Most of the possible input states to a responsive machine usually do not lead to doable outputs. The limited number of input states that lead to doable outputs must be identified and evoked by a responsive machine through its environment. So, the continued operation of a responsive machine system is based upon there being a continuous supply of these *familiar inputs* to the machine from its environment, and doable outputs from the machine to its environment.

Familiar region

Thus, a responsive machine must contrive with the environment to produce the particular set of input conditions for which it has doable outputs. The limited set of familiar inputs for which it has stored doable outputs represents the *familiar region* of the environment, as shown in Figure 2.22.

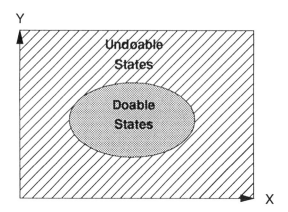

Figure 2.22. The *familiar region* of a machine is that part of the environment for which the machine has doable outputs that produce familiar inputs.

If it gets out of this familiar region, it may get lost. Once outside a familiar region, a responsive machine may have to establish new relations that may not be compatible with the relations it had established previously. If it cannot find its way back to its familiar region, or if no one brings it back, its original behavior may undergo a *behavioral death.* Thus, if the behavior of a responsive machine is to continue, part of its behavior must keep it within its familiar region.

Ongoing behavior

The primary objective of a responsive system is to establish a set of input and output states that allow the system to keep going. That is, Unit A must produce doable outputs that become inputs to Unit B, such as an environment, for which Unit B has doable outputs. In turn, these outputs must become the doable inputs to the original Unit A for which Unit A has doable outputs. If these conditions are not met any time, the system's established behavior may cease to occur, the system may stop, or its behavior may become entirely random. This condition can occur in a computer program with "indirect addressing," where the address of a program step is determined by the contents of a memory register. If the register comes up with a number that is not part of the program, the program will "blow away" into a useless loop, jump around destroying useful data, or mercifully go to the "end of file."

This relation between machine and environment is extremely important. It represents the typical machine/environment relation in all animal and machine systems. In production machine systems, specific inputs and outputs are associated with making "good" parts. If something changes to interrupt this chain of events, the system may jam, seize up, or shut down automatically. For example, production equipment may have some way of measuring the acceptability of the parts it produces. If these parts get out of tolerance, the machine is usually designed to shut down.

For an empirical machine to "run by itself," it must "learn" doable outputs for each familiar input. These doable outputs must lead to additional familiar inputs for which the machine has had experience and knows what the next doable output is.

So, the operation of a responsive machine system is based upon producing a continuous sequence of acceptable (doable) behavior. The purpose of the empirical controller is to develop and enlarge this reservoir or *repertoire* of doable behavior that allows it to act continuously without *interference*, interruption, or even assistance from outside sources. The result is *self-sustaining behavior,* in contrast to *operator assisted behavior.* This repertoire of doable behavior is responsible for the intelligence, usefulness, and success of the machine.

Machine memory

This repertoire of doable behavior must be recorded and stored within a responsive machine for use when each input occurs. This machine *memory system* must store each doable output state that occurs with each active input state. The amount of memory required is determined by the number of these input/output relations and the complexity of each input and output state. The size of memory in most machine controllers is limited by the cost of the memory hardware and the cost of programming this hardware. So what can be done to limit the size of the memory while still obtaining useful behavior?

Closed line of behavior

A discontinuous responsive machine can only store a finite number of successful input/output relations. This makes it imperative that the behavior of the machine returns to a state that has occurred before and that it repeat its behavior to keep running for long periods of time. Thus, the system must form a closed line of behavior that eventually returns to some previous state to keep running.

If each output produces an input for which there is a doable output, a system with closed line of behavior will keep running indefinitely. It will stop only if it contains hidden variables that change value, if it changes its modulus of behavior, or if it is interrupted by other outside influences. Obviously, the complexity of the line of behavior of a responsive machine is limited by the number of input/output relations that can be stored in its memory.

Closed line of behavior having time as a variable

It is difficult to draw a closed line of behavior on a two-dimensional surface with time included as a variable, as shown in Figure 2.23.

Goals

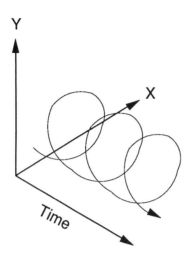

Figure 2.23. A closed line of behavior would appear as a *spiral* if time were included as a variable.

The number of revolutions of the spiral indicates how many times a machine executes a closed line of behavior. Counting the revolutions of a closed line of behavior indicates how many parts a production machine has made or how many times a truck has completed a delivery route.

Closed line of behavior with the time variable omitted

The time variable can be omitted to reduce the difficulty in representing closed lines of behavior, as shown in Figure 2.24, though the time variable is a fundamental part of any line of behavior. The simplest closed loop is a circle. Discovering a closed loop of behavior may be difficult because it may be obscured by the time variable. For example, it may occur to us for a long time that we repeat some things every day.

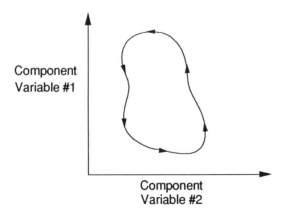

Figure 2.24. A closed line of behavior with the time variable omitted appears as a *closed loop* of behavior.

Probabilistic empirical relations

In many cases, a specific output may not occur for a given input, and some outputs may occur more frequently than other outputs for a given input. Thus, the controller of an empirical machine must remember the probability of each possible input/output relation and determine what output has the highest *historical probability* of being carried out for each input condition it encounters.

Therefore, an empirical controller must experience input/output relations many times to acquire an accurate value of the relative probability of each possible input/output relation. Every time an empirical controller goes around a closed loop of behavior, it increases the accuracy of its record of the probabilities of the relations it experiences. We experience the same effect when we become more certain about the route taken to a given location every time we follow that route successfully. Thus the empirical controller may have to establish a closed loop of behavior before it can establish an accurate understanding of the relative probabilities of the input/output relations available in its environment.

The greatest challenge of an empirical controller is to stay within the limits of its knowledge so that it can preserve the closed loop of behavior that allows it to operate without outside assistance. Much of this book deals with the process that allows an empirical controller to keep its behavior going without intervention from an outside influence.

Principle of closure

Closure, meaning the formation of repeating patterns of behavior, is a far-reaching concept. We see it in our "daily routines" and the repetitions of our weekly, yearly, and generation "cycles" of activities. Without *routines*, we would encounter too many different experiences to be able to establish accurate empirical knowledge about all of them, causing us to become "lost." Closure is a prerequisite for the continued existence of all empirical systems. Animals and plants must establish closure in their food/energy cycle to survive. They must find ways to obtain food; so they have the energy to find food. Closure is a fundamental requirement in economic systems. For example, spending must lead to earnings so that there can be more money available to produce earnings. Even computer programs run in closed cycles and loops. A program will "crash" when it calls for a file or address that does not exist, is not a correct part of the program or is otherwise undoable. For example, a program may jump to an address in an unfamiliar region from which it cannot return.

Note, however, that the computer still exists after a crash. If data and program files remain free of errors, the program still exists and can be run again. The behavior associated with a given run may get "blown away," but the program and the computer may still exist. However, if left alone (nobody reinitializes and "boots up" the computer), the program and the computer may not operate. If a computer crashes while it is operating machinery, the machinery and/or its product may be badly damaged. In this case, the continued existence of given control behavior may be more important than the continued existence of the computer itself. Closed lines of behavior with a limited number of states may be easier to protect against crashes.

This is because the machine can be given the required fault correction behavior for all of the possible faults.

We have shown that an empirical controller has to find out what it can do and what it cannot do. This is the basis of the Empirical Algorithm. The controller also has to discover behavior that causes it to receive inputs for that it has these doable outputs. So, closure is achieved when the empirical controller discovers and incorporates outputs that can be carried out (doable outputs), when these doable outputs cause inputs with which it has experience (familiar inputs), and when these familiar inputs cause another set of doable outputs leading to a closed line of behavior.

Complex line of behavior

Unique methods of survival in a wild environment have evolved in animal such as elephants. They must consume large quantities of vegetation for nourishment. This forces them to keep moving to new areas. They also cannot go for more than three days without large quantities of water. So, they travel in groups that include older members who have considerable knowledge about the location of food and water. The survival of the animals in these groups may depend upon the memory of the mature members as to the location of water holes and sources of food.

Features of behavior

The paths that foraging animals take in the wild can be called *complex lines of behavior*. They make decisions to follow a *main line of behavior* and *alternative* sublines, and must find *convergent sublines* if drawn off the main line of behavior by a *divergent subline of behavior*. These *features of behavior* are shown in the complex line of behavior shown in Figure 2.25.

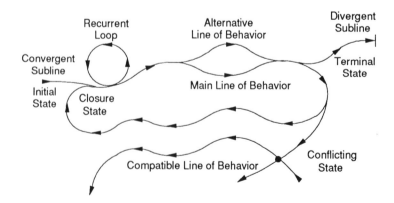

Figure 2.25. A *complex line of behavior* represents the sequence of actions typical of animal and machine behavior.

This diagram may represent the behavior of elephants in the wild, a piece of machinery, a computer program, an organism, or an empirical machine. A line of behavior is seldom a simple loop or a straight line from initial to terminal state. More often a line of behavior consists of branches, subloops, alternative paths, and multiple entry/exit states.

Closure of a complex line of behavior

Though a complex line of behavior may appear to wander, it must return to familiar points if it is to be used by empirical organisms. Many groups of animals in the wild return to specific water holes, pastures, and river crossings at particular times during in the year. Their line of behavior is closed because they return to previous points and retrace portions of their previous paths. The experienced gained by this behavior gives them a competitive edge over animals that forage at random.

The behavior of most organisms is infinitely more complex than that shown above. It may be difficult to recognize the closed or repeating nature of their behavior. However, closure is an essential requirement for the existence of empirical systems with a finite decision making ability.

If a responsive machine is to produce this kind of behavior, it must determine the *probability of success* of different options at each decision point, store this information in some kind of memory, and then search this memory and select those options representing the highest probability of success at a given decision point.

For these decisions to be valid and useful, a machine or an animal must experienced these decision points enough times to establish an accurate assessment of the real probability of success of each option.

Another requirement of a responsive machine or animal is that it find its way back to these decision points. Thus, the behavior of an animal or a machine must include correct decisions about finding its next water hole or recharging station, and must include behavior that causes them to return to the limited number of previously used water holes or recharging stations.

Elements of a complex line of behavior

A complex line of behavior is made up of *elements of behavior* such as the main line of behavior, divergent sublines, convergent sublines, recurrent loops, parallel lines of behavior, alternative sublines, coincident lines of behavior, compatible lines of behavior, initial states, active states, inactive states, terminal states, inaccessible states, accessible states, and intersecting lines of behavior. These elements of behavior cover the range of possible living and nonliving control behavior. These features describe and define how a controller and its environment can perform, and how one controller/environment system can interact with another.

Main line of behavior

The *main line of behavior* (Fig 2.26) can be defined as the core or basic behavior of a machine or organism. It generally allows the machine or organism to meet its primary need to survive or accomplish some overall objective. For example, the desired trajectory of a rocket may be considered its main line of behavior. By following its main line of behavior, the rocket will arrive at its target.

Main Line of Behavior

Figure 2.26. The *main line of behavior* leads toward some useful or desired accomplishment.

The main line of behavior may be hard to identify. Much of the behavior of a system may be taken up with deviations from and corrections to the main line of behavior. We may find it difficult to identify the main line of behavior leading toward our goals.

Initial state

An *initial state* (Fig. 2.27) is the starting place for a line of behavior. An initial state is usually an *accessible state* that has special features that make it useful and desirable as the starting point of a line of behavior.

Line of Behavior

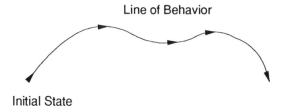

Figure 2.27. The *initial state* is the first of a set of states that occur in a given sequence.

Initial State

The "home" position of a piece of machinery, the recharging station of a mobile robot, the presence of an unfired cartridge in a gun, and a runner in the "ready position" on a track are examples of initial states. A line of behavior may have many initial states. The more initial states it has, the more it can *interact* with other systems that can turn on or activate these initial states.

Terminal state

A *terminal state* (Fig. 2.28) is the end point of an active line of behavior. It implies that behavior cannot continue beyond that point or that the line of behavior becomes inactive beyond that point.

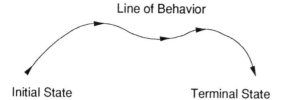

Line of Behavior

Initial State   Terminal State

Figure 2.28. A *terminal state* may prohibit a line of behavior from obtaining closure.

A terminal state may be a point where a rocket hits its target, where a machine breaks down, where a computer program reaches a "stop" or "end" command, or a search reaches an "end of file" status. A terminal state may be caused by the system entering a cul-de-sac from which it cannot emerge, such as falling off a cliff or even running out of energy in a situation where it cannot recharge itself. In general, if a line of behavior reaches a terminal state, some outside agent must restart the line of behavior at an initial state if the system can still operate. For autonomous systems acting alone, a terminal state signifies the end or death of the behavior of that system.

Closed line of behavior

Most machines and organisms operate in a closed line of behavior, as shown in Figure 2.29. If a machine or organism can produce only a limited number of states, it must establish a closed line of behavior to operate over a period of time by itself.

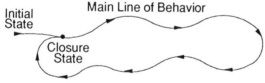

Initial State   Main Line of Behavior
Closure State

Figure 2.29. A *closed line of behavior* contains no terminal states or end points.

Divergent sublines

Many events can take us away from our main line of behavior. Falling, getting lost, experimenting with new behavior, and traveling to new places are examples of *divergent sublines* of behavior (Fig. 2.30).

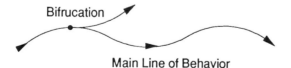

Bifrucation

Main Line of Behavior

Figure 2.30. A *divergent subline of behavior* takes the system states away from the main line of behavior.

The sudden application of a load, the interruption of a control signal, the loss of power, or the appearance of an additional control unit within a control system usually cause divergent sublines of behavior. The point of divergence or bifurcation in the line of behavior occurs when one or more variables in a transition in the main line of behavior assume a value other than their previous values. This transition acts like a switch that can direct the line of behavior one way or the other. If the changing variable is a hidden variable, the line of behavior will appear to bifurcate according to some probability that can be determined only by repeated observations.

Convergent sublines of behavior

A *convergent subline of behavior* causes the system states to merge with the main line of behavior, as shown in Figure 2.31.

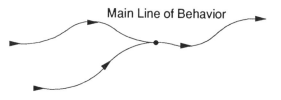

Main Line of Behavior

Convergent Subline of Behavior

Figure 2.31. A *convergent subline of behavior* restores the system to its main line of behavior.

To preserve closure and maintain the existence of a line of behavior, there must be a convergent line of behavior for each divergent line of behavior and initial state. The more convergent sublines there are, the more likely the system can withstand disturbances, perturbations, and other interferences. A machine with many convergent lines of behavior is more *robust* than a machine with a single line of behavior.

Recurrent loop

A line of behavior may diverge and converge in a *recurrent loop* that causes the behavior to return to the same or a preceding state after some period of time, as shown in Figure 2.32.

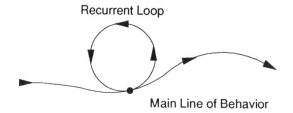

Recurrent Loop

Main Line of Behavior

Figure 2.32. A *recurrent loop of behavior* can be used to keep the main line of behavior from advancing.

A recurrent loop can be used to change the time in that an event occurs in the main line of behavior. For example, an airplane may be required to circle an airport until it is given permission to land. A recurrent loop can be used to "stall" or "kill time" in many other ways. Drumming one's fingers while waiting for someone, saying "ah" while talking, or a conditional loop in a computer program are examples of recurrent sublines of behavior.

Parallel line of behavior

When the values of variables in two or more lines of behavior differ only by a fixed constant, they may form *parallel lines of behavior* (Fig. 2.33). Parallel lines of behavior reinforce the main line of behavior by providing behavior that is similar even under different conditions. An example of parallel behavior occurs when different types of instruments in an orchestra play the same melody in different octaves or play in another form of harmony.

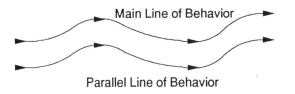

Main Line of Behavior

Parallel Line of Behavior

Figure 2.33. A parallel line of behavior may strengthen the main line of behavior.

Alternative sublines of behavior

An *alternative subline of behavior* (Fig. 2.34) is a combination of divergent, parallel and convergent lines of behavior. It allows a system to get past a point on the main line of behavior using a slightly different set of variables and/or values. If used often enough, an alternative line of behavior may become part of the main line of behavior.

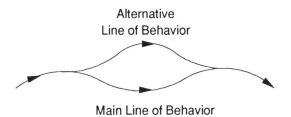

Alternative
Line of Behavior

Main Line of Behavior

Figure 2.34. An alternative line of behavior is like a detour of the main line of behavior.

Coincident line of behavior

When two or more systems operate upon the same set of variables in the same way simultaneously, they share a *coincident line of behavior*. They may operate without interferences using coincident and/or compatible lines of behavior. If two or more lines of behavior share identical states simultaneously, they are coincident, as shown in Figure 2.35.

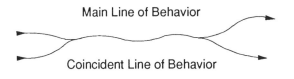

Main Line of Behavior

Coincident Line of Behavior

Figure 2.35. Two or more systems can exist in the same environment simultaneously if they operate on a coincident line of behavior.

Members of an orchestra have a coincident line of behavior when they play their instruments in unison. Two or more people pulling together on a rope share a coincident line of behavior. This type of behavior may allow a group of individuals to create a greater influence than a single individual acting alone. This may allow them to overcome physical obstacles in their environment or allow them to dominate the behavior of another group.

Compatible line of behavior

The other kind of behavior that can permit two or more systems to operate together upon the same set of variables is called a *compatible line of behavior* (Fig. 36).

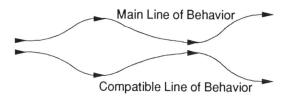

Main Line of Behavior

Compatible Line of Behavior

Figure 2.36. A *compatible line of behavior* occurs when two or more systems share common variables at different times without interfering with one another.

Two groups of elephants may follow the same river for a distance. If they travel that route at nearly the same time as another group of elephants, they may temporarily exhaust the food supply. However, their behavior may become coordinated to avoid arriving at the river at the same time. Or the two groups also may stay on the opposite side of the river so as not to interfere with each other. Compatible behavior is based upon finding ways to share common variables without interference.

Intersecting lines of behavior

If two or more different lines of behavior have identical states at the same time, they have *intersecting lines of behavior* (Fig 2.37). How and when the behavior of one system meets the behavior of another can be critical for the success or even existence of both. Since two or more objects cannot occupy the same space simultaneously, great care must be taken at these times to ensure that the behavior of one system does not interfere with the behavior of another. An intersection between two roads represents the potential for disaster that can occur when lines of behavior intersect. If an automobile on each road arrives at their intersection simultaneously, the modulus of behavior of both automobiles may be changed significantly.

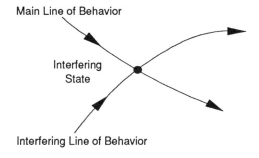

Main Line of Behavior

Interfering State

Interfering Line of Behavior

Figure 2.37. Intersecting lines of behavior create the need for specific behavior that reduces the potential for unwanted interferences.

If a group of elephants arrives at a point on their forage path that lies on the forage path of another group, all of the vegetation may have been consumed. If they arrive at the same point after a time delay long enough to allow the vegetation to recover, there may be no interference. So the relative timing of their lines of behavior may be critical to the outcomes of their behavior. In other cases, it may be important for two or more systems to be at the same place at the same time. For instance, an aircraft and a refueling tanker must meet at nearly the same place at the same time if the refueling operation is to take place successfully.

Active and inactive lines of behavior

The *active line of behavior* changes with time. An *inactive line of behavior* remains constant, as shown in Figure 2.38. Some lines of behavior may be active at one moment in time and be inactive at other moments in time.

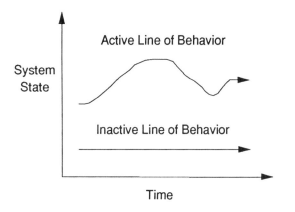

System State

Active Line of Behavior

Inactive Line of Behavior

Time

Figure 2.38. At any given time, there may be *active and inactive lines of behavior*.

Accessible and inaccessible states

Some states in a line of behavior may be *accessible* and amenable to change by an outside influence while others are *inaccessible* and unalterable (Fig. 2.39). For example, the course and trajectory of a rocket may be changed during the (accessible) burn period, but cannot be changed during the (inaccessible) ballistic portion of its flight.

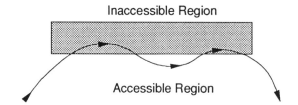

Figure 2.39. A line of behavior is made up of *accessible and inaccessible states.*

A production machine such as a *stamping press* has inaccessible states that the operator cannot alter once a machine cycle is initiated. The die stroke is an example of the inaccessible portion of the machine cycle. The operator can exercise control over a piece of machinery only while it is operating in its accessible region and may have no influence upon the machine while it is operating in its inaccessible region.

Map of behavior

All of the different possible lines of behavior can be combined into a drawing that looks like the road map shown in Figure 2.40. The actual line of behavior carried out by a driver is a given route taken by the driver. Each intersection is a potential transition. The number of intersections encountered in a line of behavior is equal to the number of transitions in that line of behavior. The specification of that line of behavior is the set of directions taken at each intersection.

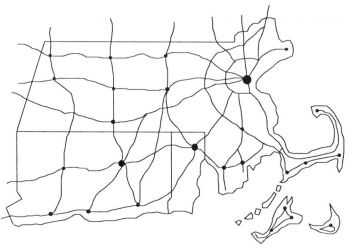

Figure 2.40. A *map of behavior* shows all of the possible routes that a driver can take.

Some regions like the towns on the islands are inaccessible unless a ferry is available. Some roads, like the roads that lead to the beaches, lead to terminal states. The main highways between cities are main lines of behavior for most drivers. Some drivers take many alternative subroutes each day while other drivers follow the same routes every day and follow other routes only when they become lost.

Information in the map of behavior

The information in these lines of behavior is determined by the number of inter-sections and the variety of choices available at each intersection. If many choices are available, the probability of selecting a given choice at random is small. Thus, more information needs to be stored in a machine that has to deal with many possible choices. This information can be embodied in circuit elements that represent the specific choices produced by a memory controlled machine (see Part II).

Self-sustaining behavior

The statements in Table 2.3 summarize some conditions that must be met for a machine to establish a self-sustaining line of behavior. The conditions that allow a machine to produce self-sustaining behavior also apply to all living systems.

Table 2.3 Requirements for self-sustaining behavior

---

1. To be self-sufficient, a machine must establish closure. It cannot have terminal states in its active line of behavior, or it must be connected to an environment, or another control system or systems that will restart its active initial states if it reaches a terminal state. This effectively reestablishes closure.
2. A machine may have many terminal states, but can maintain closure only if these terminal states lie in inactive lines of behavior.
3. The behavior of two or more machines will not interfere if their lines of behavior do not share common states at the same time.
4. The primary mission of an autonomous empirical system is to develop behavior that excludes any terminal states.
5. If a system reaches a terminal state, it must be restarted by an outside source of behavior at some initial state. Such a system can be classified as *semi-autonomous.*

The purpose of the study of empirical control is to develop machines that can spontaneously develop self-sustaining behavior.

Modes of behavior

It is possible to make generalizations about the way that machines operate. For example, a clock repeats behavior in a *routine* manner every 12 hours. A spring attempts to return to its original length when displaced in either direction in a *stable* manner. A ship may attempt to turn away from a straight course without any action from the rudder in an *unstable* manner. These *modes of behavior* are determined by the modulus of behavior of each machine.

Routine behavior

A machine that does the same thing repeatedly like a clock or punch press acts according to a closed state-determined line of behavior, as shown in Figure 2.41.

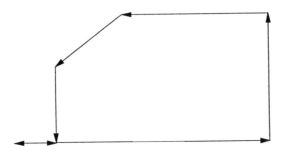

Figure 2.41. The behavior space of a machine producing *routine behavior* has a single closed line of behavior.

Most automatic industrial machines operate in a routine or repeating line of behavior.

Stable behavior

If direction of motion of a machine is changed suddenly, it will attempt to continue in its original direction, producing a divergent line of behavior. A restoring force is required to reposition the machine in the new direction as, shown in Figure 2.42.

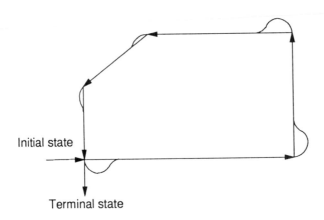

Initial state

Terminal state

Figure 2.42. Stable behavior is produced when the restoring force brings the system state back to its main line of behavior within a limited time and with a limited divergence.

The divergent line of behavior may have to be kept small to allow the machine to operate properly. The size of the divergent and convergent lines of behavior can be kept small by applying a large restoring force and by keeping the mass of the machine as small as possible.

Unstable behavior

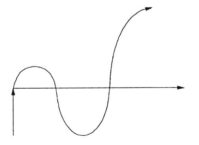

Figure 2.43. Unstable behavior occurs when the system state keeps getting further from its intended line of behavior.

If the restoring force is too large or is applied for a period that is too long, the machine may overshoot its line of behavior, as shown in Figure 2.43. Depending upon the mass of the machine, the amount of restoring force, and the amount of energy consumed in friction, a machine element may return to its intended line of behavior in a stable manner, fail to reach its intended line of behavior, or continue to oscillate to points farther from the intended line of behavior in an unstable manner.

Negative force behavior

A machine may be used to maintain its system states at or close to some stable position. Variations in the load on these systems, or other interferences, may cause a movement of the line of behavior away from the set point. This is represented by a divergent subline of behavior as shown in Figure 2.44.

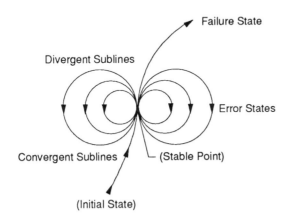

Figure 2.44. The *behavior space of a machine with negative force behavior* is characterized by a group of sublines that converge on some position for every divergent line of behavior.

To return to the original state, a stable machine must determine how far the system state has moved from the stable point and then produce a convergent subline *toward* the stable point for each divergent subline. In general, negative force behavior leads to a closed line of behavior.

Negative direct force behavior

A machine can produce a force toward its point of stability that increases or decreases as it is moved away from this point. The end of a linear spring produces a force toward its *rest position* that is proportional to its displacement from this point, as shown in Figure 2.45.

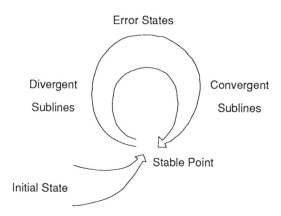

Figure 2.45. The *behavior space of a machine with negative directly proportional force* shows that an object can be drawn into a stable point with an ever increasing force the further it is displaced from that point.

Since the direction of the force produced by a spring is in the opposite direction of the displacement, and is in proportion to the displacement, a spring produces *negative directly proportional force* (NDPF) behavior. A nonlinear spring produces *negative direct force* behavior according to some power of the displacement.

Negative inverse force behavior

A machine also can produce a *negative inversely proportional force* (NIPF) if it attempts to arrive at some point with ever increasing force the closer it gets to that point (Fig. 2.46).

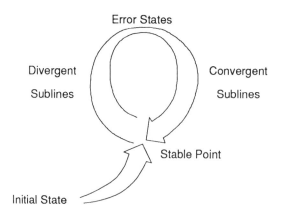

Figure 2.46. The *behavior space of a machine with negative inversely proportional force* shows that an object can be drawn into a stable point with an increasing force the closer it gets to that point.

Objects having a negative inversely proportional relation are not strongly attracted when they are far apart. However, once they are brought close together they attempt to stick together. This property makes them useful as detents and door latches. Some machine elements consisting of sets of masses, electric charges, or magnets are attracted with a *negative inverse force* behavior according to some power of the displacement shown by the equation $F = m/r^2$.

Detents

A very useful machine element is a device holds a constant position around a *stable point* until it is subjected to a force that exceeds a predetermined level, as shown in Figure 2.47.

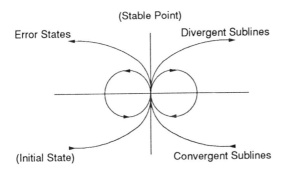

Figure 2.47. A *detent* uses negative inverse force behavior to hold a nearly constant position until it is subjected to a force above some predetermined level.

A detent can be used as a safety device that holds machine elements together when the machine operates normally, but allows the elements to *break away* if the machine malfunction.

Bistable machine

A machine that operates with negative inverse forces may have two or more stable points, as shown in Figure 2.48.

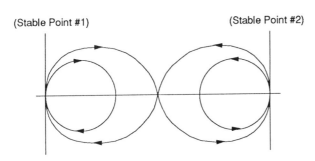

Figure 2.48. The behavior space of a machine with *negative inverse force behavior with two stable points* shows that the machine will operate about one point or the other.

A larger force may be required to move the system state from one *stable point* to the other than is involved in movements about each stable point. If a machine has two stable points, it is said to be a bistable machine.

Applications for machines with negative inverse forces

If a system with negative inverse force has two or more stable points, it will attempt to stay at the last location in that it has been placed. This property makes a unit with negative inverse force useful as a memory device or as a multiposition switch, as shown in Figure 2.49.

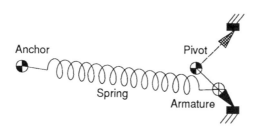

Figure 2.49. A *toggle mechanism* is a bistable machine that produces negative inverse forces.

A *toggle* is a machine that operates according to a negative inverse force between two stable points. A toggle can be moved easily in one direction or the other when it is midway between its stable points, but energetically moves toward one point if it is moved closer to that point. Since a toggle mechanism attempt to remain at its point of stability, it can be used as a *binary memory device* if both positions are used as values of a variable (see Appendix A). If a machine has three stable points, it may be used as a *trinary memory device*. Most two-position switches are *toggle switches* that are either *normally open* or *normally closed*. A *multiposition switch* is a device that operates with negative inverse forces, and has as many stable points as switch contact positions.

Positive force behavior

Some machines tend to produce sublines of behavior in the direction away from an unstable point using *positive force behavior*, as shown in Figure 2.50. For example, a ships will always turn to the left or right although its rudder is held in the center position. Some airplanes will turn (pitch) nose up if the pilot takes his or her hands off the controls. In general, positive force behavior leads to a loss of control and open lines of behavior if uncorrected.

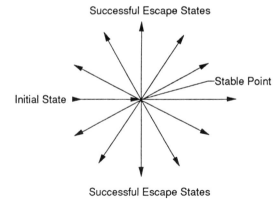

Successful Escape States

Initial State ▶

Stable Point

Successful Escape States

Figure 2.50. The *behavior space of a machine with positive force behavior* is characterized by a group of sublines that diverge from some point of instability.

Positive direct force behavior

If a machine attempts to move away from a *point of instability* with a force that increases according to its displacement from that point (Fig. 2.51), it is using positive direct proportional force behavior. A positive direct proportional force can become very destructive because it becomes more energetic as it continues. It starts small and gets stronger. A "runaway" engine with a stuck throttle is an example of a machine operating with *positive direct force*. It keeps accelerating, producing more power until it blows itself apart.

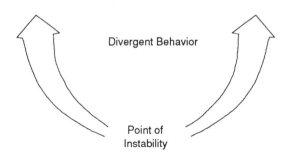

Divergent Behavior

Point of
Instability

Figure 2.51. The *behavior space of a machine with positive directly proportional force* is characterized by a set of divergent sublines of behavior that extend away from a point of instability with an ever increasing force.

Systems with positive direct *square* force such as chemical explosions are even more explosive than systems with a positive direct force because the rate at that they react doubles for every 10°C, and the explosion causes the temperature of the fuels to increase. Thermonuclear "chain reactions" may operate at even higher powers depending upon the nature of the reaction.

Geysers are also positive direct force events. Once hot gases start to escape from a vent, the pressure within the fissure may fall, causing the release of even more superheated steam. Volcanos are also examples of positive direct force events. The flow of hot magma may increase the size of a fissure allowing more magma to flow.

These *positive direct force* (PDF) processes produce sudden, powerful, and concussive releases of energy that are usually highly destructive. If a device is expected to produce an explosion, but for some reason remains at its point of instability, that device fails to operate properly and that point is a *failure state*.

Positive inverse force

A machine also can produce *positive inversely proportional force* (PIPF) behavior
(Fig 2.52). A positive inverse proportional force acts in the same direction as the
displacement from a given point but becomes smaller as it gets away from that point.
PIPF is manageable. Once the machine gets away from its point of instability, the
force level on the machine drops off.

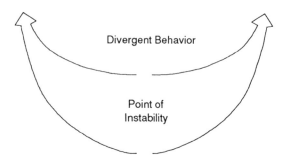

Divergent Behavior

Point of
Instability

Figure 2.52. The *behavior space of a machine
with positive inversely
proportional force* shows
that it produces a force
that is directed away
from a critical point.
This force decreases as
the machine state moves
away from that point.

Positive inverse proportional force becomes very energetic only near its *critical
point.* Overrunning clutches and magnetic suspension (repulsion) systems are
examples of machines that produce *positive inverse forces* or even *positive inverse
square forces.*

Self-energizing mechanisms

The features of positive force behavior are sometimes used to accomplish useful
tasks. For instance, the cam toggles in an overrunning clutch use a positive inverse
force to jam or lock the housing of the clutch to the clutch body in one direction
while letting the housing run free in the other, as shown in Figure 2.53.

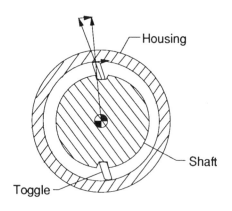

Housing

Shaft

Toggle

Figure 2.53. An *over-
running clutch* uses a
positive inverse force to
lock up the clutch in one
direction of rotation only.

Rotation of the housing in one direction moves the toggle away from the housing. Rotation in the other direction moves the toggle against the housing. If the tangent of the angle is less than the coefficient of friction, trying to rotate the housing more in that direction will cause the toggle to bear harder against the housing and the force due to friction will always be greater than the applied force. Thus, the holding force is energized by the applied force. This is an example of a positive inverse proportional force in that the force due to friction increases as the toggle is forced against the drum. It reaches a maximum as the toggle approaches the upright position and suddenly reverses direction as the toggle goes over center and diminishes as it moves away from center.

Positive inverse proportional explosions

The explosion of a compressed air cylinder and the failure of a dam are examples of positive inverse force behavior in that they start out very intense and diminish in intensity as the compressed gas or contained water is released. These events differ from chemical or thermonuclear explosions, which are positive direct force events that become more intense as the explosion continues due to the increase in temperature or radiation levels during the explosion.

Chase/escape behavior

One machine can control another machine. For example, a ship auto-pilot can be connected to the ship's rudder so that it attempts to keep the ship on a straight course. The auto-pilot produces negative force behavior by acting upon the ship in the opposite direction in which it attempts to turn. The ship produces positive force behavior in that it attempts to move away from the heading established by the auto-pilot. Thus, the ship attempts to *escape* from the heading of the auto-pilot, and the auto-pilot attempts to capture and maintain the heading of the ship on the desired course.

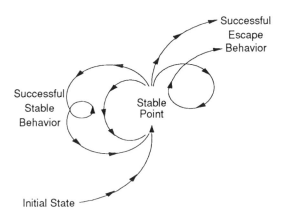

Figure 2.54. Chase/escape behavior evokes a large expenditure of action and energy.

When a *chase machine,* producing a negative force behavior, attempts to catch an *escape machine,* producing a positive force behavior, large amounts of action and energy may be evoked in the resulting *chase/escape behavior* (Fig. 2.53). If the chase machine uses negative direct force behavior and the escape machine uses positive inverse force behavior, the closer the chase machine comes to the escape machine, the harder the escape machine works to get away. The further the escape machine gets away, the harder the chase machine works to catch up.

The action of a system with chase/escape behavior may be endless in that there may be no obvious terminal condition when the two machines are of nearly equal strength, speed, and intelligence. The line of behavior carried out by these machines may be *cyclical* because the two machines probably will come together and then separate. Their behavior also may be *chaotic* in that small variations may result in large changes in the resulting line of behavior.

The motions of the planets around the sun may be considered chase/escape behavior in that the centripetal force created by the rotation of each planet around the sun acts against the gravitational forces created by the mass of the sun and its planets. Chase/escape behavior is the basis of many *sports* and *games* such as football, basketball, and baseball in which one team tries to get the ball while the other team tries to keep the ball away from them. The behavior of the machine/environment systems presented in this book consists mainly of some form of chase/escape behavior. For example, an auto-pilot attempts to hold a ship on course and the ship usually attempts to swing off course. In summary, most interesting, dynamic, and useful machine and living systems operate with chase/escape behavior.

Chapter 3
Networks of Machines

We have seen how a set of sensor variables can be connected to a set of actuator variables by a single *standalone machine*. Two or more machines can be connected to these sensor and actuator variables, forming *networks of machines*.

Compound machines

So far, we have discussed one kind of machine: the sensor/actuator machine. However, there are three more basic machine configurations: the actuator/sensor machine, the sensor/sensor machine, and the actuator/actuator machine. Each machine is useful in particular situations, and two or more of these machines can be connected to the same set of sensor and actuator variables forming a *compound machine*.

Sensor/actuator machine

Most responsive machines are considered *sensor/actuator machines* in that they sense some conditions in the environment and then produce some actions that may affect the conditions in the environment according to the modulus of behavior of the environment, as shown in Figure 3.1.

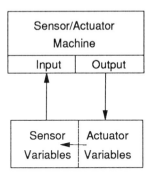

Figure 3.1. A *sensor/actuator machine* converts values of the sensor variables in the environment to values of the actuator variables in the environment.

An example of a typical sensor/actuator machine is a home heating system that senses room temperature and then operates a furnace according to the sensed temperature and the setting of the thermostat.

Actuator/sensor machine

The environment is usually considered an *actuator/sensor machine* because the environment converts the actions of a machine into new environmental conditions that are then sensed by the machine. However, another actuator/sensor machine can be connected to the environment or act in place of an environment, as shown in Figure 3.2.

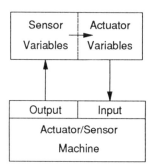

Figure 3.2. An *actuator/sensor machine* has sensors connected to the output (action) variables of another machine, and actuators connected to the input (sensor) variables of the other machine.

An actuator/sensor machine converts the action variables of a system into a new set of values of the system's sensor variables. For example, an actuator/sensor machine converts the heat produced by a furnace to some value of a thermometer. This actuator/sensor machine may consist of a blower and air ducts that carry the heat produced by a furnace to the room containing a thermostat.

Sensor/sensor machine

A *sensor/sensor machine* (Fig. 3.3) is connected only to the sensor variables of a system. The values of these variables may be transformed by a sensor/sensor process in the environment, or by another sensor/actuator/sensor process.

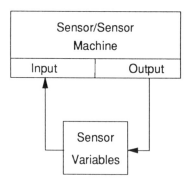

Figure 3.3. A *sensor/sensor machine* produces a new input state in response to the current input state without directly influencing the system's actuators.

A digital computer may be considered a sensor/sensor machine in that it performs a transformation on its current program statement to establish the next program statement. If the program statement number is considered a sensor value, a digital computer operates upon the next statement number or a statement number determined by some conditional command based upon the current sensor value. So it proceeds through a sequence of sensor values.

Actuator/actuator machine

An *actuator/actuator machine* (Fig. 3.4) produces a new output state in response to the current output state. It does not rely upon the system's sensor variables to carry out its behavior.

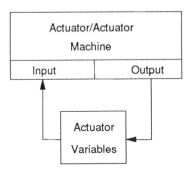

Figure 3.4. An *actuator/actuator machine* produces a new actuator value for each current actuator value.

A punch press may be considered an actuator/actuator machine in that it produces an output die position and velocity based upon a transformation of its current die position and velocity. For the sake of the simplicity of this discussion, most of the control systems presented in this book consist of sensor/actuator machines connected to an actuator/sensor environment.

Compound machine

A compound machine is created when machines of two or more of the different basic configurations defined are connected to the same set of sensor and actuator variables, as shown in Figure 3.5. The behavior of a compound machine has great durability because the system states are produced by more than one responsive device. The sensor/sensor machine can provide a continuing set of new input states if the connection to the environment sensor variables is lost. The actuator/actuator machine can position the actuators without sensor information, leaving the sensors free to deal with other information. The actuator/sensor machine also can keep the sensor/actuator machine going in case the connection to the environment is lost. Even though many real machines are similar to compound machines, we will continue to deal with the basic sensor/actuator machine with the understanding that the concepts that are developed around the sensor/actuator machine can be extended to compound machines if the need arises.

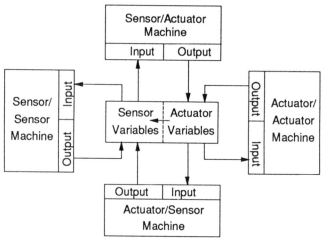

Figure 3.5. A *compound machine* can be used to keep a system operating even when disconnected from an environment.

Superimposed machines

The sensor and actuator variables of a standalone machine or a compound machine
can be connected to two or more machines. These machines can be arranged in
parallel, forming a network of *superimposed machines.*

Standalone machine

We have viewed a machine as a single entity connected to an environment or another
machine serving as an environment.

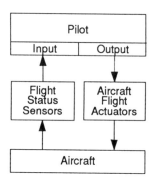

Figure 3.6. A *standa-
lone machine system* with
two or more variables
may have to deal in a
unique way with each
combination of values of
all the variables.

A *standalone machine* can produce a closed line of behavior in any mode of
behavior. For example, a pilot may maintain straight and level flight like an
auto-pilot. However, the pilot also may have to navigate and deal with equipment
malfunctions, as shown in Figure 3.6. As the number of variables increases, the task

may become more difficult. The pilot may have to find ways to deal with each combination of values for each variable, and may have to establish priorities about which variable to deal with first.

Diverse network of machines

A *diverse network of machines* is made up of a set of separate machines. A set of separate machines may provide far better control behavior than a single unit acting alone. For example, a system consisting of a pilot and copilot (Fig. 3.7), may provide better control behavior than a single individual who attempts to carry out many tasks simultaneously.

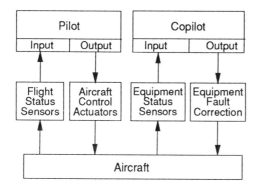

Figure 3.7. A diverse machine system is made up of separate machines each of which carries out its own control task.

Individuals in a diverse system may produce successful behavior by dealing with their own variables. The success of the system as a whole is based upon each individual successfully carrying out their individual *control tasks*. For example, if the copilot gives the pilot the correct course, the pilot maintains that course, and the copilot overcomes any equipment malfunctions, the flight may be carried out successfully.

Superimposed network

As stated, a responsive machine can continue to operate only if another machine or an environment provides a continuous supply of new input states in response to its output states. If a machine does not produce output states that lead to new input states, the behavior of the machine will cease. The machine and its environment are considered connected in *series,* since action flows in sequence through each unit. However, states that cannot be supplied by one machine can be supplied by another machine connected in *parallel* to the same or similar set of variables. When two or more machines are connected in parallel to the same or similar set of variables, they are said to be *superimposed*. This configuration may allow for a division of labor that permits two or more units to work more effectively than a single unit working alone.

Individuals superimposed upon a single input and output variable

Two or more individuals can be connected to a single input and output variable. For example, the pilot and the copilot may attempt to maintain a compass heading by sensing the direction error signal and operating upon the rudder, as shown in Figure 3.8.

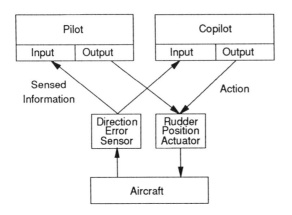

Figure 3.8. A superimposed system is made up of two or more individuals or machines connected to the same input and output variables.

When two or more machines are connected to the same variables, the behavior of the machines must be precisely adjusted to avoid conflicting behavior. For example, the pilot and copilot in a *superimposed system* must both attempt to fly the aircraft in the same direction and at the same altitude and airspeed.

Superimposed machine/attendant system

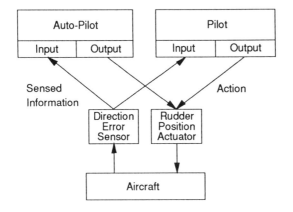

Figure 3.9. A machine and attendant can work together successfully on the same set of variables to accomplish a control task.

A machine and an *attendant* also can work together upon the same set of variables, as shown in Figure 3.9. For example, a pilot and an auto-pilot can work together to control the direction of an aircraft. The auto-pilot can be used to maintain control except when there is extreme turbulence or if traffic requires a temporary course change by the pilot.

Superimposed machines

The attendant in the system shown above also may be replaced by another machine, as shown in Figure 3.10.

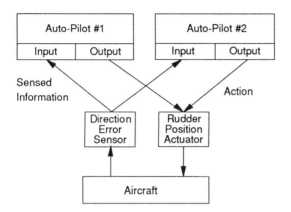

Figure 3.10. A system composed of *superimposed machines* consists of two or more machines connected to the same or similar set of variables.

For example, the pilot mentioned may be replaced by a remote control unit or a preprogrammed flight control unit. Each unit can be programmed to act upon or ignore each value or each variable. The system will work if the units produce a closed line of behavior. This can occur if each unit produces the same or nonconflicting values of the same variables at each moment in time, if each unit operates upon different values of the same variables, and/or if each unit operates upon different variables.

Coincident behavior

When different control units produce the same values of the same set of variables, they produce *coincident behavior*. Many control processes will accept coincident behavior. For example, the driver of a car and a *cruise control unit* can both operate the accelerator pedal if both attempt to maintain the same speed. Two or more people may sing the same melody in a chorus. These redundant behaviors may allow a unit to be withdrawn without interrupting the control process.

Value field selection

Many control systems will allow different control units to operate upon different values of the system variables. *Value field selection* occurs when two or more control units work on a control task by operating upon different values of the same sensor and actuator variables. Each unit can operate upon those values that are most appropriate for each unit. For example, an auto-pilot may deal with small deviations in directions, and a pilot may deal with large deviations. When different control units operate upon different values of the same variables, they are involved in a *division of labor based upon value field selection.* This process may allow each unit to be less complex, and provide better control behavior than a single system that has to be designed to deal with all the values of the controlled variables.

Individuals superimposed upon multiple input and output variables

Two or more individuals or machines also may operate upon the same set of *multiple variables* if their behavior is adjusted accordingly. For example, the pilot and copilot of an aircraft are expected to work together on the same set of controls (Fig. 3.11).

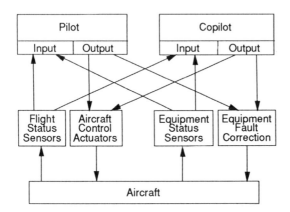

Figure 3.11. When *machines* or individuals are *superimposed on multiple variables*, the opportunity exists for each individual to deal with a different set of sensor and actuator variables.

Although there are many opportunities for the pilot and copilot to produce conflicting behavior when both attempt to deal with the same set of variables, they can establish behavior that is more desirable than a single pilot working alone. For example, the pilot or the copilot may attempt to correct a deviation in direction, attitude, or airspeed that is overlooked by the other.

Dispersion

Many control systems are made up of sets of separate variables in which combinations of values of the set have no particular significance. *Dispersion* occurs when the variables made active by one line of behavior differ from the variables made active by another line of behavior. When different control units operate upon different variables involved in a control process, they produce a *division of labor based* upon dispersion. This process also may allow each unit to be less complex, and provide better control behavior than a single system, which has to be designed to deal with all the controlled variables simultaneously. For example, the pilot and copilot may divide the control task such that the pilot monitors the instrument panel and the copilot maintains a lookout for other aircraft.

Roles of superimposed machines

When two or more machines and/or individuals are connected in parallel to the same or similar set of variables, each may play different roles according to the needs of the task environment.

Backup machine

When two or more machines and/or individuals operate upon the same set of variables and produce the same behavior, one or the other may be removed from the system without influencing the behavior of the system. This allows each to serve the *role* as a backup machine (Fig. 3.12) if the other unit is removed from the system.

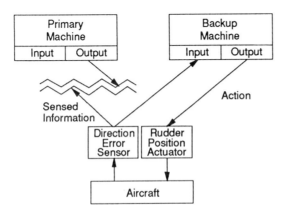

Figure 3.12. A *backup machine* must produce the same behavior as the primary machine and can maintain the behavior of the system if the primary machine is removed from the system.

The backup machine is expected to be able to produce all the control behavior that existed before units are moved from the system. The behavior of the backup machine is based primarily upon its being able to produce coincident lines of behavior.

Contingency unit

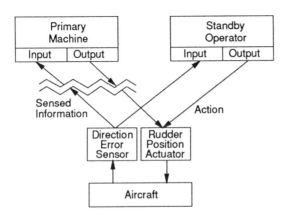

Figure 3.13. A *standby operator* deals with only a few specific states and lets the primary machine deal with most of the system states.

In some situations, an attendant or a machine may deal with only a few specific contingency situations (Fig. 3.13). When an operator or a machine unit acts only in a limited set of circumstances, it assumes the role of a *contingency unit*. For example, the operators of most automatic machines do not participate in the operation of their machines except to correct malfunctions. A contingency unit usually comes into action when critical values of a control variable go above or below some predetermined level. Thus, the behavior of a contingency unit is based primarily upon value field selection. The contingency unit also may deal with one set of variables at critical times and another set of variables at another time based upon dispersion.

Cooperating units

Thus, two or more machine units may divide the control process into separate tasks either through dispersion or value field selection. These units work on different values of the variables to which they are connected, or they may work on different variables producing separate lines of behavior that contribute to the closure of the behavior of the system.

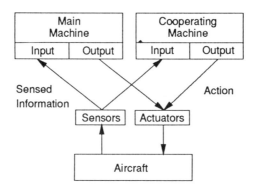

Figure 3.14. Cooperating machines may supply some behavior needed by the other units to produce successful behavior.

If one unit supplies some of the input and output states needed to maintain the existing line of behavior of the system, it is called a *cooperating unit* (Fig. 3.14). In some cases, a system cannot function without a cooperating unit. For example, it is very difficult for a land surveyor to work alone. A cooperating unit can be a co-worker, copilot, or some kind of supporting machine unit.

Structured superimposed networks of machines

Units can be superimposed upon different groups of input and output variables, forming an *assorted superimposed network*. These groups can be selected such that some units are part of a small group of variables, while other units are part of a larger group of variables. This type of *structured superimposed network* provides opportunities for diverse behavior to occur within the small groups and unitary behavior to occur within the large groups.

Assorted superimposed networks

The superimposed units shown earlier are connected to the same input and output variables. However, each unit may be superimposed upon a different set of input and output variables (Fig. 3.15).

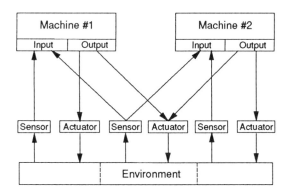

Figure 3.15. In an *assorted superimposed network*, each unit is connected to a different set of input and output variables.

Each unit has the opportunity for producing behavior by acting upon the variables to which it is connected. Each machine can be designed to operate successfully with its own set of variables and divide the task behavior involved with the variables they have in common.

Structured superimposed network of machines

Another way in which two or more machines can be connected to multiple variables is to have some machines connected to variables specific to each machine and have another machine connected to all the variables (Fig. 3.16).

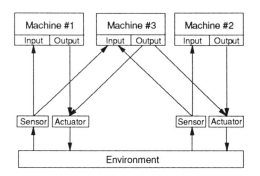

Figure 3.16. In a *structured superimposed network of machines*, some machines are connected to a few variables only, and the other machines are connect to more of the variables.

The machine connected to the larger set of variables may have to establish some method of sharing behavior with the other units. A systematic method of superimposing units may be required if the system has many variables.

Multilevel structured superimposed networks of machines

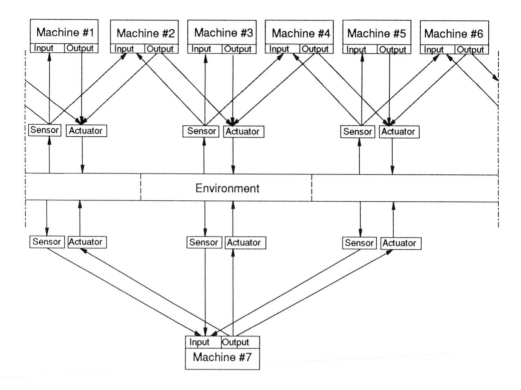

Figure 3.17. A multilevel structured superimposed network is made up of three or more over-lapping networks.

A *multilevel structured superimposed network* comes about by organizing the input/output variables into multiple groups. For example, one unit in a structured network may include all the variables in the set, and other units may include only some variables in the set (Fig. 3.17). The organization of the federal, state, and local governments in the United States is an example of a multilevel structured super-imposed network in that all three levels of government levy taxes, support education, provide law enforcement, etc., but do so with different groups of people within the country. Each level competes for the tax dollars from each individual and competes to provide services to each individual. Yet, each unit includes a different set of citizens, and each level attempts to provide different types of service.

Division of labor in a structured superimposed network

To avoid direct competition, each level in a structured superimposed network can act upon values of the service variables. For example in the U.S. system of law enforcement, the federal level concentrates upon crimes of groups of individuals involved in illegal interstate activities, states attempt to maintain order on their highways, and local police forces deal more with domestic disputes among individuals.

Parallel versus series organization

The structured superimposed network is different from a true hierarchical system of government. In a hierarchical system of government like that found in the former Soviet Union, the local governments collect taxes and pass some on to higher levels of government. The higher-level state and federation governments pass on some of this revenue to the highest-level central government. This kind of *series organization* (see hierarchical networks) is typical of many religious, social, and service organizations where dues are collected at the local level and passed on to regional and national organizations.

Series and parallel networks of machines

Up to now, we have viewed networks of machines as units connected in parallel to the sensor and actuator variables in the environment. However, machines can be connected in *series* in which the output of one machine is connected directly to the input of another machine. When machines can be connected in series, they can be arranged in networks that may allow more variables to be connected successfully than may be possible in a single individual or machine unit acting alone. Machines can be connected in series and in parallel to form a *hierarchical network*, which allows a single unit to mediate all condition/action relations. Machines also can be arranged in a *nodal network* so than more than one machine can mediate condition/action relations.

Direct series connection

We have seen how the output variables of one machine can be connected to the input variables of another machine through a given environment. The output variables of a machine also can be connected directly to the input variables of another machine in a *direct series connection* (Fig. 3.18).

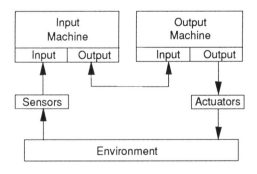

Figure 3.18. A *direct series connection* allows the output variables of one unit to influence the input variables of another unit without an intervening environment.

The first machine may transform the values of the sensor variables into a set of values that make it easier for the second machine to operate. For example, the first machine may filter out high and low values allowing the second machine to operate in a narrower bandwidth.

Direct convergent network

Two or more machines can be connected directly to a single machine, forming the *direct convergent network* shown in Figure 3.19.

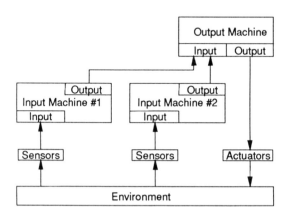

Figure 3.19. A *direct convergent network* is formed by connecting the output variables of two or more machines directly to the input variables of another machine.

In a direct convergent network, the values of multiple sensor variables can be combined and transformed by a single machine. This reduces the size of each machine compared to a single machine, which has to deal with all the sensor variables.

Direct divergent network

One machine can be connected directly to two or more other machines, forming the *direct divergent network* shown in Figure 3.20.

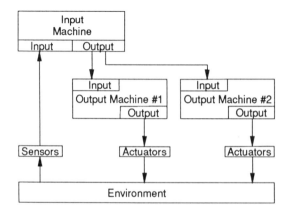

Figure 3.20. In a *direct divergent network*, action flows from one machine to two or more machines.

A direct divergent network increases the number of actuators that a single unit can control. It can use smaller output units than a single machine that controls all the actuators.

Direct hierarchical network

Machines can be arranged in a set of *assemblies* and *subassemblies* in a *hierarchical network*. In a system where sensor variables are collected by one set of machines and the actuator variables are collected in another set of machines, sensor information in a hierarchical network converges upon a central unit and then diverges to the actuators (Fig. 3.21).

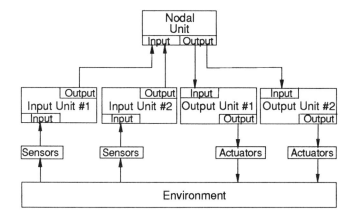

Figure 3.21. In a *direct hierarchical network of machines*, all the signals that determined behavior must pass through the single nodal unit.

Most machine assemblies, businesses, military organizations, books, and data retrieval systems are based upon a hierarchical network. This kind of organization allows the nodal unit to exercise some influence upon all sensor/actuator relations. This kind of network also reduces complexity by minimizing the complexity of each unit, limits the number of connections to and from other units, and provides an unambiguous method of identifying the activities of each unit and subunit.

Direct nodal network

A *nodal network* is similar to a hierarchical network except that each unit is connected to two or more other units. This allows two or more parallel paths for information to flow from the sensor units to the actuator units (Fig. 3.22).

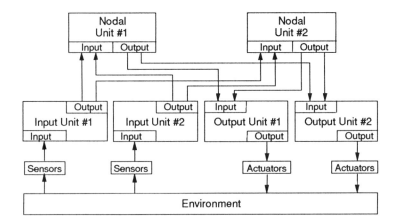

Figure 3.22. In a *direct nodal network of machines,* two or more machines connected to sensor machines may directly influence two or more machines connected to actuator variables.

When the outputs of the sensor machines are connected to more than one nodal unit, and each nodal unit is connected to more than one actuator unit, alternative paths are made available for action to flow from the sensor variables to the actuator variables. Many communications, transportation, electrical, and water distribution systems are based upon the nodal network. This kind of network allows many different pathways for effects to move from input to output variables.

Superimposed networks

Two or more networks can be connected to a common set of sensor and actuator variables, forming the *direct superimposed networks* shown in Figure 3.23.

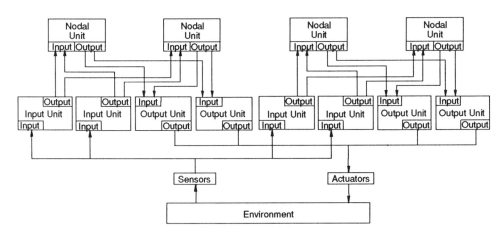

Figure 3.23. Direct superimposed networks allow more parallel paths for information to flow from sensors to actuators.

Each network can be considered a machine unit that acts in parallel with other networks.

Network of networks

Networks also can be connected in series and in parallel, forming the *direct network of networks* shown in Figure 3.24. Each network *superunit* allows the complete machine assembly to incorporate more variables and produce a wider variety of behavior than its subassemblies.

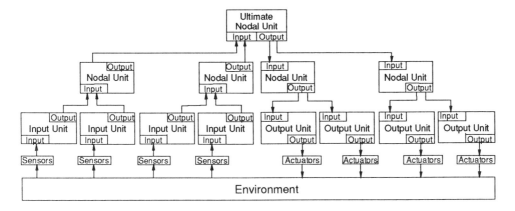

Figure 3.24. A *direct network of direct networks* forms a superunit that can deal with a greater variety of behavior than a single large unit acting alone.

Unitization

When two or more machines are connected to other machines in a network, they form a new machine made up of subunits. This process of creating assemblies made up of subassemblies is called *unitization*. All complex man made systems consist of assemblies and subassemblies. *Unitized systems* can be taken apart, repaired, and even redesigned with a minimum disruption of the rest of the system. Likewise, the behavior of parts of a machine made up of subassemblies can be changed without disturbing the behavior in other parts of the machine assembly. This process of unitization can provide a greater variety of behavior using units that are less complex than a single standalone unit. Thus, a machine system made up of a set of machine subassemblies may provide more effective behavior than a single large machine acting alone.

Mobile machines

If a machine is free to move around within an environment, it may be connected to different environmental variables. A *mobile machine* may be connected to different controlled machines. This greatly increases the opportunity to organize machine units according the task environment. For example, a group of mobile machine units may be put together to deal with a specific set of variables that inherently form a single system. Or they may be separated to deal independently with separate variables. They also may be placed in different environments so they can deal separately with separate tasks.

Selection of networks

The networks of machines presented in this study may include structured, hierarchical, or nodal networks. Each type of network may work best in different situations. When specific input and output variables can be coupled, a structured network with its competing and overlapping units provides opportunities to identify and incorporate new relations between input and output variables. A hierarchical network may work best when each combination of sensor values requires a unique combination of actuator values that can be mediated by the central machine unit. A nodal network allows for more diverse relations between input and output variables. The next chapter shows how the behavior of standalone machines and machines used in networks can be controlled by special elements that allow the behavior of these machines to be easily modified.

Chapter 4
Controlled Machines

We have seen how machines can behave in many different ways. *Control systems* can be designed that perform the same way as specific machines. Moreover, these *controlled machines* can be programmed to provide a variety of behavior without changing their basic structure. This programming can be carried out before or after they are put into use. An empirical controlled machine also can establish its own behavior without being programmed by learning to take advantage of the opportunities for action in the task environment. To show the design of controlled machines that can learn from experience, it is desirable to discuss several different types of controllers. They are as follows:

1. Directive controllers,
2. Governing controllers,
3. Positive feedback controllers,
4. Adaptive controllers,
5. Discontinuous responsive controllers.

Although each is suited for different applications, they show the evolution of control technology from the simplest systems to advanced empirical responsive controllers like the animal brain.

Definition of control

Since the subject of this book is a new kind of control system, it is important to discuss what is meant by "control," "controllers," "control systems," and "controlled machines."

Control

The definition of *control* is: to *perceive* existing circumstances, and to make a *decision* about the outcome of an event based upon these perceptions; to exercise *authority* over, to *direct* or command; to *regulate* or maintain the value of a variable within limits; or to check or *verify* by comparison with a duplicate or standard. So when we speak of control in this book, we mean a process that includes some or all of the following steps:

1. To *perceive* the state of existing conditions in preparation to making a decision and taking action.

All control systems must sense some aspect of their environment to produce useful control behavior and use this information to make decisions about what actions to carry out. For example, a household thermostat senses temperature and decides whether to turn the furnace on, off, or leave it in its current state. In some controllers, very little use is made of the sensing function. For instance, a numerical controller for a milling machine may send a great deal of position information to the machine it is controlling without receiving much information from it. Other controllers, such as a thermostat, must continuously sense input conditions.

2. To *decide* what the perceptions mean before taking action.

For some controllers, such as ourselves, the decision-making function is very important. We may decide to drive a car and decide where we want to go. We also may decide how fast we want to go and the route we wish to take. We also may decide that the circumstances do not warrant our involvement. We may choose to ignore those situations for which other people are responsible. In other controllers, such as the thermostat, there are few choices and little variety in the decision making process.

3. To exercise *authority*. If a controller has authority, it can carry out decisions that it makes. If a robot controller decides to pick up an object and is given the authority to do so, it can cause the robot arm to move in the required manner.

There are circumstances when an action should not be carried out. The arm of a robot may be about to hit an expensive piece of equipment, or a maintenance worker may be in the robot's work space. In these cases, the controller may be denied the authority to carry out some planned action. If it is working with another controller, it may decide that a workpiece lies in the work space of the other robot. In this case, in must abstain from acting upon that workpiece. Determination of authority is an important element of many control systems. For instance, we may observe a speeding car and decide that the driver is violating of the law. But we also may determine that it is not our responsibility to arrest the offender. A furnace temperature control system may sense a high room temperature, but may not have the authority to turn on the air conditioner in the winter. In a more general sense, someone may decide what the set point of the thermostat is supposed to be and the thermostat may simply carry out that decision. The individual who determines the set point is the controller with the authority to control temperature.

4. To determine, *direct*, or command. If a controller can direct behavior, it can give instructions about what is to be done under given circumstances.

The controllers described in this book will determine the motion of their actuators, and direct or command the actions of other devices. Action is the product of most control systems. A thermostat turns on the furnace. An auto-pilot turns the rudder of a ship to point the ship in the desired direction. A timer turns on an oven at some desired time. It is difficult to imagine a control system that does not attempt to evoke some kind of action. A burglar alarm monitors a region and sends an alarm only if a dangerous or targeted event occurs. Even if nothing untoward happens, the controller is still doing its job.

5. To *regulate* or maintain the value of a variable within limits. This is done by comparing the chosen results with the actual results and making a correction that brings the two closer together. The first step in the process of regulation is to check or *verify* the actual results against a duplicate or standard established previously.

This *negative feedback process* is used by all *regulators* and is usually built into their design. However, some controllers can *learn* or be programmed to produce behavior that brings actual results closer to the desired results. One interpretation of the term *verify* is given by the example of the function of the *comptroller* in a business organization. The comptroller may be responsible for creating and maintaining a budget or *standard cost* and a statement of planned revenue. The job of that office is then to verify that money is being received and spent as planned by comparing the *standard* to *actual* expenses and costs, and to signal when plans are not being met. This is basically a regulation or control function. Another use of a standard is in *controlled experiments* in scientific research where experiments are carried out using a *control variable* and an *unknown variable*. The results of each test are compared to see how the unknown variable changes with respect to changes in the control variable. This is intended to eliminate the effects of random or uncontrolled *hidden variables* that may have an influence on both variables.

Controllers

There are *controllers* around us everywhere. A thermostat that controls the temperature of a room or house is a simple *temperature controller*. We may have a controller in an oven that not only controls temperature, but also controls when the oven is turned on and off. We use automatic speed controllers such as "cruise controls" in our automobiles and automatic direction controllers such as auto-pilots in boats and airplanes. A digital computer can be programmed to control machinery or the flow of materials in a manufacturing plant or can be programmed to act as a cruise controller or auto-pilot. We become controllers when we drive a car or fly an airplane.

Control systems

When a controller and the device that it is controlling are put together, they form a *control system*. Thus a thermostat and furnace form a *temperature control system*. The cruise controller, car, and road form an *automobile speed control system*. An auto-pilot and an aircraft form an *automatic aircraft direction control system*. The pilot and airplane form a *manual flight control system*.

Controlled machine

A control system usually has sensors, a control unit, and actuators. For example, a temperature control system has a bimetal temperature sensor, an operating switch, and a furnace. An industrial robot system has sensors, a control unit, and actuators, as shown in Figure 4.1.

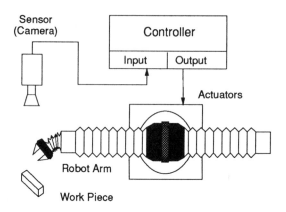

Figure 4.1. An *industrial robot arm* is a memory controlled machine that can be programmed to behave in many different ways.

A *controlled machine* is a machine that has been divided into three parts: sensors, a controller, and actuators. Controlled machines can act like any kind of kinematic machine such as a pendulum or a servomechanism. For example, the engines used in most late model automobiles use microcomputers to control their timing and choke settings, a function performed before by cams and linkages. The microcomputer/engine system is a computer controlled machine. Controlled machines can be viewed as control systems as well. For example, microcomputers are used to control heating and cooling systems. In that capacity, the sensor/microcomputer/actuator system is viewed as a computer based temperature control system and the microcomputer is viewed as a *temperature controller.*

System variables

A controller determines the *value* of one or more of the *system variables.* Thermostats control *temperature.* Cruise controllers regulate *speed.* Auto-pilots control *direction* of a vehicle. Temperature, speed, and direction are the *controlled variables.* A controlled machine has *sensor variables*, a controller, and *actuator variables.* In most cases, the actuator variables are the controlled variables.

System state

At any given time, every variable in a control system has a specific value even if the value is zero. A statement defining a set of sensor or actuator variables and their values at a given moment in time is called a *system state*. A *multivariable state* consists of two or more variables and their values. The sensor variables and/or the actuator variables can be used to define the system state.

Control behavior

Like a machine, all control systems have a characteristic way of behaving. For instance, the time setting of the oven timer specifies what it will do. The list of instructions of a numerical control machine specifies the sequence of movements of its *x/y* table and cutter. The "stiffness" and set point of a servo-controller determines how it will behave when displaced from its equilibrium position. These behavioral characteristics refer to the *modulus of behavior* of these controllers. *Control behavior* can be defined in the following four ways:

1. A transformation: The list of each state that could be produced by a controller in response to each condition,
2. A line of behavior: The list of each state that actually occurs in response to a set of conditions,
3. A circuit or machine design: The specification of a device that defines the size, shape and relation of its elements,
4. A mathematical model: The equation that defines set of states of the dependent variables that occur in relation to the independent variables.

Some controllers determine the state of a system by virtue of their predetermined or *a priori design* and construction, as described in Part I. Other memory based responsive controllers described in Part II can be *programmed* beforehand to produce specific output states for each input state likely to be encountered. Part III describes a memory based responsive controller that can independently establish and continue to produce a unique set of states after it is built and put into service *a posteriori* by the *empirical process*. The selection of this set of states by the controller represents a knowledge or understanding of what is possible within the controller/environment system.

Open- and closed-loop control

There are two broad categories of control: open-loop control and closed-loop control. In *open-loop control*, most of the control information flows from the controller to the actuator or output with little or no information sent back to the controller about the state the output. So an open-loop controller does not check to see if the actions called for are carried out. It does not use the results of its actions to create new actions. It simply sends out a control signal based upon its modulus of behavior based upon initial conditions. Most event timers, machine sequence controllers, and motor controllers without position or velocity feedback use open-loop control. Most clocks are open-loop controlled machines because they produce a sequence of states based upon a predetermined speed setting. If the desired clock speed is not achieved, the speed *parameter* must be adjusted manually to bring the clock to the desired speed. Many mechanisms in the nonliving world act like open loop controlled machines because they behave according to how the laws of physics govern their mass, position, velocity, and structure. They seldom sense the conditions in their surroundings and alter their behavior accordingly.

In *closed-loop control*, information about the actual output flows back to the sensors of the controller and is compared to some chosen standard or *set point* condition. Decisions about subsequent control activities are based upon this comparison. The flow of information back to the controller as the actual output of the controller is called *feedback*. People act as closed-loop controllers when they drive an automobile. They sense their position on the road (feedback signal) and make the corrections needed to keep their car in the position that they want (set point). Most controllers that act as regulators check to see if the action they call for is carried out. A thermostat continuously checks the temperature of the room. If the furnace does not bring the room up to the desired temperature, the thermostat will continue to call for heat. When a person drives a car, he or she turns to the right when the car is too far to the left. Since the correction is always in the opposite direction of the error from the set point, this kind of closed-loop behavior is called *negative feedback control*.

When we attempt to control the speed of our car, we adjust the position of the accelerator to what we think is the correct setting. We then check the speedometer at some later interval to see if we are in fact going at the desired speed. The difference between the speedometer reading and our wanted speed is called the *error signal*.

All *regulators* use closed-loop control with negative feedback to maintain a system state at or close to a set point. It is the basis of auto-pilots and servomechanisms. It is also the basis of *homeostasis*, the principle behind the regulators within our body that keep essential variables within safe limits.

The purpose of the comptroller of a business or government organization is to complete the information loop between planned and actual spending and earnings. The function of the comptroller is to keep the assets of an organization within the planned limits. This function is carried out using a closed-loop process using negative feedback.

There is a tendency to think that all control systems rely on closed-loop control and the negative feedback process. However, some controllers use *positive feedback*. For example, a rabbit turns in the direction that it is displaced from its pursuer. It hopes this maneuver will take it further from its pursuer. This is an example of *positive feedback control* because the prey attempts to increase, not decrease the error signal.

The empirical controllers described in this book are not closed-loop controllers in the usual sense. They do not necessarily rely on the negative feedback process described. They normally use open-loop control in that they send out control signals based upon their internal states and do not attempt to reduce the difference between the desired sensed conditions and the actual sensed conditions. However, they do monitor their actions and use this information to alter their modulus of behavior. They can *learn* to provide positive or negative force behavior to meet the needs of the task environment. They learn to produce behavior typical of any one of the five types of control systems listed in Table 4.1.

Table 4.1. Five types of control systems

1. Directive controllers (open loop devices).
2. Governing controllers (closed loop, negative feedback, point seeking predetermined devices).
3. Positive feedback controllers (closed loop, positive feedback, point avoidance, predetermined devices).
4. Adaptive controllers (closed loop, usually negative feedback, self-adjusting predetermined devices).
5. Discontinuous responsive controllers (no set point, memory based predetermined or empirical devices).

Each type of controller contributes a greater understanding of the control process:

- The *directive controller* shows that useful behavior can be produced without a feedback loop.
- The *governing controller* demonstrates the usefulness of holding the value of a controlled variable within specific limits using negative feedback.
- The *positive feedback controller* shows the usefulness of changing the value of a controlled variable so that it avoids values around a set point through positive feedback instead of negative feedback.
- The *adaptive controller* shows that it is possible to design a feedback controller that can automatically improve its performance.
- The *predetermined discontinuous responsive controller* shows how actuators can be programmed to behave at specific moments according to the discrete values of sensors.
- The *empirical discontinuous responsive controller* shows how a controller can establish behavior by itself that appears intelligent and mindful of its surroundings.

Our concern is how the *empirical discontinuous responsive controller* shown in Part III produces intelligent behavior by itself through the discovery of the natural relations in its environment, and how it incorporates this information into its behavior to assure the continued existence of its behavior.

Directive controllers

The simplest type of controller does nothing more than tell the device it is controlling what to do. A directive controller directs the action or activities of some device in a particular way based upon its design or memory without regard for the conditions in its environment. Thus, it is called an open-loop controller.

Delay timer

A *delay timer* (Fig. 4.2) is an example of a directive controller. This controller simply turns a device "on" for a period determined by a value recorded in the controller. This type of controller is called a *directive controller* because it directs the operation of the device (actuator) it is controlling.

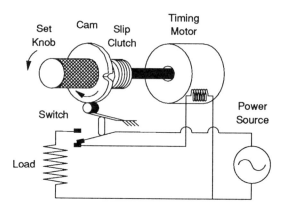

Figure 4.2. A *delay timer* causes a switch (output) to close after a predetermined time delay.

The modulus of behavior of a delay timer is specified by a single number — the time delay setting. The directive controller specifies action or influence based upon its own internal memory and decision making process. In the case of cam delay timer, the internal memory device is called a *cam*. The cam is mounted upon a shaft that is turned by a clock mechanism. The position of the cam can be adjusted by the operator through the time set knob so that it operates the output switch at the desired time of day or after a predetermined delay. The cam timer is a mechanical memory device that stores a predetermined delay and produces an output when a predetermined time has expired. A synchronous motor usually provides the time base of the controller.

Modulus of a timer

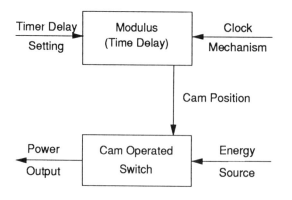

Figure 4.3. The *diagram of a delay timer control system* shows that its behavior is specified by a single number.

In some cases, the modulus of behavior of a control system can be specified by a single number like the modulus of elasticity or a spring constant. The modulus of behavior of a directive controller, like the delay timer shown in Figure 4.3, is also specified by a single number: the time delay set into the controller. If more complex behavior is required, such as a requirement that the timer close a switch for a predetermined period, then an additional number is required to specify the modulus of behavior of an *interval timer* shown in Figure 4.4.

Interval timer

An *interval timer* works in much the same way as a delay timer, except that a second cam is mounted on the timing motor shaft that can be displaced from the position of the first cam. The gap between the two cams determines the length of time over which the switch contacts remain closed.

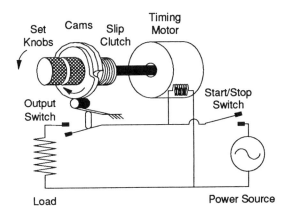

Figure 4.4. An *interval timer* uses two cams to operate a switch after a specific time delay and for a particular period.

The position of the first cam can be set on the shaft to operate the switch at a particular time of day. If more information is involved, such as when to turn the switch "off," a second cam can be added alongside the first cam and rotated to keep the contacts of the output switch closed for a specific period. Thus, more information is required to specify the modulus of behavior of an interval timer than a delay timer.

Oven timer

An *oven timer* (Fig. 4.5) may include even more information, such as temperature settings and controls that allow the user to select when these different temperatures are supposed to occur.

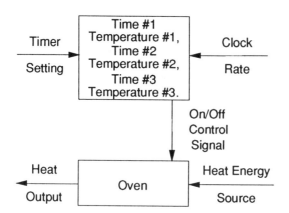

Figure 4.5. An *oven timer* may store six or more values.

As more information is incorporated in the control system, its modulus of behavior becomes more complex, and provision must be made for the input and storage of more information. In some cases, a microcomputer is used to produce the operating features of a simple oven timer.

Elements of directive controllers

The oven timer is shown because it embodies the five main features of directive control systems:

1. A place to introduce desired behavior information (set knob);
2. A memory or information storage device (cam);
3. An indexing or timing mechanism (timing motor);
4. A switch or amplifier operated by the controller that releases energy (switch);
5. An external source of energy (electrical power source).

Very complex open-loop control behavior can be achieved using multiple cam controllers made up of these five basic parts.

Numerically controlled machine

Another typical directive control system is a *numerically controlled milling machine* (Fig. 4.6) used to cut shapes into solid materials. The behavior of this machine is determined by the cutting instructions, generally called the *parts program,* stored in a computer memory, on paper tape, or on other media read by the controller.

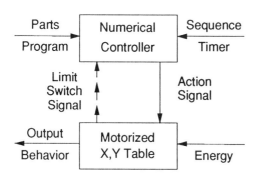

Figure 4.6. A *numerically controlled machine* produces behavior based upon prerecorded information.

The directive controller relies primarily upon the predetermined parts program rather than sensed values and observations to produce behavior. Many numerical control machines use a feedback signal to inform the controller where each axis is at any time. This is generally done to improve positional accuracy. But the predetermined parts program is used to determine the position of each axis in each step in an open-loop manner.

Modulus of a numerical controller

The modulus of behavior of a directive controller (Fig. 4.7), like a numerical control machine, is specified by the list of instructions needed to cut a particular shape and is as complex as the shape of the parts to be cut.

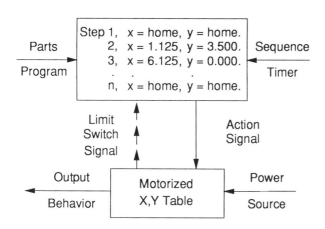

Figure 4.7. The *modulus of behavior of a numerical control machine* is a list of actions to be carried out in sequence.

The statements in the lists are sequential because the parts must be cut in a particular sequence. In a completely open-loop system, a clock or time base can be used to provide the signal needed to advance to the next statement. If the controller "knows" when it has completed a step, the time base may provide a predetermined time delay before starting the next step. There may be more than one list stored in the controller. Each list may contain the information needed to cut different parts.

Tape recorder

A *tape recorder* is an example of a directive controller since it reproduces predetermined behavior such as speech and music. A signal may tell the tape drive motor
to stop when the tape is nearly unwound from one of its two spools. A directive
controller such as a tape recorder is represented by Figure 4.8.

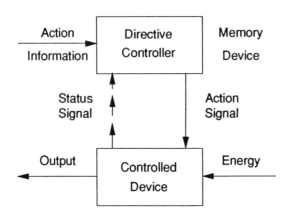

Figure 4.8. A *directive
controller* sends prede-
termined action informa-
tion to the device that it
is controlling, and uses
only a limited amount of
sensor information.

A directive controller receives a limited amount of sensor status information such
as a start/stop signal, the name or location of the stored file, and signals from limit
switches indicating that the machine has come to the limit of its movement in a given
direction.

Governing controllers

A major improvement in control occurred with the invention of the *governing
controller* near the end of the 1700s. The directive controllers used until this time
could do little more than carry out predetermined behavior like a cuckoo clock.

Governors

The earliest governing controllers that found widespread use was the centrifugal
speed "governor" (Fig. 4.9) invented by James Watt (1736—1819) and used to
control the speed of steam engines. It operates in a different way than the directive
controllers described here in that the governing controller uses information about
its output to change the way it performs.

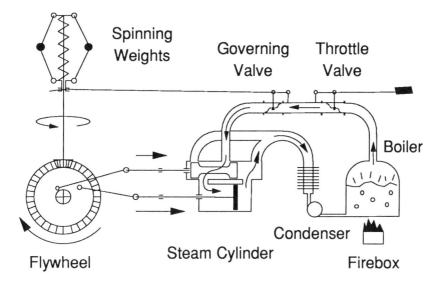

Figure 4.9. A *centrifugal speed governor* uses the force created by spinning weights on its output shaft to control the flow of steam to the engine.

A pair of weights is attached to the engine shaft through gearing so that the weights fly outward as the speed of the shaft increases. This movement causes the throttle to close, thus limiting the shaft speed to some predetermined value.

Centrifugal governor

The centrifugal speed governor can maintain a nearly constant output speed, in spite of changes in the load on the engine, by measuring the difference between the engine speed and the desired speed and producing an error signal that is used to bring the actual speed closer to the desired speed.

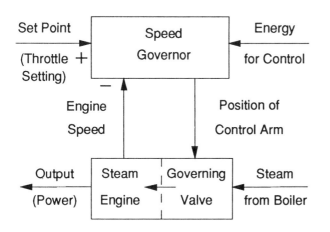

Figure 4.10. A *diagram of a centrifugal speed governor*, shows how the sensed output speed is used to control the throttle setting.

The governor is designed to create an opposing relationship between the output speed of the engine and the amount of steam allowed to reach the engine, as shown in Figure 4.10. For example, as the output speed increases, the governing valve is closed. If an increase in load on the engine reduces the engine speed, the governing valve is opened.

Negative feedback

The centrifugal speed governor senses the output of the device it is controlling and decides how to control the input of this device based upon the modulus of behavior of the controller. The modulus of behavior of the centrifugal speed governor shown is determined by the mass of its spinning weights, the stiffness of the restoring spring, and the linkage to the throttle valve. A *governing controller* can be represented by the diagram in Figure 4.11.

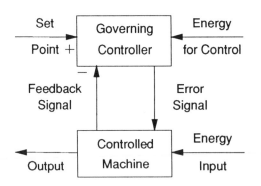

Figure 4.11. A *governing controller* uses information from the device it is controlling and its set point to influence the behavior of the device it is controlling.

A governing controller specifies an action or influence upon the device it is controlling based upon the deviation of the output of this device from a desired *set point*. A governing controller produces an *error signal*, which is the difference between the desired result and the sensed result, and uses this error signal to return the output of the device it is controlling to the value of the set point. This opposite relationship is called *negative feedback* because the polarity of the feedback signal is made *negative in relation to the set point*. Thus, the error signal will be positive if the feedback signal is smaller than the set point and the error signal will be negative if the feedback signal is larger than the set point.

Behavior space of a regulator

The value of the controlled (machine) variable of a governor always attempts to return to the set point, which may be chosen by the operator of the system. This kind of control system is also called a *regulator* or *servomechanism*. A *map of the behavior of a regulator* (Fig. 4.12) shows lines of behavior converging upon the set point for every line of behavior that diverges from the set point.

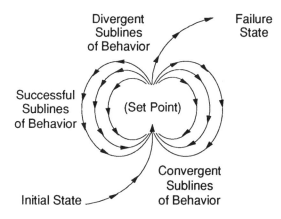

Figure 4.12. The *behavior space of a regulator* converges upon its set point.

Since a regulator always acts in the opposite direction of its deviation from the set point, and usually acts with a force or influence that increases as the deviation increases, it is said to use *negative direct feedback* (NDF). It tries to return to the set point with a force or influence that may be directly proportional to the deviation from the set point or proportional to some power of the deviation. A regulator can operate indefinitely without changing its modulus of behavior if it can produce a convergent line of behavior for each divergent line of behavior, forming a closed behavior space around the set point.

Air pressure regulator

The *air pressure regulator* (Fig. 4.13) is a common regulator found in most machine shops and science laboratories. It is used to maintain a constant selected pressure in a compressed air line regardless of the pressure of its source such as a compressor, and regardless of atmospheric pressure.

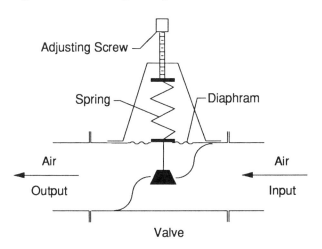

Figure 4.13. An *air pressure regulator* produces a nearly constant output pressure despite fluctuation in the pressure source or in the ambient air pressure.

The output pressure is sensed by the flexible diagram, which opens the air valve when the pressure drops, and closes the air valve when the pressure increases. An air pressure regulator may use *negative directly proportional feedback* (NDPF).

Voltage regulator

The *voltage regulator* (Fig 4.14) in an automobile works much in the same way as an air pressure regulator. The voltage produced by a generator varies according to engine speed. However, the voltage required to maintain the charge of the battery and to operate the lights and other accessories must stay within narrow limits.

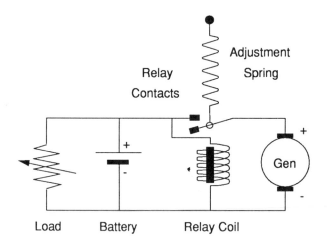

Figure 4.14. A *voltage regulator* maintains a constant predetermined voltage despite fluctuations in the voltage of the generator and changes in the load.

The electromagnetic relay coil creates a larger magnetic field as the voltage increases across the relay coil. This magnetic field produces a force on the relay switch arm. If this force is greater than the force produced by the adjustment spring, the switch will open, disconnecting the generator from the battery and the load. This will cause the voltage across the coil to drop back to the battery voltage, which is insufficient to keep the switch open. Thus, the switch will rapidly open and close, limiting the voltage to some predetermined desired level determined by the tension in the adjustment spring.

Thermostat

Another example of a regulator is a simple household *thermostat* (Fig. 4.15), which can be called a *temperature regulator*. If the sensed temperature of the room drops below the setting (set point) of the thermostat by a certain amount, as sensed by the bimetal heat sensor, the thermostat switch turns the furnace on. This causes the room to get warm enough to cause the thermostat switch to turn off the furnace.

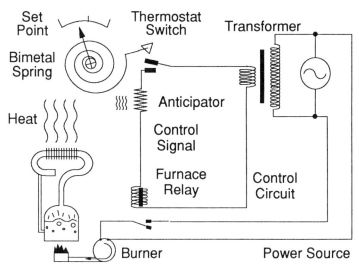

Figure 4.15. A household temperature control system maintains a constant temperature by turning on the furnace when the room temperature drops below the desired set point.

However, there is a small problem with this temperature control system. The heaters in the room must operate at a higher temperature than the set point to warm the room. If they are turned off when the room reaches the desired temperature, they will continue to heat the room for a period of time. This may cause the temperature of the room to overshoot — to get too hot. This problem can be solved by designing the circuit so that the electrical current from the signal that turns the furnace on also flows through an *"anticipator"* resistor. The heat generated in this resistor warms the bimetal temperature sensor even before the room gets warm. So the thermostat turns the furnace off before the room reaches its desired temperature and the heat remaining in the heaters continues to heat the room to its desired temperature. The wattage of the resistor is chosen to provide the best control performance. If the wattage of the anticipator is too low, the temperature of the room will overshoot. If it is too high, the anticipator will turn off the furnace too soon. In this case, the thermostat will keep cycling the furnace on and off, and the room may not come up to the desired temperature.

Diagram of a household temperature control system

The behavior of this predetermined controller is established by its design and the value of its components. This system can be represented by the diagram in Figure 4.16. In the cases of the household thermostat, the feedback loop is not specifically built into the control system. It appears as a path of immediate effects through the room environment. This temperature controller is called a thermo*stat* because it maintains a constant or *stat*ionary temperature.

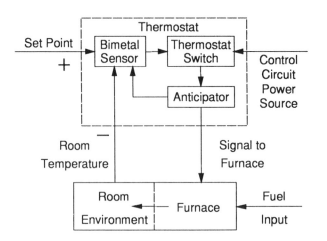

Figure 4.16. The *diagram of a household temperature control system* shows that it is made up of a thermostat, a furnace, a room environment, and a path for signals to flow between them.

Fluid position servo controller

Most auto-pilots, power steering units, and other position control systems are regulators based upon the design of the *fluid position servo controller* system shown in Figure 4.17.

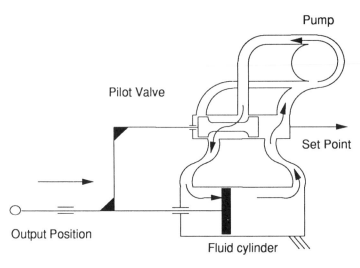

Figure 4.17. A *fluid position servo controller* uses a pilot valve to direct a pressurized fluid to one or the other side of a piston in a cylinder.

If sufficient force is applied to the piston to cause it to move, the spool in the pilot valve moves off its inlet and outlet ports. This allows a fluid, pressurized by a pump, to enter the cylinder in an attempt to restore the piston to its original position. Viewed in this way, the position servo controller is a *posistat* since it holds an object in a *stati*onary *posi*tion. Moving the pilot valve also causes the spool to uncover its inlet and outlet ports, which forces the piston to move in the direction in which the pilot valve was moved. Once the piston moves to the new position of the pilot valve, the ports are closed and no more fluid can enter on either side of the piston. A small force applied to the pilot valve may cause a large force to be produced by the piston if the pilot valve can move freely, if the diameter of the pilot valve is small in comparison to the output cylinder, and if sufficient pressure is generated by the fluid pump. Thus, the posistat can act as a force amplifier.

Position servo-actuator

A *mechanical differential* can be added to the posistat to eliminate the need to move the pilot valve. One side gear of the differential is connected to the piston rod, the other side gear to the set point, and the center gear representing the output of the differential to the spool of the pilot valve, forming the *servo-actuator* shown in Figure 4.18.

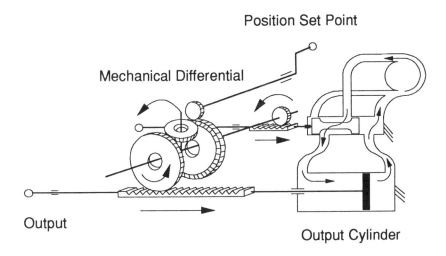

Figure 4.18. A *position servo-actuator* allows a small force to position an object that requires a much larger force to move.

The position of the center gear of a differential represents the sum of or difference between the positions of its two side gears. If its two side gears are rotated in the same direction, the position of the center gear represents the sum of their positions. If the two side gears rotate in the opposite direction, the position of the center gear represents the difference between their positions. The gears in a position servo are arranged so that the differential subtracts the output position of the amplifier from the set point position. The difference created by the subtraction process (represented by the position of the center gear) controls the offset of the spool in the pilot valve. When the output position and the set point position are the same, there is no offset of the spool and no fluid flows into or out of the output cylinder.

Hydraulic power steering unit

A fluid position servo-actuator can be used as the *hydraulic power steering unit* found in most automobiles (Fig. 4.19).

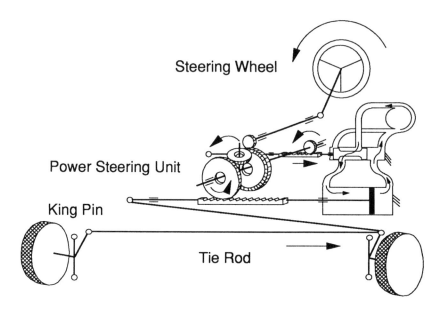

Figure 4.19. A *power steering unit* uses a position servo-actuator to aim the front wheels in the direction called for by the steering wheel.

A movement of the steering wheel moves one side gear of the differential, causing the center gear and the spool in the pilot valve to be displaced. This error signal causes fluid to flow into the output cylinder, producing a restoring force on the tie rods through the output piston. If this force is sufficiently large, it causes a movement of the output piston. This turns the wheels, the center gear of the differential, and the spool in a direction that reduces the error signal in the pilot value. When the error in the pilot valve is eliminated, the output piston is no long caused to move. Only a small force is required to position the servo valve, but a large force can be produced by the output piston. This property makes the position servo-actuator useful as a power steering device, naval gun positioner, industrial robot actuator, etc., where behavior requiring large forces must be controlled by the limited forces that can be produced a human operator.

Electric servo amplifiers

The same amplifying process can be carried out using an electric motor, mechanical differential, position transducer, and electrical amplifier shown in Figure 4.20.

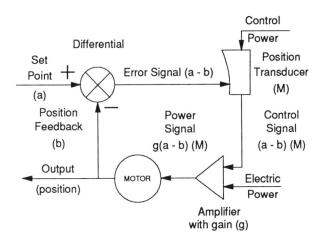

Figure 4.20. An *electric position servo amplifier* uses information about the position of an electric motor to bring the motor closer to the position of a set point.

The servo control system attempts to position the output shaft of the *electric motor* at some place determined by the position of the set point (*a*). Thus, the position of the output shaft of the motor is the controlled variable (*b*). The feedback signal consisting of the value of the controlled variable is carried to a *mechanical differential*, which automatically subtracts the feedback signal from the set point, creating an error signal (*a* - *b*). A *position transducer* converts the output of the differential into an electrical signal, which is amplified and used to drive the electric motor. The motor moves toward a position determined by the set point according to the modulus of behavior of the controller and the modulus of behavior of the motor and the *electrical amplifier*.

Power source

Every servo amplifier has a *power source* that supplies the power to drive the output of the system. The purpose of the control system is to harness this power by the correct amount and in the correct direction. In the case of the steam engine, the power comes from the boiler in the form of pressurized steam, which is throttled by the speed governor. In the case of the hydraulic servo position amplifier, the power comes from a hydraulic pump and is controlled by the servo valve. The power needed to drive the motor in an electric servo system comes from the electrical power source and is controlled by the input to the electrical amplifier.

Gain

The degree of amplification, or *gain*, is the ratio of the amount of power produced by the servo control system to the amount of power required to change the value of the set point. In the hydraulic servo, the gain is usually considered equal to the force of the output piston divided by the force it takes to move the pilot valve. In hydraulic systems, the gain can be as much as 1000 to 1. In an electronic amplifier, the gain is usually considered the ratio of the output voltage to the control voltage. In some electronic amplifiers, the gain can be over 10,000 to 1.

Stiffness

A position servo amplifier may maintain a nearly *stati*onary output position if the set point remains unchanged even if the output shaft is subjected to significant loading. However, a governing controller like a servo amplifier can be considered to act like a spring. It produces some restoring influence if displaced from its set point. The ratio of force applied to the output shaft to the deviation from its set point can be specified by a modulus of behavior just like the *stiffness* of a spring. Usually, a large force is produced by a small deviation. Therefore, the output shaft will stay close to the position called for by the input shaft regardless of the force loading on the output shaft.

Response time

Every servo system requires a certain amount of time to reach the position called for by its set point. It takes a certain amount of time for a furnace to heat a room to the desired temperature. In the hydraulic servo, it may take a significant amount of time for the hydraulic fluid to enter the actuator cylinder and move the piston to the position required by the pilot valve. Using a pump of a given capacity, the larger the cylinder, the longer it takes to fill the cylinder to move the piston a given amount. However, a large force requires a large cylinder diameter for a given fluid pressure. Thus, the output shaft of a servo controller that can produce a large force may require a significant *response time* to move to its output position.

Inertia

The piston and output shaft of a hydraulic position servo have to be made strong enough to withstand the forces applied by the load and the hydraulic fluid. A large rotor may be required in a electric servomotor to produce sufficient torque. Thus, the machine output elements may have considerable *inertia*. The inertia of an output element may increase the response time of the servo system and create the possibility that the output device will *overshoot* the set point. This may cause the position of output element to oscillate around the set point instead of coming to rest at the desired position. The load attached to the output of a servo amplifier also may influence the response time of the system and the tendency of the system to overshoot.

Damping

An oscillation of the output shaft can be reduced or eliminated by connecting a motion damper to the output element that produces a retarding force in proportion to the velocity of the output element. If the force produced by the damper is larger than the force produced by the inertia of the output element due to its acceleration and deceleration, the oscillations will disappear or not occur in the first place. This *damping force* must be kept as small as possible to provide a short response time and to allow more of the force of the servo to act upon the load.

Force to inertia ratio

The damping force and thus the response time can be kept to a minimum when the servo actuator has a high *force to inertia ratio.* Thus, much of the effort that goes into the design of a servomechanism is directed toward providing an output element that has a low inertia and can produce a high output force.

Dead band

There may be some movement of the pilot valve that does not cause fluid to flow into the cylinder. Also, there may be some movement of the output shaft that does not cause the pilot valve to move. These inaccuracies create a *dead band* of small displacements in which there is no clear relation between input and output positions. This dead band must be smaller than the smallest increments of motion (resolution) required by the control task.

Modulus of a servo controller

The modulus of behavior of an electric servo controller can be specified by the term *M* in the equation shown in Figure 4.21.

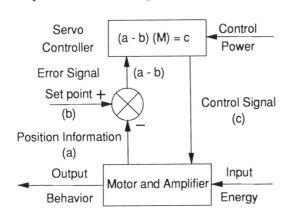

Figure 4.21. The *modulus of a servo controller* determines an output control signal based upon the difference between its set point and the actual output.

The modulus of the servo amplifier is built into the controller by the selection of the gain, motor torque, and ratio of motor position to error signal. The ratio of position error to restoring force can be linear, and the equation used to represent its behavior can be a single number like the spring constant. If the behavior of a servo amplifier is nonlinear, a more complex equation is required to specify its modulus of behavior.

Computer model

A computer can be used to compute the value of the control signal to the output element of a servo control system based upon the value of the error signal (Fig. 4.22). This control signal can take into account the values of the different parameters listed and produce results that minimize their negative effects.

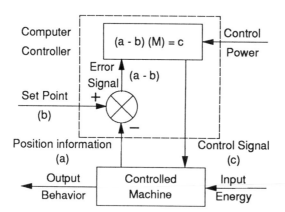

Figure 4.22. A computer model of a servo controller can be changed to minimize the negative effects of the mechanical parameters.

This type of controller can be used in a flight simulator. For example, a computer can generate the values of pitch, roll, direction, velocity, location, and altitude based upon the operators' control behavior and a model of the behavior of an aircraft defined by the equations for its modulus of behavior. If the modulus of behavior of a controller can be specified by an equation, its control behavior can be computed very efficiently.

Governing controller as a list

The modulus of behavior of a governing controller also can be specified by a list of statements describing how the controller responds to each error signal, as shown in Figure 4.23.

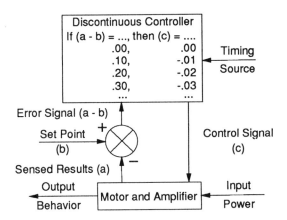

Figure 4.23. The behavior of a servo controller can be defined by a *list of responses* to error signals.

These statements are in the form: "If error signal. . ., then output. . .." This list of responses to error signals can be made extensive enough to allow a discontinuous responsive controller to behave as effectively as a conventional servo system or as a model of a servo system generated by a computer using equations for its modulus of behavior.

Gun servo system

A gun position servo system operates with negative feedback when it creates a force on the gun that is proportional to the error signal with the sign reversed (Fig. 4.24). If the error in aim is to the right, the gun must move to the left. Thus, the gun will turn closer to a target corresponding to its set point, but may overshoot if it approaches the target at too fast a rate.

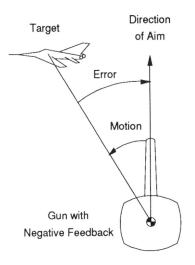

Figure 4.24. A gun position servo system with negative feedback creates a force on the gun toward the target.

To reduce overshoot, these servo control systems usually operate with *negative position and velocity feedback*. The position of the gun is monitored by a position sensor and the turning velocity of the gun is monitored by a tachometer. The torque on the gun is modified to make the velocity proportional to the magnitude of the position error signal. Thus, if the position of the gun is far from the desired direction, it can move quickly. If it is near the set point, it must go slowly. This reduces the tendency of the gun to overshoot the set point.

Chase behavior

If the set point of a regulator moves around, the regulator attempts to chase the set point. For example, the target of the gun servo system shown may take evasive maneuvers to avoid being hit by the gun. In this case, the gun servo produces *chase behavior*. In general, the chase behavior becomes more active the more that the position of the set point is varied. However, the response time of the regulator must be shorter than the time of the movements of the set point in order for the regulator to "catch" its set point.

Negative direct feedback control

In most cases, a gun servo must move the gun rapidly if it is pointing far away from its target. In this case, it is desirable to use some form of negative direct feedback, which causes a large force one the gun when it is pointing away from its target, as shown in Figure 4.25.

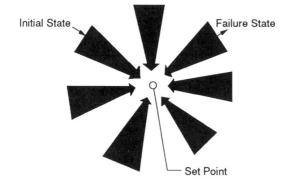

Figure 4.25. Negative direct feedback creates a larger force on the gun toward the target as the error increases.

However, negative direct feedback does not create a large aiming force when the gun is close to the target. Thus, the gun may overshoot or miss the target because of an offset due to friction.

Negative inverse feedback control

To aim the gun with more accuracy, the gun controller can use negative inverse feedback, which causes a large force on the gun when it is close to its target, as shown in Figure 4.26.

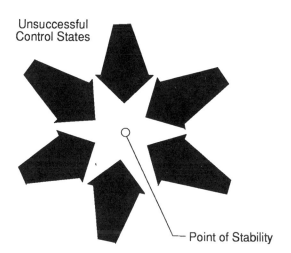

Unsuccessful
Control States

Point of Stability

Figure 4.26. Negative inverse feedback creates a larger force on the gun when it is close to its target.

Negative inverse feedback causes the gun to zero in and stay rigidly fixed upon its target. In some cases, combinations of negative direct and negative inverse feedback are used to create optimum (nonlinear) control behavior.

Positive feedback controllers

It is possible to hook up a governing controller backward so that its feedback is positive rather than negative. This usually results in undesirable and sometimes disastrous behavior. In the case of the thermostat with positive feedback, as the room gets colder, the contacts of the thermostat move farther apart. This causes the room to continue to cool. The result is that the furnace is effectively shut off. If the contacts are closed, the room will continue to heat, which will drive the contacts closer together. This causes the room to continue to get hotter. The result is that the furnace will stay on until its temperature reaches the setting of the high limit switch on the furnace. In the case of a servo system with positive instead of negative feedback, the output element will accelerate away from its set point in an explosive manner until some component in the system breaks or burns out. Thus, most closed-loop controlled machines use negative feedback. However, some useful control applications are based upon positive feedback.

Gun servo with positive directly proportional feedback

A gun also can be hooked up with positive feedback. If it is aimed to the left of the target, the gun will try to move farther to the left, as shown in Figure 4.27. If a control system with *positive directly proportional* feedback (PDPF) is used, the gun will try harder to move away from the target as the error increases. This creates an explosive burst of force and power in a direction away from the target, resulting in unstable "runaway" behavior.

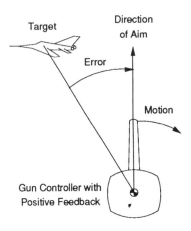

Figure 4.27. A controller with *positive directly proportional feedback* will point away from the target at an ever-increasing speed.

Positive directly proportional feedback is highly undesirable and destructive in any kind of governing control system and is particularly dangerous in a system with high gain because of the large amount of energy it evokes, as shown in Figure 4.28.

Behavior space of a control system with positive directly proportional feedback

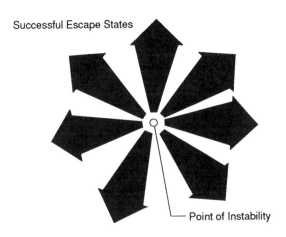

Figure 4.28. The *behavior space of a control system with positive directly proportional feedback* shows system states attempting to escape from the set point with an ever-increasing force.

However, a gun with positive feedback has an interesting feature. It does everything it can to point away from the target. This feature may be very desirable in training exercises so as to reduce the chance of accidentally shooting down friendly airplanes. If this is not the case, it behooves the target to move away from the direction in which the gun is aimed using some form of positive feedback. If the target knows the direction in which a gun is aimed, and considers this direction as its set point, its best way to move to avoid being hit is to use positive feedback. The behavior it follows in staying away from its set point may be called *escape behavior*.

Positive inversely proportional feedback

It is possible to design a gun control system with positive feedback that is not as likely to self-destruct as one with positive directly proportional feedback. It is possible to use *positive inversely proportional feedback* (PIPF) in which the torque on the gun is always away from the target but decreases in proportion to the distance from the target as shown in Figure 4.29.

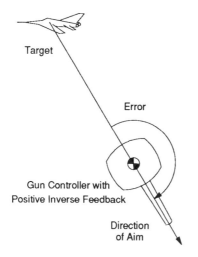

Figure 4.29. A gun controller with *positive inversely proportional feedback* moves away from the target with a decreasing force and comes to rest pointing 180° from the target.

In other words, the torque on the gun is directed away from the target, increases as it gets closer to the target, reaches a maximum when it is on the target. It reverses direction when it gets on the other side of the target, carrying it away in the other direction, where its torque decreases as it gets farther from the target, as shown in Figure 4.30.

Behavior space of a control system with positive inverse feedback

Positive inversely proportional feedback will cause the gun to point away from the target, and will not create extreme demands for power unless the gun is close to the target. Since it tries to move away from this point, the power demand diminishes quickly as the gun moves away from the target.

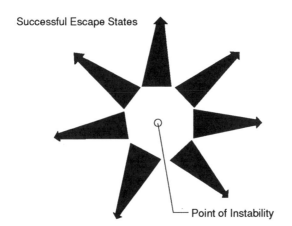

Successful Escape States

Point of Instability

Figure 4.30. The behavior space of a *control system with positive inversely proportional feedback* shows the set of system states that could be produced in its attempt to escape from a set point with an ever-diminishing force.

Constructing the gun control system so that it can be switched from negative directly proportional feedback to positive inversely proportional feedback can be very useful. In times of peace or in practice maneuvers, the gun can be placed in positive inversely proportional mode to reduce or eliminate the possibility of hitting any target and then switched to negative directly proportional feedback in actual combat.

Escape behavior

The target can also act with positive or negative feedback. Once the target moves away from the point at which the gun is aimed, the risk of being hit is reduced. However, with positive direct feedback, the target would continue to move away with ever-increasing force. A better strategy for the target is to escape with positive inverse proportional feedback, in which the target moves away from the aiming point with a diminished effort the farther it gets away from the aiming point Escape behavior, using positive inverse proportional feedback is employed by animals, fighter pilots, and players of many sports when they are being chased and do not want to get caught. The closer a pursuer comes to the prey, the more violently the prey must accelerate or change course. But once the prey gets away, the it may ease up on its effort to escape so as not to wear itself out.

Chase/escape behavior

One view of behavior is that positive feedback activities are the source of disturbances and action, and negative feedback activities tend to reduce these disturbances or limit their effects. Many sports are based upon an arrangement in which one individual or team acts in an offensive (positive feedback) mode at a given time and the other individual or team acts in a defensive (negative feedback) mode (Fig. 31). The team in the offensive mode usually "has the ball" and attempts to move the ball to some goal location. The team in the defensive "goes after the ball" and attempts to keep the offensive team from reaching the goal. Players in the offensive mode usually employ *escape behavior* using positive feedback to move the ball away from the defender, while players in the defensive mode employ *chase behavior* using negative feedback behavior to get to the ball.

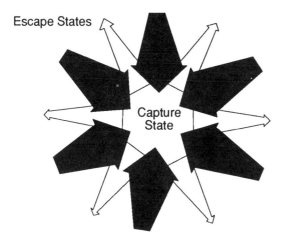

Escape States

Capture
State

Figure 4.31. The behavior space of *control systems with positive and negative feedback* may show perpetual and chaotic behavior while the moduli of behavior of the units in the system remain constant.

Placing individuals or control systems in a *chase/escape mode of behavior* is a natural way of evoking large amounts of action and energy. With the escape unit using positive inverse proportional behavior, the closer the chase unit gets to the escape unit, the harder the escape unit tries to get away. With the chase unit using negative proportional behavior, the farther the escape unit gets away, the harder the chase unit works to catch up. In a typical chase/escape encounter in the wild, the closer a predator gets to its prey, the harder the predator may work to make the kill. Thus, the predator may use negative inverse proportional or even negative inverse square behavior. The prey may use positive inverse or even positive inverse square behavior to avoid capture. Using this mode of behavior, the prey does not try to get away when the predator is far away, but makes a feverish attempt to get away if the predator comes close. The motion of the steam engine shown earlier is based upon chase/escape behavior in that the position of the pilot valve (set point) is continuously moved by the flywheel crank to a position that causes the steam piston to chase after it.

Chase/escape behavior is a *perpetual process* in that there is no obvious terminal condition if the units are nearly equal. It is *cyclical* because the two units probably will come together and then separate, and it can be *chaotic* in that small variations in initial conditions may result in large changes in behavior. Control systems that include units with positive feedback behavior create more opportunities for action than systems that act as regulators, which tend to become inactive around a set point. Thus, most of the applications of empirical control systems discussed in this book involve this type of chase/escape behavior. For example, a ship will seldom travel in a straight line. In most cases, a ship will attempt to turn to the left or right. The more it turns, the more it may attempt to turn. This unstable behavior is based upon positive feedback. An auto-pilot is a regulator that attempts to maintain the course of a ship in a straight line. Connecting an auto-pilot to a ship's rudder creates a control system based upon chase/escape behavior.

Adaptive controllers

There are other governing controllers that keep objects *stat*ionary such as a gyro*stat*, which keeps a ship from rocking, and a rheo*stat*, which maintains a constant voltage or current. These controllers use the same procedure, which is to measure or sense the value of a variable and then do something to restore the value to some predetermined set point.

The term *homeostasis* is often applied to living systems to explain their ability to maintain values within critical limits. This process employs more than the negative feedback used by a governing controller in that the correction effort may involve a change in the way in which the negative feedback is applied through a change in the modulus of behavior of the control unit.

For instance, if an organism becomes dehydrated (experiences a decrease in the concentration of water in the body), it attempts to find and drink water. If this behavior is successful, the concentration of water in the body is increased and the dehydration is reduced. This response is typical of a regulator. However, an animal accustomed to an abundant supply of water may find itself in an environment having little water. The organism may not find enough water to satisfy its need through the normal regulation behavior it had employed. This shortage of water may force the body to change the way in which it uses its limited supply of water. For example, the body may allow the urine to become more concentrated. In this case, the underlying parameters of its control system are changed. These automatic changes in the parameters of a control system may be viewed as *homeostatic changes* in the controller rather than as corrective (regulative) behavior. Thus, homeostatic and *adaptive behavior* are distinguished from regulatory behavior in that they involve automatic changes in the parameters such as the set point, gain, response time, damping, or dead band in a controller, as compared to automatic changes in controlled variables caused by the regulation process using negative feedback.

Species may change physically in response to long term changes in environment through the process of *evolution*. This process of *adaptation* also may be viewed as a change in the fundamental parameters of the species. This type of behavior is also called "adaptive" behavior since it allows the organism to "survive" in situations beyond the normal range of behavior with which a regulator or even homeostatic changes are expected to cope.

In summary, adaptive behavior and homeostasis often refer to changes in the parameters of the feedback process that change the way in which the controlled variables respond the set point. Although most *adaptive systems* operate with negative feedback, an adaptive system with positive feedback could be designed to change the way it attempts to move away or escape from a set point.

Adaptive control

In some cases, it may be difficult or even dangerous for a regulator to try to maintain the control variable at a given set point. Leaving an automobile's cruise control set at 125 mph could cause the transmission to overheat. A sensor on the transmission can be set up so that it kicks back the cruise control setting if the temperature exceeds a certain value.

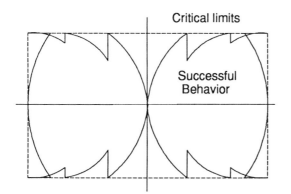

Critical limits

Successful Behavior

Figure 4.32. The *behavior space of an adaptive controller* shows how its modulus of behavior changes in a predetermined manner when it exceeds certain limits.

This process of automatically changing the set point or the values of the other control parameters in a control system in response to changes in the environment that cannot be dealt with effectively by the existing parameters is called *adaptive control* (Fig. 4.32).

Homeostat

The values of the controlled variables in a regulator change in response to external conditions though its modulus of behavior may not change. However, the modulus of behavior of an adaptive controller can be made to change in response to external conditions. This type of behavior can be shown by the *homeostat*[1] in Figure 4.33. Whenever the voltage exceeds or falls below a specific value, pointer is driven against a set of contacts at the full scale or zero position. This energizes a solenoid operated *step switch* connected to a *resistance box* that is connected in series with the meter windings and the voltage source. This either increases or decreases the sensitivity (modulus of behavior) of the meter. This process of *self-adjustment* will continue until the pointer is not driven off scale. Thus, the *range of the meter* is automatically changed to accommodate the input voltage. This design feature may reduce the risks of exceeding the electrical current-carrying capacity of the windings in the meter. (The design of a homeostat given by W. Ross Ashby (1952) is made up of a moving coil galvanometer instead of a moving magnet motor and includes a means of reversing polarity if the pointer is driven against the zero position. It also includes a means of breaking and making the solenoid circuit to change the value of the resistance box if the pointer remains on the zero or full scale contacts.)

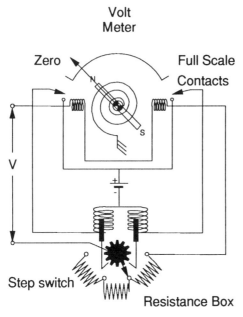

Figure 4.33. The modulus behavior of a *homeostat* changes automatically according to the conditions in its environment.

Homeostat

Adaptive clock

An *adaptive clock* can be constructed so that its rate is changed slightly whenever its time is corrected. If the clock has to be advanced, its rate is slightly speeded up. If the clock has to be set back, its rate is slightly retarded. A clock of this type is said to be adaptive because its modulus of behavior (running speed) is adjusted automatically while in use.

Adaptive controls

Adaptive controls are also used to operate metal cutting machines so that the cutting tool feed rate is reduced if the temperature of the cutting tool exceeds critical limits, or if excessive cutting force is encountered. Changes in these relations can come about by the same process used to adjust the speed of an adaptive clock. An operator can slow down the feed rate if the tool gets too hot, and the controller can incorporate these parameter changes in its modulus of behavior. However, the ability to make these changes must be built into the design of the machine controller in much the same way that an adaptive clock must be designed to speed up or slow down whenever its time is corrected.

Adaptive versus empirical behavior

The adaptive clock seems to learn from experience. Its accuracy improves over time as it incorporates information about the correctness of its rate each time it is reset.

However, this learning process is built into the clock. It has no other option than to decrease its rate if set back and increase its rate if advanced. If some other behavior is required, such as running fast during the day and running slow at night, provision would have to be made in its design to provide these features. This differs considerably from the discontinuous responsive machines shown in this book, which are design to produce whatever behavior is required of them.

Chapter 5
Discontinuous Responsive Controllers

We have seen in Chapter 2 how a machine like a spring can respond to its environment in a continuous manner. We have also seen how control systems can be made to behave like most machines. In this chapter we will see how a discontinuous responsive controller can be connected to sensors and actuators to produce a controlled machine that can act like any kind of control system or conventional machine. This discontinuous responsive controller responds to its environment in a discontinuous manner by sampling sensed conditions at specific moments in time and producing some specific action for each sensed condition based upon predetermined information stored in its memory. In some cases, the predetermined discontinuous controlled machine shown in Part II may be difficult to program. Its design can be modified so that it can program itself in an empirical manner, as shown in Part III.

Modulus of behavior of a discontinuous controller

Our goal is to design a controlled machine that can act in place of a person as a standby controlled machine, or even act like a person in some cases. What class of controller are we? Are we a directive, governing, positive feedback, or adaptive controller?

Sometimes we act like a directive controller. For instance, when we play music from memory or recite poetry or lines from a play, we act like a numerical control machine because we replay information stored within ourselves. However, there is no direct way to install musical information into a musician like installing a parts program into a numerical control machine.

Sometimes we act like a governing controller. This occurs when we steer our car for instance. As governing controllers, we visually sense the position of the car on the road and make decisions about how to turn the steering wheel to maintain the position of the car in the proper position in the road. In this case we act like an auto-pilot. When we regulate the speed of our car, we act like a speed governor or cruise controller. When we control the temperature of the inside the car by adjusting the vents and heater lever, we act like a thermostat. When we try to run away from a pursuer or play most sports, we may act like a positive feedback controller.

We act like an adaptive controller when we change the way we operate upon the variables we attempt to control. For example, we may get angry with someone after we find that it does not work to reason with them.

However, we do not have a set point that is accessible for adjustment, nor do we have a discernible means of generating an error signal. There is no way for someone or something to reach in and set a value for us to seek or escape from. We behave in many different ways by producing different sets of responses to external conditions. We can act like different kinds of controllers because our brain is a *discontinuous responsive empirical controller.*

Modulus of behavior

We are primarily interested in creating successful behavior in discontinuous responsive controllers. Their behavior is specified by the relation between the inputs to the controller and their outputs. This relation has been defined as the controller's *modulus of behavior*. This definition is represented by the diagram in Figure 5.1.

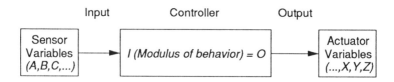

Figure 5.1. The *modulus of behavior of a controller* specifies the relation between inputs to the controller and the outputs it produces.

In our discussion so far, the modulus of behavior (*M*) has been defined as the factor that determines the output when multiplied by the input.

Modulus of behavior as a list

The modulus of behavior of a discontinuous machine was shown to consist of a list of actions taken by the machine for each sensed condition. The modulus of behavior of any discontinuous responsive controller also consists of a list of "If . . . , then . . . ." statements that relate possible input conditions to output actions (Fig. 5.2). The controller must contain enough memory to store these "If . . . , then . . . ." statements.

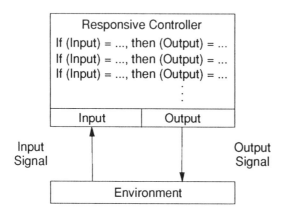

Figure 5.2. The *modulus of behavior of a discontinuous responsive controller* consists of a list of "If . . . , then . . . ." statements.

The "If . . . ," statements are the input conditions in the environment of the controller, and the "then . . . ." statements are the output actions of the controller. The list stored in a memory based controller is similar to the list in a memory based governing controller except that the "If . . . ," statements are the error signals in a governing controller rather than environmental conditions.

Modulus of behavior of a discontinuous directive controller

A directive controller sends out more information than it receives. Thus, the list of a directive controller has more "then . . . ." statements than "If . . . ," statements, as shown in Figure 5.3.

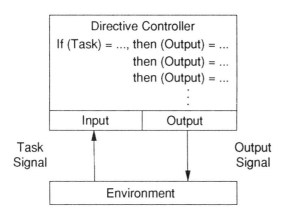

Figure 5.3. The *modulus of behavior of a directive controller* is a list having many "then . . . ." statements relative to the number of "If . . . ," statements.

A typical list of a directive controller is as follows: If the On/Off Switch is in the On position, then . . . , then . . . , then . . . . If the On/Off Switch is in the Off position, then do nothing.

Regulation behavior with no set point

A discontinuous responsive controller may have no built-in mechanism to compare sensed conditions to a set or reference point. A discontinuous responsive controller can act like a negative feedback controller (regulator) by producing enough convergent sublines of behavior to keep it at or close to a "goal state." This stable regulating behavior is shown on the map of possible behavior in Figure 5.4.

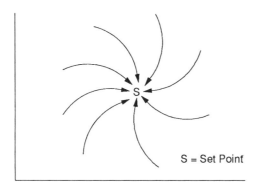

Figure 5.4. A *diagram of negative feedback behavior* shows how the action of a regulator converges upon a single point.

Most of the system's behavior consists of convergent sublines of behavior. If a discontinuous responsive controller can produce convergent sublines for each accessible initial state likely to be encountered, it can maintain a stable behavior space around a selected point.

Responsive controller with positive feedback behavior

A discontinuous responsive controller can produce *positive feedback behavior* as well as negative feedback (regulation) behavior. The controller can be programmed to avoid certain objects or situations rather than seeking them. A map of positive feedback behavior is shown in Figure 5.5.

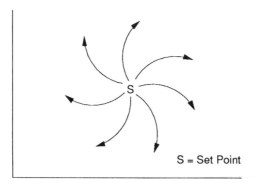

S = Set Point

Figure 5.5. A *diagram of positive feedback behavior* consists mainly of divergent sublines of behavior that are programmed into a discontinuous responsive controller.

Positive feedback behavior creates an explosive growth in behavior that is similar to the burst of energy that occurs when a servo system is hooked up incorrectly with positive feedback rather than negative feedback. These positive feedback behaviors must be restrained or modified at some point to keep the units producing these behaviors from destroying themselves.

Escaping from holes

As stated, positive inverse feedback provides a safe and often successful escape strategy. As stated earlier, it is less violent and destructive than positive direct feedback behavior. A discontinuous responsive controller can be programmed to produce escape behavior (Fig. 5.6). For example, a mobile robot with a discontinuous responsive controller will have to produce positive inverse feedback behavior to escape from a hole. As the robot nears the slope at the edge of the hole, it must begin to struggle uphill away from the center of the hole. The farther it gets into the hole, the more energy it must put into its struggle to escape. If its effort to escape increases faster than the slope of the hole, it may escape successfully.

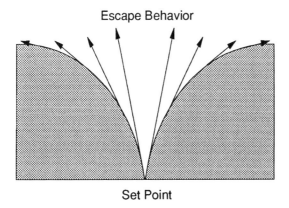

Escape Behavior

Set Point

Figure 5.6. A discontinuous responsive controller using escape behavior could climb out of a hole.

Chase/escape behavior

The behavior of a regulator usually settles down at its set point. If the system is not disturbed, the system will remain inactive. A discontinuous responsive controller with regulation behavior may produce only a relatively few states around a given point. If the environment does not demand change, the controller may not produce change. However, a controller with positive feedback behavior will create an explosive burst of new states if even slightly displaced from its point of instability. If one controller acts with negative feedback behavior toward a second controller that acts with positive inverse feedback behavior, a great deal of action is evoked from both controllers, as shown in Figure 5.7.

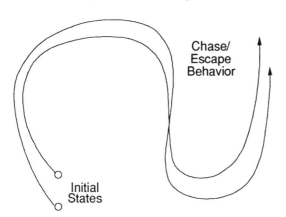

Chase/
Escape
Behavior

Initial
States

Figure 5.7. Control systems in a chase/escape configuration create a condition of ceaseless and energetic activity.

The pursuer and pursued may interact in unexpected ways. This configuration is highly dynamic and evokes large amounts of action and energy. This unstable form of behavior plays an important role in the study of most control systems. As memtioned, most ships do not want to go in a straight line and attempt to diverge from the desired course. A helmsman must be constantly alert for these tendencies to yaw, and be prepared to bring the ship back on course. A ship and an auto-pilot system is an example of a chase/escape configuration. Though the ship attempts to diverge from the desired course, a discontinuous responsive controller can be programmed to bring the ship back on course.

Homeostasis

An adaptive controller may randomly change its control parameters whenever the values of its critical variables exceed critical limits. This may lead to the development of a stable behavior that excludes these critical states. This kind of stable behavior also can be produced by a discontinuous memory based controller. Its modulus of behavior can be made to change until it stays within critical limits, as shown in Figure 5.8.

Critical Limits

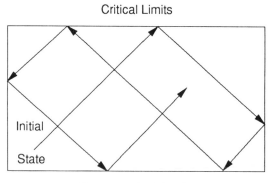

Figure 5.8. A discontinuous responsive controller with homeostatic behavior must change the way it behaves when a variable reaches a critical value.

Homeostatic Behavior

A memory based controller can produce *homeostatic behavior* by entering a new subline of behavior whenever the values of its variables exceed critical limits. These new behaviors can be created whenever a discontinuous responsive controller encounters interferences that prohibit its existing line of behavior from being carried out.

Discontinuous responsive controlled machine

Thus, the *discontinuous responsive controlled machine* shown in Figure 5.9 is neither a directive nor a feedback controller, but has some features of both. It produces behavior by responding with a unique action to each different sensed condition in its environment. This discontinuous behavior is based upon its internal memory rather than a continuous predetermined structure or a mathematical equation that relates input variables to output variables. A discontinuous responsive controller is closed loop because it uses sensed conditions to make decisions. However, it need not have a set point (it does not need to compare sensed conditions to desired conditions and create an error signal). So it is not a governing controller such as the thermostat, regulator, and servo controller described, though it can act like them.

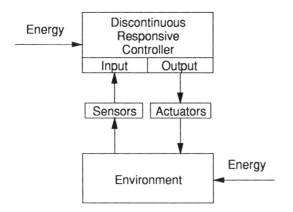

Figure 5.9. A *discontinuous responsive controller* creates behavior by producing action events based upon sensed conditions and information stored in a set of memory elements.

The discontinuous responsive controller also can act like a directive controller in that it sends out commands based upon its internal information states. But it is not an open-loop controller because it relies heavily upon input information about the conditions in its surroundings to decide what commands it sends out. It can chase or run away from other units like negative and positive feedback controllers, though it does not have a built-in mechanism that predisposes it to do so.

Predetermined discontinuous responsive controllers

Most of the controlled machines used today are programmed before they are put into use. These predetermined controlled machines usually require considerable reprogramming (debugging) after the results of the initial programs are tested.

Programming a responsive controller

The modulus of behavior of a controlled responsive machine can be established is several ways:

1. Built into the design of the controller.
2. Programmed into the controller after it is built.
3. Established after the system is in use.

A considerable effort is often needed to determine the requirements and specifications of a control program before it is written and implemented in hardware. In many cases, the modulus of behavior of a complex controlled machine has to be modified after it has been put into use.

Modulus built in

The modulus of behavior is usually built into a continuous governing controller in the same way that the spring constant is built into a spring. The responsiveness of the governor, thermostat, and other regulators is usually predetermined by the design of the hardware used in these control systems. For example, the modulus of a thermostat is built into its design through the selection of the bimetal sensor and resistance and placement of the anticipator resistor. In the case of the thermostat, the selection of the wattage dissipated by the anticipator resistor is an example of an *a priori* design decision that determines the control characteristics of the system. If the wrong resistor value is chosen, that component in the control system must be changed physically to correct the error.

Modulus programmed before use

When a computer is used as a controller, simulator, or model, its modulus of behavior is usually created by a *predetermined program*. A computer used in this way operates upon the equations and logic statements within its program to produce the desired behavior. Usually, these equations and the program are determined *prior* to the operation of the computer control system.

In the case of a numerical control machine, the list of movements needed to cut parts is usually created *before* the machining operation commences, and is based upon an analysis of the desired shape of the parts to be cut and the operating characteristics of the machine. In some applications, a numerical control machine does not require any physical modification to produce different parts — just a change in the parts program.

Faults

Most production machines operate in a repeating line of behavior. However, hidden variables may lead to variations in this behavior. In automatic machinery, variations in behavior are usually considered as *faults*. Usually, faults are actions that stop a machine from operating, such as parts jamming in a chute, or sensors indicating that parts are out of tolerance. So a fault makes the machine unable to repeat its behavior. Some faults lead to "crashes," and other faults lead to "crunches." Crashes in a computer program lead to terminal states like "End of File," while crunches may change the modulus of behavior of the machine because linkages are bent or teeth are stripped from gears. The successful operation of automatic machinery is based upon how well they deal with faults. Faults can be dealt with in the following ways: fault elimination, fault detection, fault identification, fault action, fault correction, and fault avoidance.

Fault elimination or "debugging" is the usual method of improving the modulus of behavior of a machine system. These faults can be caused by poor design, design errors, and parts made not to specification or assembled incorrectly. Usually, more time is spent debugging a machine system after it has been assembled than is spent designing and building it. Sensors may be added to machine systems that detect errors or faults. For example, a *fault detection* switch placed at the home position of an actuator can determined if the actuator moves to its normal *home position* in a "home check" maneuver. If the actuator fails to home properly, the switch will indicate that something is wrong and some corrective action can be taken. If many different faults can be detected, it may be helpful to identify which fault or faults have occurred. This *fault identification* information can be used to troubleshoot the system by showing the location and nature of each fault. In some cases, it is possible to create some appropriate action once a fault has been identified. For example, if an actuator does not home properly after the first try, it may be a good idea to try again instead of just shutting down the system. This *fault action* behavior can be programmed as part of the behavior of the machine system. In other cases, it may be possible to correct specific faults. For example, if it is determined that parts are jammed in a chute, it is possible to activate a thumper to dislodge the parts. This *fault correction* behavior also can be part of the modulus of behavior of the machine system. An even more desirable method of dealing with faults is to sense the conditions leading to faults (*fault anticipation*) and take corrective action before the fault occurs. For example, the number of parts leaving a chute can be subtracted from the number of parts entering a chute. If this difference exceeds a certain number, a jam has occurred somewhere in the chute. Then, a thumper can be activated to dislodge the jam. This *fault avoidance* behavior, along with the fault action and fault correction behavior, make a machine more reliable, robust and self-sustaining.

Self-correcting behavior

A memory based controlled machine may contain a repertoire of behavior that allows it to overcome the effects of unwanted actions. Most of the behavior in a complex system with many potential faults may consist of *self-correcting behavior*. The purpose of this self-correcting behavior is to sustain an existing line of behavior despite obstacles. Thus, much of the effort that goes into designing and developing a successful machine systems may be in developing *self-sustaining behavior*.

Closure of discontinuous responsive behavior

Most of our behavior is also based upon repeating many elements of our behavior. We get up in the morning and go to work, etc. However, hidden variables in the environment or within ourselves may cause variations in our behavior. For example, our bus may not show up on time, or we may decide to sleep a little longer. We usually learn to cope with these hidden variables. For example, if our bus does not show up, we may attempt to hail a cab. The alternative subline of behavior of getting to work by cab allows us repeat most of our daily work routine, allowing the closure of our main line of behavior.

Self-sustaining discontinuous responsive behavior

Likewise, the goal of a machine designer is to create machines that can produce and maintain a closed line of behavior. To obtain closure, a discontinuous memory based machine must have ways to deal with all likely faults. This is done by establishing convergent sublines of behavior for each possible divergent subline of behavior. To continue its behavior, a discontinuous memory based machine must perform the three requirements listed in Table 5.1.

Table 5.1. Requirements for self-sustaining behavior

---

1. Each response must be carried out successfully.
2. Every fault must be dealt with successfully.
3. The behavior of the system must stay within those conditions for which it has developed this successful behavior.

Discontinuous machines can produce only a limited number of possible responses to a limited number of possible sensed conditions. This means that a line of behavior cannot be infinitely long. Thus, a discontinuous responsive machine must act in a closed loop of behavior to keep it from running out of responses.

Self-supporting behavior

Part of the behavior of a discontinuous machine must be directed toward keeping the machine within the conditions for which it has developed successful behavior. This *self-perpetuating behavior* acts to preserve the existing behavior of the machine. For example, the limited memory in a discontinuous responsive machine acting as a temperature regulator may be able to hold a nearly constant temperature for small variations in temperature, but may be unable to operate over a wide range of temperatures. Its effort to maintain a constant temperature within its operating range usually protects it from encountering the wider temperature variations for which it lacks adequate responses.

Limitations of predetermined responsive controllers

It may be difficult or impossible to foresee all the possible faults that may be encountered in a complex machine system, or to determine what course of action should be taken in a given machine for any anticipated faults. It may be far easier to let a machine attempt to operate in a given environment and then determined what corrective actions need to be taken as each fault arises. These corrective actions can be incorporated in the modulus of behavior of the empirical responsive controllers shown in the next section, leading to a machine that can develop self-sustaining and self-supporting behavior.

Empirical discontinuous responsive controllers

The purpose of this book is to describe an *empirical discontinuous responsive controller* that can learn to maintain variables within limits like a feedback controller though it may not have a means of comparing its output to an accessible set point.

This new empirical discontinuous responsive controller can act as an open-loop numerical control machine by producing behavior based upon its internal states though it relies heavily upon sensed conditions to determine what it does. It can change the way it operates like an adaptive controller, work alongside other responsive controllers, and set the desired values of other regulators. So how does an empirical discontinuous responsive controller establish and produce this kind of successful behavior?

A posteriori changes in behavior

In the case of a servo controller, the gain and sensitivity can be adjusted after the controller is placed in service. In fact, algorithms have been developed that automatically adjust these parameters based upon the actual performance of the controller system. These self-adjusting servo controllers are called *adaptive controllers* because they change automatically in response to their surroundings. The *homeostat* used as a self-adjusting volt meter is an example of a device that can perform these automatic after-the-fact modifications to its modulus of behavior. When the modulus of behavior is developed while the control system is in service, it is called an *empirical process*, as compared to the *predetermined process* involved when the modulus of behavior is created prior to the use of the system.

The behavior of living beings can be specified by their modulus of behavior. Much of the behavior of some species is genetically predetermined as in *a priori* or *instinctive behavior*. Much of the behavior of other species is learned through experience. In the case of human beings, some of our behavior is instinctive, and some is learned through experience. We also have a special ability to alter our behavior through a cognitive process involving language and thought. In general, we optimize our performance after we become involved in a given environment and after we see the effects of our behavior. This is an example of the *a posteriori* creation of control behavior through the empirical process.

Lead through programming of a directive controller

Usually the programmer of a numerically controlled machine prepares a predetermined list of actions that the machine is expected to carry out. This list is then stored in a directive controller. An exception to this practice is the *lead through method of programming*, in which an operator causes a machine to perform the actual movements needed for a machine to produce a part while the directive controller monitors and records these movements in sequence in a *"learn mode."* Then the directive controller causes the machine to reproduce these movements in a *"playback mode"* to make additional parts.

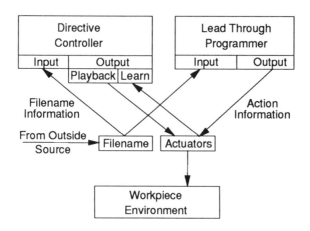

Figure 5.10. Lead through programming permits an operator to create a "parts program" by executing the actual motions required to produce a particular part.

The lead through method of programming is an example of an *a posteriori* process because the modulus of behavior of the control system is created after a machine has been put into service and while its is in use (Figure 5.10). The desired motions produced by the operator are recorded in a specific file by the directive controller in the sequence in which the motions are produced. The operator need only select that file in order for the controller to execute that sequence of motions and produce the particular part.

Lead through programming of a discontinuous responsive controller

A discontinuous responsive controlled machine can be programmed in the same *lead through method of programming* (Fig. 5.11). The operator can produce the desired machine output in a learn mode for each input condition that the operator expects the controller to encounter. Then the controlled machine can produce these desired outputs in the playback mode as each input occurs.

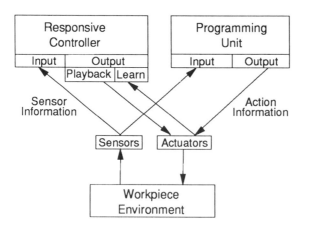

Figure 5.11. Programming a discontinuous responsive controller requires that an output state be specified for each input state as each input state is encountered.

For each sensed "If" statement encountered, there must be an output "then" statement provided. In most cases, the best way to program a discontinuous responsive controller is for the programmer to examine each "If" condition as it occurs and determine the best "then" action to be taken in the actual setting of the control system. As each input/output relation is produced, new "If" conditions may be exposed for which good "then" statements have to be supplied. These unusual condition/action relations might not be anticipated before a system is put into use. Thus, the *a posteriori* method of programming may provide a higher-quality modulus of behavior than the *a priori* method of programming a discontinuous responsive controller.

Transform conditioning

The learn and playback modes can be combined in a discontinuous responsive controller such that the controller alternately plays back the learned action for each sensed condition and then learns the actual action. If the actual action is the same as the action produced by the controller, the behavior of the machine remains unchanged. If the programmer can produce a different actual response, the controller will learn the behavior produced by the programmer. In this case, the controlled machine and the programmer may both attempt to produce the output behavior for given sensed conditions. If both produce the same actions for given sensed conditions, then the modulus of behavior of the controller will stay the same. If the programmer produces different actions for sensed conditions than the machine, and the actions of the programmer are made to prevail, the behavior of the empirical discontinuous responsive controller will change until it corresponds to the modulus of behavior of the programmer. Since the signal sent to the controller represents the actual output behavior (transform) that takes place for each sensed condition (operand), it is called the *transform conditioning signal* (Fig. 5.12).

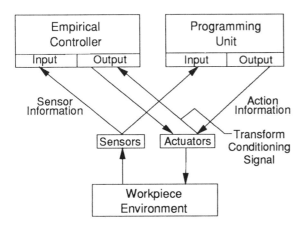

Figure 5.12. A discontinuous responsive machine with transform conditioning will learn to produce the actual output states that are produced after each sensed condition.

In some cases, a discontinuous responsive machine with transform conditioning may not need to be programmed. Initially, the condition/action relations stored in its memory will cause it to produce some indefinite behavior. The restrictions and opportunities for action in the environment will determine what actions can be carried out by the machine. This in turn will mold the modulus of behavior of this *empirical machine* to reflect the physical laws within its environment and the desires of its programmers.

Interaction

When the environment of a memory based controller is connected to a second controller or a machine operator, the responses of the second controller may contribute to the continued existence of the behavior of the first controller by causing the environment to produce the sensor conditions for which the first controller has doable outputs. Two or more control systems are said to *interact* when each supplies the information needed by the other to produce successful behavior (Fig. 5.13).

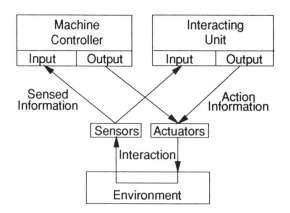

Figure 5.13. Interaction occurs between two or more control units when one unit helps the other unit act according to a successful line of behavior.

In some cases, the *interaction* of one control unit upon the other is essential for the continued existence of a line of behavior. Removing one control unit may terminate the behavior of the other unit. For example, the operator of a production machine may be able to dislodge a jammed part, allowing the behavior of the machine to continue. If the machine operator is not present during the operation of the machine, the machine may not function properly.

Interference

On the other hand, the behavior of a second controlled machine or an operator can impede the behavior of a first controlled machine. This *interference* may not allow the behavior of the first machine from being carried out. For example, the operator of a production machine may push the emergency stop button when he or she detects an impending jam that may cause the machine to break (Fig. 5.14).

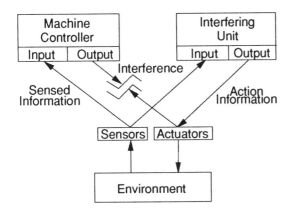

Figure 5.14. An *interference* occurs when the behavior of one control system prohibits the actions of another from being carried out.

Though an interference may restrict the behavior of a responsive control system, it may lead to the development of more successful behavior. For example, the operator of a production machine with an empirical controller may teach the empirical machine a fault avoidance step by activating the emergency stop switch prior to a potentially disastrous jam.

Goals in the study of empirically controlled machines

The purpose of this book is to describe the design and uses of a discontinuous responsive machine that can learn from experience. The next section of this book describes a controlled machine whose behavior can be established *a priori* by a programmer acting directly upon the memory in its controller. This *predetermined controlled machine* can act as a standalone machine, a standby machine that can help human operators complete tasks, and a cooperative machine that can work with other machines to produce successful behavior. Then in Part III, an *empirically controlled machine* will be described that can produce this kind of useful behavior *a posteriori* without a programmer having *direct access* to the controller's memory, or someone building behavior into the machine. The design of this empirically controlled machine is derived from the predetermined controlled machine shown in Part II and is based upon an *indirect access* to the controller's memory through the system's sensors and actuators in a process of transform conditioning. Part IV describes how groups of these empirical machines can work together more effectively that a single machine working alone, and Part V describes the applications and principles of the behavior of these empirical machines.

PART II
PREDETERMINED CONTROLLED
MACHINES

Part I showed machines that are designed by putting together the set of machine components required to do a specific task. Part I also showed how control systems can be designed to reproduce the behavior of most machines. The behavior of these *controlled machines* can be determined beforehand in a *predetermined* manner or established *empirically* while they are in use. Also, these machines can be controlled by *absolute control matrices* that make all-or-nothing decisions using *absolute memory elements,* or these machines can be controlled by *conditional control matrices* that make decisions by comparing the relative values of a set of *conditional memory elements.*

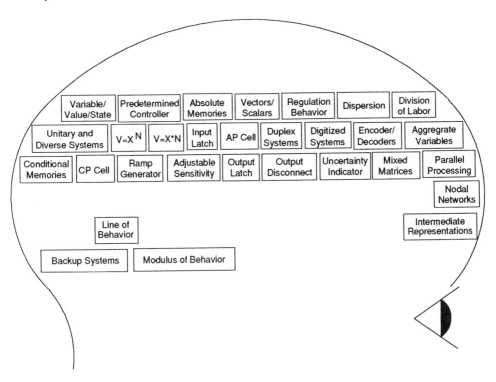

Part II describes the design of *absolute* and *conditional predetermined controlled machines* that can be programmed to produce useful behavior before they are put into use. Part III shows the design of *absolute* and *conditional empirically controlled machines* that can establish useful behavior after they are put into use in much the same way that we learn from experience.

Chapter 6
Passive Absolute Predetermined Machines

A discontinuous responsive machine can be designed that has a modulus of behavior defined by a *passive memory matrix*. The memory matrix can be programmed *a priori* to produce specific behavior by connecting each input state believed to be relevant, to the output state believed to be desirable for that input state. If this programming is carried out before the machine is put into use, and the machine produces behavior according to the discontinuous state (DS) ontology, it may be called a *passive absolute memory controlled DS machine*.

Passive absolute memory controlled machines

As shown in Part I, the information needed for a responsive machine to produce a given modulus of behavior is contained in an input/output transformation. Each input/output relation can be recorded as an *absolute connection* at the intersection of each input state on one axis and a given output state on the another axis of a predetermined connection matrix. Each connection is absolute because it either exists or does not exist. The set of connections then represents the modulus of behavior of the machine.

Passive connection matrix

Each input state of a state-determined transformation can be connected to any output state by the *passive connection matrix* shown in Figure 6.1.

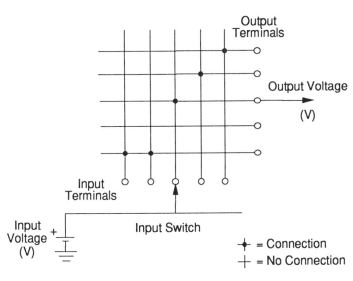

Figure 6.1. A *passive connection matrix* can transform a specific input state into a specific output state.

A *connection* can be made at the *intersection* of a given *input terminal* and a given *output terminal*. Thus, each intersection is used to designate a specific *input/output relation*. If a connection is made, and its input terminal is energized, then its output terminal is energized. Each connection is *absolute* because it exists or does not exist and the activity of an output terminal does not rely upon conditions at the other intersections. The set of connections in a matrix defines the *modulus of behavior* (U_1) of the matrix. The number of intersections is equal to the number of input terminals multiplied by the number of output terminals and is equal to the *potential variety* of the memory matrix. A passive connection matrix is the simplest kind of *memory matrix*.

Modulus of behavior of a connection matrix

The modulus of behavior of a connection matrix can be represented by a *truth table,* showing the locations of the connections between the input and output terminals (Fig. 6.2).

Output State

0	0	0	0	1	(4)
0	0	0	1	0	(3)
0	0	1	0	0	(2)
0	0	0	0	0	(1)
1	1	0	0	0	(0)
(4)	(3)	(2)	(1)	(0)	

Input State

Figure 6.2. A truth table can be used to represent the modulus of behavior of a connection matrix.

Since each connection in the connection matrix represents an *absolute state,* it is either present or it is not present. Thus, each intersection contains a one or a zero depending upon whether a connection is made or not made at that intersection. A *one* indicates that a connection is made, and a *zero* indicates that no connection is made.

Rules of a connection matrix with a single input and output variable

The memory matrix with a single input variable and a single output variable shown must follow the rules in Table 6.1:

Table 6.1. Rules for a connection matrix

---

1. Only one input state can exist at a given moment in time. Thus, only one input terminal can be energized at a given time.
2. Only one output state can exist at a given moment in time. Thus, there can be only one connection in each input column.
3. Since two or more input states can produce the same output state, there may be two or more connections made on each output row.

These rules are based upon the premise that a variable can have only one value at a given time. (Each intersection is not a binary variable because of the constraints imposed by the rules of a connection matrix. A binary variable must be able to assume either one of two values and must be free to assume either value regardless of the values of other variables. However, each intersection has one output terminal, and only one intersection in a set of intersections connected to a given input terminal can be made active at a given time. See Appendix A.)

Connection matrix as a scalar memory matrix

The connection matrix can be considered a *scalar memory matrix* because it connects one input variable to one output variable. A scalar memory matrix is defined by the number of values of its input variable (number of input states). The connection matrix shown contains five input states, of which only one can be made active at a given moment. Thus, it can be called a *pentary memory matrix*. (See *number systems* in Chapter 30.) This connection matrix can have up to five active output terminals, each representing the value of one of its five input terminals.

Permanent connection matrix

The connection matrix can be made up of two crossing sets of parallel insulated wires with one set connected to the input terminals and the other set connected to the output terminals. A permanent connection can be made by melting through the insulation at the intersection of a given pair of intersecting wires and welding the two conductors together. It is called a *welded wire connection matrix* because it is made up of an array of conductive and nonconductive intersections. The welded insulated wire matrix was widely used to connect components in high-density circuits prior to the development of integrated circuits. Because the connections are made by welding each conductor together, the modulus of behavior of the matrix is "hard wired" and cannot be changed without making and breaking welds. Thus, it is called a *permanent connection matrix*.

Conductive pin connection matrix

The connections between the input and output terminals can be changed by inserting or removing *conductive pins* that fit into sockets at the intersection of each input and output terminals, as shown in Figure 6.3.

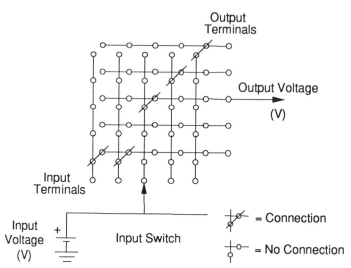

Figure 6.3. A conductive pin matrix can be used to connect any input terminal to any output terminal.

Each pin can be easily removed and placed at another intersection to reprogram the memory matrix. The rules of a connection matrix still apply. Placing a conductive pin in the sockets at an intersection makes that intersection conductive. Removing the pin makes that intersection nonconductive. Because the pins can be inserted and removed easily, the conductive pin matrix is called a *programmable connection matrix*.

Diode pin matrix

Diode pins are sometimes used to make connections at the intersections of specific input and output terminals in a *diode pin memory matrix* (Fig. 6.4). A diode pin is conductive in one direction of current flow and nonconductive in the other direction of current flow. They are used when a plain conductor pin would allow current to flow to other input terminals if two or more connections were made to a given output terminal *and* two or more outputs are connected to a given input terminal. Because diodes allow current to flow in only one direction, current cannot flow from an output terminal to another input terminal in a diode matrix. However, according to the rules of a connection matrix stated previously, only one output terminal can be connected to a given input terminal; so diodes are not needed where the rules of a connection matrix apply.

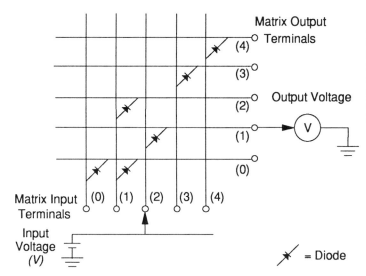

Figure 6.4. A *diode pin matrix* allows two or more connections to be made on given input and output terminals without causing current from one input terminal to flow to another input terminal.

Toggle switch

The connection between specific input and output terminals also can be made using a *toggle switch* (Fig. 6.5).

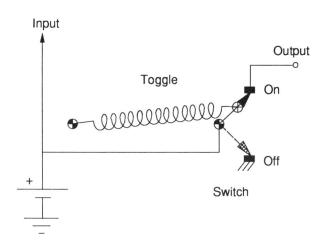

Figure 6.5. A *toggle switch* can be used as a memory device to make or break a connection between specific input and output terminals.

A toggle switch is a *bistable mechanical device* that remains in whatever position it is placed. In one position it makes a connection between its input and output terminal, and in the other position it does not make a connection between its input and output terminal. It may be viewed as a *binary memory device* if both possible output terminals are used to convey information, as in a single-pole, double-throw switch (see Appendix A), or an *absolute memory element* if only one output terminal is used to convey information as it is in a connection matrix with its single-pole, single-throw switches.

Toggle switch connection matrix

A toggle switch also can be placed at the intersection of each input and output terminal of the *toggle switch connection matrix* shown in Figure 6.6.

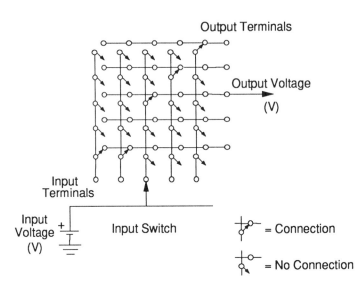

Figure 6.6. A *toggle switch connection matrix* uses a toggle switch to make or break the connection at the intersection of each input and output terminal.

A programmer can place each toggle switch in one or the other position, where it will remain until the matrix is reprogrammed. The rules of a connection matrix will apply to a toggle switch connection matrix in that only one output terminal can be connected to a given input terminal though two or more input terminals can be connected to a given output terminal. Because the modulus of behavior of a toggle switch connection matrix can be changed without adding or removing memory elements, it can be called a *programmable connection matrix*.

Passive memory controlled machine

A programmable connection matrix is called a *passive memory matrix* because the energy to produce an output state must come from an input state. A *passive memory controlled machine* (Fig. 6.7) uses a passive memory matrix to store its modulus of behavior, and can be programmed to produce a wide variety of control behavior without modifying its basic structure other than changing the connections in its memory matrix. This contrasts with a typical continuous machine such as a sewing machine, which behaves according to the overall arrangement of its mechanical structure.

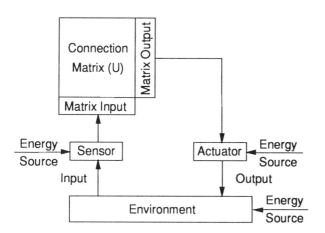

Figure 6.7. A passive memory controlled machine uses a connection matrix to specify what output state is produced by each input state.

A memory controlled machine consists of: *sensors* that convert the values of the variables in the *environment* into input states; a *connection matrix* that transforms each input state into an output state; and *actuators* that convert the output states of the control matrix into values of the variables in the environment.

Rotary position sensor

A *sensor* is used to represent a *sensor variable*. The position of the *switch armature* may be used to represent a *position variable*. The *multiposition rotary switch* creates a voltage at a *switch output terminal* that represents the value of the position variable at that moment (Fig. 6.8).

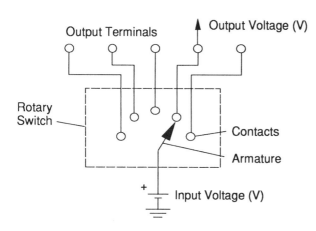

Figure 6.8. A rotary position sensor converts conditions in the physical domain of the environment, such as a shaft position, into conditions that can be understood by the control matrix.

The rotary switch can be used as a sensor in other applications. For example, a *temperature sensor* may use a *bimetal spring* to move the switch armature of a rotary switch to different terminals according to the temperature of the bimetal.

Range, resolution, and variety of a sensor

A sensor is designed to operate over a specific *range* of the high/low extreme values of the input variable it represents. It may be designed to have a certain number of distinctly different output states (values) over this range called the *variety*[2] of the sensor. Its *resolution* is its range divided by its variety and is a measure of the size (resolution) of each distinctly different state. Resolution is sometimes expressed as the percentage of the size of each state to the range of the variable. For example, the range of a multiposition rotary switch is the angle through which it can turn. Its variety is equal to the total number of contacts over this range, and its resolution is equal to the angular distance between each contact. (We could indicate the variety of a variable by placing zeros to the left of its value. For example, 000123 indicates a variable with a value of 123, a range of 0 thru 999,999, and a variety of 1,000,000 possible values.)

Electromechanical vernier actuator

A machine must produce an action that can influence the variables in the environment. Each *action* may consist of the creation of a force, motion, or both force and motion (energy) at given moments in time by an *actuator*. An action is produced by the *electromechanical vernier actuator* shown in Figure 6.9 whenever one of its *electromagnetic stator coils* is energized.

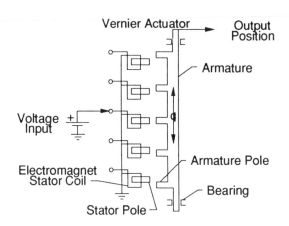

Figure 6.9. An *electromechanical vernier actuator* can convert the output state of the control matrix into a value of a position variable in the physical domain of the environment.

The *armature* of the vernier actuator attempts to align one of its *poles* with the *electromagnetic stator coil* that is energized at a given time. The output of the actuator is the position of the armature along the center line of the armature. The actuator can be designed to produce a given average force during a step motion, energy per step (force times distance of step), and power per step (energy divided by time of step).

Range and resolution of an actuator

An actuator also acts over a specific *range* of the output variable it represents and has a *variety* equal to the number of different output positions. In the example of the electromechanical vernier actuator, its range is equal to the distance between the position of the armature in its extreme position in one direction and the position of the armature in its extreme position in the other direction. Its variety is equal to the number of stator or armature poles, which determines the number of distinct stopping places of the armature between its two extreme positions. Its *resolution* is equal to the distance the armature moves between two adjacent poles and is equal to its range divided by its variety. Resolution is also expressed as the ratio of each step to the range of the actuator.

Simple memory controlled machine

The rotary position sensor and the electromechanical vernier actuator can be connected to a connection matrix, forming the simple *memory controlled machine* shown in Figure 6.10.

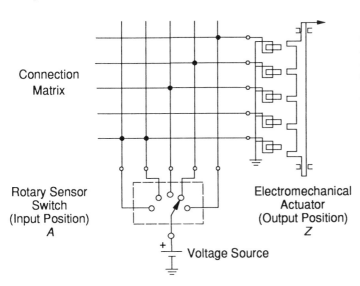

Connection Matrix

Rotary Sensor
Switch
(Input Position)
A

Electromechanical
Actuator
(Output Position)
Z

Voltage Source

Figure 6.10. A simple *memory controlled machine* can produce any of its possible output states for any of its possible input states.

The location of each connection in the connection matrix determines the output state (actuator position) of the memory controlled machine for each input state (sensor position). The simple memory controlled machine shown here makes a *step* change in its output state whenever the input state changes. The number of input/output relations is determined by the variety of the sensor and actuator variables and the number of connections in the matrix.

Parallel search

The selection of an output state is determined by simultaneously searching all the possible output states in the matrix for the particular output state that is connected to the particular input state. Because the search of all the output states is carried out in *parallel* in a connection matrix rather than in *sequence*, as it must be in a digital computer, the process of selecting an output state in a connection matrix is called a *parallel search* rather than a *sequential search*.

Discontinuous state machines

A controlled machine must sense the operand state at a specific moment in time and produce a transform state after some time delay t_d to meet the requirements of the discontinuous state ontology explained in Part I. This transform state must occur within another specific time period t_p after the delay. This requires that the connection matrix be connected to a specific *matrix driver circuit* that can produce a time delay between the arrival of specific sensed conditions and the production of an output event.

Output time delay

The discontinuous state ontology requires that a given output state be produced within a specific time period after the start of a given transition cycle. A controlled machine must have a *cycle switch* that starts and ends each transition cycle, and a *delay relay* that initiates the output action only after a specific period of time from the start of the transition cycle, as shown in Figure 6.11.

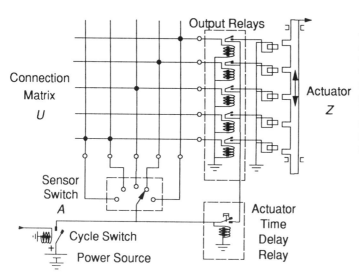

Figure 6.11. The *transition cycle timing circuit* determines the start and end of the transition step cycle and establishes a specific time delay between the start of the cycle and the production of an output state at the end of the cycle.

Operation of the transition cycle timing circuit

The connection matrix transition cycle timing circuit operates in the sequence of events listed in Table 6.2.

Table 6.2. Operation of the transition cycle timing circuit

1. The cycle switch is closed by a voltage signal from some outside timing source. This produces a voltage at the input terminal in contact with the sensor switch armature. This energizes the matrix output terminal connected by a closed toggle switch to that matrix input terminal. This closes the contacts of only one output relay. Also, when the cycle switch is closed, the delay relay starts to "time out." (The dashpot on the armature of the delay relay retards its closure for some period after its coil is energized according to the damping value of its dashpot.)
2. When the armature of the delay relay finally closes, it establishes a voltage on one of the contacts on each output relay. However, only the actuator coil connected to the closed output relay is energized.
3. After some further time delay, the cycle switch is opened by the outside timing source, marking the end of the transition cycle. The delay relay and the active output relay are "reset" (opened) when the voltage to their coils is removed.

The output relays can "amplify" the output signal from the control matrix since its output power comes through the time delay circuit and does not have to pass through the connection matrix. This allows more current to pass through the selected stator coil than may be available through the connection matrix. This may allow the actuator to position itself with more force.

Actuator brake

The actuator should not be allowed to move during the period between transition cycles when the cycle switch is open. It is usually desirable to have the actuator remain in the last position in which it was placed between transition cycles. This can be accomplished by the *actuator brake circuit* shown in Figure 6.12.

The time delay relay remains "open" between transition cycles and when it is "timing out." During this time, the actuator brake disconnect relay is in its normally closed position, allowing the power source to energize the actuator coil through the actuator brake contacts at the location where the actuator armature and stator are aligned. This creates a force that attempts to hold the actuator in this position. When the actuator time delay relay closes, it opens the brake disconnect relay, which releases the actuator armature and allows the output relay that is energized by the connection matrix to reposition the actuator armature according to the new input state and the modulus of the connection matrix.

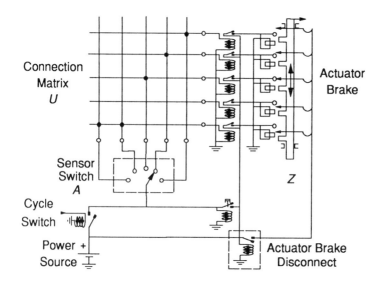

Connection
Matrix
U

Actuator
Brake

Figure 6.12. An *actuator brake circuit* attempts to hold the actuator in whatever position it was placed in the last transition cycle until it is repositioned in the next transition cycle.

Sensor
Switch
A

Z

Cycle
Switch

Power +
Source

Actuator Brake
Disconnect

Input latch

Since only one input state is allowed during each transition cycle, an *input latch circuit* (Fig. 6.13) must be connected to each matrix input terminal to maintain the voltage on the first input terminal energized in a transition cycle.

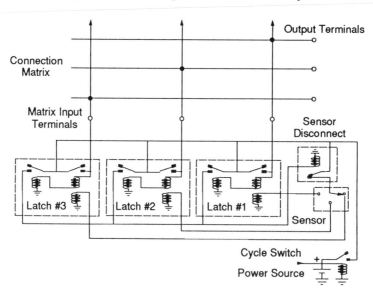

Output Terminals

Connection
Matrix

Matrix Input
Terminals

Sensor
Disconnect

Latch #3   Latch #2   Latch #1

Sensor

Cycle Switch

Power Source

Figure 6.13. An *input latch circuit* maintains the voltage at the first input terminal that is energized in a transition cycle, and disconnects the sensor switch so that no other input terminals can be energized during the remainder of that transition cycle.

The input latch also must send a voltage signal to a sensor disconnect relay that shuts off the voltage to the sensor switch, making it impossible for the sensor switch to energize any other input terminal during that transition cycle. The input latch is "reset" when the voltage is removed from the input latch circuit by the opening of the cycle switch at the end of a given transition cycle. The latch circuit is like a *short-term memory* that holds the initial input conditions constant for the duration of the transition cycle. An input latch creates a discontinuous state by *discretizing* a continuous variable.

Controlled machine with one input and one output variable

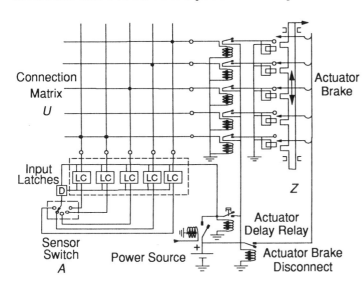

Figure 6.14. A DS passive memory controlled machine can convert each value of an input variable into a given value of an output variable at specific moments in time with a given time delay between the input and output events.

A sensor switch, input latch circuit, a connection matrix, a time delay circuit, the electromechanical actuator, and brake circuit can be connected to form a scalar passive memory controlled machine (Fig. 6.14) based upon the discontinuous state ontology. A passive memory controlled machine can produce a position value of the output device after a specific time delay from the occurrence of a given position value of the input device. It is assumed that the electromagnets are strong enough to overcome the usual frictional forces that may be encountered. However, they may not be strong enough to overcome some *interfering forces* applied by an outside load or by a human operator who may wish to locate the actuator armature at another location during the transition cycle.

Passive memory controlled ship auto-pilot

One possible mode of behavior of a machine is that of a regulator such as a thermostat, air pressure regulator, or an auto-pilot. A *ship auto-pilot* is an excellent example of a regulator with one input variable (direction error) and one output variable (rudder position). A ship auto-pilot also may require two regulators: a direction control regulator and a rudder position servo-amplifier.

Diagram of a ship auto-pilot

As stated in Part I, a controlled machine must be connected to some environment to operate. The environment must supply new input states for each output state for the machine to operate. A controlled machine can be used as a *ship auto-pilot* (Fig. 6.15) if it contains a direction error sensor, a control matrix, an electromechanical vernier actuator, and a position servo-actuator.

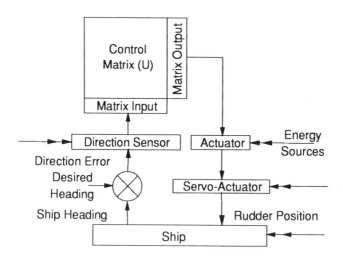

Figure 6.15. A *ship auto-pilot* must produce the correct rudder position after some time delay for each specific direction error state encountered to provide successful regulation behavior.

In the example of the ship direction auto-pilot shown, the set point is the compass direction selected by the navigator. The ship may attempt to veer from a straight course because of its positive feedback behavior, and the ship auto-pilot can be programmed to act like a negative feedback controller (regulator) by connecting each significant input terminal to the output terminal that represents the appropriate regulating rudder response for this input condition. The modulus of behavior of the ship auto-pilot is expressed in the truth table of its connection matrix.

Direction error sensor

A special sensor that produces a direction error state is required in a ship auto-pilot. This *direction error sensor* translates each deviation from a set direction into a specific error state. Each error state is represented by a given sensor terminal (contact), as shown in Figure 6.16.

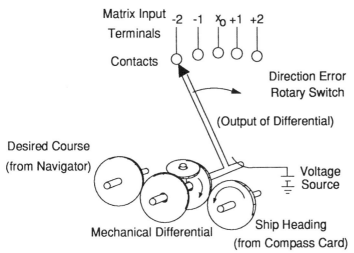

Figure 6.16. A *direction error sensor* contacts a specific terminal according to the difference between the direction in which the ship is heading and the desired course.

Since the compass card always points to magnetic north, changes in the heading of the ship cause changes in the position of the side member of the *mechanical differential* geared to the compass card. The other side member of the differential is positioned according to the direction of the desired course. The difference between these two positions determines the position of the planet gear, which moves the sensor switch contacts to a specific input matrix terminal representing a given direction error.

Servo-actuator

The output of the vernier actuator shown earlier may not produce enough force to position the rudder of a large ship. A greater force can be created by using the vernier actuator to position the servo-valve *(set point)* of a *servo-actuator* (Fig. 6.17).

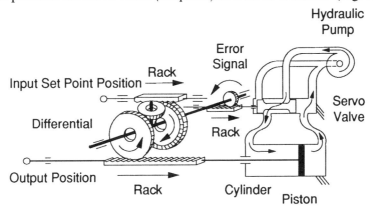

Figure 6.17. A *position servo-actuator* creates an output having a greater force and/or displacement compared to its input force after some time delay related to the response time of the servo system.

A displacement of the servo-valve, requiring only a small force, causes the hydraulic unit to pump a quantity of fluid under pressure into the cylinder on one side of the piston. This produces a force on the connecting rod. If this force is sufficient to move the load far enough to close the servo-valve, the flow of fluid into the cylinder is stopped, and the piston remains in the position determined by the original displacement of the servo-valve. The time required to move the load is determined by the time it takes for the fluid to fill the volume of the cylinder displaced by the moving piston.

Behavior of the ship auto-pilot

The ship auto-pilot shown can act as a regulator and keep the ship on the desired course if the following conditions are met:

- The compass has the required accuracy and repeatability.
- The error sensor has sufficient resolution.
- Connections are made at the appropriate locations in the control matrix.
- Correct time delays are provided between input and output states.
- The servo-actuator has the appropriate resolution, response time, and force amplification.

It is also assumed that all the elements of the control system are state determined so that their behavior is repeatable. A *memory controlled ship auto-pilot* can keep the ship on course if it is programmed to produce the rudder correction required to bring the ship back on course for each direction error signal.

Size of the control matrix

The size of the control matrix is determined by the product of the variety of the input and output variables, which is equal to the number of different possible input/output relations. The number of connections made in the matrix is equal to the number of actual input/output relations programmed into the matrix. In some cases, the control system designer can accurately estimate the variety of the input and output variables. For example, the resolution of the error sensing device in the ship auto-pilot shown may not have to be greater than 5% of its range (though the absolute direction sensing accuracy may have to be better than 5% in this application). Thus, no more than 20 input states may need to be discerned. Likewise, the rudder control mechanism may not require a resolution greater than 5% of its range; so no more than 20 rudder different positions may need to be produced. Thus, the matrix may need to establish $20 \times 20 = 400$ possible input/output relations requiring 400 intersections. If a unique output is specified for every input state, then 20 of these 400 possible connections are actually made.

Localization of control

Since each input/output relation is defined by a specific connection at a specific location, the process of producing behavior in a connection matrix is based upon an extreme *localization of control*. If for some reason a connection cannot be made at a given intersection, that input/output relation cannot exist. This contrasts with the more *distributed control* shown later in the *duplex controller*, in which a given input/output relation is represented by any two intersections that can be connected in many different places.

Nonsquare matrices

In some applications, many different inputs may require only a few outputs, and in other applications only a few input values may require many alternative output values. Since the cost of a predetermined control system is basically determined by the number of intersection memory switches it contains, considerable savings may be obtained by determining *a priori* whether the number of input or output values can be reduced.

Null functions

In some cases, the value of an output variable may remain constant for different values of the input variable (Fig. 6.18).

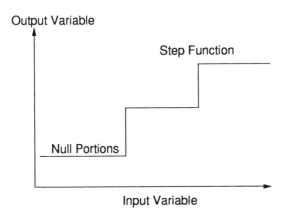

Figure 6.18. In a *null function*, the value of the output variable may not change for certain values of the input variable.

    Those values of the input variable over which the output variable does not change can be considered as a *null function*. According to the rules of a connection matrix, a null function can be represented by a set of connections from a set of the input terminals to a given output terminal. In the discontinuous state ontology, one null function may change to another null function in an abrupt *step function*.

Nonsquare input matrix

Whenever an actuator variable contains a null function, two or more connections
are made to a given output terminal, the number of output terminals may be reduced,
and fewer memory switches may be required to produce behavior. The control
matrix for the ship auto-pilot can be made into a nonsquare *input matrix* when null
functions are present and if more resolution is required for the input variable than
the output variable (Fig. 6.19).

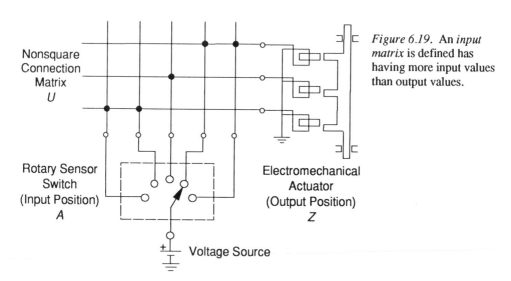

Figure 6.19. An *input matrix* is defined has having more input values than output values.

The input matrix shown has an input variety of 5 terminals and an output variety
of 3 terminals, and requires 15 intersections. If the input variable has a resolution
of 5%, it requires an input variety of 20 terminals. If the output variable has a
resolution of 20%, it requires an output variety of 5 terminals. Therefore, the number
of intersections in this input matrix is $20 \times 5 = 100$ possible connections in contrast
to the 400 intersections required in a square matrix with a resolution of 5% on each
axis.

Nonsquare output matrix

In other cases, the output variable may have higher resolution than the input variable.
The nonsquare *output matrix* shown in Figure 6.20 can be designed to provide the
same behavior as a square matrix but uses fewer memory switches. The output
matrix is drawn in a position turned 90° from the position of the input matrix

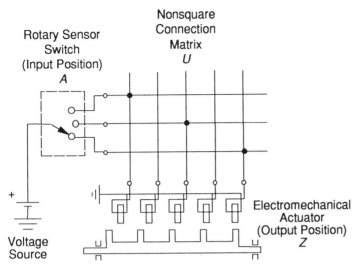

Rotary Sensor
Switch
(Input Position)
A

Nonsquare
Connection
Matrix
U

Figure 6.20. An *output matrix* has more output values than input values.

+
Voltage
Source

Electromechanical
Actuator
(Output Position)
Z

If the input variable has a resolution of 20% and the output variable has a resolution of 5%, the number of intersections is $5 \times 20 = 100$ potential connections. The rules of the connection matrix still hold in that only one output terminal can be connected to a given input terminal. This means that some output terminals in an output matrix have no connections made to them, as shown, and must remain inactive.

Matched input/output variety

The number of memory switches in a matrix and thus the number of possible choices of input/output relations is equal to the number of input values times the number of output values. The *maximum data* (see Chapter 30) can be stored in a matrix with a given number of memory switches when the number of input values equals the number of output values. This is explained by the following theorem:

For any values of a and b, where $a + b = k$, then $a \times b$ = maximum for a given value of k when $a = b$.

However, applications with big differences between the number of input or output values do occur and can be dealt with by nonsquare matrices.

Inefficient matrix

When the number of input/output relations is small in comparison to the number of input and output values, most of the intersections in a square matrix are wasted. This situation is shown on the square matrix in Figure 6.21.

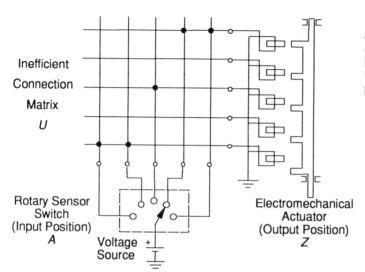

Figure 6.21. An *inefficient connection matrix* has more input or output values than input/output relations.

Inefficient
Connection
Matrix
U

Rotary Sensor
Switch
(Input Position)
A

Voltage
Source

Electromechanical
Actuator
(Output Position)
Z

For example, if the error direction sensor and the rudder actuator of the auto-pilot shown have high resolutions, yet only a few sensor values are needed to produce a few actuator values, then most of the intersections in the control matrix are not used in the control process. This problem is solved using the duplex shown in the next section.

Passive memory controlled duplex machines

In some cases, the variety of the input and output variables of a control system greatly exceeds the number of transitions required in the line of behavior of the control system. In this case, the number of intersections in the control matrix can be reduced by dividing the matrix into an input matrix and output matrix, and connecting the two with as many *intermediate states* as the expected number of transitions in its line of behavior, forming a *passive memory controlled duplex machine.*

Passive duplex controller

In the ship auto-pilot described, many input states may produce the same output state. For example, a small deviation from the set point may require a "1/2 rudder" displacement, and any greater deviation may require "full rudder" displacement in the opposite direction of the error. In this case, the size of the control matrix can be reduced by using two matrices: an *input matrix* and an *output matrix*. These matrices can be connected by an *intermediate variable* having a number of values equal to the number of different input/output relations required, as shown in Figure 6.22.

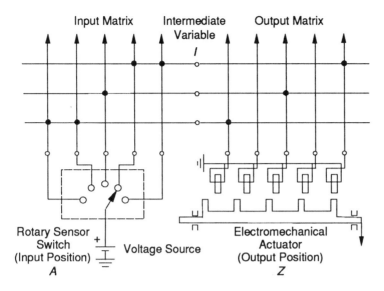

Figure 6.22. A *passive duplex controller* uses two matrices and two connections to specify each input/output relation.

In the example given above, there are five input values and five output values. Each value of the intermediate variable represents a unique input/output relation. If the number of different input/output relations is 4, then each matrix contains $5 \times 4 = 20$ intersections, resulting in a total of 40 intersections. (If a single matrix were used, it would have $5 \times 5 = 25$ intersections. The advantage of using a duplex control matrix over a single matrix appears when there are more input and output states than the number of required transitions. For example, if there were 20 input and 20 output states and only 5 significant input/output relations, a duplex control matrix would have only 200 intersections compared to 400 in the single matrix.)

Null functions in a duplex controller

When the output of a controller unit does not change over certain values of its sensor variables, each *null function* can be represented by a given value of the intermediate variable of a duplex controller. Thus, when the behavior of a duplex controller results in a null function, it does not use additional intermediate values, as shown in Figure 6.23.

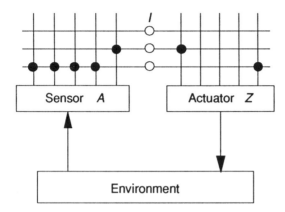

Figure 6.23. A null function can be programmed as a *single value of the intermediate variable* in a duplex controller.

According to the rules of a connection matrix, two or more input values can be connected to one intermediate terminal in a duplex controller. This single intermediate value, representing the multiple input values, can then produce a single constant output value. This leaves the intersections connected to the rest of the intermediate terminals available to produce different behavior based upon other values of the input variable.

Passive duplex controlled DS machine

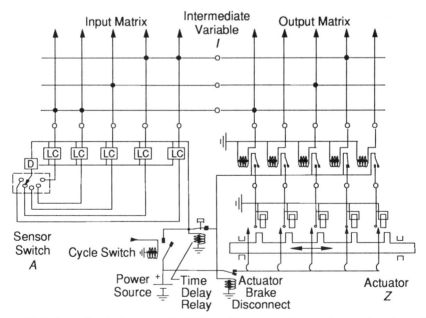

Figure 6.24. A *passive duplex controlled DS machine* can produce a given action after a fixed time delay from the start of a transition cycle based upon the value of the input variable at the start of the transition cycle.

A passive duplex controller can be made to operate according to the DS ontology by incorporating a cycle switch, input latches, a time delay relay, and an actuator brake, as shown in the *passive duplex controlled DS machine* in Figure 6.24. The passive duplex controlled DS machine can be used as a real machine in any scalar application having only one input variable and only one output variable. The resolution of the input and output variables can be increased by adding more input and output terminals, and the number of transitions can be increased by adding more intermediate terminals. A duplex controlled machine with high resolution sensor and actuator variables may require fewer intersections than a machine with a single control matrix, particularly if there are a relatively few transitions required in its line of behavior.

Passive duplex ship auto-pilot

The passive duplex control matrix can be used to specify the input/output relations of a ship auto-pilot in place of a single control matrix, as shown in Figure 6.25.

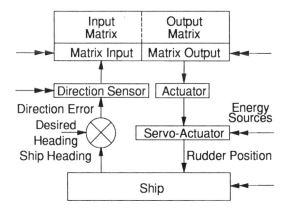

Figure 6.25. A *passive duplex ship auto-pilot may provide adequate control behavior using smaller matrices than a control system with a single control matrix.*

The duplex controller uses smaller matrices than a single controller when the number of input/output values is large in comparison to the number of required transitions. For example, the direction error sensor and actuator may both have a 5% resolution, resulting in 20 input and 20 output states. This would require a single matrix having $20 \times 20 = 400$ intersections. If there are only 5 active input/output relations, the duplex control matrices would require $20 \times 5 = 100 \times 2 = 200$ intersections instead of the 400 intersections required in a single matrix.

Distributed versus local control

Since the selection of a specific output state for a given input state in a duplex controller is determined by connections at two separate locations, the process of producing behavior in a duplex system is based upon *distributed control*. This contrasts with the *local control* in a single matrix, where information stored at one location determines behavior. Since a given transition can be represented by any value of the intermediate variable, the loss of a given intersection in either matrix does not cause an irretrievable lost of a specific transition as it would in a single matrix.

Diverse systems

So far, we have considered the design of a passive control system having only one input and one output variable. However, most control systems consist of many input and output variables. For example, a ship may need a crosswind sensor and a bow thruster besides the error direction sensor and rudder controller to maintain good direction control at low speeds in high wind conditions. Thus two or more separate control systems may be required to accomplish a given control task. When two or more separate control units are organized to deal with separate variables and each can determine behavior by itself, they form a *diverse control system*. In most cases, a diverse control system is the simplest, least expensive, and easiest system to fabricate and program. Failure to identify the diverse relations in a control task correctly compounds the difficulty in designing, building, and programming the control system.

Crosswind compensation system

A strong crosswind striking the bow of a large ship may blow the ship off course. At slow speeds, the rudder may not be able to correct for this effect. A *crosswind sensor* can be installed on the bow of a ship to measure the velocity of the crosswind, and a *bow thruster* can be installed in the bow of the ship, which acts against the force created by the crosswind (Fig. 6.26).

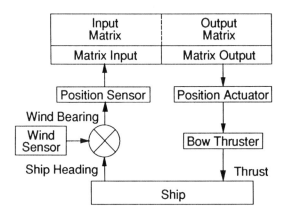

Figure 6.26. A *crosswind compensation system* has a crosswind sensor, a bow thruster, and a control unit. It produces a given bow thruster action for each measured crosswind condition.

The crosswind sensor measures the component of the wind velocity perpendicular to the center line of the ship by measuring the airflow through a tube mounted crosswise on the bow of the ship. The bow thruster is a motor-driven propeller mounted on a shaft in the bow of the ship that acts perpendicular to the center line of the ship and can create a thrust to the left or the right of the center line. The crosswind sensor and bow thruster can work together to keep the bow of the ship from being blown downwind by any condition of crosswind.

Diverse control system

In some cases, two or more separate control units can work together to solve a control problem. For example, the direction auto-pilot and the crosswind compensation system shown in Figure 6.27 may operate independently to provide better control behavior than either system working alone.

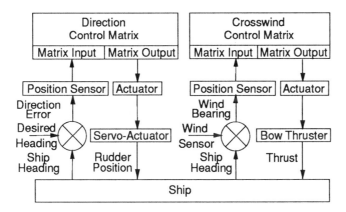

Figure 6.27. In a *diverse ship direction control system,* each control unit deals with a separate set of sensor and actuator variables.

An independent crosswind compensation system can be programmed to provide a bow thruster action based upon the relative direction and strength of the wind. The action of the bow thruster may counteract the force of the wind against the bow of the ship, allowing the direction auto-pilot to keep the ship heading in the desired direction.

Diverse organization of a control system

When a given input variable lacks a consistent or meaningful relation with a given output variable, it should not be placed in the same control system. Instead, it should be placed in another control unit, as shown in Figure 6.28.

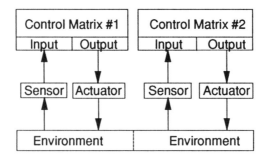

Figure 6.28. A *control system with a diverse organization* is made up of separate units, each of which connects specific input variables to specific output variables.

This set of sensors, actuators, and their control matrices creates a control system with a *predetermined diverse organization* in that specific variables are connected to specific controllers before the system is put into use.

Variety of a diverse system

The number of sensor or actuator terminals in a diverse system is equal to the *sum* of the number of possible values (x_i) of each variable in the set of (n) variables. The *variety* (V_d) of this diverse system is represented by the following *sigma equation of variety*.

$$V = \sum_{i=1}^{n} x_i. \tag{6.1}$$

A set of diverse sensor and actuator variables may be said to form a *sigma system*[3] when the variety of the input or output variables is found by summing the variety in each unit. If each variable of the sigma system has the same number of values (x), then the variety (V_d) of the set of (n) input or output variables is as follows:

$$V = x \times n \text{ (states).} \tag{6.2}$$

This *simple equation of variety of a diverse system* is more useful in many applications than the sigma equation [Eq. 6.1]. It does no harm to assume that all variables in a system have the same number of values. This is the case in all number systems. For example, each "place" in our decimal number system is a variable having 10 values. The simple equation of variety [Eq. 6.2] demonstrates and clarifies many interesting and important ideas in the theory of information (see Chapter 30).

Number of intersections in a diverse control system

The number of intersections in a diverse system is equal to the number of intersections in all the control matrices in the control system. If the number of values of all the variables are the same, then the potential system variety (V) is shown by Equation 6.3.

$$V = x^2 \times n. \tag{6.3}$$

If the ship auto-pilot can be a diverse system such that the rudder is positioned within 5% by the 5% position error sensor discussed earlier, and a bow thruster is controlled within 5% by a 5% cross wind sensor, then $x = 20$ and $n = 2$; so the diverse control system would contain $2(20 \times 20) = 800$ intersections. If duplex control units are used, each of which can produce 10 transitions, then the number of intersections would still be $4(10 \times 20) = 800$ intersections. (The duplex controllers do not provide a reduction in the number of intersections in this case, although they would if fewer transitions were required or if the variety of the input and output variables were greater than 20 values).

Results as the sum of behaviors

The heading of a ship may be determined by the sum of the action of the rudder and bow thruster, as shown in Figure 6.29.

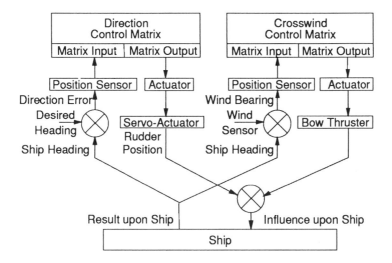

Figure 6.29. When actions are the result of the *sum of behaviors* of two or more control systems, the complexity of each system may be reduced.

The action of the crosswind compensation system may reduce the work load of the direction auto-pilot control system by reducing the need for large rudder corrections in strong wind conditions. Likewise, the action of the rudder may reduce the range over which the bow thruster must act to maintain a straight course. When two or more control units can work together to reduce the complexity of behavior required of each unit, they are practicing a successful *division of labor* that allows them to produce more complex behavior than a single unit working alone.

Null behavior in a diverse system

Two or more diverse control systems also may divide the control task into separate lines of behavior. Then, each can produce *null behavior* while the other produces *active behavior*. The rules of the connection matrix allow each connection matrix to store its null behavior on a single intermediate terminal, thus reducing the size of both matrices, as shown in Figure 6.30.

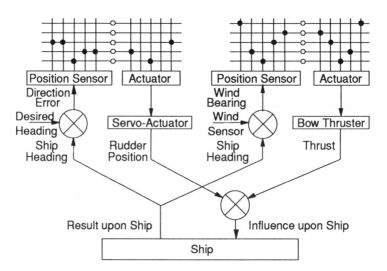

Figure 6.30. *Null behavior in a diverse system* allows each control unit to be less complex than a single unit that can deal with all the action states.

For example, a constant action of the bow thruster in one direction or the other depending upon wind direction may reduce the magnitude of the rudder responses required to keep the ship on a straight course. This may reduce the size of both duplex controllers in the auto-pilot system. When two or more control units divide the control task into separate lines of behavior that use different variables, they are practicing a *division of labor* based upon *dispersion*.

Diverse behavior and parallel control

The ship auto-pilot and the crosswind compensation system can act independently as separate systems and still produce satisfactory behavior when working together. The *diverse behavior* of each unit may indirectly help the other unit maintain successful control behavior. A diverse control system that operates successfully is an example of *parallel control.*

Multiple controllers connected to a single input and output variable

Until now, we have considered the design and application of single standalone passive memory controlled machines operating upon separate variables. However, two or more controllers connected to the same variable may share the work load of a control task more effectively than a single control unit acting alone. Two or more control units can be connected to the same input variable and to the same output variable by means of a differential or a *summing actuator.* These units will not interfere with each other if the sum of their actions results in the desired behavior. In one case, only one unit may produce the desired behavior, while the other units produce a constant (null) behavior for all conditions. In another case, different units may operate upon different values of the sensor variable using *value field selection.* These methods of obtaining a *division of labor* may reduce the workload of each control unit.

Summing behavior

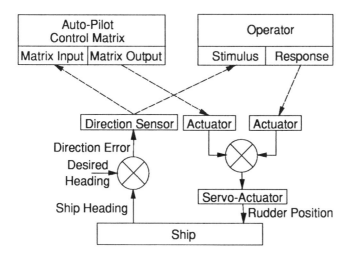

Figure 6.31. Two or more control units can operate upon the same input and output variables if their actions are summed in a *control system with an output differential.*

If the helmsman does not have confidence in the behavior or the reliability of the ship auto-pilot discussed, the helmsman may choose to steer the ship while the ship auto-pilot is in operation. So long as the auto-pilot produces the desired behavior, the helmsman does not need to produce behavior. If the auto-pilot fails to produce the required action, the helmsman can provide the required correction by acting through the *output differential* shown in Figure 6.31. The combined behavior may increase the *reliability* of a control system through *redundancy,* or may reduce the work load of both the auto-pilot and the helmsman. For example, the auto-pilot may produce small corrections for small direction errors, and the helmsman may provide the large corrections required for large direction errors.

Sigma actuator

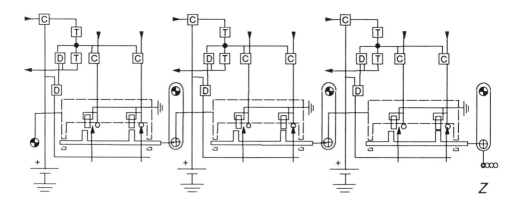

Figure 6.32. A *sigma actuator* produces an output position equal to the sum of the outputs of each of its vernier actuators.

A *sigma actuator* is made up of two or more vernier actuators connected in series, as shown in Figure 6.32, and can replace the output differential shown. The vernier actuators of two or more control units can be connected in series forming the sigma actuator. If one vernier actuator in a sigma actuator does not change its position (producing a null function), it acts like a fixed link in the set of adjustable links.

Summing the behavior of two or more controllers

The outputs of two or more controllers can be connected to a single actuator variable by a sigma actuator, as shown in Figure 6.33. When two or more controllers are connected to a sigma actuator, each controller may contribute to the output behavior or may refrain from producing behavior.

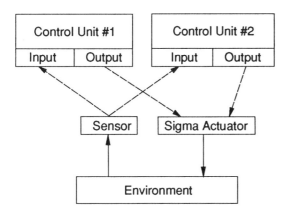

Figure 6.33. Control units connected to a sigma actuator produce behavior that is equal to the sum of their individual behaviors.

When two or more control units are connected to a sigma actuator, both units may contribute to the output of the actuator. This may increase the range and variety of the actuator variable without increasing the complexity of each control unit.

Division of labor based upon value field selection

The complexity of the single variable duplex ship auto-pilot shown earlier can be reduced if the helmsman or another controlled machine share the work load of the control task. For example, the duplex ship auto-pilot may be able to deal effectively with the small rudder corrections needed to maintain a straight course. However, if a new course is required in a totally new direction, the helmsman may wish to supply the large direction error values needed to establish a new course heading and continue to control the ship until the direction error is reduced to some smaller value. In the example given, the duplex controller can deal with the small error values involved in keeping the ship going in a straight line and can ignore the larger error values if the operator deals with the large error values and does not get involved in small error behavior. This *value field selection* reduces the size of duplex controller and the workload of the operator. Value field selection may allow two or more cooperative controllers to accomplish a control task that is too complicated for one controller acting alone.

Division of labor and parallel control

When the control of a variable is determined by control units that are connected in parallel with the variable, and they can divide the control task according to the opportunities for value field selection. This process of producing behavior is also based upon *parallel control*. Parallel control allows a *division of labor* that may decrease the size requirements of each parallel controller, and may reduce the size of the overall control system in some cases.

Chapter 7
Coded Unitary Passive Controllers

The scalar control units described so far consist of a single input variable and a single output variable. For example, the error signal and rudder position of the direction auto-pilot shown earlier are both single variables, and the separate direction auto-pilot and crosswind compensation systems were used together to form a diverse system. However, it may be desirable for the rudder responses also to be determined in part by the crosswind sensor, and the bow thruster responses also to be determined in part by values of the direction error sensor. This can be done in a single passive coded unitary controller made up of two input variables and two output variables.

In this configuration, the rudder response of an auto-pilot may be different when there are different amounts of crosswind for a given direction error, and the bow thruster response may be different when there are different error signals for a given crosswind. In this case, each *combination* of crosswind and direction error may require a unique *combination* of rudder and bow thruster response. When different combinations of sensor variables must produce a different combination of actuator values, the sensor and actuator variables form a *unitary system*.

If a system state has two or more variables, there are two different ways in which the system can be defined: It can be defined as a set of *diverse systems* in which the condition or action of the system is best described by a set of separate and independent sensor and actuator variables, or it can be defined as a single *unitary system* made up of two or more sensor and actuator variables where each combination of values of its sensors may produce a unique combination of values of its actuators. The control system will work properly only when the correct method of *organization* is identified, and the system is made to operate accordingly.

Unitary systems

The separate ship auto-pilot system and crosswind compensation system shown earlier can be combined into one system, forming a *multivariable direction control system*. The combined system contains two sensor variables, two actuator variables, and one control matrix. Each combination of values of the sensor variables may have to determine different combinations of values of the actuator variables in a unitary manner to provide satisfactory control behavior.

Variety of a unitary system

Since each input or output state in a unitary system is made up of a specific combination of values of its component variables, the *potential variety V_u* (number of possible input or output states) in this unitary system is equal to the *product* of the number of values of all the variables. When there are n variables and x_i equals the number of values of the ith variable, the variety of the set of unitary input or output variables is shown in the following *pi equation of variety:*

$$V = \prod_{i=1}^{n} x_i. \tag{7.1}$$

Therefore, a set of input or output variables in a unitary system forms what may be called a *pi system.*[4]

If the number of values of each variable is the same, as in all number systems, then the potential variety (V_u) of this unitary system can be written as:

$$V_u = x^n \text{ (states)}. \tag{7.2}$$

This *simple equation of variety of a unitary system* is more useful in many applications than the *pi equation* [Eq. 7.1]. For example, all number systems are unitary systems with variables that have the same number of values. The variety (V) of any number system with a specific number of variables (n) and a *number base* with a specific number of values (x) is given by Equation 7.2, the simple equation of variety of a unitary system (see Chapter 30 for more information on number systems).

Unitary system viewed as a vector

The set of sensor variables and the set of actuator variables in a unitary system may be considered components of a *sensor vector* and an *actuator vector* because each different combination of values of their *component variables* results in a unique value of the *resultant* (sensor and actuator) *variable.* The set of separate sensor and actuator variables in a diverse system may be considered *separate vectors,* and the set of sensor and actuator variables in each separate system may be considered the components of each separate sensor and actuator vector.

Variety of unitary and diverse systems

The *potential variety* of a unitary system (x^n) is much higher than the potential variety of a diverse system ($x \times n$) when both are made up of the same number of variables (n) and both have the same number of values (x), and when x and n are greater than two. This is because there are many duplications in the results of the multiplication of different combinations of values of x and n due to the commutative law of multiplication, but fewer duplications in the results of the exponentiation of x^n for different combinations of values of x and n. (The only duplication that comes to mind in exponentiation is 2^4 and 4^2, which are both equal to 16.)

Unitary decoding

A device that can create all x^n unitary states for n variables having x values is called a *decoder.* For example, if the input to an electrical decoder has two variables, and each variable has two values, then the decoder would produce an electrical signal at one of $2^2 = 4$ output terminals, as shown in Figure 7.1.

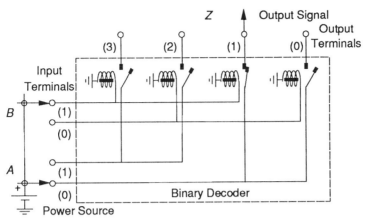

Figure 7.1. A *binary decoder* creates a signal at a given output terminal of the output variable Z, which represents a unique combination of one of two values of the input variables A and B.

If the output variable of a decoder is considered a vector, the input variables of a decoder can be considered the components of this vector. The decoder converts the code represented by the values of the set of input (component) variables into a specific value of the single decoder output (vector) variable. An electrical decoder is a common logic device and is often used to translate a code into the value of a single variable, such as converting a digital code into the position of an electron beam on a cathode ray tube. A digital-to-analog *(D/A) device* is considered a decoder in that the input digital word (usually a multi-digit number) is made up of a set of variables, each having a set of values based upon the base of its number system, and the output is a single analog variable. However, the decoders in this book usually convert multiple discrete input variables into a single discrete output variable.

Decoder truth table

A decoder using a number base of two is called a *binary decoder*. The modulus of behavior of a two-variable, two-value (binary) decoder can be defined by the *binary decoder truth table* in Figure 7.2:

Binary Decoder Truth Table			
Input Variables		Output Variable	
A	B	Z	
0	0	0	
0	1	1	Output Value
1	0	2	
1	1	3	

(Input Values)

Figure 7.2. A *binary decoder truth table* shows which value of the single output variable represents each combination of values of its input variables.

Thus, each combination of binary values of the two input variables is represented by one of the four possible values of the single output variable.

Unitary encoding

Each value of a single input variable also can be converted back into a unique combination of values of a set of output variables by an *encoder*. If the input variable of an encoder has V values, then it must have n output variables each having x values, where $V = x^n$, if all the input values are to be represented by the output code. The *binary encoder* represented by the circuit in Figure 7.3 has two values of each output variable.

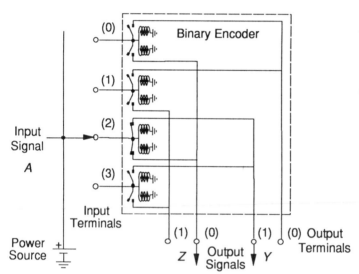

Figure 7.3 A *binary encoder* creates a specific combination of one of two values of its output variables Y and Z for each value of its single input variable A.

The input variable of an encoder can be considered a vector, and its output variables can be considered the components of this vector. In the example shown, the four values of the input (vector) variable are converted into binary values of the two output (component) variables. Encoders are also common electromechanical devices that are often used to provide a digital display of the value of a single analog variable, such as shaft position, temperature, pressure, or flow. An analog-to-digital *(A/D) device* may be considered an encoder because an analog signal is a single variable that the encoder represents by a set of digital variables. However in this book, an encoder is considered a device that converts the values of a single discrete input variable into values of multiple discrete output variables.

Encoder truth table

The modulus of behavior of this two-variable binary encoder can be defined by the *binary encoder truth table* in Figure 7.4.

Binary Encoder Truth Table		
Input Variable	Output Variables	
A	Z	Y
0	0	0
1	1	0
2	0	1
3	1	1

Figure 7.4. A *binary encoder truth table* shows each combination of values of the output variables produced for each value of the single input variable.

(Input Value on the left of the data rows; Output Values on the right of the data rows)

Thus, the four values of the single input variable can be represented by the four combinations of binary values of the two output variables.

Diode encoder

A diode can be used in place of each relay used in the binary encoder shown previously, forming the *diode encoder* shown in Figure 7.5. A diode encoder will be used throughout this book because it shows the operation of an encoder in the simplest way. Whenever a diode is used, a relay could be substituted that has the power source connected to its armature contact and its coil in the manner shown in the relay encoder shown earlier.

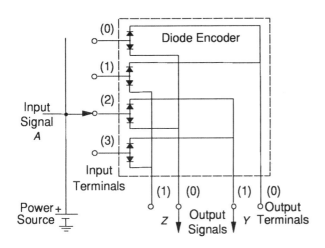

Figure 7.5. A *binary two-variable diode encoder* produces a unique combination of binary values of its two output variables for each value of its input variable.

Code translation

Note that a *decoder* converts the combinations of values of a set of unitary variables into values of a single variable, while an *encoder* converts the values of a single variable back into a combination of values of a set of unitary variables. A decoder and an encoder can be connected by a single scalar connection matrix, forming the *code translator* shown in Figure 7.6.

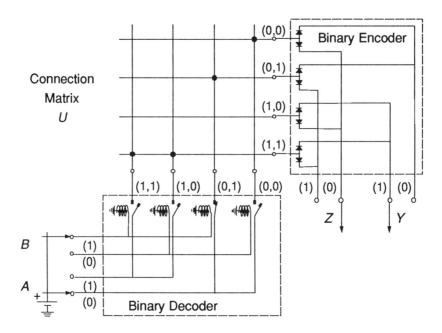

Figure 7.6. A *code translator* has a decoder, a scalar connection matrix, and an encoder. It can be used to translate one code, represented by the values of one set of variables, into another code, represented by the values of another set of variables.

According to the way in which the output of the decoder is connected to the input of the encoder by the intervening connection matrix, an encoder and a decoder can be used to translate one set of numbers into another set of numbers in a given number system or to transform the values of the component variables of one vector variable into values of the component variables of another vector variable in a unitary manner.

Code transformation truth table

The modulus of the code translation or vector transformation can be represented by the *transformation truth table* shown in Figure 7.7.

Transformation Truth Table				
Input Variables		Output Variables		
A	B	Z	Y	
0	0	0	0	
0	1	1	0	
1	0	0	1	
1	1	1	1	

Input Values / Output Values

Figure 7.7. A *transformation truth table* makes it possible to translate one code into another code or transform one vector into another vector.

A transformation truth table contains the information about the relation of one code to another code or one vector to another vector. If one code is considered an input code, and the other code is considered the output code, the truth table defines the modulus of behavior of the code transformation system.

Unitary decoding machine

The input variables of a decoder may be connected to a set of sensor switches, and the output of the decoder can be connected to a vernier actuator, forming the *unitary decoding machine* shown in Figure 7.8.

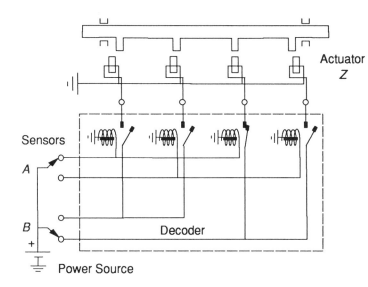

Figure 7.8. A *unitary decoding machine* can be used to produce a unique actuator position for each combination of values of a set of sensor variables.

The actuator will assume a unique position depending upon the settings of the sensor switches. If there are n sensor switches, each having x positions, the actuator must produce a variety of x^n different positions.

Unitary encoding machine

The input variable of an encoder can be connected to a single multiposition sensor switch, and the output can be connected to a set of actuators, forming the *unitary encoding machine* shown in Figure 7.9. The set of actuators will assume a unique combination of positions for each position of the sensor switch. If there are n actuators, each having x number of output positions, then the sensor switch must have a variety of x^n terminals to create every possible combination of actuator positions.

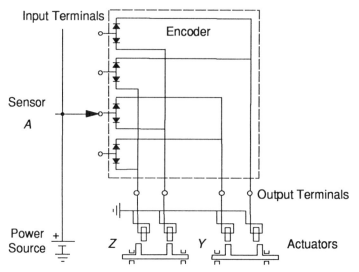

Figure 7.9. A *unitary encoding machine* has a single multiposition sensor switch and two or more output actuators.

Coded unitary passive machines

A *coded unitary passive machine* having multiple input and output variables can be created by connecting a set of sensors to the input of a decoder, connecting a set of actuators to the output of an encoder, and connecting the decoder and encoder to a passive scalar control matrix, as shown in Figure 7.10.

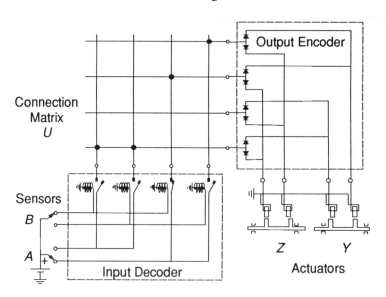

Figure 7.10. A *coded unitary passive machine* can have multiple input and output variables. Each combination of sensor values can be programmed to produce a unique set of actuator values.

A coded unitary machine can deal effectively with an environment organized as a vector. For example, this controller could be used as a predetermined ship auto-pilot having two or more input variables such as direction error and crosswind sensor, and two or more output variables such as rudder position and bow thruster. A particular unitary output state is produced whenever a particular unitary input state is encountered.

Number of intersections in a coded unitary controller

The number of intersections in the matrix of a coded unitary machine is equal to the number of matrix input states times the number of matrix output states. Thus, the number of matrix input states in a unitary controller is $V_I = x^n$, where n is the number of input variables and x is the number of values of these variables, assuming all the input variables have the same number of values. The number of matrix output states is $V_O = x^n$, where n is the number of output variables and x is the number of values of these variables, assuming all the output variables have the same number of values. Thus, the number of intersections is equal to $V_I \times V_O$. In some cases, the input variety (V_I) may equal the output variety (V_O), resulting in a square matrix. If the number of input variables equals the number of output variables, and all have the same number of values, the number of intersections is equal to $(x^n)^2$. (The maximum number of intersections and thus the maximum potential variety for a given number of terminals occurs under these conditions.) In the example given above, $x = 2$, $n = 2$, so $x^n = 4$, and $(x^n)^2 = 16$ intersections.

However, two or more input states may produce the same output state, although only one output state can be produced by a given input state. Therefore, there may be less actual output variety (V_O) and fewer matrix output terminals than the number of input states (V_I). This situation causes some output terminals and matrix intersections to remain inactive and in a sense to be wasted.

Modulus of behavior of coded unitary machine

The *modulus of behavior of a coded unitary machine* is defined by the connection matrix truth table shown in Figure 7.11. Note that a unique combination of values of the multiple input variables produces a unique combination of values of the multiple output variables through a single connection. This is an example of an *extreme localization of control,* since the responsive behavior of multiple variables is determined by a single connection at only one location. If the connection at a given intersection cannot be made because of faulty contacts, that transition becomes *unavailable.* The connections in the matrix of a coded unitary machine must conform to the rules of a connection matrix.

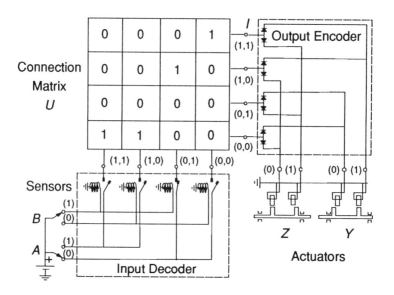

Figure 7.11. The *modulus of behavior of a coded unitary machine* can be changed by changing the connections in its control matrix.

Coded unitary DS machine

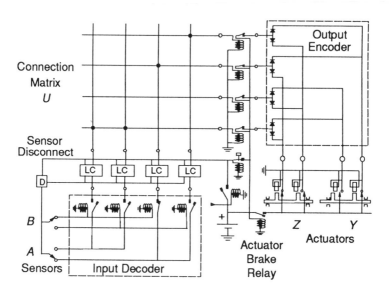

Figure 7.12. A *coded unitary DS machine* creates a specific output state after a specific delay and for a specific period after the occurrence of a specific input state.

A cycle switch, time delay circuit, input latches, and actuator brake circuit can be added to the coded unitary machine shown to create a control system that produces a specific output state that must occur within a specific period after the input state according to the discontinuous state ontology, as shown in Figure 7.12. The input and output states of a *coded unitary DS machine* may include two or more variables. Each input and output state is a specific combination of values of the two or more variables.

Operation of a coded unitary DS machine

A coded unitary DS machine operates in the sequence of steps given in Table 7.*1*.

Table 7.1. Operation of a coded unitary DS machine

---

1. The cycle switch is closed at the beginning of a transition cycle. The actuator delay relay starts to time out and the input decoder energizes an input terminal in the connection matrix based upon the combination of values of the sensors.
2. The selected input terminal is latched "on" and the voltage to the sensors is disconnected so that no new sensor values can register during the current transition cycle.
3. The connection matrix selects a matrix output terminal. This causes the encoder to energize some combination of encoder output terminals according to the value of the encoder input terminal selected by the connection matrix.
4. When the actuator delay relay times out, the actuator brake is released and the selected encoder output terminals energize the selected actuator terminals for while the cycle switch is closed. This allows the actuators to move to the selected positions within this period if nothing physically prohibits them from doing so.
5. At the end of the transition cycle, the cycle switch is opened. This energizes the actuator brake, which holds the actuators in whatever position they are in at the end of the transition cycle, and the latches, output relays, and disconnects are returned to their initial condition.

A coded unitary DS machine may provide excellent control behavior in a system with few variables with few values in which each combination of input values requires a unique combination of output values.

Passive coded unitary ship auto-pilot

Each combination of values of the direction error and crosswind sensors may require a unique combination of rudder position and bow thruster values to maintain adequate control of a ship under all circumstances. Unitary control can be obtained by decoding the sensor variables and encoding the output of the control matrix, as shown in the *passive coded unitary direction control system* (Fig. 7.13). A passive coded unitary ship direction control system can deal effectively with multiple input and output variables if every possible combination of input values requires a specific different combination of output values.

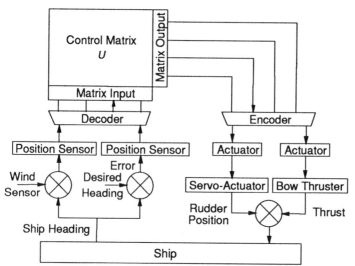

Figure 7.13. A passive coded unitary direction control system produces a unique combination of actuator values for each combination of sensor values.

Number of intersections in a coded unitary auto-pilot

If the two input variables of the auto-pilot shown earlier consist of an error signal and a crosswind sensor, both with a resolution of 5%, then $n = 2$, and $x = 20$, and $x^n = 400$ required matrix input states. If the rudder and the bow thruster also operate with 5% resolutions, the number of matrix output states is equal to 400. The scalar connection matrix for this system requires $400 \times 400 = 160,000$ intersections, of which a maximum of only 400 intersections may be connected. If the sensor and actuator variables were separate and diverse, the scalar matrix would require $400 + 400 = 800$ intersections, of which 40 intersections would be connected. The number of intersections compared to the number of connections in the unitary controller is excessive. The size of the scalar matrix in this coded unitary machine can be reduced from 160,000 intersections to 8000 intersections by using a coded unitary duplex controller shown in the next section.

Coded unitary passive duplex controllers

A unitary input decoder and output encoder can produce every possible combination of values of the input and output variables. However, only certain combinations values of the input and output variables may be required to provide adequate control behavior, and the other possible combinations may not require unique responses. For example, unique bow thruster and rudder positions may be required for large direction error values and high crosswind conditions only. So many possible transitions may be disregarded. The size of the coded unitary control matrix can be reduced by using a duplex control matrix and including only the number of intermediate states needed to represent the number of active transitions.

Passive machine with an input decoder

The two input variables can be decoded and connected to a passive connection matrix, forming the *passive machine with input decoder* shown in Figure 7.14.

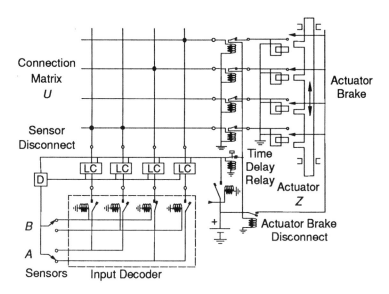

Figure 7.14. A *passive machine with input decoder* produces a unique value of its output variable for specific combinations of values of its input variables.

Any one of the possible combinations of input values can be connected to a given output terminal. If every possible combination of input values must produce a different output value, a square matrix would be required.

Passive machine with an output encoder

The output of a passive connection matrix can be connected to an encoder that produces different combinations of values of its actuator variables for each output value of the connection matrix, forming the *passive machine with an output encoder* (Fig. 7.15). Any one of the possible combinations of output values can be connected to a given input terminal. If every possible combination of output values must be produced, a square matrix would be required.

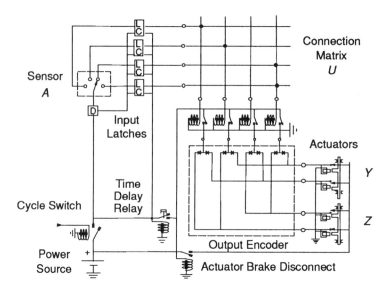

Figure 7.15. A *passive machine with an output encoder* can produce a unique combination of values of its actuator variables for each value of its input variable.

Passive machine with a coded unitary duplex controller

A *passive machine with a coded unitary duplex controller* (Fig. 7.16) can be created by connecting an input matrix with a decoder to an output matrix with an encoder by an intermediate variable (I) with as many value terminals as expected active output states. If the number of expected transitions involving a different output state is equal to 10, and both of the input and output variables have a 5% resolution, then the number of decoder output terminals and encoder input terminals is $20^2 = 400$. The number of intersections in one matrix is equal to 400 (terminals) $\times$ 10 (transitions) = 4000 intersections. Thus, both matrices required 8000 intersections together as compared to the 160,000 intersections in a single control matrix with the same resolution. Connections would be made in 20 of these 8000 intersections if there are 10 different output states. Many more transitions could be made using these 8000 intersections if many input states produce the same output states. Since two connections are required to complete a connection from the sensor variables to the actuator variables, the duplex system uses a *distributed selection process* in contrast to the local selection process used in the coded unitary machine using a single matrix.

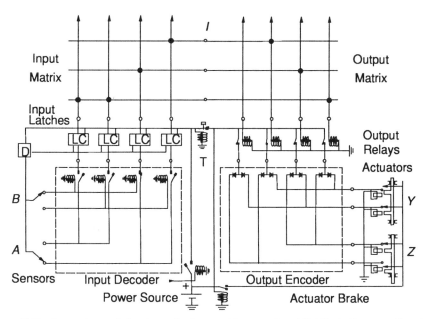

Figure 7.16. A *passive coded unitary duplex machine* uses two (distributed) connections to create each transition and uses only as many intermediate values as the number of expected different output states.

Coded unitary passive duplex direction control system

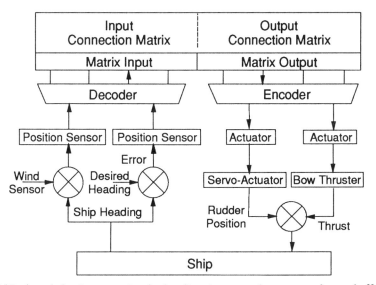

Figure 7.17. A *coded unitary passive duplex direction control system* can be used effectively in any application having two or more input and output variables that act like vectors, and when there are few required input/output relations in comparison to the input/output variety.

A duplex controller can produce control behavior with fewer intersections than a single matrix when a relatively few active transitions are required. Thus, the *coded unitary passive duplex ship auto-pilot* shown in Figure 7.17 can be used more effectively in most applications than the coded unitary ship auto-pilot using a single matrix. A duplex controller is essential in any coded unitary application with input and output variables that have a significant number of values and a relatively few active transitions. However, enough intermediate states must be provided to include all the required combinations of values of the output variables.

Problems with coded unitary systems

The coded unitary system requires a matrix input terminal for each combination of values of its input and output variables. Since many of these combinations may not be needed to produce adequate control behavior, much of the system hardware is wasted. Also, some of these unneeded combinations of sensor values may actually occur randomly within a real control task. Each *irrelevant* combination of sensor values may have to be programmed to produce benign results. This greatly increases the programming effort and may result in an impractical control system except in a very small system where nearly all the possible combinations of input and output variables are meaningful.

Intrinsic organization

If the variables in a control system are separate and distinct, where the meaning or significance of each variable is independent of the values of the other variables and combinations of values have no significance, then the variables may form a *diverse* system. If each combination of values of a set of variables produces unique results, where the significance of the value of one variable is dependent upon the value of another variable, then the variables form a *unitary system*. One of the difficult tasks facing the designer of a predetermined control system is to determine this *intrinsic organization* of the task environment.

The successful design of a control system having multiple sensor and actuator variables is based upon correctly identifying the related input and output variables, and organizing them into units according to these relations. The failure to organize the units into the correct unitary and/or diverse assemblies leads to a system that does not work well, requires too much memory, and is unnecessarily expensive.

Problems using a coded unitary controller in a diverse environment

If the crosswind does not influence direction control and the rudder does not need the bow thruster to maintain a steady course, the coded unitary controller shown previously can be programmed to produce a specific output for each value of one sensor variable and *any* value of the other variable. This is done by programming the control matrix to produce specific values of at least one output variable and constant values of the other output variable for each value of the given input variables and *every* value of the other input variable. This results in what appears to be diverse behavior of one or more of the input and output variables. However, producing this result requires an extensive programming effort, requires a much larger program, and costs much more to build than a set of diverse units. It would be much simpler to reconfigure the system into separate diverse systems.

Problem with committed terminals

Another problem with the coded unitary controller is that too many matrix input and output terminals may be committed to values that may not be used. For example, $20^2 = 400$ sensor decoder output terminals and 400 actuator encoder input terminals are required for a pair of input and output variables where each variable has a resolution of 5%. If the number of expected transitions is equal to 10, then only 10 of the 400 matrix input terminals and a maximum of 10 of the 400 matrix output terminals would be used in a line of behavior. This coded unitary duplex controller with 10 transitions would require 8000 intersections, of which only 20 intersections would be used if each input state produces a unique output state. A diverse duplex system with the same number of variables, values, and number of transitions would require only 800 intersections, of which 80 would be used to determine behavior.

Problems using diverse controllers in a unitary environment

If a set of diverse controllers were used in a unitary environment, they could not identify the specific combinations of values needed to obtain specific unitary states. For example, the ship auto-pilot system may require a different rudder correction for a given crosswind condition when a given bow thruster value is applied. This kind of *coordinated behavior* is not possible in a diverse system.

Incorrectly connected diverse units

Illogical and unsuccessful control situations would also arise if separate diverse controllers were connected to an inappropriate set of variables. For example, the crosswind sensor may not hold a ship on a straight course at low speed if it is connected to the rudder, and the bow thruster may not be able to maintain the ship on a given course using the direction error signal. An improperly configured diverse system may consistently make gross mistakes.

Selection of a unitary or diverse system

A unitary control system is required if the behavior of one input variable has a bearing upon the meaning of another input variable, or the behavior of one output variable has a bearing upon the behavior of another output variable. But if the value of each input or output variable is not related to the values of the other variables, a diverse system is desirable in most cases, and may be required in some cases.

A coded unitary duplex controller may be impractical even in a unitary environment with many sensor and actuator variables because too many matrix intersections may not be used. So another method must be found to create a multivariable controller that does not commit terminals to combinations of values of the input or output variables and can form unitary or diverse relations of its input and output variables.

Superimposed coded unitary machines

If the system designer can identify the diverse and unitary variables in a system, the designer can put together the unique set of unitary and diverse units needed to connect the diverse and unitary variables. If the control system hardware must be configured to include every possible unitary and diverse relation, the system may require more hardware than is practical in any system except the simplest.

Combined network

If the system designer does not know whether a given set of sensor and actuator variables is intrinsically unitary or diverse, the designer may wish create a unit made up of unitary and diverse subunits connected to the same set of variables, as shown in the Figure 7.18.

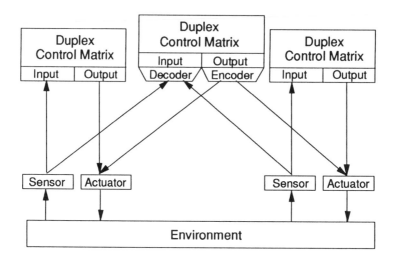

Figure 7.18. Sensor and actuator variables in a *predetermined combined network* are connected to both unitary and diverse units.

A *combined network* allows unitary relations to be programmed in the unitary units and diverse relations to be programmed in the diverse units. A properly designed combined network, having unitary variables connected to unitary controllers and diverse variables connected to diverse units, may use far fewer intersections and be easier to program than a single coded unitary duplex controller with enough intermediate states to handle the unitary and diverse relations.

Number of intersections in a combined network

The number of intersections in a combined network is equal to the number of intersections in a purely diverse system plus the number of intersections in a purely unitary system. Thus the number of intersections may be excessively large if there are more than a few unitary variables.

Composite network

In a *composite network*, subunits of all possible levels of unitary and diverse organization are superimposed upon a given set of variables, as shown in Figure 7.19. The programmer may program those subunits that conform to the intrinsic organization of the task environment and bypass the rest of the subunits with null or nonconducting intersections.

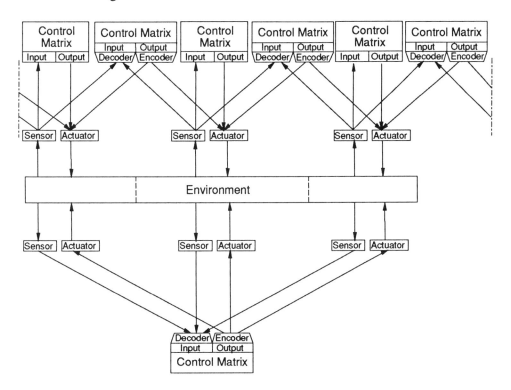

Figure 7.19. A *predetermined composite network* is made up of subunits having every possible level of unitary or diverse relation.

The number of intersections in a composite network increases rapidly as the number of variables increases. For example, a composite network of n binary variables requires approximately n^{10} intersections. This is impractical in all but the smallest possible system having only a few variables with only a few values. However, a properly designed control system, consisting of the subset of a composite network having just the correct number and level of unitary and/or diverse subunits, may be practical in some applications. But what can be done to deal effectively with a control system with many unitary and diverse variables?

Chapter 8
Active Switch Machines

It is not possible to construct a single control matrix with two or more input or output variables using a passive matrix except by using the unitary coding and decoding techniques shown in the last chapter. However, by using an *active switch matrix*, it is possible to construct an *active switch multivariable input matrix* that can have two or more input variables and one output variable without using a decoder. It is also possible to construct an *active switch multivariable output matrix* with one input variable and two or more output variables without using an encoder. Then the multivariable input matrix can be connected to the multivariable output matrix by an intermediate variable, forming an *active switch multivariable duplex controller* that can handle multiple input and output variables. These active input and output matrices also can be combined into a single *active switch monolithic control matrix*. These monolithic control matrices can be connected by a set of intermediate variables, forming an *active switch universal duplex controller* that can be programmed to produce unitary or asynchronous diverse behavior.

Active switch matrices

An *active switch matrix* is made up of a distribution of three-terminal *active switches* within a matrix of conductors instead of a distribution of *conductive pins* at the intersections of specific input and output terminals in the passive connection matrix shown previously. The active switch matrix also can replace the two-terminal *passive switches* at the intersection of each matrix input terminal and each matrix output terminal in the passive switch matrix shown previously. Power to drive the matrix output terminals of an active switch matrix comes from a separate *matrix power source* connected to all the active switches. Two conditions must be met for an active switch matrix to produce an output signal at one of its matrix output terminals: An active switch must be connected to the latched input terminal, and the matrix power source must produce a voltage to the output enable terminal of this active switch. An active switch multivariable input matrix can produce an output based upon the value of two or more input variables, and an active switch multivariable output matrix can produce values of a set of output variables based upon the value of a single input variable.

Passive switch

Until now, we have dealt with passive connection matrices in which the power needed to activate their output terminals comes from their input terminals. A passive pin matrix is consists of a matrix of conductors with a socket at each intersection. It is programmed by placing a conductive pin at the intersection of each desired input/output relation. A passive toggle switch matrix consists of a matrix of conductors with a *passive toggle switch* at every intersection. It is programmed by manually placing each switch in an open or closed position. Power flows from the input terminal of a passive switch to an output terminal only when the switch is closed and a voltage is applied to its input terminal.

Active switch

However, an *active switch* has three terminals: It has an input sense terminal, an output enable terminal, and an output terminal, as shown in Figure 8.1.

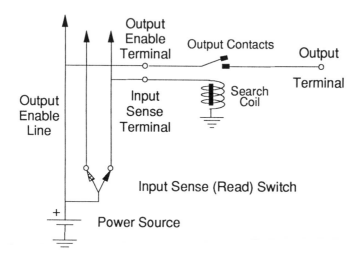

Figure 8.1. An *active switch* can be opened or closed by a control signal that is separate from the source of its output power.

An active switch is usually an *active component* like a *relay, vacuum tube,* or *transistor.* Power flows from the *output enable terminal* to its *output terminal* only when a voltage is applied to its output enable terminal, and sufficient current flows from the *input sense terminal* through its *input search coil* to close its *output contacts.*

Active matrix driver

An active switch can be placed at the intersection of a given output terminal and a given input terminal, causing that output terminal to be energized whenever that input terminal is energized, as shown in Figure 8.2. Once the cycle switch is closed by the cycle signal from an outside voltage source, the output enable bus and the selected input sense line are energized. The selected input sense line is the one connected to the sensor terminal making contact with the switch armature. This causes a voltage to appear at the matrix output terminal connected to the active switch that is connected to the selected input sense line. A voltage at a given matrix output terminal energizes its stator coil and causes a pole on the actuator armature to attempt to align itself with that stator coil.

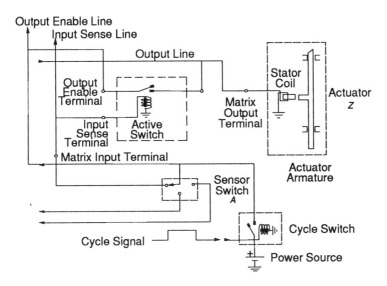

Figure 8.2. An *active matrix driver* provides the power source to drive the output enable and input sense lines of an active switch matrix.

Input latch for an active switch matrix

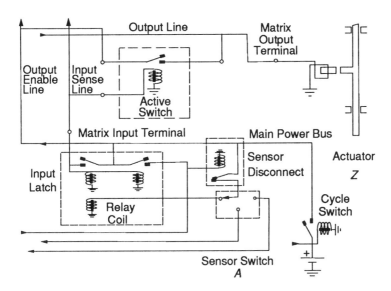

Figure 8.3. The *input latch for an active switch matrix* allows only one input sense line to be energized during a transition cycle.

According to the DS ontology and as in a passive memory matrix, only one matrix input terminal and only one matrix output terminal can be energized during a given transition cycle in an active matrix. Therefore, some provision must be made to keep a second input terminal from being energized if the armature of the sensor switch is moved during the transition cycle. This can be accomplished by the *input latch for an active switch matrix* shown in Figure 8.3.

Once an outside transition cycle signal closes the cycle switch, the relay coil is energized in the input latch connected to the sensor switch terminal selected by the sensor switch. This causes both contacts in the selected input latch to close. This energizes its input sense line, the sensor disconnect, and causes current to continue to flow through the other two latch coils while the cycle switch is closed. This maintains the voltage on the latched input sense line and prohibits any other input sense line from being energized for the remainder of that transition cycle regardless of any changes in the position of the sensor switch.

Actuator delay circuit for an active switch matrix

According to the DS ontology, a time delay is required between the occurrence of a given input state and the production of a given output state. This delay can be provided by the *actuator delay circuit for an active switch matrix* shown in Figure 8.4.

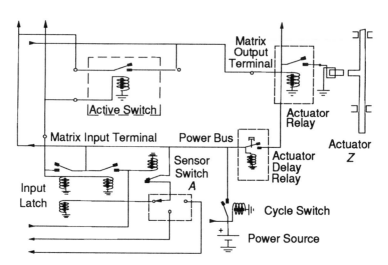

Figure 8.4. An *actuator delay circuit for an active switch matrix* creates a delay between the beginning of the transition cycle and the production of the selected actuator state.

Once the outside transition timing signal closes the cycle switch, the coil of the actuator delay relay is energized. However, the armature of this *delay relay* cannot close immediately because it is fitted with a *damper* that restricts it rate of closure. The amount of *damping* can be selected to provide a long or short delay. Once the relay armature closes, the input contacts of all the actuator relays are energized. However, only the actuator stator coil connected to an active switch energized by the input latch can be energized.

End-of-cycle circuit for an active switch matrix

In a real machine, some time may be required to allow the actuator to move to its selected position. In some cases, a relatively long time may be allowed for the actuator to position itself and in other cases a very short time may be required. This *actuator response time* can be specified by the *end-of-cycle circuit for an active switch matrix* shown in Figure 8.5.

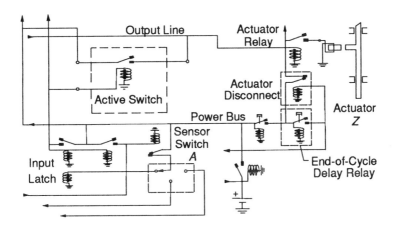

Figure 8.5. The *end-of-cycle circuit for an active switch matrix* determines how much time the actuator has to complete its movement.

The actuator delay relay energizes the coil of the end-of-cycle delay relay when it "times out." This causes the end-of-cycle delay relay to start timing. Once it times out, it activates the actuator disconnect, which cuts off power the actuator relay. This ends the period during which the control matrix can position the actuator. If the actuator has not completed its movement by this time, it is left stranded in some indeterminate position. The end-of-cycle signal is also sent to the outside transition timing source, indicating that the transition cycle for that matrix is over. The outside timing source can then end that transition cycle by suspending the voltage to the cycle switch. This eliminates the voltage to all the components in the matrix circuit. This allows the latches and timers to return to their initial positions.

Actuator brake circuit for an active switch matrix

When the transition cycle signal opens the cycle switch at the end of the transition cycle, power to every element of the matrix is terminated, including power to the actuator. However, the actuator must be held in the selected position or in whatever position it was left until an active switch attempts to reposition it in the next transition cycle. The position of the actuator can be maintained between cycles by the *actuator brake circuit for an active switch matrix* shown in Figure 8.6.

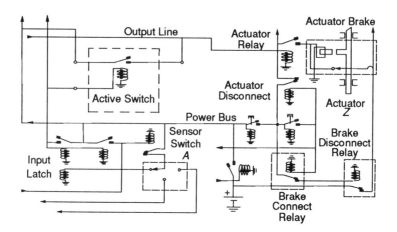

Figure 8.6. The *actuator brake circuit for an active switch matrix* attempts to hold the actuator armature in a fixed position between transition cycles.

The *actuator brake* must be released when the actuator delay relay has timed out and before the end-of-cycle delay relay has timed out. During this period, the actuator brake must be free to move under the influence of the actuator coils. The brake is disengaged by energizing the brake disconnect relay with the output of the actuator delay relay acting through the brake connect relay. When the end-of-cycle delay relay times out, it energizes the brake connect relay. This opens the brake disconnect relay and allows the actuator brake to re-engage and attempt to hold the actuator in a fixed position until an active switch attempts to position the actuator during the next transition cycle. The end-of-cycle relay also energizes the coil of the actuator disconnect so that an active switch and the actuator brake cannot operate on the actuator simultaneously.

Active switch DS Machine

As few as two input terminals can be connected to two output terminals by two active switches, forming the *active switch DS machine* shown in Figure 8.7. In the active switch matrix shown, the input value (0) produces the output value (1), and the input value (1) produces the output value (0). Since the input variable of the control matrix contains two values, it can be called a *binary control matrix*.

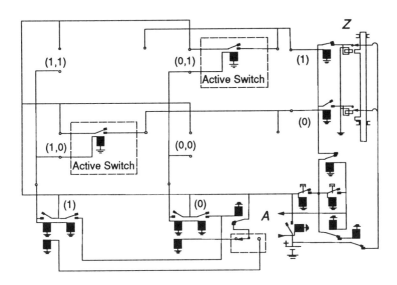

Figure 8.7. An *active switch DS machine* can connect any value of one input variable *A* to any value of one output variable *Z*.

Operation of an active switch DS controlled machine

An active pin DS controlled machine operates according to Table 8.1.

Table 8.1. Operation of an active pin DS controlled machine

---

1. An outside timing source closes the cycle switch at the start of a transition cycle.
2. This energizes the matrix output enable power bus and the matrix input terminal selected by the sensor switch.
3. The input latch disconnects the sensor switch and maintains the voltage at the selected matrix input terminal.
4. A voltage is produced at the input sense terminals of the active pin connected to the selected matrix input terminal. (According to the rules of a connection matrix, only one active pin can be connected to a given sense terminal.) This closes the output contacts of this switch, allowing the voltage from the output enable bus to appear at the matrix output terminal connected to switch.
5. After a delay determined by the actuator delay relay, the actuator brake is released and the output relay connected to this matrix output terminal produces a voltage that drives the actuator to a position determined by the position of the energized output relay.
6. When the end-of-cycle relay times out, the brake is re-engaged, power to the output relays is disconnected, and a signal is sent to the outside timing source indicating that the transition is complete.
7. At the end of the transition cycle, the outside timing source opens the cycle switch. This disconnects the matrix power source, releases the input latch and restores all the other disconnects and delay relays to their normal (initial) positions.

Thus, an active pin controlled machine produces the same results as a passive controlled machine in this single variable control system except that the power to drive the actuator relays does not come through the matrix input terminals.

Application of an active switch control matrix

The number of input terminals of the active switch control matrix can be increased to match the number of values of the direction error sensor, and the output terminals can be increased to match the number of values of the pilot valve of the rudder servo-actuator in the ship auto-pilot shown earlier, forming the *active switch ship auto-pilot* shown in Figure 8.8.

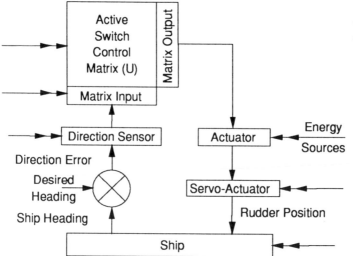

Figure 8.8. A *ship auto-pilot using active switches* can be programmed to provide adequate control behavior if enough input and output terminals are provided.

An active switch must be connected to each specific value of the direction error sensor likely to be encountered and to the specific rudder position required to hold the ship on course.

Active switch duplex control systems

As shown before, the number of intersections in a control system can be greatly reduced by connecting a matrix containing the input variable to a matrix containing the output variable by an intermediate variable having only as many values as expected input/output relations, forming an *active switch duplex controller*.

Scalar active switch input matrix

A *scalar active switch input matrix* (Fig. 8.9) can be used in place of a passive input matrix in any application having one input variable and one output variable, or in a coded unitary controller with multiple input and output variables.

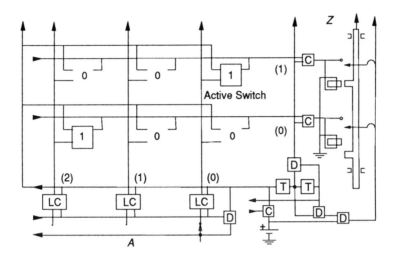

Figure 8.9. A *scalar active switch input matrix* isolates or separates the input sense power source from the output enable power source and contains a single input variable and a single output variable.

Though an active input matrix is more complex and therefore more expensive than a passive input matrix, isolating the output enable power source from the input sense power source increases the design flexibility of the electrical circuit, allowing it to handle multiple input variables in a manner to be shown. In the scalar input matrix given, the input value $A(0)$ produces the output value $Z(1)$, input value $A(1)$ produces no output, and input value $A(2)$ produces output value $Z(0)$. Because it has three values of its input variable, it may be said to be a *trinary input matrix* (see Chapter 30). The number of values of its input and output variables can be increased indefinitely, as indicated by the arrows, to match the variety of any sensor and actuator variable.

Scalar active switch output matrix

The conductive pins in a scalar passive output matrix can be replaced with active switches, forming the *scalar active switch output matrix* shown in Figure 8.10. In the scalar output matrix given, the input value $A(0)$ produces the output value $Z(1)$, input value $A(1)$ output value $Z(2)$, and output value $Z(0)$ will not work in the matrix conformation shown. Because it has three values of its output variable, it may be said to be a *trinary output matrix* (see Chapter 30). The number of values of its input and output variables can be increased indefinitely, as indicated by the arrows, to match the variety of any sensor and actuator variable. The active output matrix usually has more values of its single (scalar) output variable than values of its single input variable and operates in the same as active switch input matrix rotated 90°.

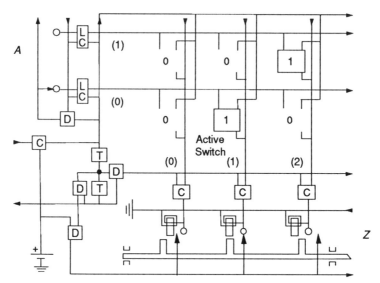

Figure 8.10. A *scalar active switch output matrix* uses active memory cells to connect a few values of the single input variables to many possible values of its output variable.

Scalar active switch duplex DS machine

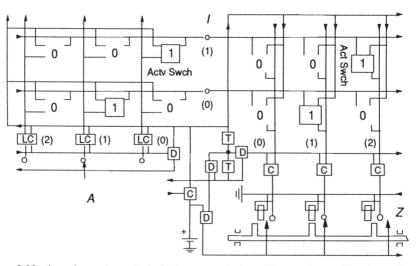

Figure 8.11. A *scalar active switch duplex controlled machine* can be used in place of a scalar passive duplex controlled machine whenever it is desirable to isolate the input and output terminals of the control matrices.

Since each unique output state can be represented by a value of the intermediate variable *I* in the *scalar active switch duplex DS machine*, only as many intermediate terminals are required as there are output states in an expected line of behavior. Thus, the number of intersections in the active scalar matrices shown previously can be reduced, when there are significantly fewer transitions than input and output states, by using the active switch duplex controller shown in Figure 8.11. A scalar active switch duplex controller is made up of a scalar active switch input matrix connected to a scalar active switch output matrix.

A scalar passive duplex controller may be the preferred design for most predetermined duplex controlled machines with a single input and output variable. However, the design of the active duplex controller leads to the design of a duplex controlled machine that can deal with two or more input variables and two or more output variables, as will be shown.

Operation of a scalar active switch duplex machine

A scalar active switch duplex machine operates according to Table 8.2.

Table 8.2. Operation of a scalar active switch duplex machine

---

1. An outside timing source closes the cycle switch at the start of a transition cycle.
2. This energizes the output enable power bus and the matrix input terminal selected by the sensor switch.
3. The input latch disconnects the sensor switch and maintains the voltage at the selected matrix input terminal.
4. A voltage is produced at the input sense terminals of the active switch connected to the selected matrix input terminal.
5. This causes a current to flow through the relay coil in this active switch, causing the output contacts of this active switch to close. This allows the voltage of the matrix power source to appear at the value of the intermediate variable connected to the energized active switch.
6. This value of the intermediate variable then causes a current to flow through the relay coil in the active switch in the output matrix connected to this intermediate terminal. This causes the output contacts of this active switch to close, allowing the voltage of the matrix power source to appear at the matrix output terminal connected to this second selecting active switch.
7. After a delay determined by the actuator delay relay, the actuator brake is released, and the output relay connected to this matrix output terminal produces a voltage that drives the actuator to a position determined by the location of the active output relay.
8. When the end-of-cycle relay times out, the brake is re-engaged, the voltage to the output relays is disconnected, and a signal is sent to the outside timing source that the transition cycle is over.
9. The outside timing source then releases the cycle switch. This disconnects the matrix power source, releases the input latch and restores all the other disconnects and delay relays to their normal positions.

Thus, a scalar active switch duplex DS machine produces the same results as a scalar passive duplex controlled machine in this single variable (scalar) control system.

Application of a scalar active switch duplex controller

A duplex controller can have more input and output values with fewer memory cells than a single matrix if fewer than half as many input/output relations are required than the number of values of its input and output variables.

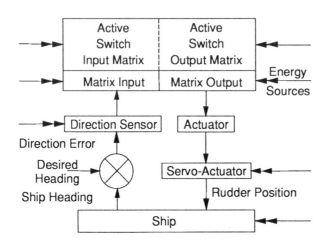

Figure 8.12. A *ship auto-pilot using a scalar active switch duplex controller* can provide adequate control behavior using the least number of intersections in a system with 5% resolution and up to 9 transitions.

For example, if the direction error sensor in the ship auto-pilot shown in Figure 8.12 has 20 values, and the rudder servo-actuator has 20 values, for a resolution of these variables of 5%, and only 5 transitions are required to maintain adequate control, then the duplex controller would require only 200 intersections in contrast to a single matrix that would require 400 intersections. The scalar active switch duplex controller is useful as a controller in any system with a single input variable and a single output variable with relatively few values, and where a relatively few transitions are required in the line of behavior of the control system.

Active switch multivariable input matrices

The use of active switch switches allows two or more single-variable matrices to be connected to form an *active switch multivariable input matrix* with multiple input variables without using an input decoder, and a single output variable. The active switch multivariable input matrix is different from a coded input matrix, which can produce all the possible combinations of values of its input variable in a unitary manner. The active switch input matrix can produce outputs for only those combinations of values that are programmed into the matrix.

Active switch multivariable binary input matrix

The design of the active switch matrix allows two or more input variables to be included in a single multivariable input matrix. For example, two or more single-variable active switch matrices can be connected in series with one matrix power source, as shown in Figure 8.13.

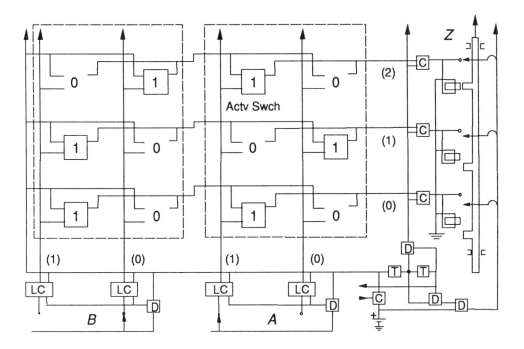

Figure 8.13. An *active switch multivariable binary input matrix* contains two or more binary input variables (*A* and *B*), one matrix power source and one output variable (*Z*).

An *active switch multivariable binary input matrix* can produce an output value that represents a unique combination of values of its input variables in a unitary manner. For example, output terminal Z(0) is energized when input terminals A(1) *and* B(1) are energized, output terminal Z(1) is energized when input terminals A(0) *and* B(1) are energized and output terminal Z(2) is energized when input terminals A(1) *and* B(0) are energized. A given output can be produced at a given output terminal in a multivariable input matrix with two input variables only when two active switchs connected to this output terminal are made active by the positions of both sensor switches.

Operation of an active switch multivariable input matrix

An active switch multivariable input matrix operates according to Table 8.3.

Table 8.3. Operation of an active switch multivariable input matrix

---

1. An outside timing source closes the cycle switch at the start of a transition cycle.
2. This energizes the matrix enable power source and one input sense terminal connected to a sensor switch representing each input variable in each matrix.
3. Each input latch disconnects each sensor switch, as each matrix input terminal is energized, and maintains the voltage at the selected matrix input terminals.
4. A voltage is produced at the input sense terminals of all the active switches connected to the selected matrix input terminals.
5. This causes a current to flow through their relay coils, causing their output contacts to close. If there is an active switch in both matrices connected to a latched input terminal and to a given output terminal, a current can flow from the matrix power source, across the matrix, to that output terminal.
6. After a delay determined by the actuator delay relay, the actuator brake is released and the output relay connected to this matrix output terminal produces a voltage that drives the actuator to a position determined by the location of this output relay.
7. When the end-of-cycle relay times out, the output relay is disconnected, the actuator brake is re-engaged, and a signal is sent to the outside timing source indicating that the transition is over.
8. The outside timing source then opens the cycle switch. This disconnects the matrix power source, releases the input latch, and restores the disconnects and delay relays to their *normal* positions.

Thus, a multivariable input matrix can produce a unique output value of its single output variable for each unique combination of values of its multiple input variables. The active switch multivariable input matrix can be used in any control application with two or more input variables and one output variable that can be programmed before it is put into use.

Distributed control in an active switch multivariable input matrix

Since two or more memory elements are required to produce a given output state in a multivariable input matrix, this process of using two or more input variables in parallel to control an output variable is called *distributed control*.

Unitary relations in an active switch multivariable input matrix

A multivariable input matrix with x^n output terminals can provide unique *unitary* output values for every combination of n input variables with x values. For example, the active switch multivariable input matrix shown can produce four unique outputs, each representing one of the four possible combinations of the two values of its two input variables.

Synchronous diverse behavior in an active switch multivariable input matrix

The multivariable input matrix also can provide a unique *diverse* output value based upon the value of a given input variable despite the values of the other variables if it is programmed according to the *conformation* shown in Figure 8.14.

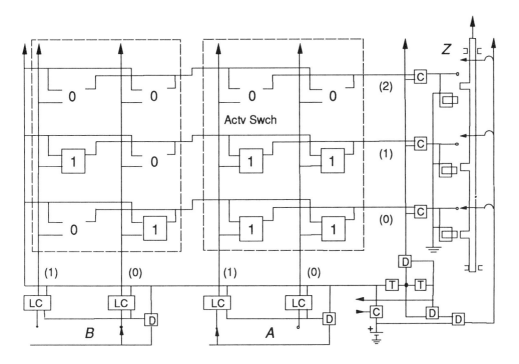

Figure 8.14. An *active switch multivariable input matrix in a synchronous diverse conformation* can disregard one or more if its input variables.

For example, it can produce a given output value Z(0) whenever its input terminal B(0) is energized, though either value of the other input variable A(0) or A(1) is energized as shown. Thus, a variable like A is ignored by the matrix whenever active switches are connected to all of its input terminals. By adding more input variables, other diverse relations can be established that occur at different (synchronous) times.

Active switch trinary multivariable input matrix

The number of values of each variable in a multivariable input matrix can be increased indefinitely to match the variety of the sensor and actuator variables. For example, the number of values of the input variables can be increased from two in an active switch binary multivariable input matrix to three in the *active switch trinary multivariable input matrix* shown in Figure 8.15.

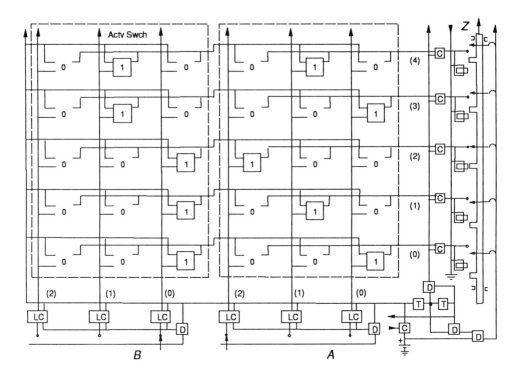

Figure 8.15. An *active switch trinary multivariable input matrix* can produce a unique response to more combinations of values of its input variables than a binary input matrix with the same number of input terminals.

The *trinary* number system is more "efficient" than the binary number system in terms of the amount of variety it can generate with a given number of matrix intersections. For example, a 6-variable binary input matrix requires 12 input terminals, and can identify and one of $2^6 = 64$ different input states. A 4-variable trinary input matrix also requires input terminals, but can identify any one of $3^4 = 81$ different input states. A 3-variable quadrary input matrix also requires input terminals, but can identify any one of only $4^3 = 64$ different input states, the same as a binary system. The improvement of a trinary system over a binary or quadrary system increases as the number of variables increases. The number of values of the input variables can be increased indefinitely to correspond to the variety of the sensor and actuator variables, although the "efficiency" of the matrix decreases as the number of values increases above 3 terminals per variable.

Active switch multivariable input matrix control system

The multivariable input matrix can be used to connect two or more input variables to a single output variable in a unitary or synchronous diverse fashion. For example, an active switch multivariable input matrix with at least twenty values of its input and output variables can be used to connect the direction error sensor and the crosswind sensor in the ship direction control system shown earlier to a single rudder position actuator, forming the *active switch multivariable ship direction control system* shown in Figure 8.16.

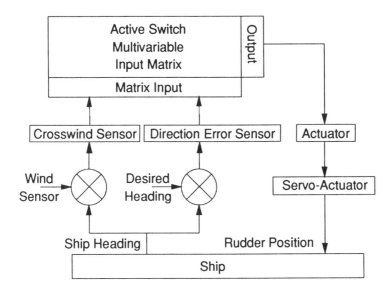

Figure 8.16. An *active switch multivariable input matrix ship direction control system* can use one or all its input variables to control one output variable.

If both the crosswind and direction error signals are needed to maintain the course of a ship, the matrix can be programmed to produce unique rudder actuator values for each combination of values of the crosswind and direction error sensors in a unitary manner. If the crosswind has no influence upon the amount of rudder correction required to keep the ship on course, active switches can be placed in the crosswind matrix for all values of crosswind, eliminating it as a controlling variable in a synchronous diverse manner.

Active switch multivariable output matrices

One input variable can be connected to two or more output variables in an active switch multivariable output matrix. This *active switch multivariable output matrix* can be used in any predetermined control system with one input variable and two or more output variables without using an output encoder, and can be programmed to produce unitary or synchronous diverse behavior.

Active switch binary multivariable output matrix

The design of an active matrix also allows two or more output variables to be connected to a single input variable by connecting two or more single-variable active control matrices in *parallel* with the input variable, forming the *active switch binary multivariable output matrix* shown in Figure 8.17.

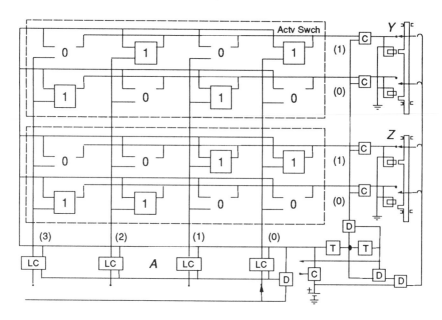

Figure 8.17. An *active switch binary multivariable output matrix* has one input variable and uses a separate single-variable active matrix for each output variable.

A multivariable output matrix can produce a unique combination of output values for each value of its input variable. For example, two output terminals $Y(1)$ and $Z(1)$ are energized when input terminal $A(0)$ is energized, and two output terminals $Y(0)$, and $Z(0)$ are energized when input terminal $A(1)$ is energized. Notice that the active switches connected to a given output variable are divided in the horizontal direction in a multivariable output matrix, whereas the active switches connected to a given input variable are divided in the vertical direction in a multivariable input matrix.

Rotated active switch multivariable output matrix

It is desirable to rotate the multivariable output matrix 90° (Fig. 8.18) to view it in its position as the output matrix of a duplex controller.

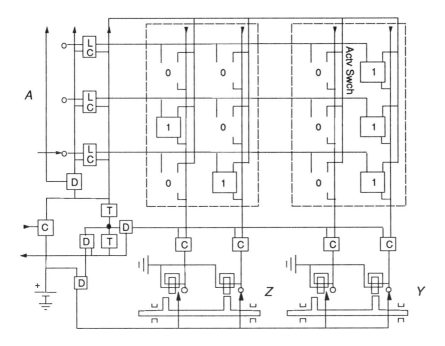

Figure 8.18. A *rotated active switch multivariable output matrix* shows the arrangement of latches, connects, disconnects and active switches as they appear in the output matrix of a multi-variable duplex controller.

The time delays occur in the vertical direction in a rotated multivariable output matrix, and the submatrices are divided in the vertical direction like the submatrices in a multivariable input matrix. However, the submatrices in a multivariable output matrix are still connected in parallel in contrast to the submatrices of a multivariable input matrix, which are connected in series.

Operation of an active switch multivariable output matrix

An active switch multivariable output matrix operates according the steps shown in Table 8.4.

Table 8.4. Operation of an active switch multivariable output matrix

---

1. The outside timing source closes the cycle switch at the start of a transition cycle.
2. This energizes the matrix power bus and the one matrix input terminal representing the value of the input variable at the beginning of the transition cycle.
3. The input latch disconnects the sensor switch and maintains the voltage at the selected matrix input terminal.
4. A voltage is produced at the input sense terminals of the active switches in both matrices connected to the selected matrix input terminal.
5. This causes a current to flow through the relay coils in all of these active switches, causing their output contacts to close. This causes the voltage of the matrix power supply to appear at the matrix output terminals connected to these active switches in both the matrices.
6. After a delay determined by the actuator delay relay, the actuator brake is released and the output relays connected to these matrix output terminals produce a voltage that drives both actuators to positions determined by the position of the energized output relays.
7. When the end-of-cycle relay times out, the brake is re-engaged, the output relays are disconnected, and a signal is sent to the outside timing source indicating that the transition cycle is over.
8. When the outside timing source releases the cycle switch, the matrix power bus is disconnected from the matrix power source, the input latch and other disconnects are released, and the delay relays are reset.

Thus, a multivariable output matrix produces a unique combination of output values in a given transition that is determined by the value of its input variable. The active switch multivariable output matrix can be used for any control application with one input variable and two or more output variables that can be programmed before being put into use.

Synchronous diverse behavior in a multivariable output matrix

A multivariable output matrix can provide unique *synchronous diverse* output values based upon the value of its input variable. That is to say, it can produce a set of output values on one or more of its output variables and produce constant values on one or more of its other output variables for each value of its input variable. In the example shown, the multivariable output matrix can produce an *active set of values* of its output variable Y for each value of its input variable A, and produce a *constant value* at its output variable Z. This requires that connections be made to one output terminal in the submatrix representing the variable Z. Submatrix Z is then said to be a *null matrix*.

Unitary relations in a multivariable output matrix

A multivariable output matrix with x^n input terminals also can provide unique *unitary* output values for every combination of its n output variables if each variable has x values. For example, the binary multivariable output matrix shown can produce all four combinations of the two values of its two output variables, as shown in Figure 8.19.

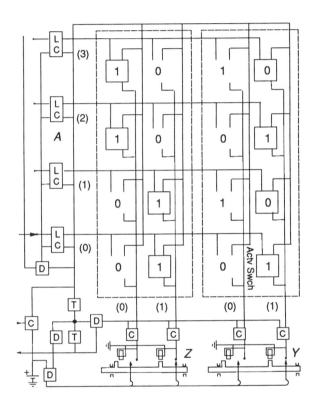

Figure 8.19. An *active switch multivariable output matrix in a diverse conformation* produces a constant value of one or more of its output variables.

Thus, an active switch multivariable output matrix can be programmed to produce either unitary relations between its input and output variables, or can be programmed to produce a constant value of one or more of its output variables for some values of its input variable, and produce a constant value of other output variable for other values of it input variable in a synchronous diverse manner.

Active switch trinary multivariable output matrix

The number of values of the output variables can be increased indefinitely. For example, the number of values of each variable can be increased from two in the binary output matrix shown to three in the *active switch trinary multivariable output matrix* shown in Figure 8.20.

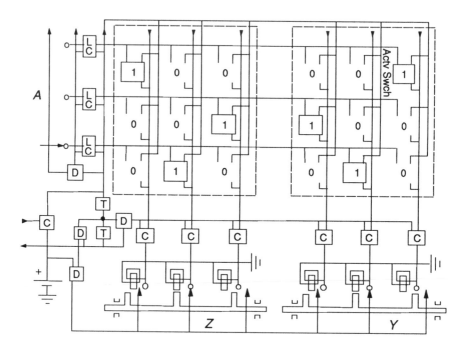

Figure 8.20. An *active switch trinary multivariable output matrix* contains three values of each output variable.

Like the multivariable trinary input matrix shown previously, the multivariable trinary output matrix can produce the greatest number of different output states for a given number of intersections. The number of values of the output variables can be increased to match the variety of the output devices, though a number of values greater than three results in a less efficient use of the matrix.

Active switch multivariable output matrix control system

An active switch multivariable output matrix can be used to control two or more output variables from a single input variable. Thus, the crosswind sensor and direction error sensor in the ship direction control system shown earlier can be summed together in a differential and the output used as the input variable of a multivariable output matrix. Then the output variables of the matrix can be used to control the rudder servo-actuator and the bow thruster in the *active switch multivariable output matrix control system* shown in Figure 8.21.

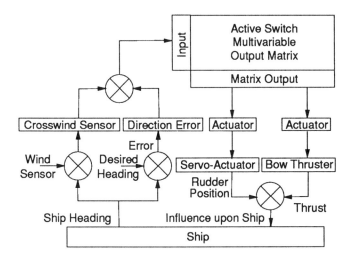

Figure 8.21. An *active switch multivariable output matrix control system* can produce multiple outputs from a single input variable.

The output matrix can be programmed such that unique combinations of values of both variables are produced for each value of the input variable in a unitary manner, or one output variable may be held constant for all values of the input variable so that the other output variable can form a synchronous diverse relation with the input variable.

Active switch multivariable duplex controllers

A control task may require two or more input variables and two or more output variables, and have a relatively few active states in its line of behavior. It also may be required to act in a unitary or synchronous diverse manner. The use of active switches allows a multivariable input matrix to be connected to a multivariable output matrix, forming an *active switch multivariable duplex control matrix* that can handle multiple input and output variables in a unitary or synchronous diverse fashion, and can produce as many transitions as are needed in its active line of behavior.

Active switch multivariable duplex control matrix

The output variable of an active switch multivariable input matrix can be connected to the input variable of an active switch multivariable output matrix, as shown in Figure 8.22, forming an *active switch multivariable duplex controller* that can handle multiple input and output variables without input or output coding.

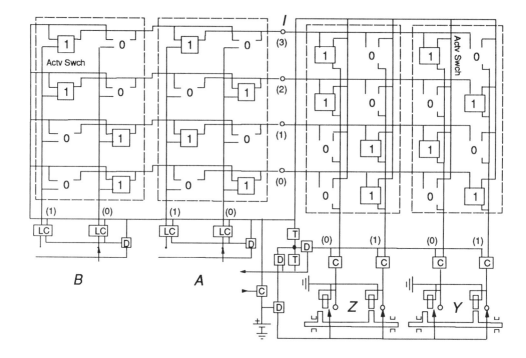

Figure 8.22. An *active switch binary multivariable duplex controller* can connect multiple input variables to multiple output variables in some predetermined number of unitary or synchronous diverse relations.

An *active switch multivariable duplex controller* can be programmed to produce unitary or *synchronous diverse relations* between its input and output variables. It can do so with a minimum number of intersections because it does not commit input or output terminals to specific combinations of values, as does a coded system, and can be designed to produce some predetermined number of relations according to the number of intermediate states designed into its matrices.

The active switch multivariable duplex controller shown can produce a unique combination of values of its output variables for each unique combination of values of its input variables in a unitary conformation. For example, input values of $A(0)$ and $B(0)$ produce output values of $Y(1)$ and $Z(1)$, and input values of $A(1)$ and $B(0)$ produce output values of $Y(0)$ and $Z(1)$ etc. in a unitary fashion.

Operation of an active switch multivariable duplex controller

A given intermediate terminal in a multivariable duplex controller can produce a unique combination of output values of multiple output variables for each unique combination of values of its multiple input variables. An active switch multivariable duplex controller operates according to the steps shown in Table 8.5.

Table 8.5. Operation of an active switch multivariable duplex controller

---

1. The outside timing source closes the cycle switch at the start of a transition cycle.
2. This energizes the matrix enable power bus and one matrix input terminal in each input matrix connected to a sensor switch representing each input variable.
3. Each input latch disconnects each sensor switch as each matrix input terminal is energized. Each latch maintains the voltage at the selected matrix input terminals.
4. A voltage is produced at the input sense terminals of the active switches connected to the selected matrix input terminals.
5. This causes a current to flow through their relay coils, causing the output contacts of these switches to close. This allows the voltage of the matrix enable power bus to appear at the intermediate variable terminal connected to the set of active switches connected to the latched input terminals and connected in series through both of the input matrices.
6. A voltage is produced at the input sense terminals of the active switches in the output matrix connected to the selected intermediate variable terminal.
7. This causes a current to flow through their relay coils, causing the output contacts of these switches to close. This allows the voltage of the matrix power bus to appear at the matrix output terminals connected to these active switches.
8. After a delay determined by the actuator delay relay, the actuator brake is released, and the output relays connected to these matrix output terminals produce a voltage that drives the actuators to positions determined by the location of the energized output relays.
9. When the end-of-cycle relay times out, it re-engages the actuator brake, disconnects the output relays, and sends a signal to the outside timing source that the transition cycle is complete.
10. Then the outside timing source opens the cycle switch. This disconnects the matrix power source from the matrix power bus and releases the input latch and other disconnects, and allows the delay relays to reset.

The active switch multivariable duplex controller can be used in any control application with two or more input variables and two or more output variables that can be programmed in an all-or-nothing manner before it is put into use.

Synchronous diverse conformation of an active switch multivariable duplex controller

The active switch multivariable duplex controller also can be programmed so that one or more input variables operate upon less than all of the output variables at different times in a synchronous diverse manner if the matrices connected to the other variables are programmed to act as null matrices, as shown in Figure 8.23.

In the synchronous diverse conformation shown, different values of the input variable A cause different values of the output variable Z and cause constant values of the output variable Y, despite the values of the input variable B. However, input variable B cannot establish a diverse relation with output Y without the conformation of the matrix being changed. The multivariable duplex controller can be programmed to provide either unitary or synchronous diverse behavior, and can be designed with just enough intermediate states to provide the required number of input and output relations. Thus, it can greatly reduce the number of matrix intersections needed to provide satisfactory multivariable control behavior compared to the coded unitary controller.

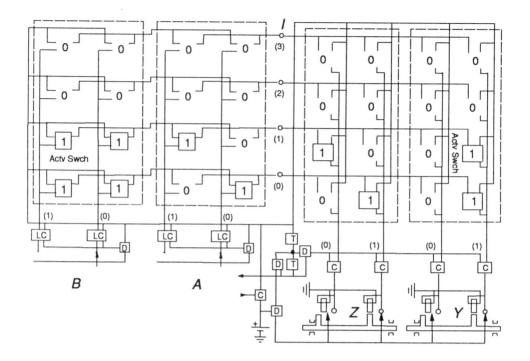

Figure 8.23. An *active switch multivariable duplex controller in a synchronous diverse conformation* can cause only one input variable to cause action in only one output variable.

Series-parallel distributed control

Because at least two active switches in the duplex system act in *series* to select an output, and because the values of two or more variables are searched in *parallel* in the input and in the output matrix to produce a given output state, the method of selecting an output in a multivariable duplex system is called *series-parallel distributed control* or *series-parallel control.*

Active switch multivariable trinary duplex controller

The number of values of the sensor and actuator variables can be increased indefinitely to match the variety of the sensor and actuator variables. The maximum variety of behavior can be produced for a given number of matrix intersections when the input and output variables have three values, as shown in the *active switch trinary multivariable duplex controller* (Fig. 8.24).

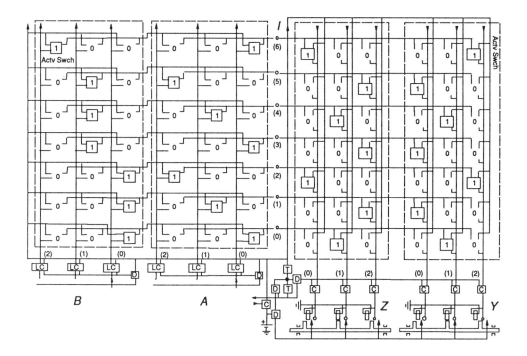

Figure 8.24. An *active switch trinary multivariable duplex controller* can connect any combination of the three values of its multiple input variables with any combination of the three values of its multiple output variables.

The active switch trinary multivariable duplex controller can be programmed to produce more different combinations of values of its output variables for different combinations of values of its input variables using fewer matrix intersections than a binary, quadrary, pentary, multivariable controller, or a controller based upon any other number system.

Active switch multivariable duplex ship direction control system

A multivariable duplex controller is the most general and useful predetermined control system for applications having a relatively few input/output relations. It can be used in any predetermined application having multiple input and output variables, such as the *active switch multivariable duplex ship direction control system* shown in Figure 8.25.

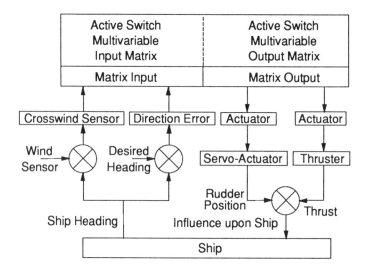

Figure 8.25. An *active switch multivariable duplex ship direction control system* requires far fewer matrix intersections than a coded unitary system with the same number of input and output variables with the same resolution.

The coded unitary duplex system shown earlier, having two input variables and two output variables, each with a resolution of 5%, and with 10 unitary transitions, requires 8000 intersections, of which only 20 are programmed with active switches. A multivariable duplex controller, having the same actual performance as a coded unitary duplex system requires only $20 \times 2 = 40$ input matrix terminals and 40 output matrix terminals with 10 intermediate terminals for a total of 800 intersections of which 40 are programmed with active switches. However, this system, with 20 values of its input and output variables, is not as efficient as a controller based upon a binary or trinary digitized system, as shown in Chapters 10 and 11.

Active switch universal controllers

One limitation of the multivariable duplex controller is that it can produce only one (synchronous) diverse relation at a given time since only one value of the intermediate variable is made active in a given transition. This limitation can be overcome by creating an active switch monolithic control matrix that can connect multiple input variables to multiple output variables in a unitary or *asynchronous* diverse manner. An active switch monolithic input matrix can be connected to an active switch monolithic output matrix, forming an *active switch universal controller* that requires only as many intermediate terminals as there are transitions in its lines of behavior. Its multiple input variables and multiple output variables can be connected in a unitary manner, or they can be connected to form *asynchronous diverse lines of behavior* that can occur at any time regardless of other diverse lines of behavior.

Active switch binary monolithic control matrix

Active switch scalar submatrices can be connected in series to a given output variable, and in parallel to a given input variable, forming the *active switch binary monolithic control matrix* shown in Figure 8.26. It can be made up of any number of input and output variables having two values, and can be programmed to produce unitary or asynchronous diverse behavior.

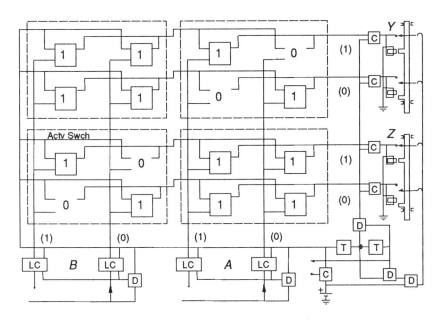

Figure 8.26. An *active switch binary monolithic control matrix* can connect multiple input variables with multiple output variables in a unitary or asynchronous diverse manner.

The active switch binary monolithic control matrix shown has two values of each input and output variable. A given input variable can be connected to a given output variable in an asynchronous diverse manner (regardless of the values of the other input variables) by inserting active switches in the intervening intersections of the other input variables, or each combination of input values can produce a different combination of output values in a unitary manner by programming all the intersections to produce a unitary code translation. (In the example given, the unitary code appears to be diverse because the variables in this particular code transformation mirror each other.)

Active switch trinary monolithic control matrix

Another value can be added to each sensor and actuator variable in the active switch binary monolithic control matrix shown previously, forming the *active switch trinary monolithic control matrix* shown in Figure 8.27.

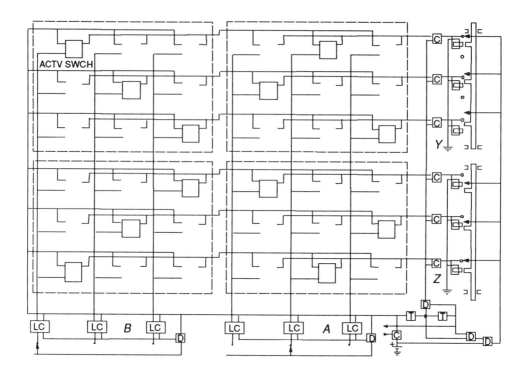

Figure 8.27. An *active switch trinary monolithic control matrix* can connect more unique sensor states to more unique actuator states than a binary monolithic control matrix with the same number of active switch intersections.

The active switch trinary monolithic control matrix shown has three values of each input and output variable. The number of values of each input and output variable can be increased to match the resolution of each sensor and actuator, although a number of values greater than three results in a less efficient matrix. The resolution of multiposition switch sensors and vernier actuators is limited to about 3% (thirty positions).

Application of an active cell monolithic control system

A crosswind sensor and a direction error sensor can be connected to a rudder servo-actuator and a bow thruster by an active switch monolithic control matrix, forming the *active switch monolithic ship direction control system* shown in Figure 8.28.

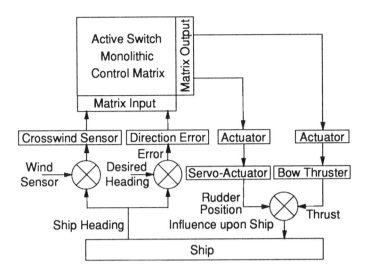

Figure 8.28. An *active switch monolithic ship direction control system* can be programmed to produce a unique combination of values of its rudder and bow thruster for each combination of values of its direction error and crosswind sensors in a unitary manner, or each sensor can be programmed to control one actuator in an asynchronous diverse manner.

The number of intersections in a monolithic control matrix increases according to the product of the total number of input values times the total number of output values. This number could greatly exceed the number of intersections in a duplex controller if only a relatively few transitions are required in the active line of behavior of a given system with many variables and/or many values.

Active switch universal duplex controller

An active switch monolithic control matrix can be used as the input matrix of a duplex controller, and another active switch monolithic control matrix can be used as the output matrix of this duplex controller, forming the *active switch universal duplex binary controller* shown in Figure 8.29. The use of monolithic input and output matrices creates two or more intermediate variables, in contrast to the single intermediate variable in the duplex controllers shown earlier.

The number of *asynchronous diverse lines of behavior* expected in a particular control problem can be used to determine the number of intermediate variables designed into the universal duplex controller. The number of transitions expected in each diverse relation can be used to determine the number of values of each intermediate variable. Thus, the number of intersections in the universal duplex controller may be fewer than the number of intersections in a monolithic controller if there are only a relatively few transitions in the expected lines of behavior.

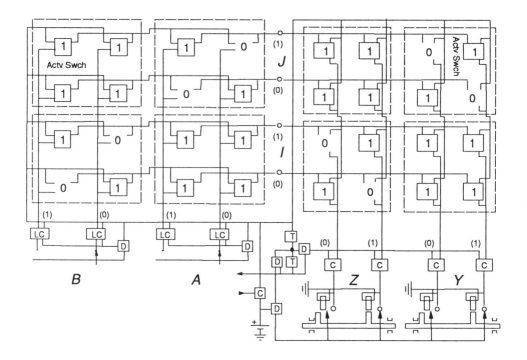

Figure 8.29. An *active switch binary universal duplex controller* can be programmed to produce unitary behavior, or be programmed to produce as many asynchronous diverse lines of behavior as it has intermediate variables.

Active switch trinary universal duplex controller

The number of input, output, and intermediate variables and the number of values of these variables can be increased indefinitely to match the number of variables, values, diverse lines of behavior, and transitions in a control task. If the active switches are connected as shown in the *active switch trinary universal duplex trinary controller* (Fig. 8.30), the maximum variety of control behavior can be produced with a minimum number of memory cells.

The number of intermediate variables can be selected to match the number of asynchronous diverse relations that the controller is expected to carry out, and the number of values of each intermediate variable can be selected to match the number of transitions required in each diverse line of behavior.

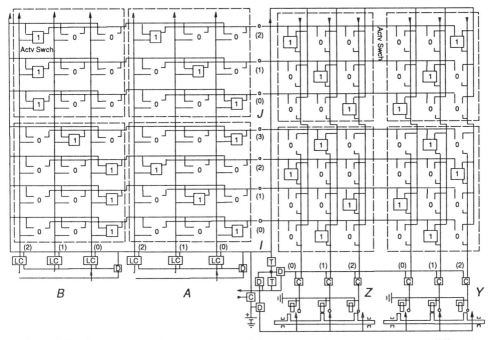

Figure 8.30. An *active switch trinary universal duplex controller* can produce more different input and output states for a given number of intersections than a binary universal duplex controller.

Application of an active switch universal duplex control system

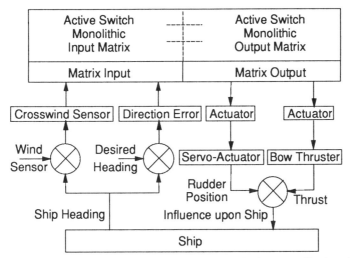

Figure 8.31. An *application for an active switch universal duplex controller* is a ship direction control system with multiple input and output variables, some of which may have to be connected in an asynchronous diverse manner.

The number of values of the input variables of the active switch universal duplex controller shown can be increased to match the variety of crosswind and direction error sensors, and the number of values of its output variables can be increased to match the variety of rudder position and bow thruster actuators, forming the *universal ship direction control system* shown in Figure 8.31. When the number of values of each variable exceeds three, the number of input and output states that can be produced by a given number of memory cells begins to fall off. A solution to this problem is to digitize the input and output variables, as shown in the Chapter 10.

Chapter 9
Absolute Predetermined Machines

The active switch control matrices shown in the previous chapter had to be programmed by physically placing active switches at the intersections of the matrix input and output terminals that represent the desired transitions. However, a *programmable memory matrix* can be programmed without physically placing or removing active switches by using *absolute predetermined memory cells* that can be set in a sensitive or insensitive state. These active predetermined memory cells can be used to construct a *multivariable input matrix* that can have two or more input variables and one output variable. An active predetermined memory matrix also can be used to construct a *multivariable output matrix* with one input variable and two or more output variables. The multivariable input matrix can be connected to a multivariable output matrix by a single intermediate variable, forming a *multivariable duplex controller* that can handle multiple input and multiple output variables. These active predetermined input and output memory matrices also can be combined into a single *predetermined monolithic control matrix*. These monolithic control matrices can be connected by a set of intermediate variables to form a *universal duplex controller* that can be programmed to produce unitary or diverse behavior.

Active absolute predetermined memory matrices

An *active predetermined memory matrix* is made up of a set of three-terminal *active memory elements* located at each intersection of the matrix input terminal and output terminals, instead of a set of active switches located only at the intersections of the input and output terminals that produce behavior. Like the active switch control matrices, power to drive the matrix output terminals of an active memory matrix also comes from a separate *matrix power source* connected to all the active memory elements. Three conditions must be met in order for an active memory matrix to produce an output signal at one of its matrix output terminals: The matrix power source must produce a voltage to the active memory elements; a voltage must be applied to a matrix input terminal; and the toggle switch of one active memory element connected to this input terminal must be in a conductive state.

Active absolute memory cell

A toggle switch can be connected in series with the search coil of the active switch such that a current can flow through the search coil only when the toggle switch is set in the "closed" position, as shown in Figure 9.1. Thus, the active memory cell shown can produce an output only when a voltage is present at its output enable terminal and its input sense terminal, and its toggle switch is set in the "closed" position, providing a low resistance in its search coil circuit.

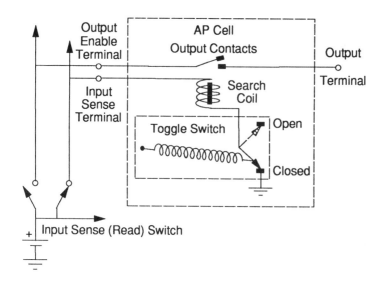

Figure 9.1. An *active absolute predetermined memory cell* (AP cell) produces an output only when a voltage is present at its output enable terminal and its input sense terminal, and a low resistance is present in its search coil circuit.

Since the toggle switch is usually placed in the "closed" or "open" position before the active memory is put into use and remains in that position unless it is intentionally moved, and since only one output terminal is used to convey information, the active memory element can be called an *active absolute predetermined memory cell* or *AP cell*.

Absolute predetermined DS Machine

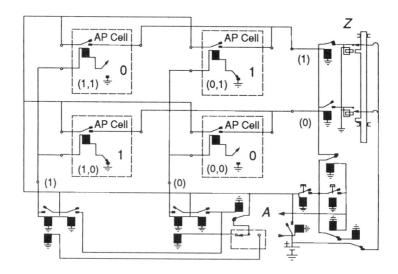

Figure 9.2. An *absolute predetermined DS machine* can connect any number of values of one input variable (*A*) to any number of values of one output variable (*Z*).

These AP cells can be placed at the intersection of every input and output terminal of a AP Cell control matrix. As few as four AP cells can be connected to the AP cell actuator brake circuit, forming the *absolute predetermined DS machine* shown in Figure 9.2. In the active connection matrix shown, the input value (0) produces the output value (1), and the input value (1) produces the output value (0). Since the input variable of the control matrix contains two values, it can be called a *binary control matrix*.

Operation of an absolute predetermined DS controlled machine

An absolute predetermined DS controlled machine operates according to Table 9.1.

Table 9.1. Operation of an absolute predetermined DS controlled machine

---

1. The cycle switch is closed at the start of a transition cycle by an outside timing source.
2. This energizes the matrix output enable power bus and the matrix input terminal selected by the sensor switch.
3. The input latch disconnects the sensor switch and maintains the voltage at the selected matrix input terminal.
4. A voltage is produced at the input sense terminals of the AP cells connected to the selected matrix input terminal.
5. This causes a current to flow through the search coil in the AP cell having a closed toggle switch. (According to the rules of a connection matrix, only one AP cell connected to a given sense terminal can have a closed toggle switch.) This causes the output contacts of this AP cell to close, allowing the voltage from the output enable bus to appear at the matrix output terminal connected to this AP cell.
6. After a delay determined by the actuator delay relay, the actuator brake is released, and the output relay connected to this matrix output terminal produces a voltage that drives the actuator to a position determined by the position of the energized output relay.
7. When the end-of-cycle relay times out, the brake is re-engaged, power to the output relays is disconnected, and a signal is sent to the outside timing source indicating that the transition is complete.
8. At the end of the transition cycle, the outside timing source opens the cycle switch. This disconnects the matrix power source, releases the input latch and restores the other disconnects and delay relays to their normal (initial) positions.

Thus, an active predetermined controlled machine produces the same results as a passive controlled machine in this single variable control system except that the power to drive the output relays does not come through its input terminals.

Application of an absolute predetermined control matrix

The number of values of the absolute predetermined control matrix can be increased to match the number of values of the direction error sensor and pilot valve of the rudder servo-actuator in the ship auto-pilot shown earlier, forming the *ship auto-pilot* shown in Figure 9.3.

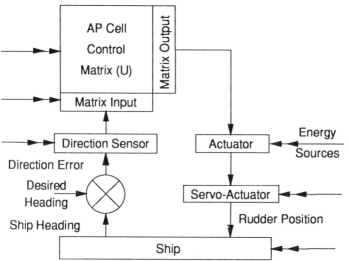

Each specific value of the direction error sensor likely to be encountered must be connected to the specific rudder position required to hold the ship on course.

Absolute predetermined duplex control systems

As shown before, the number of memory elements in a control system can be greatly reduced by connecting a matrix containing the input variable with a matrix containing the output variable by an intermediate variable having only as many values as there are expected input/output relations, forming an active *absolute predetermined duplex controller*.

Scalar absolute predetermined input matrix

A *scalar absolute predetermined input matrix* (Fig. 9.4) can be used in place of a passive input matrix in any application having one input variable and one output variable, or in a coded unitary controller with multiple input and output variables. Though an active input matrix is more complex and therefore more expensive than a passive input matrix, isolating the output enable power source from the input sense power source increases the design flexibility of the electrical circuit, allowing it to be used with multiple input variables. In the example scalar matrix given, the input value $A(0)$ produces the output value $Z(2)$, input value $A(1)$ produces no output, input value $A(2)$ produces output value $Z(1)$, and input value $A(3)$ produces output value $Z(0)$. Because it has four values of its input variable, it may be said to be a *quadrary* memory matrix (see Chapter 30). The number of values of its input and output variables can be increased indefinitely, as indicated by the arrows, to match the variety of any sensor and actuator variable.

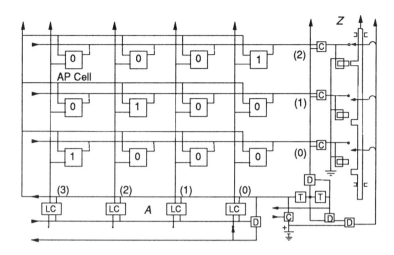

Figure 9.4. A *scalar absolute predetermined input matrix* isolates (separates) the input sense power source from the output enable power source, contains more input values than output values, and has a single input variable and a single output variable.

Scalar absolute predetermined output matrix

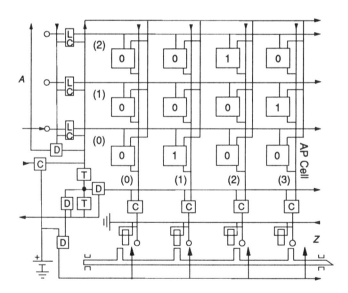

Figure 9.5. A *scalar active absolute predetermined output matrix* uses active memory cells to connect the few values of the single input variables to many possible values of its output variable.

The toggle switches in a scalar passive output matrix can be replaced with absolute predetermined memory cells, forming the *scalar absolute predetermined output matrix* shown in Figure 9.5. The AP cell output matrix usually has more values of its single (scalar) output variable than values of its single input variable, and operates the same as an active absolute predetermined input matrix rotated 90° in this case.

Scalar absolute predetermined duplex DS machine

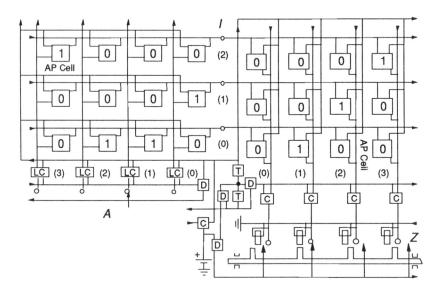

Figure 9.6. A *scalar absolute predetermined duplex controlled machine* can be used in place of a scalar active switch duplex controlled machine whenever it is desirable to reprogram the controller without adding or removing components.

The number of memory cells in the scalar absolute predetermined matrices shown can be reduced when there are significantly fewer transitions than input and output states by using the *scalar absolute predetermined duplex DS machine* shown in Figure 9.6. A scalar AP cell duplex controller is made up of a scalar AP cell input matrix connected to a scalar AP cell output matrix. A scalar passive predetermined duplex controller may be the preferred design for most predetermined duplex controlled machines with a single input and output variable. However, the design of the active duplex controller leads to the design of a duplex controlled machine that can deal with two or more input variables and two or more output variables, to be shown in the sections following this section.

Operation of a scalar absolute predetermined duplex machine

A scalar absolute predetermined duplex machine operates according to Table 9.2.

Table 9.2. Operation of a scalar absolute predetermined duplex machine

---

1. An outside timing source closes the cycle switch at the start of a transition cycle.
2. This energizes the output enable power bus and the matrix input terminal selected by the sensor switch.
3. The input latch disconnects the sensor switch and maintains the voltage at the selected matrix input terminal.
4. A voltage is produced at the input sense terminals of the AP cells connected to the selected matrix input terminal.
5. This causes a current to flow through the relay coil in the AP cell having a closed toggle switch. This causes the output contacts of this AP cell to close, allowing the voltage of the matrix power source to appear at the value of the intermediate variable connected to this selecting AP cell.
6. This value of the intermediate variable then causes a current to flow through the search coil in the AP cell in the output matrix connected to this intermediate terminal having a closed toggle switch. This causes the output contacts of this AP cell to close, allowing the voltage of the matrix power source to appear at the matrix output terminal connected to this second selecting AP cell.
7. After a delay determined by the actuator delay relay, the actuator brake is released, and the output relay connected to this matrix output terminal produces a voltage that drives the actuator to a position determined by the location of the active output relay.
8. When the end-of-cycle relay times out, the brake is re-engaged, the voltage to the output relays is disconnected, and a signal is sent to the outside timing source indicating that the transition cycle is over.
9. The outside timing source then opens the cycle switch. This disconnects the matrix power source, releases the input latch, and restores the other disconnects and delay relays to their normal positions.

Thus, a scalar absolute predetermined duplex DS machine produces the same results as a scalar passive or active switch duplex controlled machine in this single variable (scalar) control system.

Application of a scalar absolute predetermined duplex controller

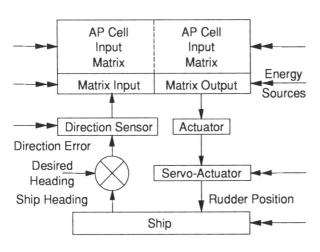

Figure 9.7. A *ship auto-pilot using a scalar active absolute predetermined duplex controller* can provide adequate control behavior in a system with 3% resolution and up to 50 transitions.

The duplex controller can have more input and output values with fewer memory cells than a single matrix if fewer than half as many input/output relations are required than the number of values of its input and output variables. For example, this duplex controller can be used in the ship auto-pilot shown in Figure 9.7. The scalar AP cell duplex controller is useful as a controller in any predetermined system with a single input variable and a single output variable, each with relatively few values and relatively few transitions. For example, a duplex ship auto-pilot with 3% resolution requires approximately 60 AP cells per transition. If 50 transitions are required, then 3000 AP cells are required. A trinary digitized duplex controller can reduce the number of AP cells to 900.

Absolute predetermined multivariable input matrices

The use of active memory cells allows two or more single-variable matrices to be connected to form an *absolute predetermined multivariable input matrix* with multiple input variables and a single output variable without using an input decoder. The multivariable input matrix can deal with multiple input variables and be programmed to produce either unitary or synchronous diverse behavior. The AP cell multivariable input matrix is different from a coded input matrix, which also can deal with multiple input variables, but only in a unitary manner.

Absolute predetermined multivariable binary input matrix

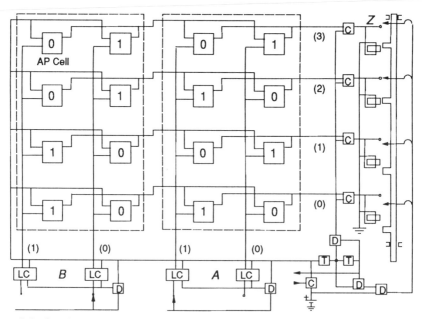

Figure 9.8. An *absolute predetermined binary multivariable input matrix* contains two or more binary input variables (*A* and *B*), one matrix power source and one output variable (*Z*).

The design of the active memory matrix allows two or more input variables to be included in a single multivariable input matrix. For example, two or more single-variable active memory matrices can be connected in series with one matrix power source, as shown in Figure 9.8. An *AP cell binary multivariable input matrix* can produce an output value that represents a unique combination of values of its input variables in a unitary manner. For example, output terminal $Z(0)$ is energized when input terminals $A(1)$ *and* $B(1)$ are energized, output terminal $Z(1)$ is energized when input terminals $A(0)$ *and* $B(1)$ are energized and output terminal $Z(2)$ is energized when input terminals $A(1)$ *and* $B(0)$ are energized. A given output can be produced in a binary input matrix only when two of its AP cells are in the conductive (1) position and both are made active by the position of both sensor switches.

Operation of an AP cell multivariable input matrix

An absolute predetermined multivariable input matrix operates according to Table 9.3.

Table 9.3. Operation of an AP cell multivariable input matrix

---

1. An outside timing source closes the cycle switch at the start of a transition cycle.
2. This energizes the matrix enable power source and one input sense terminal connected to a sensor switch representing each input variable in each matrix.
3. Each input latch disconnects each sensor switch as each matrix input terminal is energized and maintains the voltage at the selected matrix input terminals.
4. A voltage is produced at the input sense terminals of the AP cells connected to the selected matrix input terminals.
5. This causes a current to flow through the search coils in the AP cells that have closed toggle switches and are connected to the selected matrix input terminals. This causes the output contacts of these AP cells to close, allowing the voltage of the matrix power source to appear at the matrix output terminal connected to the conducting AP cells in both input matrices.
6. After a delay determined by the actuator delay relay, the actuator brake is released, and the output relay connected to this matrix output terminal produces a voltage that drives the actuator to a position determined by the location of this output relay.
7. When the end-of-cycle relay times out, the output relay is disconnected, the actuator brake is re-engaged, and a signal is sent to the outside timing source indicating that the transition is over.
8. The outside timing source opens the cycle switch at the end of the transition cycle. This disconnects the matrix power source, releases the input latch, and restores the disconnects and delay relays to their *normal* positions.

Thus, a multivariable input matrix can produce a unique output value of its single output variable for each unique combination of values of its input variables. The absolute predetermined multivariable input matrix can be used in any control application with two or more input variables and one output variable and can be programmed in an all-or-nothing manner before it is put into use.

Distributed control in an absolute predetermined multivariable input matrix

Since two or more memory elements are required to produce a given output state in a multivariable input matrix, this process of using two or more input variables in parallel to control an output variable is called *distributed control.*

Unitary relations in a multivariable input matrix

A multivariable input matrix with x^n output terminals can provide unique *unitary* output values for every combination of n input variables with x values. For example, the AP cell multivariable input matrix with the *conformation* (set of AP cell switch settings) shown can produce four unique outputs, each representing one of the four possible combinations of the two values of its two input variables.

Synchronous diverse behavior in a multivariable input matrix

The multivariable input matrix also can provide a unique *synchronous diverse* output value based upon the value of a given input variable regardless of the values of the other variables if programmed according to the following *conformation* shown in Figure 9.9.

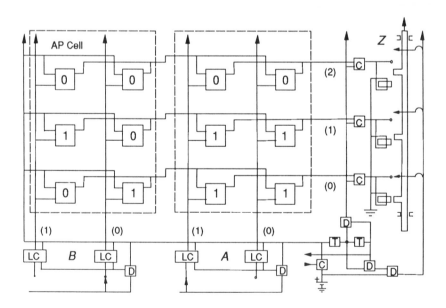

Figure 9.9. An *AP cell multivariable binary input matrix in a synchronous diverse conformation* can disregard the values of one or more if its input variables.

For example, it can produce a given output value $Z(0)$ whenever its input terminal $B(0)$ is energized, though either value of the other input variable $A(0)$ or $A(1)$ is energized as shown. Thus, a variable like A is ignored by the matrix whenever the memory cells connected to its input terminals are set in a conductive state. By adding more input variables, the matrix can be programmed so that other input variables are ignored at different times in a synchronous diverse manner.

AP cell trinary multivariable input matrix

The number of values of each variable in a multivariable input matrix can be increased indefinitely to match the variety of the sensor and actuator variables. For example, the number of values of the input variables can be increased from two in an AP cell multivariable binary input matrix to three in the *AP cell trinary multivariable input matrix* shown in Figure 9.10.

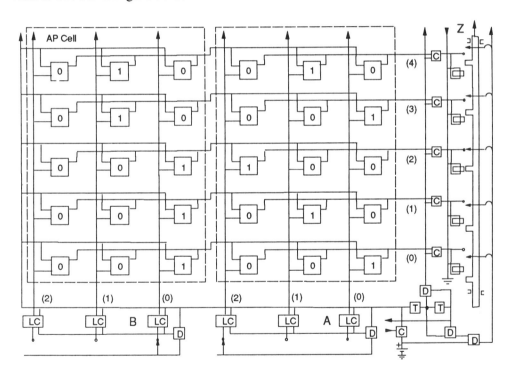

Figure 9.10. An *AP cell trinary multivariable input matrix* can produce a unique response to more combinations of values of its input variables than a binary input matrix with the same number of input terminals.

The *trinary* matrix is more "efficient" than a binary matrix in terms of the amount of variety it can generate with a given number of memory cells. For example, a 6-variable binary input matrix requires 12 memory cells per output state and can produce $2^6 = 64$ different output states. A 4-variable trinary input matrix also requires

12 memory cells per output state but can produce $3^4 = 81$ different output states. A 3-variable quadrary input matrix would also require 12 memory cells but would produce $4^3 = 64$ different output states, the same as a binary system. The improvement of a trinary system over a binary or quadrary system increases as the number of variables increases. The number of values of the input variables can be increased indefinitely to correspond to the variety of the sensor and actuator variables, although the "efficiency" of the matrix decreases as the number of values increases above 3 values per variable.

AP cell multivariable input matrix control system

The multivariable input matrix can be used to connect two or more input variables to a single output variable in a unitary or diverse fashion. For example, an AP cell multivariable input matrix with many values of its input variables can be used to connect the direction error sensor and the crosswind sensor in the ship direction control system shown earlier to a single rudder position actuator forming the *AP cell multivariable ship direction control system* shown in Figure 9.11.

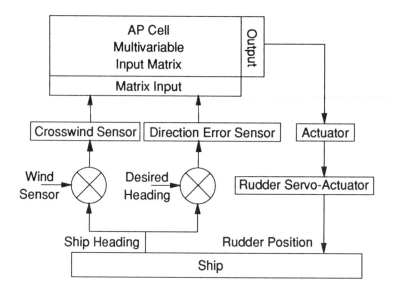

Figure 9.11. An *AP cell multivariable input matrix ship direction control system* can use one or all of its input variables to control one output variable.

If both the crosswind and direction error signals are needed to maintain the course of a ship, unique rudder actuator values can be programmed to be produced for each combination of value of crosswind and direction error. If the crosswind has no influence upon the amount of rudder correction required to keep the ship on course, 1's (connections) can be programmed into the crosswind matrix for all values of crosswind, eliminating it as a control variable.

Absolute predetermined multivariable output matrices

One input variable can be connected to two or more output variables in an absolute predetermined multivariable output matrix. This *AP cell multivariable output matrix* can be used in any predetermined control system with one input variable and two or more output variables, and can be programmed to produce unitary or synchronous behavior.

AP cell multivariable binary output matrix

The design of an active matrix also allows two or more output variables to be connected to a single input variable in a single multivariable output matrix. This is accomplished by connecting two or more single-variable active control matrices in *parallel* with the input variable, forming the *AP cell multivariable binary output matrix* shown in Figure 9.12.

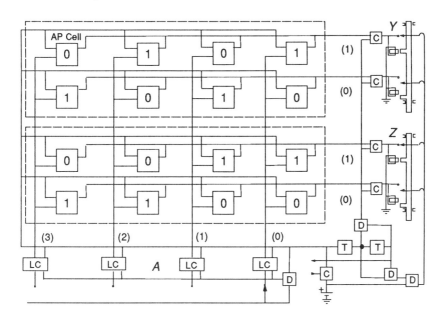

Figure 9.12. An *AP cell multivariable binary output matrix* has one input variable and uses a separate single-variable active matrix for each output variable.

A multivariable output matrix can produce a unique combination of output values for each value of its input variable. For example, output terminals $Y(1)$ and $Z(1)$ are energized when input terminal $A(0)$ is energized, and output terminals $Y(0)$, and $Z(0)$ are energized when input terminal $A(1)$ is energized. Notice that the AP cells connected to a given output variable are divided in the horizontal direction in a multivariable output matrix, whereas the AP cells connected to a given input variable are divided in the vertical direction in a multivariable input matrix.

Operation of an AP cell multivariable output matrix

An AP cell multivariable output matrix operates according to Table 9.4.

Table 9.4. Operation of an AP cell multivariable output matrix

---

1. An outside timing source closes the cycle switch at the start of a transition cycle.
2. This energizes the matrix power bus and the one matrix input terminal representing the value of the input variable at the beginning of the transition cycle.
3. The input latch disconnects the sensor switch and maintains the voltage at the selected matrix input terminal.
4. A voltage is produced at the input sense terminals of the AP cells in both matrices connected to the selected matrix input terminal.
5. This causes a current to flow through the search coils in these AP cells that have their toggle switch in the "on" position. This causes the output contacts of these AP cells to close, allowing the voltage of the matrix power supply to appear at the matrix output terminals connected to the conducting AP cells in both the matrices.
6. After a delay determined by the actuator delay relay, the actuator brake is released and the output relays connected to these matrix output terminals produce a voltage. This drives both actuators to positions determined by the position of the energized output relays.
7. When the end-of-cycle relay times out, the brake is re-engaged, the output relays are disconnected, and a signal is sent to the outside timing source indicating that the transition cycle is over.
8. The outside timing source allows the cycle switch to open at the end of the transition cycle. This disconnects the matrix power source from the matrix power bus, releases the input latch and other disconnects, and allows the delay relays to reset.

Thus, a multivariable output matrix produces a unique combination of output values in a given transition determined by the value of its input variable. The AP cell multivariable output matrix can be used for any control application with one input variable and two or more output variables and can be programmed before it is put into use.

Unitary relations in a multivariable output matrix

A multivariable output matrix with x^n input terminals can provide unique *unitary* output values for every combination of its n output variables if each variable has x values. For example, the multivariable binary output matrix with the conformation shown can produce all four combinations of the two values of its two output variables.

Diverse behavior in a multivariable output matrix

A multivariable output matrix also can provide unique *diverse* output values based upon the value of its input variable. That is to say, it can produce a set of output values on one or more of its output variables and produce a constant on one or more of its other output variables for each value of its input variable, as shown in Figure 9.13.

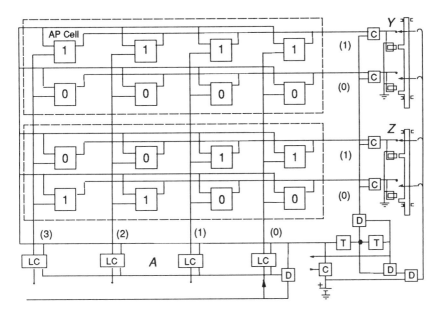

Figure 9.13. An *AP cell multivariable output matrix in a synchronous diverse conformation* contains a null function in one or more of its output variables.

In the example shown, the multivariable output matrix can produce an *active set of values* of its output variable Y for each value of its input variable A, and produce a *constant value* at its output variable Z. This requires that connections be made to one output terminal in the submatrix representing the variable Z. Submatrix Z is then said to be a *null matrix*. Thus, a multivariable output matrix can be programmed to produce either unitary behavior or a constant value of one or more of its output variables for all values of its input variable in a synchronous diverse manner.

Rotated AP cell multivariable output matrix

It is desirable to rotate the multivariable output matrix 90° (Fig. 9.14) to view it in its position as the output matrix of a duplex controller. The delays occur in the vertical direction in a rotated multivariable output matrix, and the submatrices are divided in the vertical direction like the submatrices in a multivariable input matrix. However, the submatrices in a multivariable output matrix are still connected in parallel in contrast to the submatrices of a multivariable input matrix, which are connected in series.

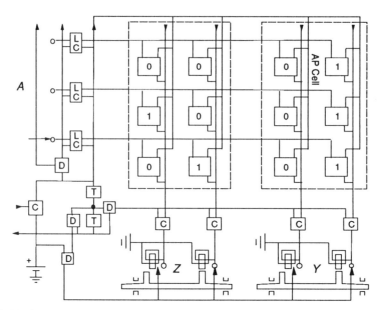

Figure 9.14. A *rotated AP cell multivariable output matrix* shows the arrangement of latches, connects, disconnects and AP cells as they will appear in the output matrix of a multivariable duplex controller.

AP cell multivariable trinary output matrix

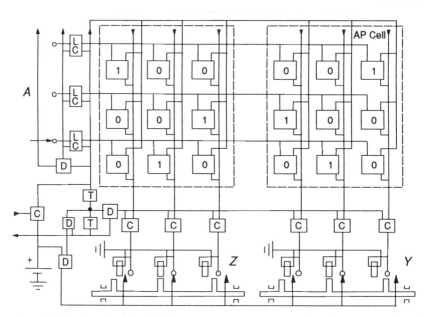

Figure 9.15. An *AP cell multivariable trinary output matrix* contains three values of each output variable.

The number of values of the output variables can be increased indefinitely. For example, the number of values of each variable can be increased from two in the binary output matrix shown to three in the *AP cell multivariable trinary output matrix* shown in Figure 9.15. Like the multivariable trinary input matrix shown previously, the multivariable trinary output matrix can produce the greatest number of different output states for a given number of memory cells. The number of values of the output variables can be increased to match the variety of the output devices, though a number of values greater than three results in a less efficient use of memory cells.

AP cell multivariable output matrix control system

An AP cell multivariable output matrix can be used to control two or more output variables from a single input variable. Thus, the crosswind sensor and direction error sensor in the ship direction control system shown earlier can be summed together in a differential and the output used as the input variable of a multivariable output matrix. Then the output variables of the matrix can be used to control the rudder servo-actuator and the bow thruster in the *AP cell multivariable output matrix control system* shown in Figure 9.16.

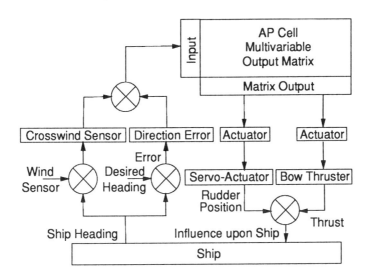

Figure 9.16. An *AP cell multivariable output matrix control system* can produce multiple outputs from a single input variable.

The output variables can be programmed such that unique combinations of values of both variables are produced for each value of the input variable in a unitary manner, or one output variable may be held constant for all values of the input variable so that the other output variable can form a synchronous diverse relation with the input variable.

Absolute predetermined multivariable duplex controllers

Many control tasks require two or more input and two or more output variables having more values than active states in its line of behavior. They also may have to be programmed in a unitary or diverse conformation. The use of active memory elements allows a multivariable input matrix to be connected to a multivariable output matrix, forming an *AP cell multivariable duplex control matrix* that can handle multiple input and output variables in a unitary or synchronous diverse fashion, and have as many transitions as are needed in its active line of behavior.

AP cell multivariable duplex control matrix

The output variable of an AP cell multivariable input matrix can be connected to the input variable of an AP cell multivariable output matrix, forming the *AP cell multivariable duplex controller* shown in Figure 9.17, which can handle multiple input and output variables without input or output coding.

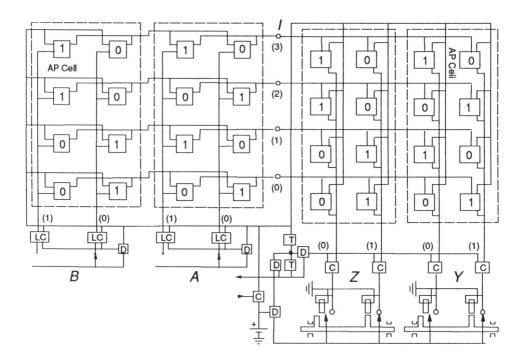

Figure 9.17. An *AP cell multivariable binary duplex controller* can connect multiple input variables to multiple output variables in some predetermined number of unitary or synchronous diverse relations.

The AP cell multivariable duplex controller shown can produce a unique combination of values of its output variables for each combination of values of its input variables in a unitary conformation. For example, input values of $A(0)$ and $B(0)$ produce output values of $Y(1)$ and $Z(1)$, and input values of $A(1)$ and $B(0)$ produce output values of $Y(0)$ and $Z(1)$ etc. in a unitary fashion. It can be programmed to produce unitary or synchronous diverse relations with a minimum number of intersections because it does not commit input or output terminals to specific combinations of values. This contrasts with a coded unitary system, which requires an input and output terminal for ever possible combination of values of all sensor and actuator variables. It can produce a limited number of relations according to the number of intermediate states designed into its matrices.

Operation of an AP cell multivariable duplex controller

A given intermediate terminal in a multivariable duplex controller can produce a unique combination of output values of multiple output variables for a unique combination of values of its multiple input variables. The AP cell multivariable duplex controller can be used in any control application with two or more input variables and two or more output variables that can be programmed in an all-or-nothing manner before it is put into use. An AP cell multivariable duplex controller operates according to the steps shown in Table 9.5.

Table 9.5. Operation of an AP cell multivariable duplex controller

---

1. An outside timing source closes the cycle switch at the start of a transition cycle.
2. This energizes the matrix enable power bus and one matrix input terminal in each input matrix connected to a sensor switch representing each input variable.
3. Each input latch disconnects each sensor switch as each matrix input terminal is energized. Each latch maintains the voltage at the selected matrix input terminals.
4. A voltage is produced at the input sense terminals of the AP cells connected to the selected matrix input terminals.
5. This causes a current to flow through the search coils in those AP cells having their toggle switches in the "on" position. This causes the output contacts of these AP cells to close, allowing the voltage of the matrix enable power bus to appear at the intermediate variable terminal connected to the set of conducting AP cells forming a series connection through the input matrices.
6. A voltage is produced at the input sense terminals of the AP cells in the output matrix connected to the selected intermediate variable terminal.
7. This causes a current to flow through the search coils in those AP cells in the output matrix having toggle switches in the "closed" position. This causes the output contacts of these AP cells to close, allowing the voltage of the matrix power bus to appear at the matrix output terminals connected to these conducting AP cells in the output matrices.
8. After a delay determined by the actuator delay relay, the actuator brake is released and the output relays connected to these matrix output terminals produce a voltage. This drives the actuators to positions determined by the location of the energized output relays.
9. When the end-of-cycle relay times out, it re-engages the actuator brake, disconnects the output relays, and sends a signal to the outside timing source indicating that the transition cycle is complete.
10. The outside timing source then opens the cycle switch. This disconnects the matrix power source from the matrix power bus, releases the input latch and other disconnects, and allows the delay relays to reset.

Diverse conformation of an AP cell multivariable duplex controller

The AP cell multivariable duplex controller also can be programmed so that the input
variables operate upon only one output variable or one input variable alone determines the values of the output variables in a *synchronous diverse manner.* Also, if
only one -input variable is needed to control one output variable, the matrices
connected to the other variables can be programmed to act as null matrices, as shown
in Figure 9.18.

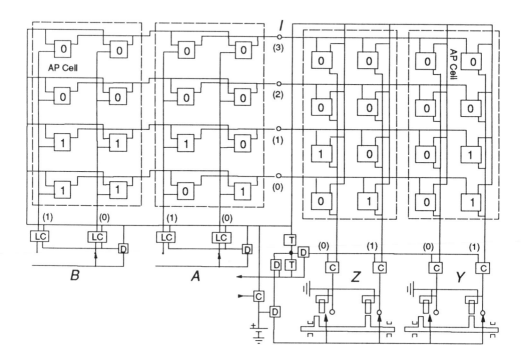

Figure 9.18. An *AP cell multivariable duplex controller in a diverse conformation* can cause
only one input variable to cause action in only one output variable.

In the synchronous diverse conformation shown, different values of the input
variable A cause different values of the output variable Z and cause constant values
of the output variable Y, regardless of the values of the input variable B. However,
input variable B cannot establish a diverse relation with output Y without the conformation of the matrix being changed. Since the multivariable duplex controller
can be programmed to provide either unitary or synchronous behavior, and can be
designed with just enough intermediate states to provide the required number of
input and output relations, it can greatly reduce the number of memory elements
needed to provide satisfactory multivariable control behavior compared to a coded
unitary controller.

AP cell trinary multivariable duplex controller

The number of values of the sensor and actuator variables can be increased indefinitely to match the variety of the sensor and actuator variables. The maximum variety of behavior can be produced for a given number of memory cells when the input and output variables have three values, as shown in the *AP cell trinary multivariable duplex controller* (Fig. 9.19).

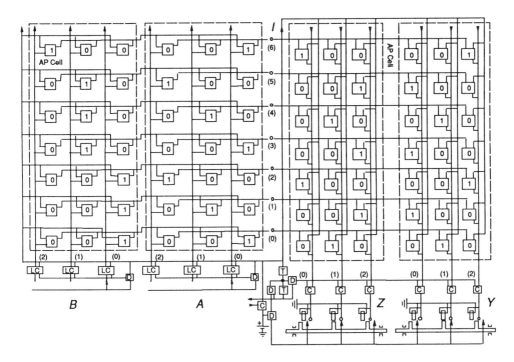

Figure 9.19. An *AP cell trinary multivariable duplex controller* can connect any combination of the three values of multiple input variables with any combination of the three values of multiple output variables.

The AP cell multivariable trinary duplex controller can be programmed to produce more different combinations of values of its output variables for different combinations of values of its input variables using fewer AP cells than a multivariable binary, quadrary, pentary, or a controller based upon any other number system.

AP cell multivariable duplex ship direction control system

A multivariable duplex controller is the most general and useful predetermined control system for any applications having a relatively few input/output relations. It can be used in any predetermined application having multiple input and output variables, such as the ship direction control system shown in Figure 9.20.

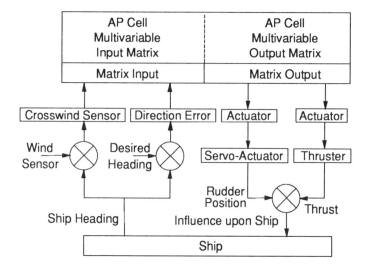

Figure 9.20. An *AP cell multivariable duplex ship direction control system* requires far fewer memory elements than a coded unitary system with the same number of input and output variables with the same resolution.

The coded unitary duplex system shown earlier, having two input variables and two output variables, each with a resolution of 5%, and with 10 unitary transitions, requires 8000 AP cells, of which only 20 can be made conductive. A multivariable duplex controller, having the same actual performance as a coded unitary duplex system requires only $20 \times 2 = 40$ input matrix terminals and 40 output matrix terminals with 10 intermediate terminals for a total of 800 AP cells. However, this system, with 20 values of its input and output variables, is not as efficient as a controller based upon a binary or trinary digitized system, as shown in Chapters 10 and 11.

Absolute predetermined universal controllers

One limitation of the multivariable duplex controller is that it can produce only one (synchronous) diverse relation at a given time since only one value of the intermediate variable is made active in a given transition. This limitation can be overcome by creating an *absolute predetermined monolithic controller* that can connect multiple input variables to multiple output variables in a unitary or asynchronous diverse manner. An *absolute predetermined universal control matrix* can be formed by connecting an absolute predetermined monolithic input matrix with an absolute predetermined monolithic output matrix. Its multiple input variables and multiple output variables can be connected by multiple intermediate variables and be programmed to produce unitary or *asynchronous diverse behavior* using only as many intermediate terminals as are required by its active lines of behavior.

Absolute predetermined binary monolithic control matrix

AP cell scalar submatrices can be connected in series to a given output variable and in parallel to a given input variable, forming the *AP cell binary monolithic control matrix* shown in Figure 9.21. It can be made up of any number of input and output variables, and can be programmed to produce unitary or asynchronous diverse behavior.

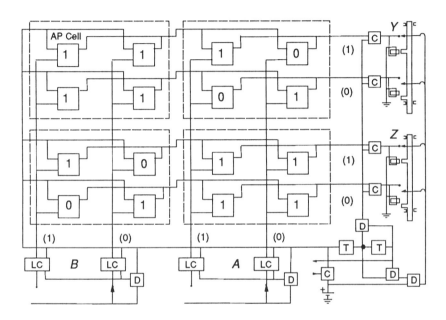

Figure 9.21. An *absolute predetermined binary monolithic control matrix* can connect multiple input variables with multiple output variables in a unitary or asynchronous diverse manner.

The *AP cell binary monolithic control matrix* shown has two values of each input and output variable. A given input variable can be connected to a given output variable in an *asynchronous diverse* manner (that allows diverse relations to be carried out at any time regardless of the states of other diverse relations by programming conductive values in the intervening memory cells), or each combination of input values can produce a different combination of output values in a unitary manner by programming the memory cells to produce a unitary code translation. (In the example given, the unitary code appears to be diverse because the variables in this particular code transformation mirror each other.)

Absolute predetermined trinary monolithic control matrix

The number of values of each sensor and actuator variable can be increased indefinitely. One more value is added to each sensor and actuator variable to form the *AP cell trinary monolithic control matrix* shown in Figure 9.22.

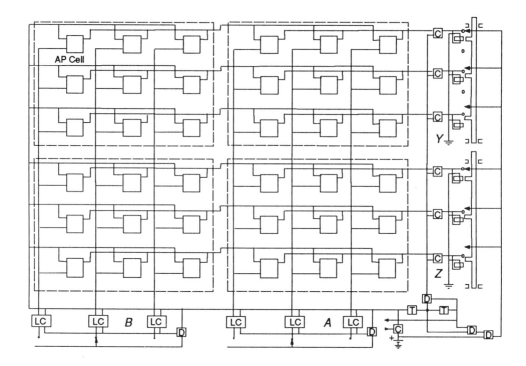

Figure 9.22. An *absolute predetermined trinary monolithic control matrix* can connect more combinations of values of sensor variables to more combinations of values of actuator variables using fewer AP cells than a binary monolithic control matrix.

It can be programmed to produce more unique transitions than for a given number of memory cells than a binary monolithic control matrix or any other monolithic control matrix with more than three values of each variable. For example, a binary monolithic control matrix with three sensor variables and three actuator variables requires $6 \times 6 = 36$ AP cells, and can connect any one of $2^3 = 8$ possible sensor states to any one of $2^3 = 8$ actuator states. A trinary monolithic control matrix with two sensor variables and two actuator variables also requires 36 AP cells, but can connect any one of $3^2 = 9$ sensor states to any one of $3^2 = 9$ actuator states.

Application of an AP cell monolithic control system

The number of values of the sensor and actuator variables can be increased to match the resolutions of a crosswind sensor, a direction error sensor, a rudder servo-actuator, and a bow thruster. These sensors and actuators can be connected by an AP cell monolithic control matrix, forming the *AP cell monolithic ship direction control system* shown in Figure 9.23.

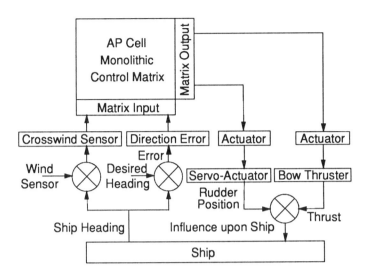

Figure 9.23. An *AP cell monolithic ship direction control system* can be programmed to produce a unique combination of values of its rudder and bow thruster for each combination of values of its direction error and crosswind sensors in a unitary manner, or one sensor can be programmed to control one actuator in an asynchronous diverse manner.

The number of memory cells in a monolithic control matrix increases according to the product of the total number of input values times the total number of output values. This number could greatly exceed the number of memory cells in a duplex controller if only a relatively few transitions are required in the active line of behavior of a given system with many variables and/or many values.

Absolute predetermined universal duplex controller

An AP cell monolithic input matrix can be connected to an AP cell monolithic output matrix by two or more intermediate variables, forming the *AP cell binary universal duplex controller* shown in Figure 9.24. The number of *asynchronous diverse lines of behavior* expected in a particular control task can be used to determine the number of intermediate variables designed into the universal duplex controller. The number of transitions expected in each diverse line of behavior can be used to determine the number of values of each intermediate variable. Thus, the number of memory cells in the universal duplex controller may be fewer than the number of memory cells in a monolithic controller if there are a relatively few required transitions in each expected line of behavior of the control task compared with the variety of its variables.

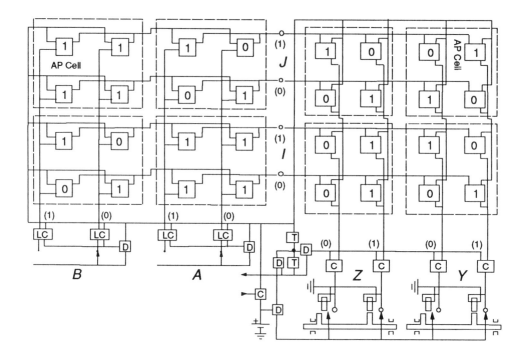

Figure 9.24. An *AP cell binary universal duplex controller* can be programmed to produce unitary behavior or programmed to produce as many asynchronous diverse lines of behavior as it has intermediate variables.

AP cell trinary universal duplex controller

The number of input, output, and intermediate variables and the number of values of these variables can be increased indefinitely to match the number of variables, resolution, transitions, and diverse lines of behavior required by the control task. If the AP cells are connected as shown in the *AP cell trinary universal duplex controller* (Fig. 9.25), the maximum variety of control behavior can be produced with a minimum of memory cells.

The number of intermediate variables can be selected to match the number of asynchronous diverse lines of behavior that the controller is expected to carry out, and the number of values of each intermediate variable can be selected to match the number of transitions required in each diverse line of behavior.

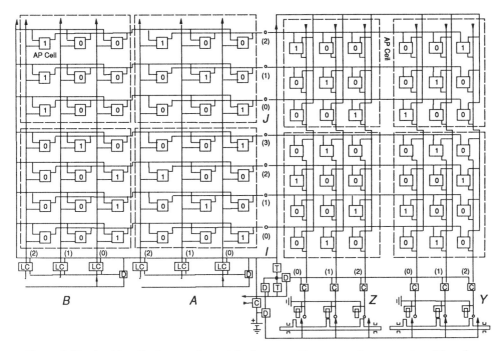

Figure 9.25. An *AP cell trinary universal duplex controller* can produce more transitions for a given number of memory cells than a binary universal controller.

Application of an AP cell universal duplex control system

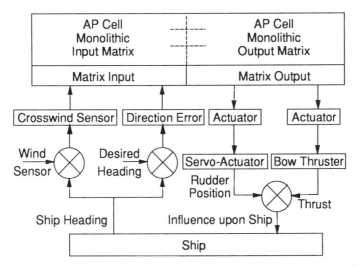

Figure 9.26. An *application for an AP cell universal duplex control system* is a ship direction control system with multiple input and output variables, some of which may have to be connected in an asynchronous diverse manner.

The number of values of the input variables of the AP cell universal duplex controller shown can be increased to match the variety of crosswind and direction error sensors, and the number of values of its output variables can be increased to match the variety of rudder position and bow thruster actuators, forming the *universal ship direction control system* shown in Figure 9.26. When the number of values of each variable exceeds three, the number of input and output states that can be produced by a given number of memory cells begins to fall off. A solution to this problem is to digitize the input and output variables, as shown in the next chapter.

Chapter 10
Binary Digitized Predetermined Machines

The design of the duplex controller can reduce the number of required memory cells. However, it is still wasteful of memory cells if the resolution of the input and output variables significantly exceeds the task requirements. This is because matrix input and output terminals are committed to values of the input and output variables that may never be used. The number of memory cells can be greatly reduced without sacrificing the *actual resolution* of the system by *digitizing* each input and output variable into a set of *aggregate variables* and using each set of aggregate variables to represent each input and output variable.

Analysis and synthesis

There are two basic processes involved in all sorting, arranging, assembling, and decision-making functions — *analysis* and *synthesis*. The first involves taking things apart — the second involves putting things back together again. These two processes are used together to create new entities. For example, a machine can be taken apart (analyzed) and then be reassembled (synthesized) into any one of many different possible new machines. This process of analysis and synthesis can be carried out by conventional electrical devices. An *encoder* can analyze the value of a variable by breaking it into its component parts and a *decoder* can put these component variables back into a single whole.

Decomposition of a variable

The first process of *analysis* consists of breaking a variable into its component or *aggregate variables*. This process makes it possible to identify the individual characteristics that make up the variable. For example, the position of a robot arm may be defined by its position along x, y, and z axes. It may be much easier for a programmer to analyze the position of the robot in its work space in terms of these orthogonal coordinates than in terms of a single space vector. For example, the programmer may determine that a robot arm must move directly down one inch in the z axis to pick up a part.

A position variable also can be broken into a set of aggregate variables by an *encoder*. A *coordinate measuring machine* produces a unique set of values representing the x, y and z coordinates for each point that it measures. A coordinate measuring machine may be considered a unitary *position vector encoder*.

Each aggregate variable also may differ from each other by some order of magnitude. For example, one aggregate variable may equal 1/10 of the value of the variable it represents, another aggregate variable may equal 1/100 of the value of that variable, and another may equal 1/1000 of that value. Then the variable can be characterized by having a large or small value of the first aggregate variable, its *most significant place* (MSP), and a high or low value of the last aggregate variable, its *least significant place* (LSP). In this case, the variable is digitized into a decimal number system by a *decade encoder*. The process of breaking a vector variable into its component parts is also called the *decomposition* of the vector.

The variable can be broken into a set of aggregate variables that differ by an order of magnitude equal to 2 instead of 10. In this case, each aggregate variable is 1/2, 1/4, and 1/8 of the value of the variable they represent, and the variable is digitized by a *binary encoder*. Also, the variable can be broken into a set of aggregate variables that differ by an order of magnitude of 3. In this case, each aggregate variable is 1/3, 1/9, and 1/27 of the value of the variable they represent, and the variable is digitized by a *trinary encoder*.

Composition of a variable

The second process of *synthesis* involves putting the aggregate characteristics of a variable together again into a single whole. For example, producing specific motions in the *x*, *y*, and *z* component axes of a robot arm results in a specific change in its position vector. Since a robot produces a unique position for each combination of values of its component position variables, a robot may be considered a unitary *position decoder*.

If each aggregate variable differs by some order of magnitude, a *decoder* can produce a single output value by adding the values of its aggregate variables. This sum corresponds to a unique combination of values of its input variables in the same way that a multi-place number is created by summing the values of each place. In a *decade decoder,* each input variable differs by an order of magnitude of 10. The aggregate variables of a *binary decoder* differ by an order of magnitude of 2. The aggregate variables of a *trinary decoder* differ by an order of magnitude of 3. The process of combining the components of a vector into a single vector through a process of synthesis is called the *composition* of a vector.

Digitization

When a variable has many values, and only a few of these value elements are important, the variable can be represented more efficiently by a set of aggregate variables rather than the variable itself. This is the reason we use *number systems*. For example, if we wish to count to 100 using a single variable, we would need 100 different numbers, value elements, or different objects to represent each of the 0 through 99 values of this variable. By using a decimal encoder, which converts the variable into a set of aggregate variables in the decimal number systems, we can represent any one of the 100 values by some combination of values of the two (decade) variables, each of which has 10 possible values. Thus, the 100 values of a single variable can be represented by only 20 different objects (value elements) belonging to the "units" and "tens" place decade aggregate variables.

The process of *digitization* works in two directions. Converting a single variable into a set of aggregate variables by an encoder is the digitization process that makes it possible to represent the variable with fewer value elements. The set of aggregate variables can then be used to reconstruct a single variable with more value elements than the aggregate variables in the digitization process using a decoder.

Local and distributed representations

Local representation occurs when a single device is used to represent a state or a value of a single variable. The properties of localization are listed in Table 10.1

Table 10.1. Properties of a local representation

---

1. Each value of a single variable is represented by a single object.
2. There must be as many objects as there are possible values of the single variable if the variable is fully represented.
3. If one of these objects is lost or destroyed, that value can no longer be represented.

The output of a decoder is a local representation because its output can have only one value at a given time. This local representation may be *fragile* because a given value may be lost (inaccessible) if its terminal does not function or is not represented.

The opposite of localization is a *distributed representation*. A distributed representation can be created by the process of digitization that uses two or more variable to represent a single variable. This reduces or eliminates the properties of localization. The properties of a distributed representation are listed in Table 10.2.

Table 10.2. Properties of a distributed representation

---

1. No one object represents a value of a single variable.
2. Many objects combine to represent a value of a single variable such that a value of a single variable is represented by a given combination of values of these objects.
3. A few objects may represent all possible values of the single variable.
4. Even if one or more of the variables representing a single variable are lost or destroyed, some characteristics of the original single variable may remain.

A distributed representation defines a value of a variable in terms of the set of values of its component parts. No two values of a distributed variable are the same unless all values of its component parts are the same. An encoder creates a distributed representation rather than a local representation of the values of a variable. The distributed representation may be more *robust* because errors in the least significant places do not introduce gross errors in the distributed value.

Binary digitized absolute predetermined scalar controllers

A discontinuous controlled machine with high-resolution sensor and actuator variables may require many memory cells to produce specific input/output relations. The number of memory cells can be greatly reduced by digitizing the high resolution sensor variables into sets of aggregate variables using input encoders (in contrast to the decoders used in the coded unitary controllers shown earlier) and digitizing the high resolution actuator variables into sets of output aggregate variables using output decoders (in contrast to the output encoders used in the coded unitary controllers shown earlier).

AP cell binary digitized scalar input matrix

The number of *inactive intersections* in an input matrix with high sensor resolution and only a few output values can be reduced by converting the sensor variable into a set of *aggregate input variables* by an encoder and using some combination of values of these aggregate variables to represent any value of the single (scalar) input variable, as shown in the *AP cell binary digitized scalar input matrix* (Fig. 10.1).

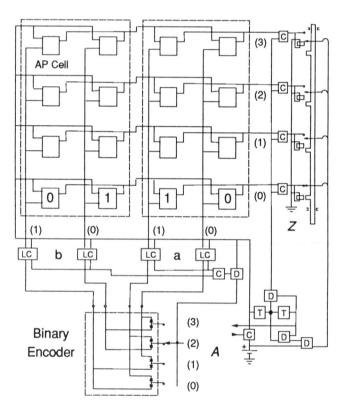

Figure 10.1. In an *AP cell binary digitized scalar input matrix*, each combination of values of the set of two binary aggregate input variables can be used to represent any one of the four values of the single input variable.

When the input variable is digitized into a set of aggregate variables, the value of the input variable is not represented by a single value element. Instead, each input value is represented by some combination of values of the set of aggregate variables. Using aggregate variables, *any* value of the input variable can be represented though only a few may be used to produce behavior. The use of aggregate input variables (*a* and *b*) is an example of a distributed representation of the input variable (*A*).

AP cell binary digitized scalar input matrix with three aggregate variables

The number of aggregate variables can be increased indefinitely if they are connected appropriately, as shown in the *AP cell binary digitized scalar input matrix with three aggregate variables* in Figure 10.2.

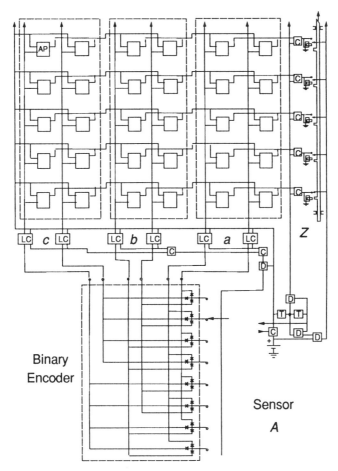

Figure 10.2. An *AP cell binary digitized scalar input matrix with three aggregate variables* can represent any one of the eight values of its input variable with only six AP cells.

The AP cell binary digitized scalar input matrix with three aggregate variables shown can represent the eight values of its input variable as compared to four values in the AP cell binary digitized input matrix with two aggregate variables shown previously. An undigitized AP cell input matrix would require eight AP cells to represent the eight values of its input variable as compared to the six AP cells required in the digitized input matrix with three aggregate variables shown.

Application of an AP cell binary digitized scalar input matrix

An AP cell binary digitized scalar input matrix can be programmed to identify any one of many possible values of an input variable such as temperature and produce an output signal to a buzzer or other annunciator to warn that the variable has reached a critical value (Fig.3).

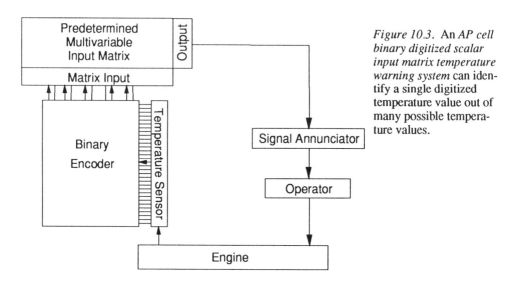

Figure 10.3. An AP cell binary digitized scalar input matrix temperature warning system can iden-tify a single digitized temperature value out of many possible tempera-ture values.

If the number of aggregate variables is increased, a digitized input matrix can identify values of a variable with greater resolution. For example, if the number of binary aggregate variables is increased to 5, the digitized binary input matrix can identify one out of $2^5 = 32$ different values and requires only $2 \times 5 = 10$ memory cells for each identification.

Binary digitized predetermined scalar output matrix

The number of inactive intersections in an output matrix with high actuator resolution and only a few input values can be reduced by using a set of *aggregate output variables* and converting combinations of values of these aggregate variables into a value of the single output variable by a decoder, as shown in the *AP cell binary* digitized scalar output matrix in Figure 10.4. When the set of aggregate output variables is *composed* by a decoder into the value of a single output variable, each output value is not represented by a single value element. Instead, each output value is represented by some combination of values of the set of aggregate variables. By using aggregate output variables, *any* one of many output values can be represented, though only a few may be used to produce behavior.

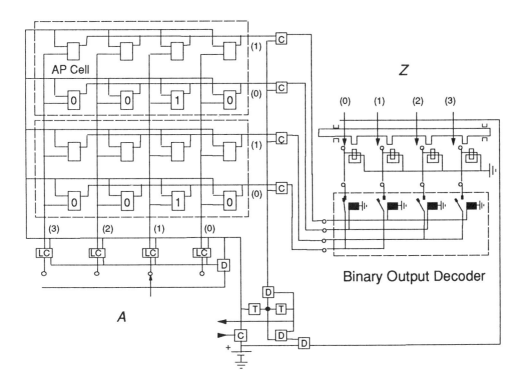

Figure 10.4. An *AP cell binary digitized scalar output matrix with two aggregate variables* can decode each combination of values of a set of two aggregate output variables into any one of four values of a single actuator variable.

Rotated AP cell digitized scalar output matrix

The digitized absolute predetermined output matrix can be rotated 90° (Fig. 10.5) to show the connects, disconnects, latches, and AP cells in the position they are placed in a digitized predetermined duplex controller. Each aggregate output variable (*a* and *b*) contains AP cells connected to the sensor variable *A*. In this binary output matrix, two AP cells connected to a given input terminal are required to be conductive to produce an output of the actuator *Z*. A unique output can be produced for each value of the sensor variable in a unitary manner or generally high or low output values can be produced for given values of the sensor variable in a synchronous diverse manner.

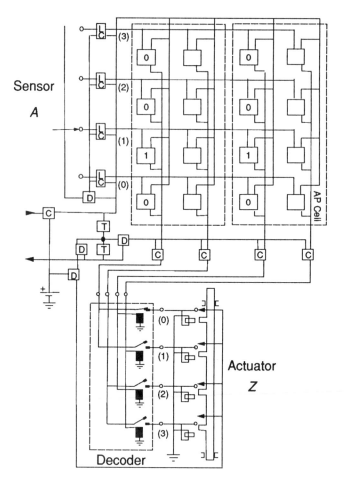

Figure 10.5. A *rotated AP cell digitized scalar output matrix* shows the location of the circuit components used in a digitized absolute predetermined duplex controller.

AP cell digitized scalar output matrix with three binary aggregate variables

The number of aggregate variables of an output variable can be increased indefinitely by using the circuit design shown in the *AP cell binary digitized output matrix* shown in Figure 10.6. Like the digitized input matrix with three binary aggregate variables, a digitized output matrix with three binary aggregate variables (*a*, *b*, and *c*) can produce 8 output values using only 6 matrix output terminals. Thus, only 6 AP cells are required for each selected actuator value in contrast to an uno.gitized matrix that would require 8 AP cells per selection.

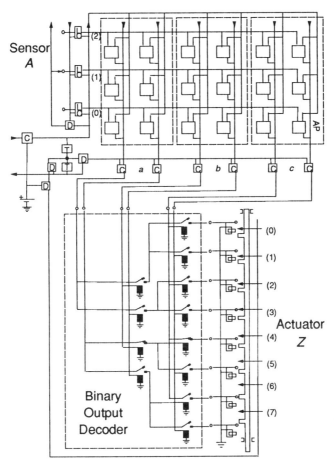

Figure 10.6. An *AP cell digitized scalar output matrix with three binary aggregate variables* can produce any one of eight output values using only six AP cells.

Application of an AP cell digitized scalar output matrix with binary aggregate variables

An AP cell binary digitized output matrix can be used to provide a *medium-resolution positioning system* (Fig. 10.7) with a relatively few memory cells. For example, if the number of binary aggregate variables were increased to 5, the output matrix could be programmed to produce any one of 32 different output values of a vernier actuator using only 10 AP cells per selection. The total number of different output values that can be programmed is determined by the number of matrix input terminals. If only a few different positions are required, as in a repetitive machine motion, then only a few matrix input terminals are required. The resolution of the vernier actuator is limited to approximately 3% (32 coils).

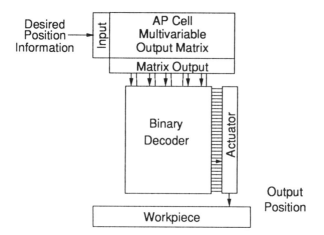

Figure 10.7. A *digitized AP cell binary scalar output matrix operating a vernier actuator* provides a much higher resolution with fewer AP cells than an undigitized output matrix.

Binary digitized absolute predetermined scalar duplex controller

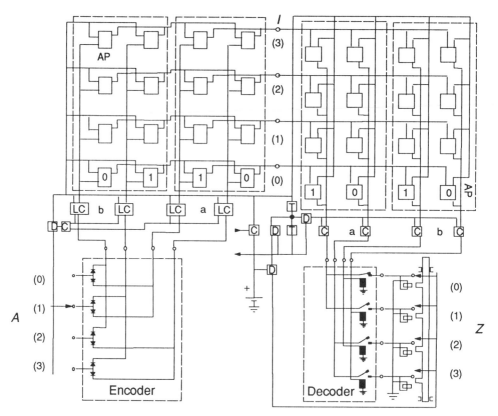

Figure 10.8. An *AP cell binary digitized scalar duplex controller* can connect values of a scalar sensor variable to values of a scalar actuator variable.

A binary digitized AP cell input matrix can be connected to a binary digitized AP cell output matrix, forming the *AP cell binary digitized scalar duplex controller* shown in Figure 10.8. When a sensor variable is digitized into a set of aggregate variables in a digitized input matrix and connected to an actuator variable in a digitized output matrix, any one of many different possible sensor values can produce any one of many different possible actuator values using fewer AP cells than a scalar duplex controller without digitized variables.

Application of an AP cell binary digitized scalar duplex system

An AP cell binary digitized scalar duplex controller can be used for any application having one sensor variable and one actuator variable with resolutions of no more than 3%, and where fewer transitions are required in its repertoire of control behavior than what is possible based upon the variety of the set of sensor and actuator variables (Fig. 10.9).

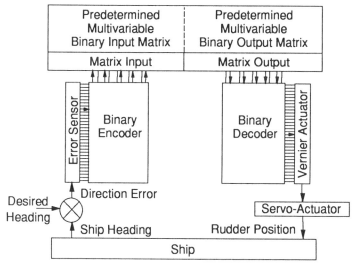

Figure 10.9. An application of a binary digitized AP cell scalar duplex controller is a ship auto-pilot that is expected to produce only a few of many possible rudder positions from a few of many possible direction error signals.

A ship auto-pilot using a binary digitized duplex controller can be programmed to provide much higher-resolution rudder positions in response to much higher-resolution direction sensor values using far fewer AP cells than an undigitized scalar duplex controller. For example, the binary digitized duplex controller shown with 5 binary aggregate variables for its sensor and actuator variables and a resolution of 3% (having 32 sensor and 32 actuator values), would require only 20 AP cells per transition. If this ship auto-pilot required only 10 transitions, it would need only 200 AP cells, of which 50 would be made conductive. An undigitized scalar duplex controller with the same 3% resolution of its sensor and actuator variable would require 64 AP cells per transition and would require 64 × 10 = 640 AP cells, of which only 20 would be made conductive.

AP cell binary digitized monolithic scalar control matrix

If many possible sensor values of a digitized scalar system must be programmed to produce many different actuator values, then the *AP cell binary digitized scalar monolithic control matrix* (Fig. 10.10) can provide the required control behavior using fewer memory cells than the binary digitized duplex controller shown previously. This is because all required transitions do not have to be represented by different values of an intermediate variable.

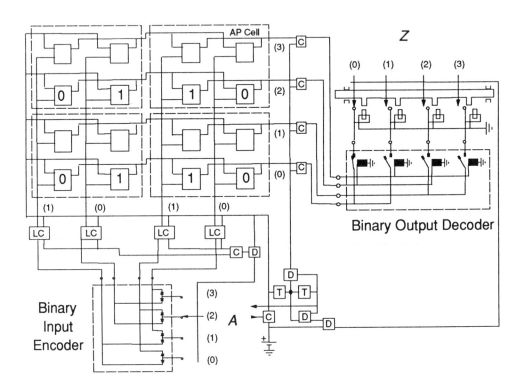

Figure 10.10. An *AP cell binary digitized scalar monolithic control matrix* can be programmed to produce different actuator positions for different sensor values by using binary aggregate variables.

If all sensor values are programmed to produce different actuator values, all the AP cells in the monolithic matrix are used. The number of memory cells in a digitized monolithic matrix is equal to the product of the number of terminals of the input encoder and the output decoder. If the resolutions of the input and output variables are the same or nearly the same, the monolithic matrix may be square or nearly square.

Application of an AP cell binary digitized monolithic scalar control system

When a controlled machine must be programmed to produce transitions for most of the values of its sensor and actuator variables in a nearly continuous function, a binary digitized AP cell scalar monolithic controller provides the most effective type of control system. For example, the temperature control system shown in Figure 10.11 may have to use every damper position to regulate the temperature of a room over the range of the temperature sensor.

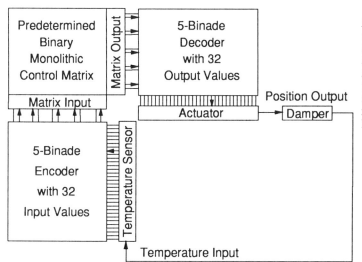

Figure 10.11. An application of a binary digitized AP cell scalar monolithic control system is a nearly continuous temperature control system using a damper to regulate the amount of heat that enters a room.

The binary digitized AP cell scalar monolithic control matrix can be programmed to produce a unique damper position for each value of the temperature sensor using a linear or nonlinear function with a resolution of around 3%. In the example shown, the monolithic controller requires ten input terminals and ten output terminals for a total of 100 AP cells. An undigitized scalar matrix would require $32 \times 32 = 1024$ AP cells. Though a binary (two value) digitized system is shown, a trinary (three value) digitized system is preferred, as will be explained later.

Distributed representation of transitions in a monolithic matrix

Since the monolithic control matrix eliminates the intermediate variable of the duplex controller, it eliminates any local representation of a transition. The *distributed representation of transitions* in a monolithic matrix makes much more efficient use of its memory cells whenever the matrix must produce many transitions because nearly all of its memory cells are used to select behavior. The digitized monolithic control matrix makes even more use of a distributed representation by translating the local values of the input and output variables into the values of the set of aggregate variables. Each transition of states of aggregate variables is represented by the distributed set of memory states in the control matrix.

Local and general characteristics of a digitized variable

The individual value elements (terminals) of a sensor or actuator variable have been referred to as *local* values because each value element represents a specific item or place in the universe. For example, everyone in the country could be assigned a different Post Office box number. Each number represents a specific Post Office box and thus a local value. If an error is made in addressing a letter, it may have little chance of reaching the desired addressee. However, by breaking up the address variable into a set of aggregate variables such as addressee, street number, Zip Code, city, and state, one or more of the values can be in error and the letter may still arrive at the desired addressee. The values of these aggregate variables represent *generalized features* such as somewhere in the state of Massachusetts and somewhere in an area with particular Zip Code rather than a particular local Post Office box.

When a variable is digitized into a set of aggregate variables, the *general characteristics* of a variable can be represented by the different aggregate variables. For example, the motion of a pick-and-place robot arm may be characterized by movements primarily in the x axis, which may be considered an aggregate or component variable of the robot's vector output variable. In another example, a high value of the most significant place of a digitized variable "generally" puts the value of the variable near the high end of its range regardless of the values of the other aggregate variables.

Stratified relations between individual aggregate variables

The general characteristics of a variable may provide a more appropriate specification of behavior than the local values in some applications. For example, the direction error sensor on the ship auto-pilot shown may provide more resolution than is required for the control tasks. In this case, the AP cells connected to all the values of the least significant places may be set in a conductive state so that only the most significant place activates the intermediate values of the duplex controller in a diverse manner. Likewise, the AP cells in the output matrix connected to the least significant places may be programmed to produce constant values, and the AP cells connected to the most significant places can be programmed to produce large actuator step motions.

In this *stratified relation* between specific aggregate variables, one or more sensor aggregate variables can then be connected to one or more of the actuator aggregate variables without regard for the values of the other aggregate variables. A digitized controller can be programmed to connect values of the most significant digits of the sensor aggregate variables to the most significant digits of the actuator aggregate variables and to ignore their less significant places. This provides "coarse" low-resolution control in a high-resolution controller. This coarse control is easier to program since only the low-resolution stratified relations need to be specified. If higher-resolution is required, relations between the values of more of the aggregate variables can be specified. The use of duplex controllers, aggregate variables, and monolithic systems increases the number of memory elements involved in the selection of a given transition and reduces the number of memory elements not involved in the selection process. This is an example of the greater use of *distributed processing* in a responsive controller.

Difference between a digitized controller and a coded unitary controller

A digitized controller is different from the coded unitary controller shown earlier, in that a digitized controller uses a sensor *encoder* to decompose the sensor variable into a set of aggregate variables and an actuator *decoder* to compose its aggregate variables into a single variable. However, a coded unitary controller uses a sensor *decoder* to compose a single matrix input variable from a set of sensor variables and an output *encoder* to decompose the matrix output variable into the set of actuator variables. A digitized multivariable controller is far more effective than the coded unitary controller.

Binary digitized absolute predetermined multivariable controllers

The number of digitized sensor and actuator variables can be increased indefinitely by increasing the number of sets of aggregate variables needed to form a multivariable controller.

Binary digitized absolute predetermined multivariable duplex controller

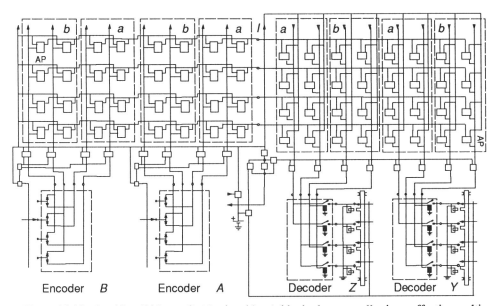

Figure 10.12. An *AP cell binary digitized multivariable duplex controller* is an effective multivariable predetermined controller if a relatively few transitions are required by the behavioral task in comparison to the number of sensor and actuator values.

If a system has more than one sensor and actuator variable, each can be digitized and combined in the *AP cell binary digitized multivariable duplex controller* shown in Figure 10.12. The digitized AP cell multivariable duplex controller can be used for any application having two or more low resolution sensor and actuator variables, and where fewer transitions are required in its repertoire of control behavior than is possible based upon the variety of the set of sensor and actuator variables. The memory matrices in the digitized multivariable duplex controller operate in the same way as the matrices in an undigitized multivariable duplex controller.

AP cell binary digitized multivariable duplex control system

The AP cell digitized multivariable duplex controller is the most general and useful programmable responsive controller and can be used in any application requiring multiple-medium resolution (3%) sensors and actuators and where a relatively few transitions are required relative to the potential variety of the sensor and actuator system, as in the *ship direction control system* shown in Figure 10.13. A digitized AP cell binary multivariable duplex control system can be programmed to produce unique combinations of values of a rudder and bow thruster from unique combinations of values of crosswind and direction error sensors. It also can be programmed to use values of only one sensor to produce values of only one actuator while holding the values of the other actuator constant in a synchronous diverse manner. Thus, different variables can be made active in different lines of behavior or both actuators can be programmed to work together to reduce the work load on a given actuator.

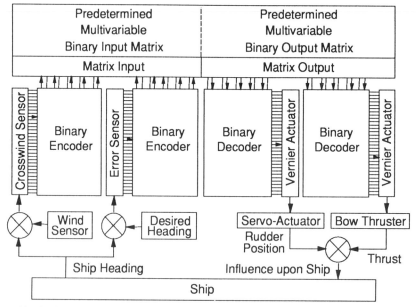

Figure 10.13. An *AP cell binary digitized multivariable duplex control system* can be programmed to produce a low resolution ship direction control system using far fewer AP cells than an undigitized system.

AP cell binary digitized multivariable monolithic control matrix

The AP cell binary digitized monolithic scalar controller can be expanded to include two or more sensor and actuator variables, as shown in the *AP cell binary digitized multivariable monolithic control matrix* in Figure 10.14. An AP cell binary digitized multivariable monolithic control matrix can be programmed to provide a unique combination of values of the actuator variables for each combination of values of the sensor variables. It also can be programmed so that a given actuator is controlled by a given sensor in an *asynchronous* diverse manner. For example, conductive values can be programmed into the set of submatrices common to input variable A and output variable Z, and conductive values can be programmed into the set of submatrices common to input variable B and output variables Y as shown. Then input variable A will control output variable Y regardless of the values of B, and input variable B will control output variable Z regardless of the values of A.

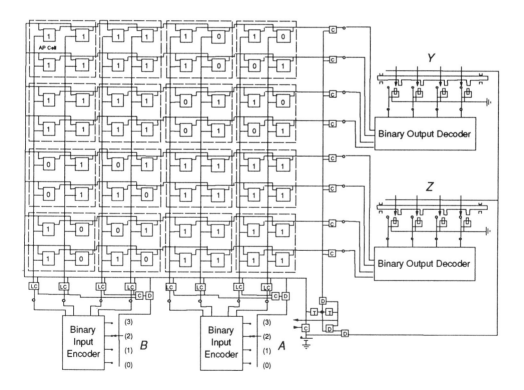

Figure 10.14. An *AP cell binary digitized multivariable monolithic control matrix* can connect multiple sensor variables to multiple actuator variables in a unitary or asynchronous diverse manner.

Application of an AP cell binary digitized multivariable monolithic controller

An AP cell binary digitized multivariable monolithic controller can be used to control the rudder actuator and bow thruster of the *ship direction control system* discussed previously, as shown in Figure 10.15.

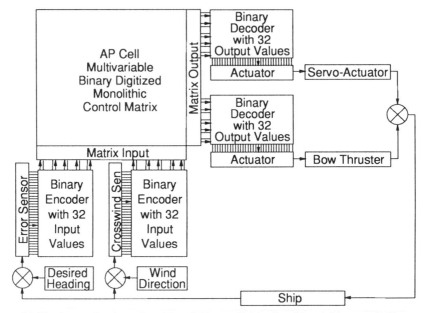

Figure 10.15. An *application of an AP cell binary digitized multivariable monolithic controller* is a multivariable ship direction control system that may require unitary or asynchronous diverse lines of behavior.

The number of variables in an AP cell digitized multivariable monolithic controller can be increased indefinitely and the resolution of these variables can be increased to match the needs of the task environment by increasing the number of aggregate variables. The number of AP cells in the control matrix increases according to the number of matrix input terminals times the number of matrix output terminals. If an additional sensor or actuator is added to the monolithic matrix, the number of additional AP cells required may be quite large. For example, the number of additional AP cells required when an additional actuator is added is equal to the number of new actuator terminals times the total number of input terminals.

AP cell binary digitized universal duplex controller

An AP cell binary digitized monolithic input matrix can be connected to an AP cell binary digitized monolithic output matrix through the intermediate variables *I* and *J*, forming the *AP cell binary digitized universal duplex controller* shown in Figure 10.16. If only a relatively few transitions are required in the control task, the AP cell binary digitized universal duplex controller can produce the required control behavior using fewer AP cells than the AP cell binary digitized monolithic controller shown previously. Also, if a new sensor or actuator is added to the system, the number of additional AP cells is determined by the number of intermediate terminals times the number of new matrix input or output terminals. This may be far less than the number of additional AP cells required when a new sensor or actuator variable is added to a monolithic control matrix.

Each intermediate variable can conduct an asynchronous diverse line of behavior that has as many transitions as the intermediate variable has values (terminals). The universal duplex controller can produce as many different asynchronous diverse lines of behavior as it has intermediate variables. The universal duplex controller can produce a single unitary line of behavior through any one intermediate variable or by programming the matrix to energize just one intermediate terminal in any intermediate variable in a given transition.

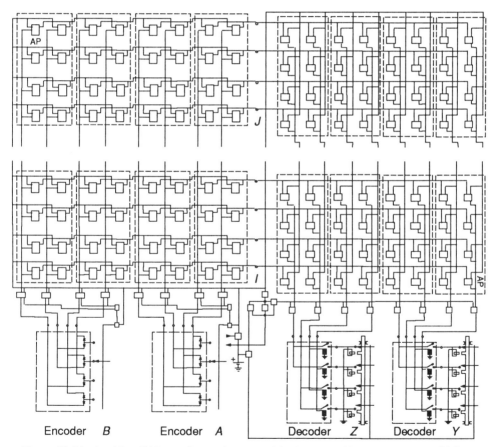

Figure 10.16. An *AP cell binary digitized universal duplex controller* can connect multiple sensor variables to multiple actuator variables in a unitary manner, or can connect any sensor variable to any actuator variable with as many asynchronous (diverse) relations as there are intermediate variables $(I, J, \ldots)$, with each relation having as many transitions as there are values of its intermediate variable.

Application of an AP cell binary digitized universal duplex controller

The direction error signal and rudder actuator in the ship direction control system shown previously may need to operate independently from the crosswind sensor and bow thruster. A *ship direction control system using an AP cell binary digitized universal duplex controller* (Fig. 10.17) can be used if a relatively few transitions are required in each separate line of behavior.

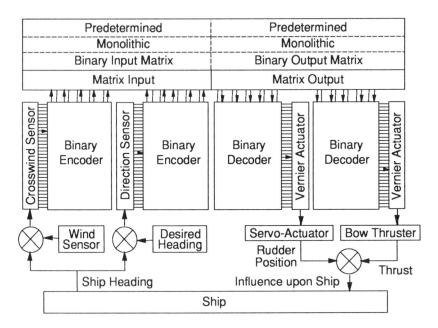

Figure 10.17. An *application of an AP cell binary digitized universal duplex controller* is a ship direction control system that requires asynchronous diverse behavior, and where few transitions are required in each diverse line of behavior relative to the variety of the sensor and actuator variables involved in this behavior.

The bow thruster can be programmed to act only in response to the crosswind sensor through a given intermediate variable by programming conductive values in all the AP cells in the input matrix connected to the direction error sensor and that intermediate variable and programming conductive values all the AP cells in the output matrix connected to the rudder and that intermediate variable. If the rudder is to respond only to different values of the direction error sensor through another intermediate variable, all the AP cells in the input matrix connected to the crosswind sensor and that intermediate variable must be programmed into a conductive state, and all the AP cells in the output matrix connected to the bow thruster and that intermediate variable must be programmed into a conductive state. If better control behavior can be achieved by using specific combinations of values of the direction error sensor and the crosswind sensor to control specific combinations of values of the rudder and bow thruster, then these unitary relations can be programmed through one intermediate variable. The AP cell binary digitized universal duplex controller is the most effective controller for most predetermined control applications that may have some asynchronous diverse lines of behavior. Moreover, it can be improved upon by using the trinary digitized systems shown later in this chapter.

Other methods of digitizing predetermined machines

Electrical encoders and decoders are not the only way of digitizing sensor and actuator variables. A position variable can be decomposed into a set of aggregate variables by a mechanical device such as a *coordinate measuring machine* or an *odometer register*. A set of aggregate position variables can be composed into a single position vector by any robotlike machine with multiple orthogonal axes such as a *motorized x, y, z table* or a *pi actuator*. These devices can be connected to a multivariable controller to obtain sensor/actuator behavior having a very high resolution, while using far fewer memory elements than an undigitized system.

Sensors as components of a vector

In some cases, the sensor and/or actuator variables intrinsically form unitary variables. For example, the sensor variables may be axis position sensors of a *coordinate measuring machine*, in which each point in a plane is specified by the x and y coordinates of a *position vector*, as shown in Figure 10.18.

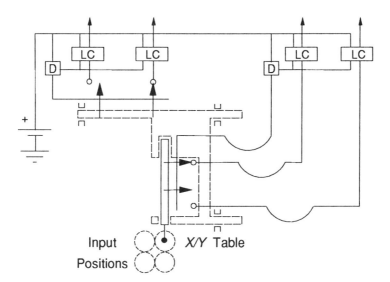

Figure 10.18. Sensors act *as components of a vector* when each combination of sensor values represents a critical input state.

When a sensor is used to measure the position of each axis of a coordinate measuring machine or a robot with a *rational coordinate system*, each set of aggregate sensor values represents a unique value of an *input space vector*. A multivariable matrix is required to establish a unitary interpretation of these sensor values. On the other hand, movements in only one axis may satisfy the needs of the task environment. In this case, the values of the other sensor axes must be ignored by programming all the memory elements connected to this axis into a conductive state.

Actuators as components of a vector

Actuators also can be connected to the axes of a *x/y* table or a robot with a *rational coordinate system* such as the orthogonal arrangement shown in Figure 10.19, where each combination of actuator values results in one unique (vector) output position.

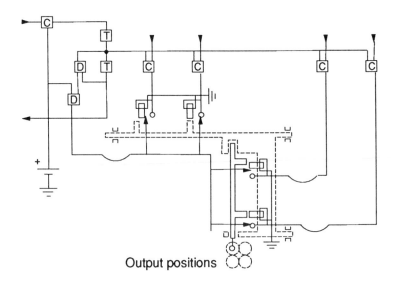

Output positions

Figure 10.19. Actuators can act as components of a vector when they are connected to the axes of a robot with a rational coordinate system.

When actuators are used to position the axes of a robot, the output matrix must produce a unique combination of output values to produce a given position of the *output space vector*. In the example given, each combination of values results in one of four possible output positions. In some cases, movement in only one axis may satisfy the needs of the task environment. In this case, the output matrix must specify the values of this critical coordinate in response to input conditions, and may have to maintain the positions of the other coordinates at a constant value. In some cases, one or more axes of a robot may be shut off and those axes locked because they are not needed or may not be working properly. A multivariable controller can be programmed to make an axis inactive by specifying the same output value for that axis in every transition.

Variety of mechanical digitized systems

A coordinate measuring machine acts intrinsically as a unitary mechanical position *encoder*, and a multiaxis robot with a rational axis system acts intrinsically as a unitary mechanical position *decoder*. The number of points (variety) that a typical coordinate measuring machine can measure or to which a robot system can go is shown in Figure 10.20.

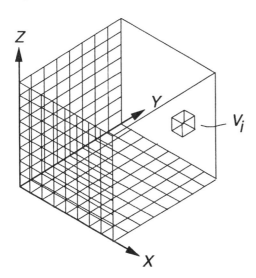

Figure 10.20. A picture of three-dimensional *variety space* shows how the number of distinct possible stopping positions on each axis determines the variety of possible output positions of a multiaxis system.

The variety of a mechanical digitizer is the same as the variety of a unitary system, which is

$$V = x^n,$$

where x is the number of distinct stopping positions of each axis, and n is the number of axes of the digitizer.

If a coordinate measuring machine or robot has three axes, and each axis has a resolution of 0.2% of its range, then $x = 500$, $n = 3$, and $V = 500^3 = 125,000,000$ possible positions. A conventional machine would require 27 two-position (binary) pneumatic actuators to generate the same variety, since $2^{27} = 125,000,000$ states. A system having one variable would require 125,000,000 positions to provide the same variety.[1]

AP cell duplex controller connected to vectors

Since aggregate variables may be considered as components of a vector variable, such as the position of an object in two- or three-dimensional space, they can be used to translate one space vector into another space vector in an AP cell multivariable duplex controller, as shown in Figure 10.21.

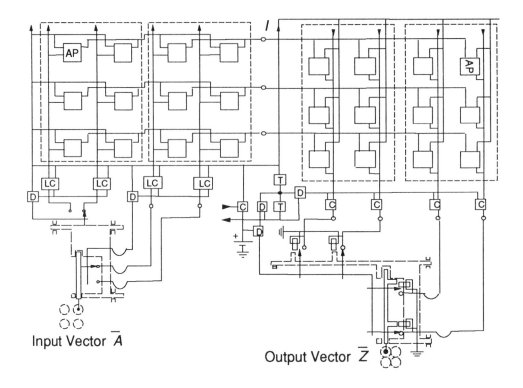

Input Vector $\overline{A}$

Output Vector $\overline{Z}$

Figure 10.21. An *AP cell multivariable duplex controller connected to vectors* can be programmed to transform one vector state into another vector state.

If a specific output vector must be produced by a specific input vector, the multivariable duplex controller can use the values of all components of the input vector to produce a specific set of values of all components of the output vector in a unitary manner. For example, the three axes of a three-dimensional position sensor (a coordinate measuring machine) can be the aggregate input variables of an AP cell multivariable duplex controller, and the three axes of a robot with a rational coordinate system can be its aggregate output variables. Each position in space measured by the coordinate measuring machine can be translated by the three-axis robot into a unique position in space.

AP cell monolithic controller connected to vector appliances

A two-axis coordinate measuring machine such as a "digitizer" and a two-axis robot such as a "plotter" also can be connected to an AP cell monolithic controller, as shown in Figure 10.22.

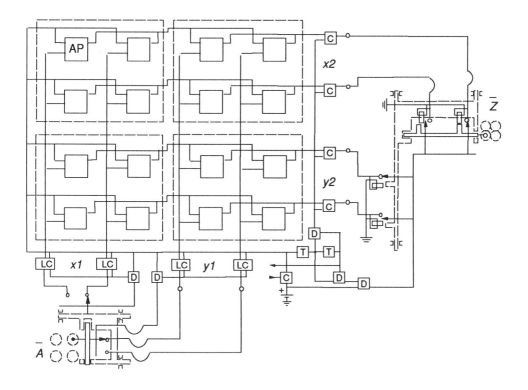

Figure 10.22. An *AP cell monolithic controller connected to vector appliances* can transform a sensor vector into an actuator vector.

A typical application for this kind of vector transformation is in a *teleoperator system* where a person is expected to position a vector appliance (actuator) in a remote location by positioning a vector appliance (sensor) at his or her location.

Stratified transformation of vectors

If the values of one component of the input vector are expected to produce the values of one component of the output vector, the controller connected to a sensor and actuator vector can establish a diverse relation between these aggregate variables. For example, the x axis of a three-dimensional coordinate measuring machine can control the x axis of a three-dimensional robot to track or follow an object moving on a conveyor in the x direction. If a controller uses just one of its three component input and output variables to produce behavior and holds the values of the other component variables constant, it establishes a *stratified transformation* of its vector variables.

Mechanical binary encoder

There are ways in which an input variable can be digitized without an electrical encoder or a coordinate measuring machine. For example, a rotary position variable can be digitized using the cams and gear train of the *mechanical binary encoder* shown in Figure 10.23.

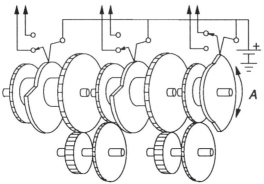

Figure 10.23. A *mechanical binary encoder* converts each input position to a set of values of its output terminals.

2:1 Reduction   2:1 Reduction

In the binary odometer register shown, each register variable (cam position) has two values (a high and low position on the cam), and the three register variables can produce $2^3 = 8$ combinations of values. Each place in a decade register would have 10 values and a 3-decade odometer register can produce $10^3 = 1000$ combinations of values (see Chapter 30).

Binary pi actuator

The output variables of a multivariable controller can be connected to a set of aggregate actuators that can produce a unique output position for each combination of values of the matrix output variables, as shown in the *binary pi actuator* in Figure 10.24.

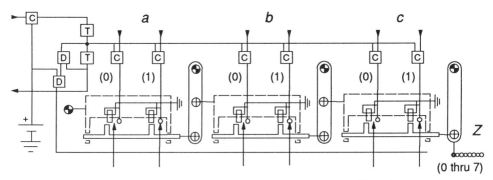

Figure 10.24. A *binary pi actuator* doubles the number of output positions it can produce with each additional aggregate actuator.

Each aggregate actuator in a pi actuator is connected in series, and each produces a displacement that is some multiple of the displacement of the aggregate actuator to which it is connected. The number of output positions that can be produced by a pi actuator is equal to the number of positions of each aggregate actuator times the number of aggregate variables. Each aggregate actuator in a binary pi actuator is designed to produce two positions and each is designed to produce twice the displacement of the actuator that drives it. This configuration causes a unique output position to be produced for each combination of values of the set of actuators. Since the number of output positions is equal to the *product* of the number of positions of each actuator, it is called a *pi actuator*. A pi actuator can have much higher resolution than a single vernier actuator and uses far fewer coils for a given resolution. For example, a pi actuator with 8 binary aggregate variables can produce $2^8 = 256$ different output positions (0.4% resolution) using only 16 coils as compared to the 256 coils in a single vernier actuator with the same resolution. (As a practical matter, the resolution of a single vernier actuator is limited to about 30 coils or a resolution of 3%, while the resolution of a pi actuator can be ten times that of a single vernier actuator.)

AP cell duplex controller with mechanical binary digitized variables

Mechanically digitized input variables can be connected to mechanically digitized output variables by an AP cell multivariable duplex controller, as shown in Figure 10.25.

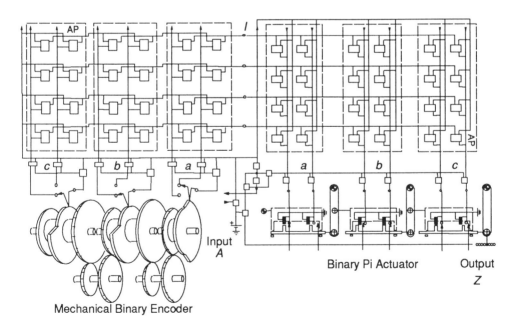

Figure 10.25. An *AP cell duplex controller with mechanically digitized binary variables* can produce high-resolution condition/action relations.

The duplex system shown can produce any one of eight possible output positions for any one of eight input positions, or it can produce a generally large or small output displacement for a generally large or small input displacement. In the configuration shown, the system can produce four of these transitions, although this number can be increased to all eight possible transitions by adding more intermediate values. If the number of aggregate input and output variables were increased to four, it could produce any one of 16 output positions for any one of 16 input positions, providing a resolution of 6.25% using only 16 AP cells per transition, as compared to the 32 AP cells per transition required in an undigitized duplex system with the same resolution.

Application of a mechanical binary digitized duplex controller

The resolution of the input and output variables of a digitized duplex controller increases dramatically as the number of aggregate variables increases without creating the need for an overwhelming number of new memory elements. For example, the resolution of a mechanical binary digitized duplex controller (Fig. 10.26) can be increased to 1 part in 256 (0.4%) by increasing the number of binary aggregate variables to 8. This binary digitized duplex controller requires only 32 AP cells per transition .

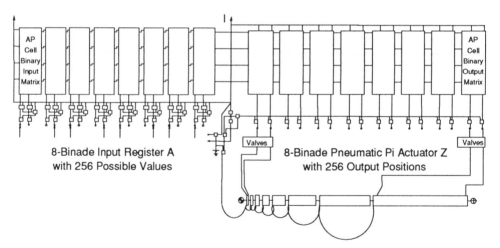

Figure 10.26. An *application of a mechanical binary digitized duplex controller* is a high-resolution linear actuator system.

Sensor switches can be replaced by a set of binary registers and latches that respond to and hold the value of the first electrical signal that arrives at each register in a given transition cycle. The set of eight binary registers forms an 8-binade input register that can produce 256 different combinations of values of its binary aggregate variables. The electrical pi actuator can be replaced by a pneumatic pi actuator, which can produce a greater force. The *pneumatic pi actuator* is made up of a set of two-position (binary) air cylinders connected in series, with each cylinder having a stroke that is twice as long as the cylinder to which it is connected. The 8-binade pneumatic pi actuator also can produce 256 different output positions.

AP cell binary monolithic controller with mechanically digitized sensor and actuator

A mechanical encoder and a pi actuator also can be connected to an AP cell monolithic controller, forming the mechanically digitized system shown in figure 10.27.

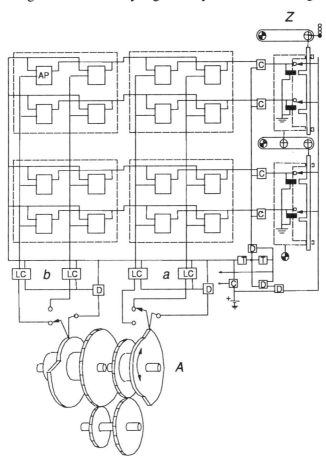

Figure 10.27. An *AP cell monolithic controller with mechanically digitized sensor and actuator* can be designed to produce many high-resolution transitions.

The number of binary variables of the mechanical encoder can be increased to eight, forming an 8-binade input register that can identify any one of 256 different input states. The number of places in the binary pi actuator can be increased to eight, forming an 8-binade pi actuator that can produce any one of 256 different output positions.

Application of an AP cell binary digitized universal controller

Two monolithic controllers can be connected by a set of intermediate variables I and J, and expanded to include two 8-binade input registers and two 8-binade pi actuators, as shown in Figure 10.28.

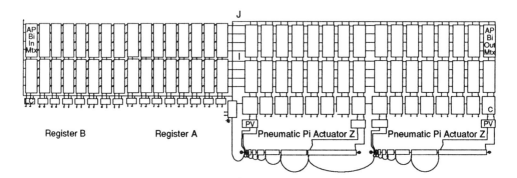

Figure 10.28. An *application of an AP cell binary digitized universal controller* is a high-resolution multiple actuator system that can produce a single vector response to a set of input variables, or produce asynchronous diverse responses to each input variable.

The input and output variety of this system viewed as a vector is $256 \times 256 = 65,536$ states, any one of which can be identified and produced by only 64 AP cells. The binary digitized systems shown can be improved upon by using the trinary digitized systems shown in the next section.

Chapter 11
Trinary Digitized Predetermined Controllers

Contrary to popular belief, a binary digitized system is not the most efficient digitizing system. A *trinary* (three-value) *digitized system* can provide a significantly higher resolution using fewer AP cells per transition than a binary digitized system. A *quadrary* (four-value) digitized system provides the same efficiency as a binary digitized system, while a *pentary* (five-value) or higher systems provide less efficiency than any of the above. Although control systems digitized with binary encoders and decoders make far more effective use of their memory cells than undigitized control units, trinary digitized controllers are even better.

Trinary digitized input matrices

A sensor variable can be decomposed into a set of trinary aggregate variables by a trinary encoder. More values of the sensor variable can be represented by trinary aggregate variables using fewer matrix input terminals than any other method of digitization.

Trinary encoder

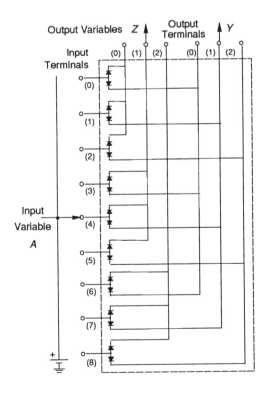

Figure 11.1. A *trinary two-variable encoder* produces an output at one of three possible output value terminals in each of its output variables when a voltage is present at an input value terminal.

The number of values in each variable in an encoder can be increased from two to three by using the method of connection shown in the *trinary two-variable encoder* shown in Figure 11.1. The number of variables (n) in a trinary encoder can be increased indefinitely. The number of input terminals in a trinary encoder is equal to 3^n. Thus, the two-variable trinary encoder shown requires 9 input terminals. A three-variable trinary encoder requires 27 input terminals and a four-variable trinary encoder requires 81 input terminals etc. The number of output terminals is equal to $3n$. Note that the two-variable trinary encoder shown has 6 output terminals and 9 input terminals, as compared to a three-variable binary encoder, which also has 6 output terminals but only 8 input terminals.

Trinary encoder truth table

The transformation between input and output variables of a trinary encoder can be shown in the *trinary encoder truth table* in Figure 11.2.

Trinary Encoder		
Input Variable	Output Variables	
A	Z	Y
0	0	0
1	1	0
2	2	0
3	0	1
4	1	1
5	2	1
6	0	2
7	1	2
8	2	2

Input Value (left label) — Output Values (right label)

Figure 11.2. A *trinary encoder truth table* shows the values of the output variables of the encoder for each value of its input variable.

The encoder can be connected internally in different ways to provide different output values for each input terminal than that shown. The code shown is based upon common number systems with most and least significant places. That is to say, when the value of a less significant place reaches its maximum value of 2, it resets to 0 and the next higher significant place increments (carries) up one value as the value of the input variable changes uniformly.

Trinary digitized absolute predetermined input matrix

A trinary encoder can be connected to a multivariable trinary input matrix, forming the *AP cell trinary digitized input matrix* shown in Figure 11.3.

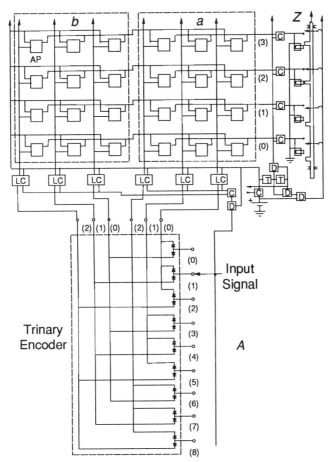

Figure 11.3. A *trinary digitized AP cell input matrix* with two trinary aggregate variables can produce more unique output states for a given number of AP cells than a binary digitized input matrix with three binary aggregate variables.

The trinary digitized sensor variable with two aggregate variables can identify nine sensor values using six AP cells. A binary digitized sensor with three aggregate variables also requires six AP cells, but can identify only eight sensor values.

Application of an AP cell trinary digitized input matrix

The number of aggregate variables can be increased indefinitely. A trinary digitized AP cell input matrix with three trinary aggregate variables can identify 27 sensor values and requires only nine AP cells per identification. Thus, a trinary digitized input matrix can be programmed to respond to only one of many different possible input values. Only as many AP cells are needed to produce this response as the number of output terminals of the digitizing encoder. This feature is very desirable in the *high-resolution temperature warning system* shown in Figure 11.4.

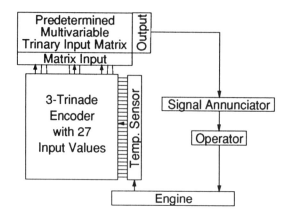

Figure 11.4. An *application of a trinary digitized AP cell input matrix* is a high-resolution temperature warning system with a trinary digitized temperature sensor that is intended to identify a relatively few of many possible values of temperature.

AP cell trinary digitized multivariable input matrix

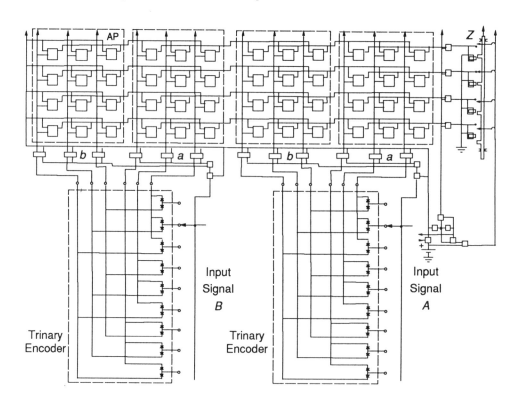

Figure 11.5. An *AP cell trinary digitized multivariable input matrix* can be programmed to produce a unique output value for any combination of digitized values of any number of input variables.

The number of trinary digitized variables also can be increased indefinitely if their AP cells are connected as shown in the *AP cell trinary digitized multivariable input matrix* in Figure 11.5. The number of different possible input values in a digitized multivariable input matrix increases dramatically as the number of aggregate variables increases. For example, the 12 AP cells in the system shown can identify any one of 81 different input states. These 12 AP cells can identify only 64 different input states in a binary digitized system with 6 aggregate variables.

Application of a trinary digitized multivariable input matrix

A trinary digitized multivariable AP cell input matrix is the preferred controller for an *absolute predetermined ship direction control system* with digitized crosswind and error direction sensors and a single rudder position output variable, as shown in Figure 11.6.

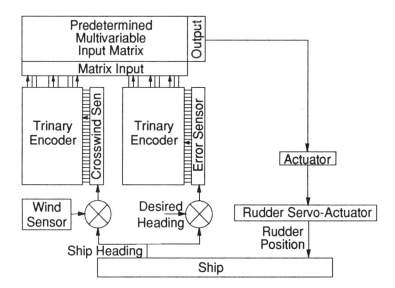

Figure 11.6. An *application of an AP cell trinary digitized multivariable input matrix* is as a ship direction control system with multiple digitized input variables and a single output variable.

The AP cell trinary digitized multivariable input matrix can be programmed to produce the output rudder position required for each combination of crosswind and direction error using sensors and an actuator with the highest resolution for a given number of AP cells per transition. For example, the two trinary digitized encoders each with 3 aggregate variables can identify any one of $3^6 = 729$ different input states using only $6 \times 3 = 18$ AP cells per identification.

Trinary digitized output matrices

The multiple output variables of an output matrix can be composed into a single actuator variables by a trinary decoder. More actuator values can be represented by aggregate output variables, using fewer output terminals than any other method of digitization.

Trinary decoder

A *trinary decoder* has three values for each of its input variables instead of the two values of a binary decoder. The trinary decoder can produce a unique value of its output variable for each combination of values of its input variables, as shown in Figure 11.7.

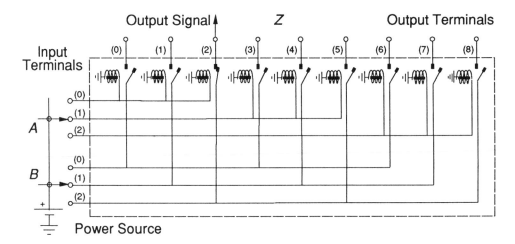

Figure 11.7. A *two-variable trinary decoder* converts each of three values of its two input variables into a value of its single output variable.

The trinary decoder can have more values of its output variable for a given number of input terminals than a binary or any other type of decoder. For example, the trinary decoder shown can produce 9 output values from its 6 input terminals as compared to the 8 output values produced by a binary decoder also with 6 input terminals.

Trinary decoder truth table

The *trinary decoder truth table* shown in Figure 11.8 is based upon the trinary decoder shown. It is hard wired according to common number systems having most and least significant places. That is to say, when the value of the less significant variable

B reaches its maximum value of 2, it resets to 0 and the next more significant variable increments up (carries) one value as the values of the output variable change in uniformly.

Trinary Decoder Truth Table		
Input Variables		Output Variable
A	B	Z
0	0	0
0	1	1
0	2	2
1	0	3
1	1	4
1	2	5
2	0	6
2	1	7
2	2	8

Input Values

Output Value

Figure 11.8. A trinary decoder truth table shows the value of the output variable produced by each combination of values of the input variables.

The number of input variables of a trinary decoder can be increased indefinitely. As the number of input variables increases, the advantage of using a trinary decoder also increases. For example, a trinary decoder with 8 input variables requires $3 \times 8 = 24$ input terminals and can produce $3^8 = 6561$ different output values, as compared to the 4096 different output values that can be produced by a binary decoder having 12 input variables with the same number of 24 input terminals.

AP cell trinary digitized output matrix

The AP cell trinary multivariable output matrix shown earlier can be connected to a trinary decoder to provide the *AP cell trinary digitized output matrix* shown in Figure 11.9. Each matrix input terminal is connected to only six AP cells. One of these AP cells in each of the two aggregate variables can be programmed to produce any one of nine possible decoder output values. The binary digitized output matrix with three aggregate variables shown earlier, also with six AP cells per transition, can produce only one of eight possible output values.

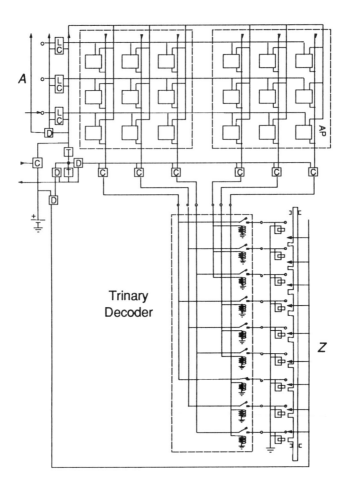

Figure 11.9. An *AP cell trinary digitized output matrix* with two trinary aggregate variables can produce more different output values than a binary digitized output matrix with the same number of AP cells.

Application of a trinary digitized AP cell output matrix

In some cases, only one of many possible output states must be produced by a given input state. For example, a *medium-resolution machine actuator* may have to move to a few precise positions. This behavior can be programmed most efficiently with the AP cell trinary digitized output matrix system shown in Figure 11.10. A trinary digitized output matrix with only 3 aggregate variables can be programmed to produce any one of 27 different output positions from any one of its matrix input terminals using only $3 \times 3 = 9$ AP cells per transition. A binary digitized system would require 5 binary aggregate variables to produce 32 output positions and 10 AP cells per transition to produce the same resolution of approximately 3%.

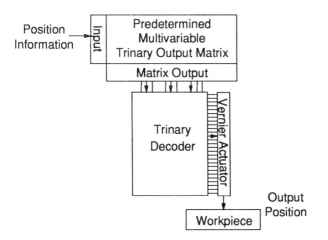

Figure 11.10. An *application of an AP cell trinary digitized output matrix* with two or more aggregate variables is a linear actuator positioning system that must produce a relatively few medium-resolution positions.

AP cell trinary digitized multivariable output matrix

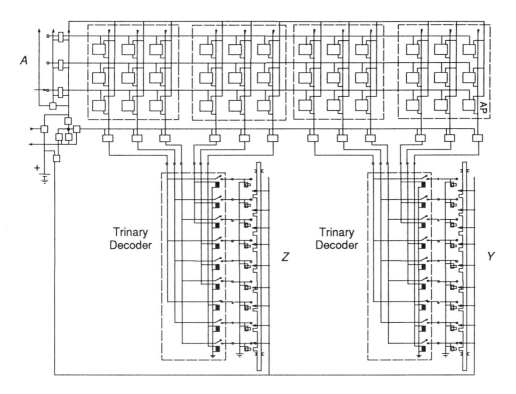

Figure 11.11. An *AP cell trinary digitized multivariable output matrix* can produce more combinations of values of two or more output variables using fewer memory cells than any other number system.

The number of aggregate output variables in an AP cell trinary digitized output matrix can be increased indefinitely, forming the *AP cell trinary digitized multivariable output matrix* shown in Figure 11.11. The AP cell trinary digitized multivariable output matrix shown can produce any one of $3^4 = 81$ different combinations of output positions using only $3 \times 4 = 12$ AP cells. A binary digitized output matrix with the same number of 12 AP cells per transition can only produce only $2^6 = 64$ different combinations of output positions.

Application of an AP cell trinary digitized multivariable output matrix

An AP cell trinary digitized multivariable output matrix can be programmed to control a medium-resolution rudder servo-actuator and a bow thruster using input values from a differential connected to a crosswind sensor and a direction error sensor, as shown in Figure 11.12.

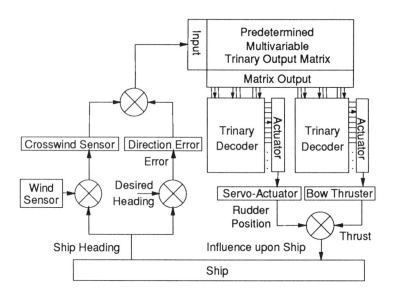

Figure 11.12. One *application of an AP cell trinary digitized multivariable output matrix* is a ship direction control system using a differential to sum the crosswind and direction error values to provide a single input variable.

After installing and testing the control system shown, it may be determined that adequate control can be achieved from the rudder and bow thruster using information from the direction error sensor alone. In this case, the crosswind sensor can be disconnected and the side member of the differential connected to it can be held constant. Also, using either the bow thruster or the rudder may provide the best direction control. Either of these control options can be achieved in the control system shown by programming either actuator to remain constant for all input conditions.

Trinary digitized systems

A trinary encoder and decoder can be used to decompose, transform, and compose a unique value of a sensor variable into a unique value of an actuator variable using fewer memory cells than any other method of digitization.

AP cell trinary digitized duplex controller

An AP cell trinary digitized input matrix can be connected to an AP cell trinary digitized output matrix, forming the *AP cell trinary digitized duplex controller* shown in Figure 11.13.

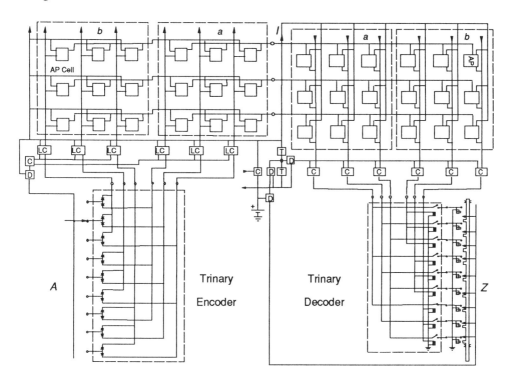

Figure 11.13. An *AP cell trinary digitized duplex controller* allows a digitized input variable to control a digitized output variable.

Using a trinary encoder with 2 aggregate variables to digitize the input variable and a trinary decoder with 2 aggregate variables to digitize the output variable, any one of 9 input values can be programmed to produce any one of 9 output values using only 12 AP cells per transition. An undigitized duplex controller would require 18 AP cells per transition to produce the same quality of machine behavior. The advantage of using a trinary digitized duplex controller increases as the variety of the sensor and actuator variables increases.

Application of a trinary digitized AP cell duplex controller

The preferred design of the predetermined ship auto-pilot discussed earlier uses the AP cell trinary digitized duplex controller, as shown in Figure 11.14.

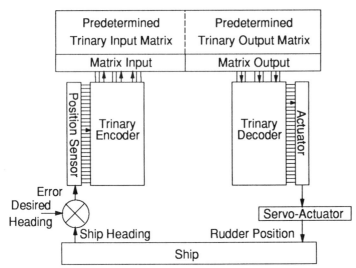

Figure 11.14. An *application of an AP cell trinary digitized duplex controller* is a medium-resolution sensor and actuator system such as a ship auto-pilot.

A trinary digitized direction error sensor with 3 aggregate variables can identify any one of 27 different input values using only $3 \times 3 = 9$ AP cells. A trinary digitized rudder servo-actuator can produce any one of 27 different output position using only 3 aggregate variables requiring $3 \times 3 = 9$ AP cells. Thus, a trinary digitized duplex controller can produce any one of 27 different output values from any one of 27 different input values using only 18 AP cells per transition. If 10 transitions are required, this digitized duplex system would require $18 \times 10 = 180$ AP cells. An undigitized square matrix would require $27 \times 27 = 729$ AP cells and an undigitized duplex controller with 10 transitions would require $2 \times 27 \times 10 = 540$ AP cells to produce the same performance.

AP cell trinary digitized multivariable duplex controller

The number of trinary digitized sensor and actuator variables in a duplex controller can be increased indefinitely, forming an *AP cell trinary digitized multivariable duplex controller* like the one shown in Figure 11.15.

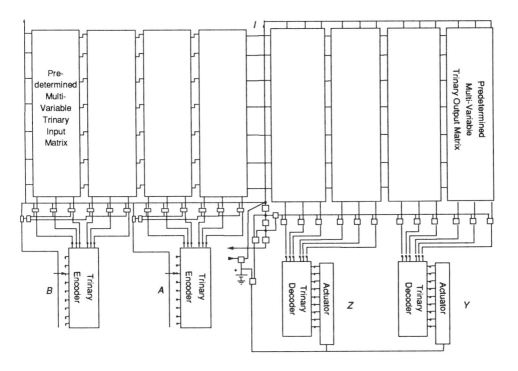

Figure 11.15. An *AP cell trinary digitized multivariable duplex controller* can be programmed to cause any combination of values of the digitized sensor variables to produce any combination of values of the digitized actuator variables.

When both input variables are digitized with two trinary aggregate variables, and both output variables are digitized with two trinary aggregate variables providing a resolution of 11% in both variables, then any one of $9 \times 9 = 81$ combinations of values of the input variables can produce any one of 81 combinations of actuator values using only 24 AP cells per transition. A coded unitary controller would require a matrix having $81 \times 81 = 6561$ AP cells. An undigitized multivariable duplex controller would require $81 + 81 = 162$ AP cells per transition.

Application of an AP cell trinary digitized multivariable duplex controller

The AP cell trinary digitized multivariable duplex controller is the most useful and desirable predetermined discontinuous responsive controller. It can be used in any application have multiple medium-resolution sensor and actuator variables, as in the *ship direction control system* shown in Figure 11.16.

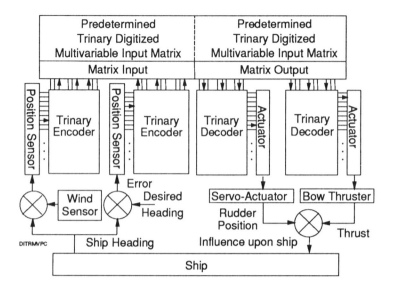

Figure 11.16. An *application for an AP cell trinary digitized multivariable duplex controller* is in a programmable multivariable ship direction control system.

The trinary digitized system can provide a higher resolution with fewer AP cells than a binary digitized system. For example, this trinary digitized system with 2 sensor and actuator variables having 3 trinary aggregate variables per sensor and actuator would require 36 AP cells per transition. This trinary system with a 4% resolution could produce any one of 729 possible combinations of output values for any one of 729 possible combinations of input values, as compared to a binary system with 36 AP cells per transition, which could produce only 512 possible combinations of input and output values.

AP cell trinary digitized monolithic controller

A trinary digitized sensor and a trinary digitized actuator also can be connected to a monolithic controller, forming the *AP cell trinary digitized monolithic controller* shown in Figure 11.17. The trinary digitized monolithic controller with each sensor and actuator variable having 2 aggregate variables requires 6 input terminals and 6 output terminals and a total of 36 AP cells. It can identify any one of 9 sensor values and produce any one of 9 actuator values. A binary digitized monolithic controller with 3 binary aggregate sensor and actuator variables also requires 6 input and 6 output terminals for a total of 36 AP cells, but can identify only 8 sensor values and produce only 8 actuator values.

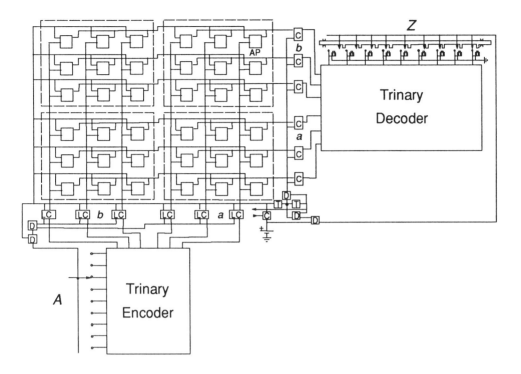

Figure 11.17. An *AP cell trinary digitized monolithic controller* can identify more sensor states and produce more actuator states than a binary digitized monolithic controller with the same number of AP Cells.

Application of an AP cell trinary digitized monolithic controller

A practical ship auto-pilot (Fig. 11.18) may require a direction error sensor and rudder actuator with resolutions of no more than 4%. This can be achieved with a trinary digitized monolithic controller that has trinary digitized sensor and actuator variables each of which has three trinary aggregate variables. The monolithic controller with trinary digitized sensor and actuator variables, each with 3 aggregate variables, requires $3 \times 3 = 9$ input terminals and 9 output terminals for a total of 81 AP cells. This digitized system can identify $3^3 = 27$ sensor values and produce 27 actuator values. This is comparable to a binary digitized system with 5 binary aggregate variables per sensor and actuator variable, which can identify 32 sensor states and produce 32 actuator states. A binary digitized system would required 2 $\times 5 = 10$ input and 10 output terminals for a total of 100 AP cells. A single scalar matrix would require $32 \times 32 = 1024$ AP cells to provide the same resolution as the binary digitized system.

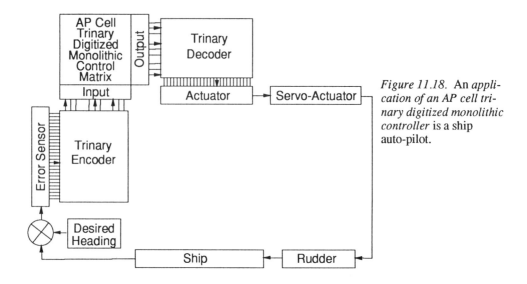

Figure 11.18. An *application of an AP cell trinary digitized monolithic controller* is a ship auto-pilot.

Application of an AP cell multivariable trinary digitized monolithic controller

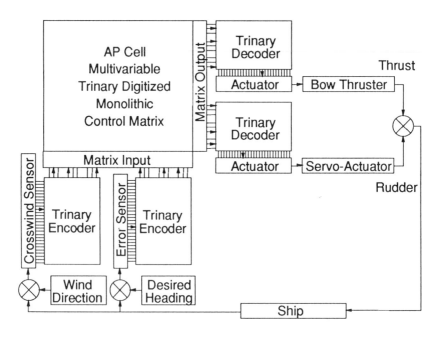

Figure 11.19. An *application of an AP cell multivariable trinary digitized monolithic controller* is a ship direction control system that can be programmed to have many unitary or asynchronous diverse lines of behavior involving multiple sensor and actuator variables.

The number of sensor and actuator variables in an AP cell trinary digitized monolithic controller can be increased indefinitely, making it possible to use an AP cell multivariable trinary digitized monolithic controller in the *ship direction control system* shown in Figure 11.19. A practical ship direction control system, with sensor and actuator resolutions of only 4%, can be achieved in a trinary digitized system using only 3 trinary aggregate variables that can identify and produce 27 states for each sensor and actuator variable. The monolithic system with 2 sensor and 2 actuator variables requires $18 \times 18 = 324$ AP cells, and can identify $27 \times 27 = 729$ different combinations of sensor and actuator states. The binary digitized monolithic system with 2 sensors and 2 actuators requires $20 \times 20 = 400$ AP cells, and can identify $32 \times 32 = 1032$ different combinations of sensor and actuator states. These additional AP cells increase the resolution from 3.7% in the trinary system to only 3.1% in the binary system.

AP cell trinary digitized universal duplex controller

Two AP cell monolithic trinary digitized controllers can be connected by a set of intermediate variables $(I, J, \ldots)$, forming the *AP cell trinary digitized universal duplex controller* shown in Figure 11.20.

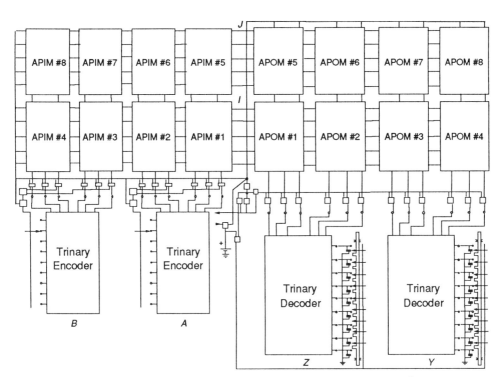

Figure 11.20. An *AP cell trinary digitized universal duplex controller* can produce as many asynchronous diverse lines of behavior between trinary digitized sensor and actuator variables as it has intermediate variables.

One or more input variable can be connected to one or more output variable through a given intermediate variable to produce an asynchronous diverse line of behavior, and other asynchronous diverse lines if behavior can be programmed to act through other intermediate variables. These intermediate variables also can produce a unitary line of behavior. The trinary digitized universal duplex controller is the most generally useful absolute predetermined digitized control system.

Application of an AP cell trinary digitized universal controller

An AP cell trinary digitized universal controller can be used in the *ship direction control system* shown in Figure 11.21 if unitary or asynchronous diverse behavior may be required.

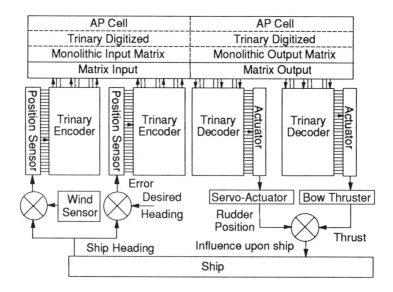

Figure 11.21. An *application of an AP cell trinary digitized universal duplex controller* is a ship direction control system that can be programmed to connect the direction error and cross-wind sensors to the rudder and bow thruster in a unitary manner, or connect just one sensor to just one actuator through one intermediate variable in an asynchronous diverse manner.

The ship control system using trinary digitized variables can produce a greater variety of behavior using fewer AP cells than a binary digitized system. The resolution of an actuator variable using a decoder and a vernier actuator is limited to approximately 3% of its range. This may be adequate for a ship direction control system, but may be inadequate for precision position control systems that may require the mechanically digitized devices shown in the next section.

Mechanically digitized trinary controllers

Just as the trinary digitized electrical encoders and decoders can provide higher-resolution sensors and actuator than binary digitized systems with the same number of memory cells, so can trinary digitized mechanical devices provide higher-variety behavior than binary digitized systems with the same number of memory cells.

Mechanically digitized trinary sensor

It may be impractical to make a sensor switch with more than a few dozen contacts. However, a single position variable can be mechanically encoded using far fewer contacts than a single multiposition switch by the *trinary mechanical encoder* shown in Figure 11.22. (The binary mechanical encoder shown earlier also reduces the number of contracts needed to represent a sensor variable, but requires more contacts to produce a given variety than the trinary mechanical encoder shown here).

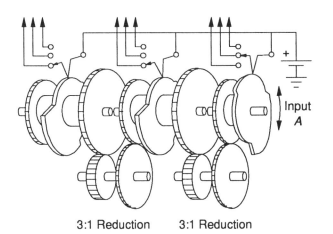

3:1 Reduction   3:1 Reduction

Figure 11.22. A trinary mechanical encoder is made up of two or more dials that are geared together through a gear ratio of 1:3, with each dial having three switch position values.

The three trinary aggregate variables, each made up of three switch positions, can produce a unique output for any one of 27 different input positions using only 9 contacts. A binary mechanical encoder with four binary aggregate variables uses 8 contacts but can resolve only $2^4 = 16$ positions. A trinary mechanical encoder with five aggregate variables requires $3 \times 5 = 15$ contacts and can resolve $3^5 = 243$ different input positions, which is adequate for most position-sensing applications. A binary encoder with 8 aggregate variables requires 16 contacts and can resolve $2^8 = 256$ input positions. A single sensor switch would require 256 contacts to provide the same resolution.

Mechanically digitized trinary actuator

A vernier actuator may not be practical in applications requiring a variety more than a few dozen values. In applications requiring a resolution higher than 3%, a vernier actuator must be broken into a set of aggregate variables, as in the binary pi actuator shown earlier. However, the *trinary pi actuator* (Fig. 11.23) makes better use of the AP cells in a control matrix and its components, such as its stator coils, than does a binary pi actuator.

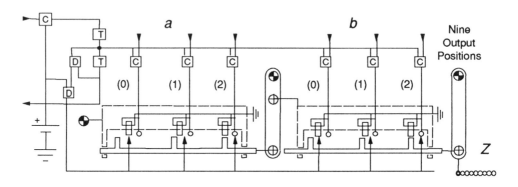

Figure 11.23. A *trinary pi actuator* is the most effective digitized actuator in terms of the number of different output positions it can produce compared to the number of actuator coils required to produce these positions.

The two trinary aggregate variables with only six coils can produce any one of nine different output positions. A trinary pi actuator with five aggregate variables, using only $3 \times 5 = 15$ coils, can produce $3^5 = 243$ different output positions, which is adequate for most precision motion control applications.

AP cell mechanically digitized trinary duplex control system

The mechanical trinary encoder and trinary pi actuator can be connected to a multivariable duplex controller made up of trinary aggregate variables, forming the *AP cell mechanically digitized trinary duplex controller* shown in Figure 11.24. The trinary system shown with two sensor and two actuator aggregate variables can produce any one of nine different output positions from any one of nine different input positions, using the same number of components and AP cells as a binary system that can produce only eight different output positions from eight different input positions. The resolution of the sensor and actuator variables can be increased greatly by increasing the number of trinary aggregate variables.

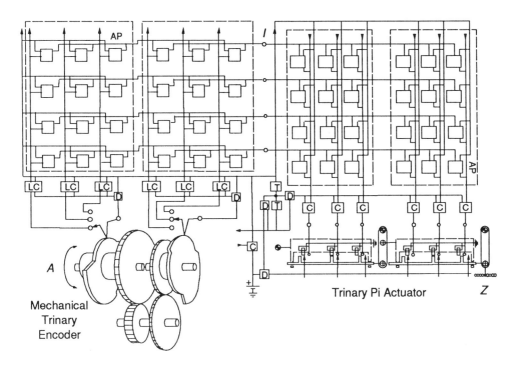

Figure 11.24. An *AP cell mechanically digitized trinary duplex controller* can create high-resolution behavior with fewer components, including fewer AP cells, than any other system.

AP cell trinary digitized high-resolution duplex control system

The number of trinary aggregate variables in the digitized duplex controller can be increased, as shown in the *AP cell trinary digitized high-resolution duplex control system* in Figure 11.25. Each trinary aggregate sensor variable can be operated by a voltage signal to any one of its three terminals. The set of five trinary sensor variables forms a *five-trinade input register*. Each trinary aggregate actuator variable can operate a *pilot valve* that causes a *three-position pneumatic cylinder* to take any one of three positions. If the cylinders are connected in series and the range of each cylinder is three times the range of the one preceding it, the five cylinders form a five-trinade pneumatic pi actuator that can produce a large force at any one of 243 possible positions. A binary digitized system with 8 binary aggregate variables can produce 256 possible positions, with a resolution of approximately the same 4%, but requires 32 AP cells per transition. The resolution of the trinary system requires $243 \times 243 = 59,049$ AP cells in an undigitized square matrix or $243 + 243 = 486$ AP cells per transition in an undigitized duplex controller.

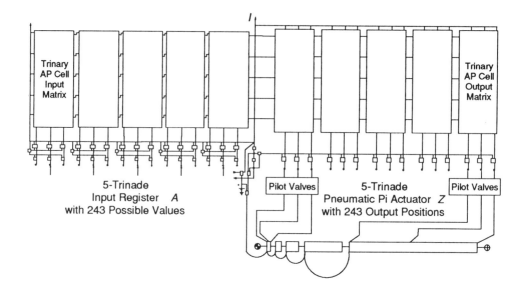

Figure 11.25. An *AP cell trinary digitized high-resolution duplex controller* with five sensor and actuator aggregate variables can produce any one of 243 possible output states for any one of 243 possible input states for a resolution of .04%, and uses only 30 AP cells per selection.

The number of sensor or actuator aggregate variables can be increased indefinitely, providing a great increase in the resolution with only a small increase in the number of AP cells. For example, a system with 8 trinary aggregate sensor variables can provide a variety of $3^8 = 6561$ input values using only $3 \times 8 = 24$ AP cells per transition. A binary digitized system with the same number of AP cells per transition requires 12 binary aggregate variables, but can produce only $2^{12} = 4096$ input values, which are 38% less values than the trinary system in this example.

The improvement in the trinary digitized system over the binary digitized increases as the number of aggregate variables increases. For example, a trinary digitized duplex system with a sensor and actuator each with 10 aggregate variables requires 60 AP cells per transition, and can produce any one of $3^{10} = 59,049$ possible actuator output values for any one of 59,049 sensor values. In contrast, a binary digitized system with 15 aggregate variables also uses 60 AP cells per transition, but can produce any one of only $2^{15} = 32,768$ possible combinations of actuator values for any one of only 32,768 possible sensor values, which are 44.5% less values than the trinary system.

AP cell mechanically digitized trinary monolithic controllers

The AP cell trinary monolithic control matrix shown earlier can be connected to mechanically digitized sensor and actuator variables, forming the *AP cell mechanically digitized trinary monolithic controller* shown in Figure 11.26.

Trinary Pi Actuator *Z*

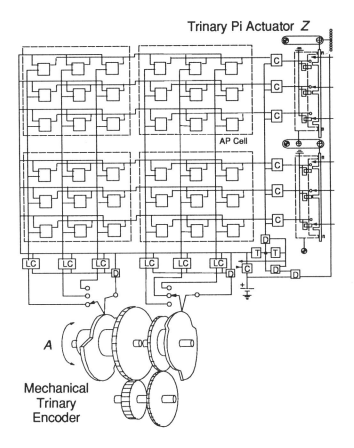

A

Mechanical
Trinary
Encoder

Figure 11.26. An *AP cell mechanically digitized trinary monolithic controller* can produce a greater variety of behavior for a given number of memory cells than a mechanically digitized binary monolithic controller.

The trinary digitized mechanical encoder can identify any one of nine different input states using only six terminals. The trinary pi actuator can produce any one of nine different output states using only six terminals. Thus, the monolithic controller can connect any one of nine input states to nine output states using only 6 × 6 = 36 AP cells. A binary digitized monolithic controller with 36 AP cells can connect only eight input states to eight output states. A scalar matrix would require 9 × 9 = 81 AP cells to connect any one of these nine input states to nine output states, whereas the trinary digitized duplex controller can connect any one of nine input states to any one of nine output states using only 6 + 6 = 12 AP cells per transition.

Trinary digitized universal controller as the basic AP cell control unit

The multiple input and output variables of an AP cell universal controller (Fig. 11.27) can be connected to any kind of sensors and actuators, digitized by encoders and decoders, vectored sensors and actuators, and other mechanical devices, and be programmed to produce unitary behavior or as many asynchronous diverse lines of behavior as it has intermediate variables.

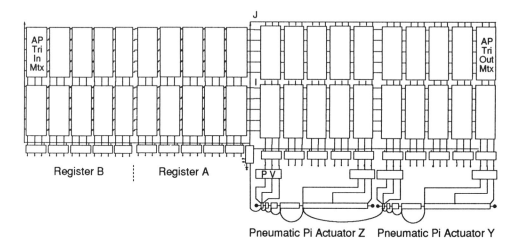

Figure 11.27. An *AP cell trinary digitized high-resolution universal controller* is the basic AP cell control unit that can be used in any predetermined control task.

For example, a universal controller with two sensor and actuator variables, each digitized with five trinary aggregate variables, providing a resolution of 0.4%, can identify any one of 59,049 possible combinations of values of both sensor variables and produce any one of 59,049 combinations of values of both actuator variables, using only 60 AP cells per selection in a unitary manner, or it can connect any one of 243 possible values of either sensor variable to any one of 243 possible values of the actuator variable through either intermediate variable I or J in an asynchronous diverse manner. A binary digitized duplex system having two sensor and actuator variables with the same 60 AP cells per transition has a resolution of only 0.8% and can identify and produce only $2^{15} = 32,768$ different possible states. A coded unitary controller requires a matrix having $59,049 \times 59,049 = 3.5 \times 10^9$ AP cells, and an undigitized multivariable duplex controller requires $486 + 486 = 972$ AP cells per transition.

Thus, the AP cell trinary universal controller can be used by itself as a *standalone controller* in almost any kind of control application that can be programmed *a priori*. For example, the system shown can be used as a high-resolution ship direction control system or as an *industrial robot controller* for vectored actuators that can be pre-programmed in a unitary or asynchronous diverse manner to produce a relatively limited amount of high-resolution control behavior.

Chapter 12
Conditional Predetermined Machines

Not all programming decisions can be absolute (all-or-nothing) events. In some cases, decisions may have to evolve gradually and many options must be compared to make a decision with a high confidence level. This requires a memory element that can be changed by degrees over time so that the results of many experiences can be averaged out and allows the results stored in each memory element to be compared. This can be done by the *conditional predetermined memory matrices* shown in this chapter, in contrast to the absolute predetermined memory matrices shown earlier.

Conditional predetermined memory cells

The absolute memory cell selects the output terminal to which it is connected based upon its stored memory value, without regard for the values of other memory cells in the matrix. However, the stored memory value of a *conditional memory cell* can be placed somewhere along a continuum of values, and all the conditional memory cells connected to a given input terminal can be interrogated to find the conditional memory cell with the highest sensitivity *relative* to the values of the other memory cells. This memory cell can then be used to produce behavior.

Current divider

In an absolute memory cell, current in the search relay coil is changed from all to nothing by changing the position of its toggle switch. However, in a conditional memory, the current in a search relay coil can be changed continuously to some value between all and nothing by changing the position of the *wiper arm* on the *variable resistor* shown in the *current divider* in Figure 12.1.

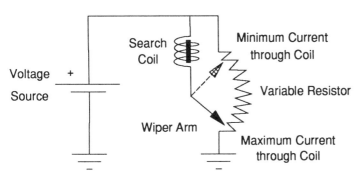

Figure 12.1. A current divider causes the current in the search relay coil to change according to the position of the wiper arm on the variable resistor.

With the wiper arm in the minimum current position, the voltage at one end of the *search coil* is the same as the voltage at the other end. Thus, no current flows through the search coil with the wiper arm in this position. With the wiper arm in the maximum current position, one end of the search coil is essentially connected to the electrical ground while the other end is at the full potential of the voltage source. Thus, the maximum current flows through the search coil with the wiper arm in this position. The current in the search coil varies according to the position of the wiper arm between its minimum and maximum position.

Relay with adjustable sensitivity

The search coil in the current divider shown can be used to close a set of *relay contacts* in the *adjustable sensitivity relay* shown in Figure 12.2.

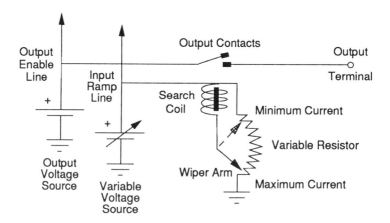

Figure 12.2. An *adjustable sensitivity relay* (ASR) can be used as a conditional memory element.

The current from the variable voltage source is split between the search coil and the variable resistor according to the position of the wiper arm. With the wiper arm in the minimum current position, very little current flows through the search coil at a given voltage. With the wiper arm in the maximum current position, nearly all the current flows through the search coil. If a voltage is present on the output enable line and a ramp voltage is applied to the input sense terminal of a sufficient voltage, the output contacts will close and a voltage will be produced at the output terminal. The voltage of the ramp required to produce an output signal is thus determined by the position of the wiper arm and the *intrinsic sensitivity* of the relay. These two factors determine the *actual sensitivity* of the relay. The actual sensitivity of an ASR can be used as the stored value of a conditional memory element by connecting a set of ASRs to a single ramped voltage source and determining which ASR is the first to produce an output signal.

Changing an adjustable sensitivity relay

The sensitivity of the adjustable sensitivity relay can be changed by moving the wiper arm of the current divider. This process is summarized in Table 12.1

Table 12.1. Changing an adjustable sensitivity relay

---

1. The control current flowing from a ramp voltage source is split between the search coil and the variable resistor. Depending upon the position of the wiper arm, more or less current will flow through the relay coil and the variable resistor.
2. If the wiper arm is in the minimum position, little or no current will flow through the relay coil and all the current will flow through the variable resistor. This corresponds to minimum sensitivity, since a large control current may not produce enough magnetic flux to close the contacts.
3. If the wiper arm is in the maximum position, all the current will flow through the search coil, and no current will flow through the variable resistor. With the wiper arm in this position, a very small current from the control terminal may be sufficient to cause the relay contact to close. This condition corresponds to maximum sensitivity.

Thus, the *sensitivity of an adjustable sensitivity relay* is dependent upon the position of the wiper arm and can be used as the value of a conditional memory element.

Output latch circuit

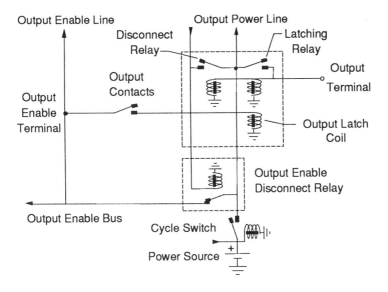

Figure 12.3. An *output latch circuit* energizes and holds a voltage at one output terminal and sends a voltage to an output enable disconnect that shuts off the output enable voltage to the memory elements in the matrix.

According to the rules of the connection matrix, only one memory element on a given input terminal in a single matrix can be set in the conductive state. However, all the conditional memories on a given input terminal may have some propensity to conduct. Therefore, once one conditional memory produces an output, the rest of the memories in the matrix must not be allowed to produce an output. This is done by the *output latch circuit* shown in Figure 12.3. Once the output contracts close in one conditional memory element in a matrix, current flows through its *output latch coil*. This closes a latching relay, which causes current from the output power line to flow through the latching coil and the coil of a *disconnect relay*. This holds the *latching relay* and disconnect relay closed and causes a voltage to appear at the output terminal and at the coil of the *output enable disconnect relay* for the remainder of the transition cycle. Thus, the voltage to the coil of the output enable disconnect relay holds the relay contacts open for remainder of the transition cycle.

Voltage ramp generator

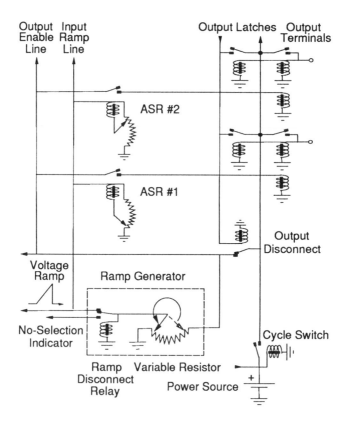

Figure 12.4. A *ramp generator* can be used to select the conditional memory with the highest sensitivity within a set of conditional memories.

The search coils of a set of conditional memory elements must be connected to a single *voltage ramp generator* (Fig. 12.4) that provides a variable voltage source. When the voltage is increased according to some *ramp curve*, the adjustable sensitivity relay having its wiper arm nearest the maximum sensitivity position will be the first to produce an output signal.

At the start of a transition cycle, the wiper arm of the ramp generator is caused to rotate by some means such as an electric motor. This causes the voltage to increase at the search coils of the conditional memories connected to the ramp generator. When the contacts of one conditional memory close, a signal appears at its output latch coil. This causes an output latch relay to close, which energizes and maintains the voltage at the output enable disconnect. This in turn prohibits a voltage from appearing at any other output latch and energizes the output terminal of that latch according to the operation of the output latch circuit. The ramp wiper arm can continue to rotate to the maximum voltage position where it activates a *ramp disconnect relay* that terminates the ramp signal for the rest of the transition cycle. The ramp wiper arm must be reset to the minimum voltage position before the next transition cycle begins. This can be done by allowing the ramp wiper arm to continue to rotate passed the maximum position around to the minimum position.

Conditional predetermined memory matrix

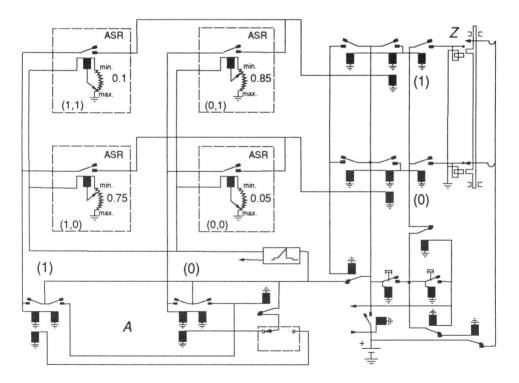

Figure 12.5. A *conditional predetermined control matrix* can be programmed by adjusting the sensitivities of its adjustable sensitivity relays.

A ramp generator, a set of output latches, and an output enable disconnect can be added to the absolute predetermined control matrix shown earlier. Also, its absolute predetermined memory cells (AP cells) can be replaced with memory elements made up of adjustable sensitivity relays, forming the *conditional predetermined control matrix* shown in Figure 12.5. Instead of having memory elements containing all-or-nothing values, the memory elements in a conditional memory matrix have values ranging from zero to one and the selected output is determined by the ASR with the highest sensitivity *relative* to the other ASRs.

Operation of a conditional memory matrix

A conditional predetermined matrix operates according to the sequence of steps listed in Table 12.2.

Table 12.2. Operation of a conditional predetermined matrix

---

1. An outside timing source closes the cycle switch at the beginning of a transition cycle. This causes a voltage to appear at the sensor switch, ramp generator, and output power line.
2. The position of the armature of the sensor switch energizes one input terminal and causes its input latch to close. This disconnects the other input latches and holds a voltage on the selected input terminal.
3. The ramp generator initiates a voltage increase that is conducted to the search coils of the conditional memory elements in the matrix.
4. The output contacts close at some voltage level in the memory element with the highest sensitivity in the selected input terminal. This energizes the output latch and the actuator coil connected to the selecting memory element.
5. The output latch also sends a voltage to the output disconnect coil, which opens the output disconnect relay for the remainder of the transition cycle. This prohibits any other memory element from producing an output for the remainder of the cycle.
6. The outside timing source opens the cycle switch at the end of the transition cycle, allowing the memory elements and circuit driver switches to return to their normal positions.

Thus, the most sensitive conditional memory element *relative* to the other memory elements connected to the same input terminal selects the one and only output allowed in a given transition cycle.

Ramp curve

The ramp curve used to interrogate a set of conditional switches must start at a voltage level just below the voltage required to close the contacts of the most sensitive relay in the set, and must rise high enough to close the contacts of the least sensitive relay in the set. In general, the *ramp curve* can be the linear ramp shown in Figure 12.6.

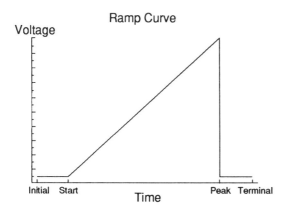

Ramp Curve

Voltage

Initial   Start   Peak   Terminal

Time

Figure 12.6. A *ramp curve* can be a linear saw tooth curve operating over a period short enough to cause at least one ASR to close before the actuator delay relay times out, and long enough to allow this ASR to disconnect the output enable line before a second ASR is closed.

The rate of rise of the ramp curve has to be determined by the range of sensitivities of the adjustable sensitivity relays and their response times. In general, the conditional matrix will work best when the response times of the ASRs and the output enable disconnects are very short.

Sensitivity indicator

If a very high ramp voltage is obtained in a given transition cycle, it indicates that none of the conditional memories connected to the active input terminal have a high sensitivity value. So the maximum voltage that the ramp obtains in a given transition cycle can be used to *indicate the sensitivity* of the selecting conditional memory element and thus the degree on confidence the programmer has in that selection.

No-selection indicator

A situation can occur in an absolute predetermined memory in which all of the AP cells connected to a given input terminal are set in the nonconductive state. Thus, no output is produce in a given transition cycle when that input terminal is made active. A similar situation may exist in a conditional predetermined memory matrix in which all of the conditional memories connected to a given input terminal may be set on the minimum value such that no memory unit can produce an output even at the maximum voltage level produced by the ramp generator. Thus, if the ramp generator reaches the value that switches the contacts of the *ramp disconnect relay*, it indicates that none of the conditional memory cells has selected an output in that transition cycle.

Nearly identical conditional memories

Another condition can occur in which two or more conditional memories connected to a given input terminal may be set to nearly the same sensitivity value. If the *output enable disconnect is made to act fast enough, and the *slope of the ramp curve* is not too steep, two or more output terminals may be energized in a given transition cycle only when there are very small differences in sensitivities between conditional memory cells. Even when two output device coils of a vernier actuator are energized simultaneously, the actuator armature is likely to assume the position of the coil closest to its last position due to the *negative inverse square force* produced by the coils upon the armature. This feature is called the *dominance of small differences*. If more than two coils are energized at exactly the same time, the armature will assume a position determined by the location of the greatest concentration of energized coils.

Programming a conditional predetermined memory matrix

A conditional predetermined matrix can be programmed by making *incremental changes* in the positions of the wiper arms in repeated runs through the desired line of behavior. After enough runs have been made, the conditional control matrix may contain the set of sensitivity values needed to produce the desired line of behavior. The advantage of using a conditional predetermined control matrix is that the line of behavior may not be changed drastically by one error in programming. The *programming error* would have to be repeated several times in most cases to change an established line of behavior.

Information stored in a conditional memory element

The adjustable sensitivity relay can store more *information* than the single value of an absolute memory because the wiper arm can be placed in any intermediate position representing some value between "0" and "1." The information stored in this *conditional memory* can be read in the following three ways:

1. As the voltage required to close the output contracts of a conditional memory.
2. As the time it takes for the contacts of a conditional memory to close after the start of a transition cycle, assuming a constant voltage ramp.
3. As the conditional memory element within the set of conditional memory elements with the highest sensitivity (the first to produce an output).

The number of values that can be stored in a conditional memory element is determined by the *resolution of the variable resistor*, the repeatability of the search relay mechanism, and the ability to measure or control the voltage ramp. Since many values can be stored in a conditional memory element, it is more of an *analog memory* device than a *digital memory device*. Because a conditional memory can store more values, it is a higher *variety* device than an absolute memory and can be used to store much more information.

Changing sensitivity of a conditional memory element

The conductive state for a given input/output relation is stored in the sensitivity level of the conditional memory element that represents the input/output relation. The conductive state of a conditional memory element is changed in a manner similar to that of an absolute memory element in that the wiper arm is moved in one direction to increase its propensity to conduct while the wiper arm is moved in the other direction to decrease its propensity to conduct. However, the propensity of the conditional memory must keep increasing (or decreasing) with repeated movements of the wiper arm while the absolute memory reaches a maximum (or minimum) with one flip of the toggle switch.

Logarithmic subtracting mechanism

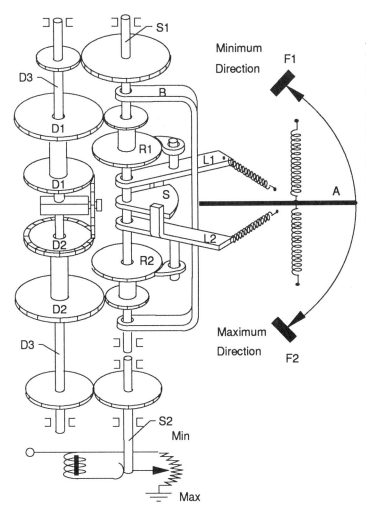

Figure 12.7. A *manually operated logarithmic subtracting mechanism* uses two levers that measure the position of the wiper arm from maximum and minimum points, and a differential that subtracts some proportion of the position of a lever from the position of the wiper arm whenever the bail is moved to the maximum or minimum position.

Obviously, the conductive state of a conditional memory cannot keep getting more or less sensitive forever. A solution to this requirement is to move the wiper arm some fraction of the distance remaining toward its maximum or minimum position. This can be done with the *logarithmic subtracting mechanism* (LSM) shown in Figure 12.7. The position of each lever L1 and L2 is determined by the position of the wiper arm through a moveable stop S mounted on the stop shaft S1. If the bail B is moved in the minimum direction by the programmer, the bail B contacts the minimum lever L2 at the position determined by the moveable stop S and pushes the minimum lever L2 to the minimum position F1. This causes a movement of the minimum ratchet R1 equal to the distance of the moveable stop from the minimum position. This motion is transmitted to the minimum subtracting member D1 of the differential, causing the output shaft of the differential D3 to turn in the same direction as D1, but through one half the distance moved by D1. This causes the wiper arm to move toward the minimum position through a rotation determined by the gear ratios between the ratchet and the differential and the gear ratio between the differential output shaft D3 and the wiper arm shaft S2. The movement of the differential shaft D3 in the minimum direction also causes the stop S to move toward the minimum position through a rotation determined by the gear ratio between the output of the differential shaft D3 and the stop shaft S1.

A motion of the bail toward the maximum or minimum position is produced by the programmer in a manner similar to flipping the toggle switch to the open or closed position in the absolute memory cell. Yet, the movement in the wiper arm toward the maximum position or minimum position must be in proportion to its original distance to the maximum or minimum position. Thus, the wiper arm is moved toward the maximum or minimum position according to whether the bail is moved toward the maximum (write maximum) or minimum (write minimum) direction, and is moved by an amount proportional to the position of the wiper from each position each time the programmer attempts to change the conditional memory. The operation of the logarithmic subtracting mechanism is similar to the operation of a *mechanical lagged demand register*[51].

Logarithmic response curve

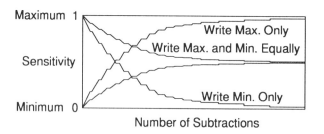

Figure 12.8. The *logarithmic response curves* of a conditional memory shows how its sensitivity must asymptotically approach its maximum or minimum value.

The action of the logarithmic subtracting mechanism produces the *logarithmic response curve* shown in Figure 12.8. If more write minimum movements occur than write maximum movements, the sensitivity of the conditional memory will reach a value closer to the minimum sensitivity. If more write maximum movements occur than write minimum movements, the sensitivity of the conditional memory will reach a position closer to the maximum sensitivity. If an equal number of write maximum and write minimum movements occur for a sufficient number of times, the conditional memory will approach the mid position.

Response curves as the sum of a geometrical progression

A *geometrical progression* is made up of the series of terms in Equation 12.1:

$$a; \ ar; \ ar^2; \ ar^3; \ \dots \ \ \ \ \ (12.1)$$

Each term is obtained by multiplying the preceding term by a constant called the *constant ratio r* in the same way that each new position of the wiper arm is determined by the wiper arm position that precedes each subtraction.

In a series of write minimum subtractions from the maximum sensitivity of 1, the differential of the logarithmic subtraction mechanism must reposition the wiper arm and its moveable stop S to a value equal to the sum of each subtraction. If the position of the moveable stop is reduced by 10% of its value in each subtraction, then $a = 1/10$; so the gear ratio between ratchet wheel and the moveable stop through the differential must be 1/10.

Since each subtraction gets progressively smaller, then $r < 1$, and the sum of these subtractions (S) approaches the Equation 12.2 as a limit, as the number of subtractions increases:

$$S = a/(1 - r). \ \ \ \ \ (12.2)$$

Since the moveable stop must arrive at its zero position after many subtractions, then, the sum of the subtractions S must be equal to 1. Therefore, $r = 9/10$.

Thus, the subtractions of the moveable stop consist of the series:

$$0.1; \ 0.09; \ 0.081; \ 0.073; \ \dots \ ; \ 0,$$

and the position of the moveable stop and wiper arm follows the series:

$$1; \ 0.9; \ 0.81; \ 0.729; \ \dots \ ; \ 0.$$

The wiper arm may be connected to the differential output shaft, the moveable stop shaft, or a separate shaft, depending upon the range of the variable resistor. Using a value of $a = 1/10$, and $r = 9/10$, the wiper arm will travel about 1/2 of the distance to zero in 6.6 subtractions, 1/4 of the distance to zero in 13.3 subtractions, and 1/8 of the distance to zero in 19.7 subtractions according to Equation 12.3:

$$S = a(1 - r^n)/(1 - r) = 1 - r^n, \ \ \ \ \ (12.3)$$

where n is the number of subtractions.

Thus, the response curves of the wiper arm are established by the values of a and r, which determine how many programming events are required to influence the sensitivity of the conditional memory significantly. The sensitivity (*S*) of a conditional memory cell represents the *geometric average* of the number of write maximum events versus the number of write minimum events.

Conditional predetermined control matrix

In an absolute predetermined control matrix, a memory cell may be changed from conducting to nonconducting leaving no trace of its former position. In a conditional predetermined control matrix, a conditional memory cell may be changed to a sensitivity level just below another cell on that input terminal, causing the second memory cell to select its action. However, if the output of the original cell proves to be more desirable than the output of the second memory cell, the *latent information* in the original memory cell can be readily identified and set to a higher sensitivity level. In fact, the sensitivity values of the set of conditional memory cells on a given input terminal can be adjusted to match the *relative desirability* of each output for that input value. Except for the ramp generator and the output latches, a conditional predetermined memory matrix works the same as an absolute predetermined memory matrix, and can be used in the same applications.

Conditional predetermined memory cell

The adjustable sensitivity relay and the logarithmic subtracting mechanism can be combined to form a *conditional predetermined memory cell* (CPM cell) that stores a *continuous value* represented by the position of the wiper arm for multiple programming efforts, as shown in Figure 12.9.

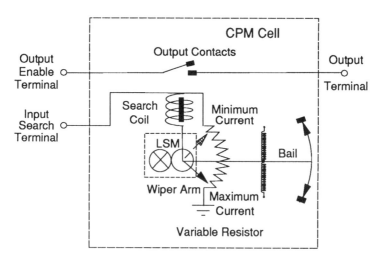

Figure 12.9. A *conditional predetermined memory cell (CPM cell)* uses an adjustable sensitivity relay (ASR) and a manual logarithmic subtracting mechanism to store a sensitivity value representing an averaged programmed sensitivity value.

The CPM cell is programmed by manually operating the logarithmic subtraction mechanism. A movement of the bail by the programmer in the write maximum direction increases the sensitivity of the memory cell. A movement of the bail in the write minimum direction decreases the sensitivity of the memory cell.

The value of a conditional predetermined memory cell is "read" by determining the threshold voltage required to close the output contacts of the search coil relay connected in parallel with the variable resistor in the adjustable sensitivity relay (ASR) in each memory cell. The higher the voltage required to close the contacts, the less sensitive the relay. The lower the voltage required to close the contacts, the more sensitive the relay. A conditional memory cell is an active memory device since a voltage must be present on the output enable terminal to produce an output voltage. When a ramp voltage is applied to the search coils of a set of conditional memory cells, the first cell to produce an output signal is identified as the most sensitive cell.

CPM cell control matrix

The absolute predetermined memory cells (AP cells) in the absolute predetermined control matrix shown earlier can be replaced with conditional predetermined memory cells, forming the *conditional predetermined control matrix* shown in Figure 12.10.

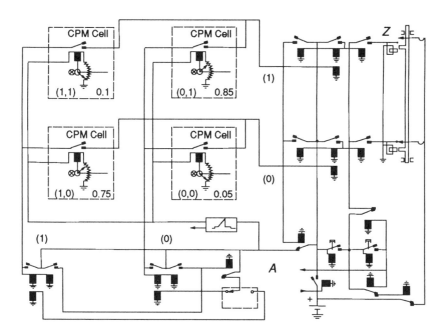

Figure 12.10. A *CPM cell control matrix* can find the conditional memory cell on a given input terminal that has been programmed more often to act than any other conditional memory cell on that input terminal.

The confidence that the programmer has for each output occurring with a given input can be recorded by the number of times the programmer manually operates the logarithmic subtracting mechanism of a CPM cell that represents each input/output relation. If the programmer operates the bail toward the maximum sensitivity position more times than the programmer operates the bail toward the minimum position relative to the other CPM cells connected to that input terminal, then the matrix will select the output terminal connected to that CPM cell when its input terminal is made active.

The conditional predetermined duplex controller

In most applications, not every value of the input variable is part of the active line of behavior of a discontinuous controlled machine. If a controlled machine has many input and output values, and a relatively few values are used, the *conditional predetermined scalar duplex controller* shown in Figure 12.11 can be programmed to provide the required control behavior using fewer memory cells than a single matrix.

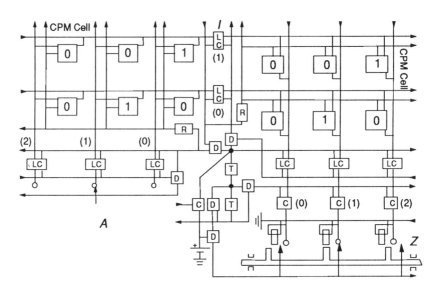

Figure 12.11. The *conditional predetermined scalar duplex controller* can be programmed to connect different values of its input variable to any value of its output variable using fewer memory cells than a single matrix if a relatively few transitions are required.

The CPM cell and the CPM cell duplex controller can be used in control systems having only a single input and output variable. In systems with multiple input and/or output variables, the conditional predetermined multivariable controllers shown in the following section must be used.

Conditional predetermined multivariable controllers

To build a multivariable conditional control matrix, the input search function must be separated from the input sense function as it is in an *active* absolute memory cell. This requires the addition of an output enable relay in each conditional memory cell and an output enable circuit in the conditional predetermined multivariable memory matrix.

CP cell

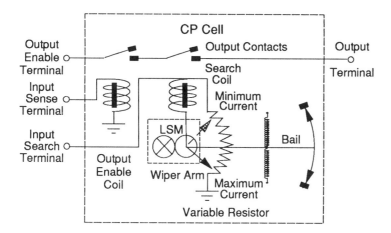

Figure 12.12. A *conditional predetermined cell* has an output enable relay that allows it to be connected in a multivariable control matrix.

An *output enable relay* can be added to the CPM cell, forming the CP cell shown in Figure 12.12. The output enable relay separates the input search function from the output enable function. This allows the *conditional predetermined cell (CP cell)* to be connected in a multivariable control matrix and mixed in with AP cells in the mixed absolute and conditional predetermined matrix shown later.

CP cell control matrix

The output disconnect in the CPM cell matrix shown earlier can be connected to a separate *output enable circuit* and the CPM cells replaced with CP cells, as shown in Figure 12.13, forming a conditional predetermined control matrix that can be used in a multivariable control matrix. To produce an output in a *CP cell control matrix*, the selecting CP cell must have a voltage present at both its input sense terminal and its output enable terminal, and its input search terminal must have a sufficient voltage to close its output contacts. By separating the output enable function from the input sense function, the CP cell can be used in a matrix with multiple input and output variables.

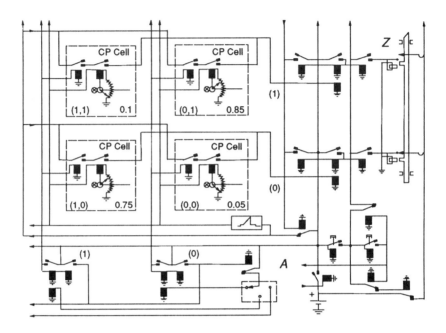

Figure 12.13. A *CP cell control matrix* has the potential of connecting multiple input variables to multiple output variables.

CP cell multivariable input matrix

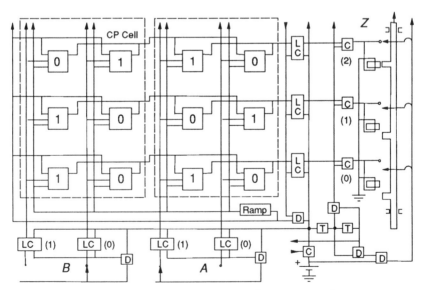

Figure 12.14. A *CP cell multivariable input matrix* can be programmed to connect any combination of values of multiple input variables with any value of a single output variable.

Two or more CP cell matrices can be combined to form a multivariable input matrix, as shown in Figure 12.14. If one input variable in a *CP cell multivariable input matrix* does not have any influence upon the value of its output variable, the CP cells connected to these variables can be set in a high sensitivity position. Then, the value of that input variable will have no influence upon the selection of an output state. If each combination values of input variables is required to produce a specific output value, the sensitivities of the CP cells in the matrix can be set to produce this result.

CP cell multivariable output matrix

The CP cell matrices shown earlier also can be combined to form the *CP cell multivariable output matrix* shown in Figure 12.15. All of the CP cells in the matrix are connected to a single ramp generator, and each set of CP cells connected to a given output variable has a separate output enable disconnect so that only one value of any output variable can be selected in a given transition cycle. Both of the conditional predetermined input and output matrices can be used as standalone controllers.

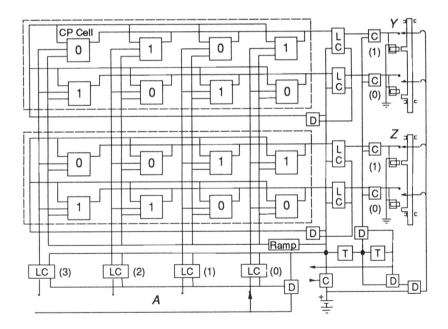

Figure 12.15. A *CP cell multivariable output matrix* can be programmed to connect each value of the input variable to any combination of values of the output variables and to hold the value of any output variable constant for all values of its input variable.

CP cell multivariable duplex controller

The conditional predetermined input and output matrices shown can be combined into the *CP cell multivariable duplex controller* shown in Figure 12.16.

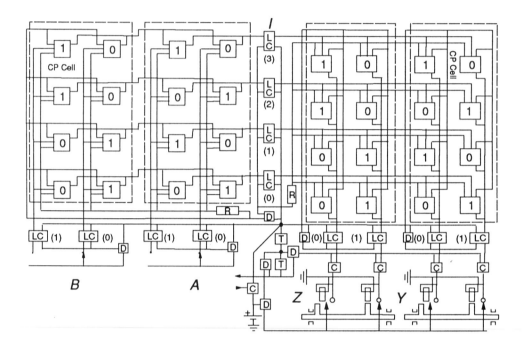

Figure 12.16. A *CP cell multivariable duplex controller* can be programmed to produce any combination of values of its output variables for each combination of values of its input variables in a unitary manner, or can be programmed to produce synchronous diverse behavior.

The duplex controller also can be programmed to produce constant values of one or more of its output variables for any value of one or more input variables. This allows the other variables to produce an active line of behavior while other output variables remain constant in a synchronous diverse manner.

CP cell multivariable duplex control system

Crosswind and direction error sensors can be connected to rudder and bow thruster actuators by a CP cell multivariable duplex controller, forming the *ship direction control system* shown in Figure 12.17. The number of input/output relations required to produce satisfactory behavior determines the number of intermediate terminals required in the duplex controller. If many intermediate terminals are required because many transitions are required in the line of behavior of the control task, then the conditional predetermined monolithic controller shown in the next section may be more useful.

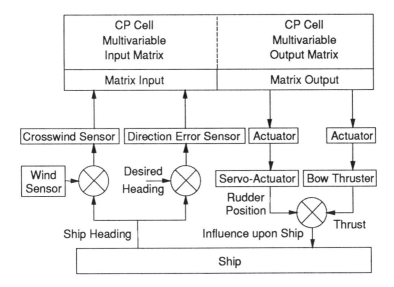

Figure 12.17. A *CP cell multivariable duplex control system* can be programmed to produce specific behavior between multiple sensors and actuators, including holding the value of specific actuators constant for any or all of the input conditions that are expected to be encountered.

Absolute predetermined mixed memory cell

An output enable relay can be added to the absolute predetermined cell (AP cell) shown earlier, forming the *absolute predetermined mixed memory cell* shown in Figure 12.18.

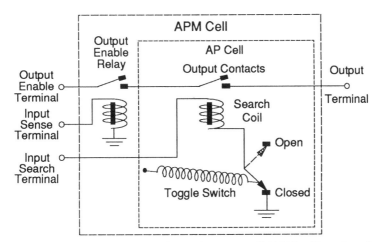

Figure 12.18. An *absolute predetermined mixed memory cell (APM cell)* can be mixed with CP cells to form a mixed predetermined control matrix that can have some behavior that is programmed absolutely, and other behavior that is programmed conditionally.

The addition of the output enable relay to an AP cell allows both kinds of predetermined cells to be mixed in a single predetermined matrix shown in the next section. This *mixed matrix* can be made into all of the other active conditional predetermined multivariable and monolithic control matrices shown in subsequent sections.

Mixed absolute and conditional predetermined control matrix

Absolute predetermined mixed memory cells can be added to the conditional control matrix shown, resulting in a *mixed absolute and conditional predetermined control matrix* (Fig. 12.19).

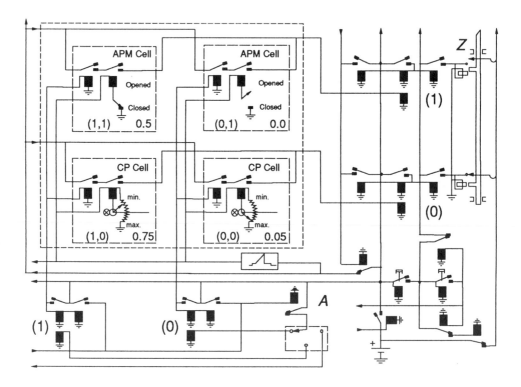

Figure 12.19. A *mixed absolute and conditional predetermined matrix* can be predisposed toward some behavior that is more persistent than other behavior.

The intrinsic sensitivity of the absolute cells can be set by design so that they operate at some percentage of the range of the ramp voltage. Then the conditional cells can be programmed above and below this sensitivity level.

A conditional cell will produce an output if its sensitivity is greater than the intrinsic sensitivity of absolute cells connected to the same input terminal. Conditional cells will not produce outputs if their sensitivities are below the intrinsic sensitivity of absolute cells connected to the same input terminal. By designing the ramp voltage required to activate the absolute cells to be at a high voltage level, the control matrix becomes more like a conditional matrix. By designing the ramp required to operate the absolute cells to be at a low voltage level, the control matrix becomes more like an absolute matrix. The matrix can be *predisposed* toward some behavior that is more *durable* than other behavior if certain cells in the matrix are absolute cells and other cells are conditional cells.

Conditional predetermined monolithic controllers

The circuit features of the conditional predetermined input and output matrices can be combined to form a single multivariable *CP cell monolithic control matrix* that can produce different combinations of values of the output variables for different combinations of values of the input variables. The monolithic control matrix is more efficient than a multivariable duplex controller when many different values of the input variables must produce many different values of the output variables.

Conditional predetermined binary monolithic control matrix

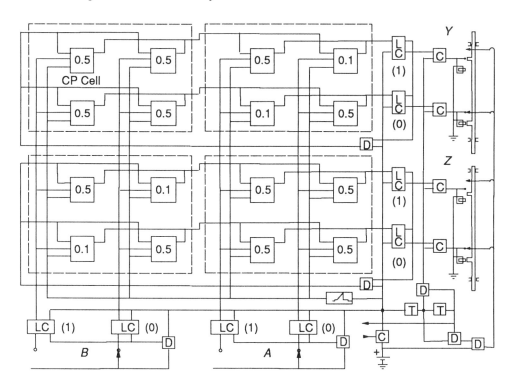

Figure 12.20. A *CP cell binary monolithic controller* uses a set of CP cell submatrices to produce a unique combination of values of the output variables for each unique combination of values of the input variables in a unitary manner.

The CP cells in the multivariable input and output matrices can be connected in a set of submatrices, forming the *CP cell binary monolithic controller* shown in Figure 12.20. The CP cell monolithic controller also can be programmed to connect specific input variables to specific output variables regardless of the values of the other input variables by programming high sensitivity values in the matrices that intervene with the asynchronous diverse relations. In the conformation shown, input variable *A* controls output variable *Y* regardless of the value of *B*, and input variable *B* controls output variable *Z* regardless of the value of *A*.

Conditional predetermined trinary monolithic control matrix

The number of values of the sensor and actuator variables of the CP cell binary monolithic control matrix shown can be increased by one to form the *CP cell trinary* monolithic controller shown in Figure 12.21. The CP cell trinary monolithic controller is the preferred embodiment for any conditional predetermined applications with a large number of variables that have a few values, and where there are many transitions its line of behavior.

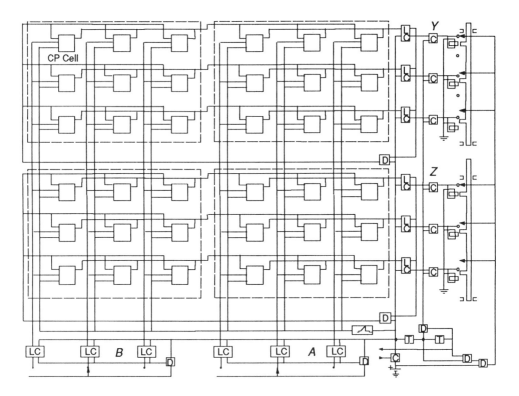

Figure 12.21. A *CP cell trinary monolithic control* matrix can produce more unique combinations of values of its output variables for more unique combinations of values of its input variables than a binary monolithic control matrix with the same number of CP cells.

CP cell monolithic control system

The crosswind and direction error sensors of the *ship direction control system* shown earlier can be connected to its rudder position and bow thruster actuators by the conditional predetermined monolithic control matrix shown in Figure 12.22.

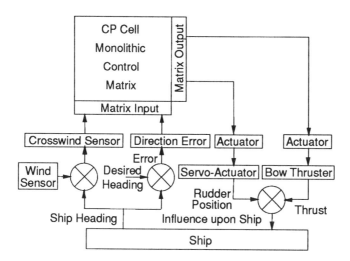

Figure 12.22. A *CP cell monolithic control system* can be programmed to produce a different combination of output values for each combination of input values.

The CP cell monolithic control matrix is particularly well suited when each of many different combinations of values of the input variables must be programmed to produce different combinations of values of the output variables. For example, 8 binary input variables can produce $2^8 = 256$ different transitions in 8 binary output variables in a monolithic control matrix using only $2 \times 8 = 16$ input and 16 output terminals for a total of only $16 \times 16 = 256$ memory cells. A duplex controller would require $16 \times 256 = 4096$ memory cells in its input matrix and 4096 memory cells in its output matrix for a total of 8192 memory cells to produce 256 transitions. Conversely, if only one transition is required, as in a temperature warning system, a monolithic matrix would still require 256 memory cells while a duplex matrix would require only $16 + 16 = 32$ memory cells to produce the same results.

CP cell universal duplex controller

A CP cell monolithic matrix can be used as the input and output matrices of the *CP cell universal duplex controller* shown in Figure 12.23. The conditional predetermined universal duplex controller can produce as many different output values for different input variables as there are values of its intermediate variables, and can be programmed to produce as many *asynchronous diverse relations* as there are intermediate variables. Thus, the conditional predetermined universal duplex controller is the most comprehensive conditional predetermined controller.

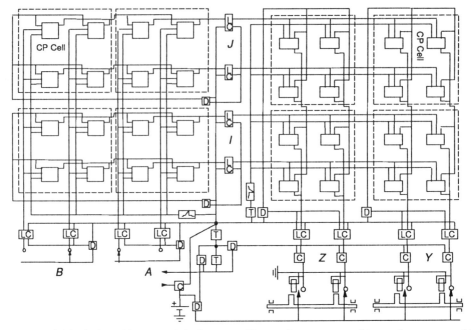

Figure 12.23. A *CP cell universal duplex controller* can be programmed to produce any combinations of values of its output variables for any combination of values of its input variables, or it can be programmed to produce as many asynchronous diverse relations as there are different intermediate variables.

Application of a conditional predetermined universal controller

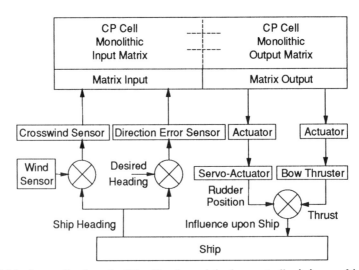

Figure 12.24. An *application of a CP cell universal duplex controller* is in a multivariable ship direction control system that may require asynchronous diverse relations.

A CP cell universal duplex controller can be used to control the multiple input and output variables of almost any control system such as the *ship direction control system* shown in Figure 12.24. The use of conditional predetermined memory cells allows the programmer to establish a distribution of memory sensitivity values over time that reflects the desirability of each input/output relation. The CP cell universal duplex controller may be the most desirable conditional predetermined controller in applications that have very low resolution. If greater resolution is required, the digitized conditional predetermined machines shown in the next chapter are required.

Chapter 13
Digitized Conditional Predetermined Controllers

The conditional predetermined input, output, duplex, monolithic, universal, and mixed control matrices can be connected to digitized input and output variables, creating digitized conditional predetermined control systems, with the trinary digitized controller being the preferred embodiment. These digitized control systems achieve the benefits of digitized absolute predetermined systems, such as dealing effectively with higher resolution input and output variables using a minimum number of memory cells and being able to connect the general or specific values of these variables if required to do so by the control task.

Binary digitized conditional predetermined controlled machines

An input encoder can be used to create a set of aggregate input variables that represent a sensor variable, and an output decoder can be used to combine a set of aggregate output variables into a value of an actuator variable. These processes allow a digitized conditional predetermined controller to operate at far higher resolution while using far fewer CP cells than an undigitized control system.

CP cell binary digitized scalar input matrix

An encoder can be connected to a CP cell multivariable input matrix, forming the *CP cell binary digitized scalar input matrix* shown in Figure 13.1). The output enable disconnect allows only one output value to be selected for any combination of values of the aggregage input variables *a* and *b*.

The CP cell binary digitized input matrix with two aggregate variables can identify any one of the four possible input states using only four CP cells. This is no better than a square scalar matrix. However, a binary digitized controller with three binary aggregate variables can produce eight different output values and uses only six CP cells for each selection. The advantage of a digitized system increases as the number of aggregate variables increases.

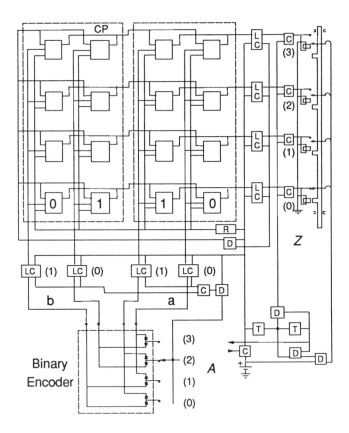

Figure 13.1. In a *CP cell binary digitized scalar input matrix*, each combination of values of the set of two binary aggregate input variables can produce any one of four values of its single output variable.

CP cell digitized scalar output matrix

A single input variable can be connected to a single actuator variable by multiple aggregate output variables as shown in the CP cell binary digitized output matrix in Figure 13.2. Each position produced by the aggregate output variables (*a* and *b*) contains the highest sensitivity CP cells connected to the latched value of the sensor variable *A*. A unique output can be produced for each value of the sensor variable in a unitary manner or generally high or low output values can be produced for given values of the sensor variable in a synchronous diverse manner.

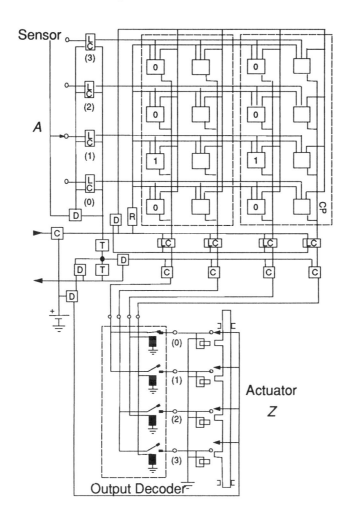

Figure 13.2. A *CP cell binary digitized scalar output matrix* uses two or more aggregate variable and an electrical decoder to position a single actuator.

CP cell binary digitized duplex controller

A CP cell binary digitized input matrix can be connected to a CP cell binary digitized
output matrix, forming the *CP cell binary digitized duplex controller* shown in Figure
13.3.

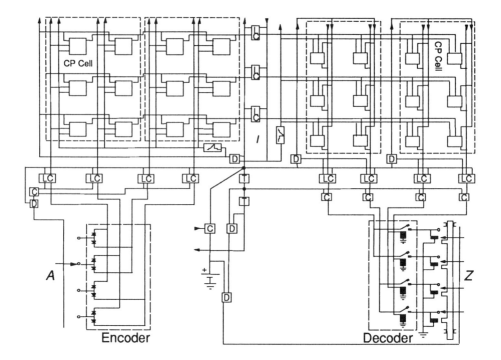

Figure 13.3. A *CP cell binary digitized duplex controller* can produce the set of digitized output values most often programmed for a limited set of digitized input values.

The number of digitized sensor and actuator variables can be increased indefinitely and the number of intermediate variables can be increased to match the number of transitions required by the task environment. The number of values of each aggregate variable can be increased from two to three, forming a trinary digitized system, which is the preferred embodiment of a digitized duplex system.

CP cell binary digitized monolithic control matrix

One example of a very useful digitized conditional predetermined control matrix is the *CP cell binary digitized monolithic control matrix* shown in Figure 13.4. The CP cell digitized monolithic control matrix is used when most of the input values are used to produce many different output values as in a *nearly continuous relation* between an input variable and an output variable.

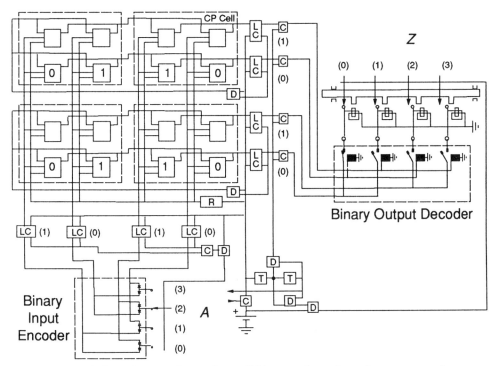

Figure 13.4. A *CP cell binary digitized monolithic control matrix* can be programmed to produce a different output value for each different input value.

Application of a CP cell binary digitized monolithic control system

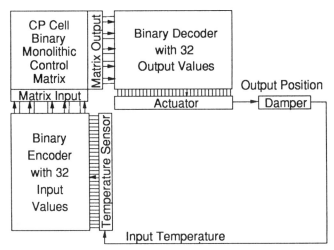

Figure 13.5. An *application of a CP cell binary digitized monolithic control system* is a temperature control system, where many different temperatures require many different output damper
positions.

A CP cell binary digitized monolithic controller can be used as a temperature control system (Fig. 13.5) that can produce many different damper positions for many different sensed values of temperature. A binary encoder with 5 aggregate variables can identify 32 different values of a temperature sensor, and a binary decoder with 5 aggregate variables can produce any one of 32 different damper positions. A monolithic control matrix with 5 binary aggregate input and output variables requires $10 \times 10 = 100$ CP cells.

CP cell binary digitized universal duplex controller

If a system requires multiple sensors and actuators, and if only a relatively few transitions of medium-resolution sensor and actuator variables are required, a *CP cell binary digitized universal duplex controller* (Fig. 13.6) can meet the needs of the control task with fewer memory cells than a digitized monolithic controller. A CP cell trinary digitized universal duplex controller is preferred in this application. The binary digitized design is shown only because it is the least complex circuit that can present the design of the complete controller in the limited space on this page.

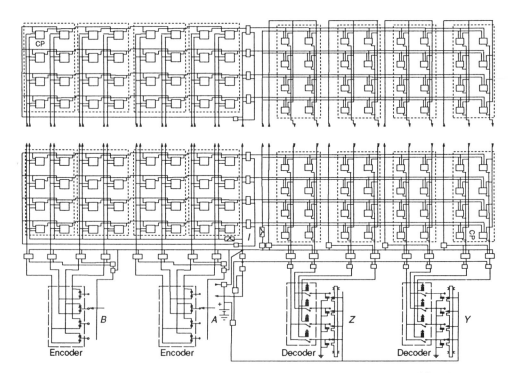

Figure 13.6. A *CP cell binary digitized universal duplex controller* can be conditionally programmed to produce unitary or asynchronous diverse relations between multiple sensor and actuator variables.

Trinary digitized conditional predetermined controlled machines

Conditional predetermined controlled machines also can be digitized using trinary (three value) encoders and decoders. These trinary digitized controlled machines have a higher resolution for a given number of CP cells than binary digitized machines.

CP cell trinary digitized input matrix

A trinary encoder can be connected to a multivariable trinary input matrix, forming the *CP cell trinary digitized input matrix* shown in Figure 13.7.

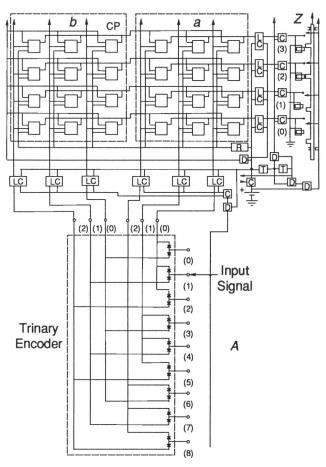

Figure 13.7. A *trinary digitized CP cell input matrix* with two trinary aggregate variables can produce more unique output states for a given number of CP cells than a binary digitized input matrix with three binary aggregate variables.

The trinary digitized sensor variable with two aggregate variables can identify nine sensor values using six CP cells. A binary digitized sensor with three aggregate variables also requires six CP cells, but can identify only eight sensor values.

CP cell trinary digitized output matrix

A CP cell trinary multivariable output matrix can be connected to a trinary decoder to provide the *CP cell trinary digitized output matrix* shown in Figure 13.8.

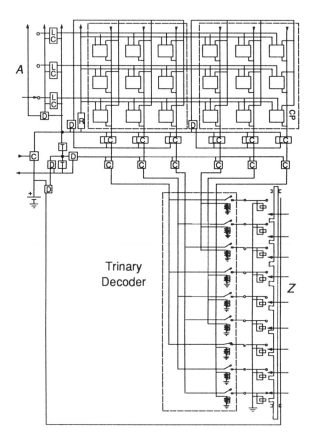

Figure 13.8. An *AP cell trinary digitized output matrix* with two trinary aggregate variables can produce more different output values than a binary digitized output matrix with the same number of CP cells.

Each matrix input terminal is connected to only six CP cells. One of these CP cells in each of the two aggregate variables can be programmed to produce any one of nine possible decoder output values. A binary digitized output matrix with three aggregate variables also requires six AP cells per transition, but can produce only one of eight possible output values.

CP cell trinary digitized duplex controller

A CP cell trinary digitized input matrix can be connected to a CP cell trinary digitized output matrix, forming the *CP cell trinary digitized duplex controller* shown in Figure 13.9.

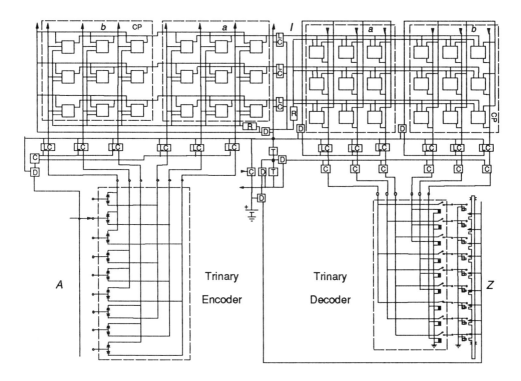

Figure 13.9. A *CP cell trinary digitized duplex controller* allows a digitized input variable to control a digitized output variable.

By using a trinary encoder with 2 aggregate variables to digitize the input variable and a trinary decoder with 2 aggregate variables to digitize the output variable, any one of 9 input values can be programmed to produce any one of 9 output values using only 12 CP cells per transition. An undigitized duplex controller would require 18 CP cells per transition to produce the same quality of machine behavior. The advantage of using a trinary digitized duplex controller increases as the variety of the sensor and actuator variables increases.

CP cell trinary digitized monolithic control matrix

Trinary submatrices can be combined to produce the *CP cell trinary digitized monolithic control matrix* shown in Figure 13.10. The binary digitized monolithic control matrix is shown for the sake of simplicity since the simplest binary system requires only four input and four output terminals and a trinary system requires six input and six output terminals. The CP cell digitized monolithic control matrix is the preferred method of creating a digitized conditional predetermined monolithic control matrix.

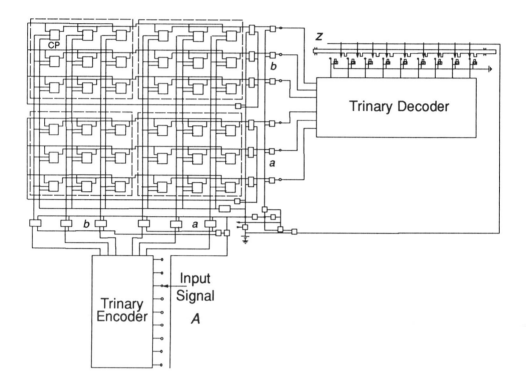

Figure 13.10. A *CP cell trinary digitized monolithic control matrix* can be programmed to produce more different output values for more different input values than a binary digitized monolithic control matrix with the same number of CP cells.

Application of a CP cell trinary digitized monolithic control system

A CP cell trinary digitized monolithic controller also can be used as a scalar temperature control system, as shown in Figure 13.11. A trinary encoder with 3 aggregate variables can identify 27 different values of a temperature sensor, and a trinary decoder with 3 aggregate variables can produce any one of 27 different damper positions. A monolithic control matrix with 3 trinary aggregate input and output variables requires $9 \times 9 = 81$ CP cells compared to a binary digitized system with nearly the same resolution that requires 100 CP cells. If the control task requires a resolution greater than 3%, a mechanically digitized system using a pi actuator to be shown may be required.

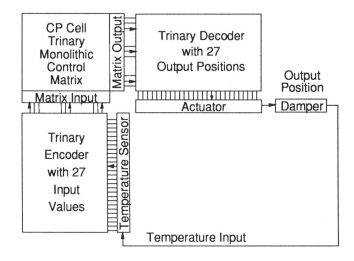

Figure 13.11. An *application of a CP cell trinary digitized monolithic control system* is as a temperature control system, where many different temperatures require many different output damper positions.

Application of a CP cell trinary digitized universal duplex controller

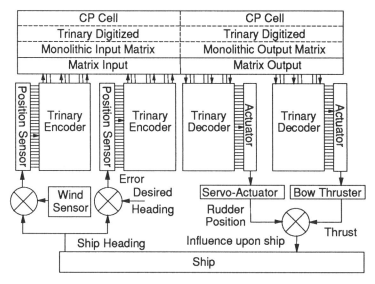

Figure 13.12. An *application of a CP cell trinary digitized universal duplex controller* is a multivariable ship direction control system that can be conditionally programmed to produce unitary or asynchronous diverse relations between multiple sensor and actuator variables.

A CP cell binary digitized universal duplex controller can be used as a *ship direction control system,* as shown in Figure 13.12. A CP cell trinary digitized universal duplex controller with three aggregate variables in each sensor and actuator can produce any one of 729 combinations of values of its actuator variables for any one of 729 combinations of values of its sensor variables using only 36 CP cells per transition. A binary digitized duplex system with 36 CP cells per transition could produce only one of 512 possible combinations of values of its sensors and actuators in each transition.

Mechanically digitized conditional predetermined controlled machines

The resolution of encoded multiposition sensor switches and vernier actuators is limited to about 3% (thirty positions). Mechanical encoders and pi actuators can be used to extend the resolution of sensor and actuator variables to any required level.

Mechanically digitized trinary CP cell duplex controller

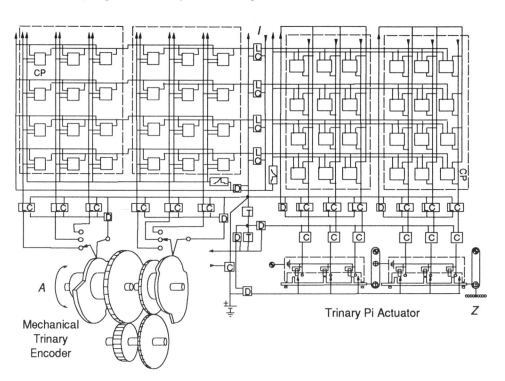

Figure 13.13. A mechanically digitized trinary CP cell duplex controller can transform high-resolution sensor values into high-resolution actuator values.

Two or more CP cell trinary input submatrices can be connected to a mechanical digitized trinary sensor, and two or more CP cell output submatrices can be connected to a trinary pi actuator (Fig. 13.13) when high-resolution control behavior is required. The number of aggregate variables of the sensor and actuator variables can be increased indefinitely to provide the required resolution. Trinary digitized sensor and actuator variables provide greater sensor and actuator variety for a given number of memory cells than a binary digitized system.

CP cell high-resolution duplex control system

The number of aggregate variables in a CP cell trinary duplex controller can be increased, forming the CP cell high-resolution duplex control system shown in Figure 13.14.

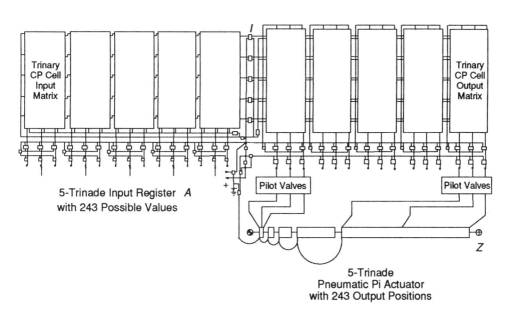

Figure 13.14. A *high-resolution trinary digitized CP cell duplex controller* can transform very high-resolution sensor values into very high-resolution actuator values.

A trinary digitized input register with five aggregate variables can identify any one of 243 different sensor values, and a trinary digitized pneumatic pi actuator can produce any one of 243 different positions. The number of intermediate values can be increased to match the number of transitions in the required line of behavior.

Mechanically digitized trinary CP cell monolithic controller

The CP cell binary monolithic controller also can be connected to mechanical binary digitized sensor and actuator variables using a mechanical trinary encoder and a trinary pi actuator (Fig. 13.15) when high-resolution control behavior is required.

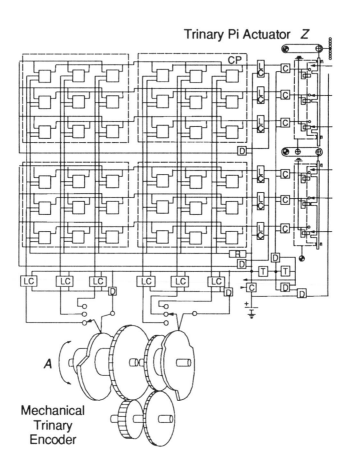

Figure 13.15. A *mechanically digitized trinary CP cell monolithic controller* can transform very high-resolution sensor values into very high-resolution actuator values.

The number of sensor and actuator variables and the number of aggregate variables can be increased indefinitely. Trinary digitized sensor and actuator variables provide the maximum sensor and actuator variety with a minimum number of CP cells.

CP cell high-resolution universal duplex control system

The CP cell trinary universal duplex controller also can be connected to mechanical trinary digitized sensor and actuator variables using a set of input registers and a trinary pi actuator (Fig. 13.16) using three-position pneumatic actuators.

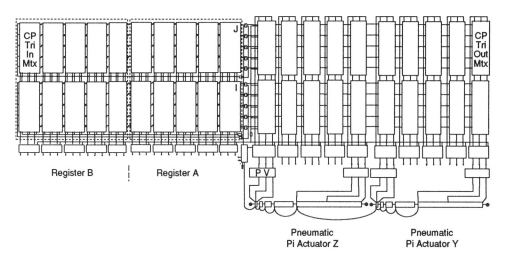

Figure 13.16. A high-resolution trinary digitized CP cell universal duplex controller can transform very high-resolution sensor values into very high-resolution actuator values in a unitary or asynchronous diverse manner.

The number of sensor and actuator variables, and the number of aggregate variables in each sensor and actuator variable can be increased indefinitely to match the required sensor and actuator variety. The number of intermediate variables can be increases to match the need for asynchronous diverse behavior, and the number of values of each intermediate variable can be increased to match the number of transitions in the required lines of behavior.

Controlling vectors using conditional predetermined matrices

A multi-dimensional position variable can be combined into a single vector state by a CP cell input matrix by using each axis of a coordinate measuring machine as its aggregate input variables. A CP cell output matrix also can produce a single vector state by using its aggregate output variables to produce a set of positions in a robot-like machine with a rational coordinate system.

CP cell binary duplex controller operating vector devices

A CP cell multivariable duplex controller can be connected to the component variables of vector devices (Fig. 13.17), allowing values of the sensor vector to be transformed into values of the actuator vector according the distribution of sensitivities programmed into the control matrix.

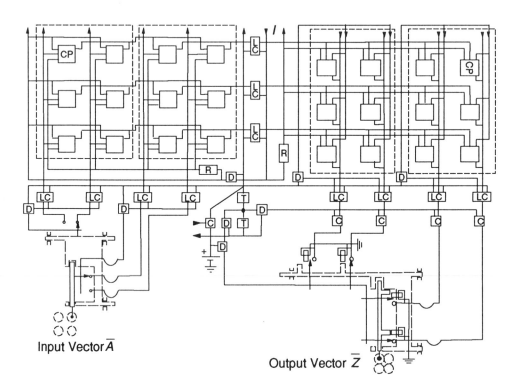

Figure 13.17. A *CP cell binary multivariable duplex controller operating vector devices* can transform a relatively few high-resolution sensor vectors into a relatively few high-resolution actuator vectors.

The number of vector variables and the number of component variables can be increased indefinitely. The number of values of each component variable also can be increased to match the required resolution. Binary component variables are shown only for simplicity. If the number of values of a component variable exceeds approximately eight, it is desirable to digitize each component variable using trinary encoders and decoders. If the number of values of each component exceeds about thirty, it is desirable to digitize the component variables with mechanical encoders and pi actuators, as shown previously.

CP cell binary monolithic controller operating vector devices

The CP cell monolithic controller also can be connected to the component variables of vector devices (Fig. 13.18), allowing values of the sensor vector to be transformed into values of the actuator vector according the distribution of sensitivities programmed into the control matrix.

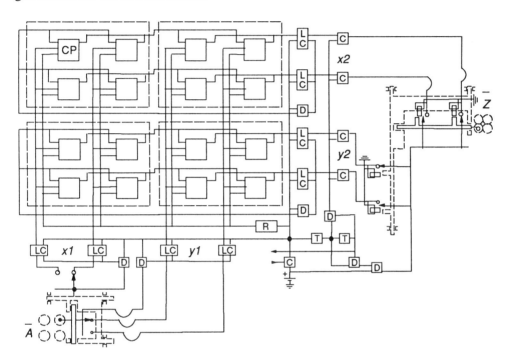

Figure 13.18. A *CP cell binary monolithic controller operating vector devices* also can transform one vector into another vector.

If only a relatively few transitions are required, the duplex controller may require fewer CP cells than a monolithic controller, particularly when there are many possible sensor and actuator values.

Limitations of a single predetermined control unit

A properly programmed trinary digitized predetermined universal duplex controller may be the most efficient controller for most unitary predetermined control tasks where nearly all its variables are active in each transition. However, the universal duplex controller becomes less efficient in terms of its use of memory cells as the task becomes more diverse and when there are many random variables present that have no bearing upon the process of selecting behavior. When individual input and output variables are not part of the memory selection process, the controller must be programmed so that the predetermined memory cells (AP cells or CP cells) connected to the values of the these variables are set in a conductive state in that part of the matrix used to produce behavior. They must conduct for all values of these input/output variables. This *wastes memory cells* (matrix geography). This inefficiency is a natural consequence of the reduction in variety of a unitary or pi system, as compared to the variety of a diverse or sigma system.

Although these problems are not fatal — these systems will still work — there are ways to reduce the number of unused memory cells. The solution to the problem is to break up a large controller into a set of smaller controllers so that each may not include as many diverse or random variables. These smaller controllers can be connected into *networks* and be programmed to act separately upon separate variables. These networks of predetermined controllers, shown in the next chapter, can be made up of absolute predetermined or conditional predetermined memory matrices or combinations of both. All predetermined memory cells will be referred to as *P cells*.

Chapter 14
Networks of Predetermined Controlled Machines

We have seen how a predetermined controlled machine and an operator can be connected to a set of input and output variables, forming a *parallel network* of control units. These control units can work together *successfully* if the *sum of their behavior* results in the desired behavior. They can establish a *division of labor* based upon *value field selection* in which each unit acts at different times on different values of its input and output variable. In some cases, value field selection reduces the work load of each unit and makes it possible for a set of simple units to produce more complex behavior than a single unit working alone. These control units also can be connected to multiple sensor and actuator variables. They may establish a division of labor based upon *dispersion* when each unit works on different variables at different times. Control units can also be connected in *series* with other units, allowing states to be transformed and combined. Control units can be placed in *series and parallel networks* to create new opportunities for diverse and unitary behavior using fewer memory cells in some cases than a single control unit working alone.

Parallel networks of predetermined controlled machines

When two or more control units are connected to a given sensor and actuator variable, they can operate together upon the actuator variable, or the control task can be divided so that each unit operates upon different values of the actuator variable (value field selection). If they are connected to multiple sensor and actuator variables, different control units can be assigned to the set of variables made active by each line of behavior when the set of variables made active by one line of behavior differs from the set of variables made active by another line of behavior (dispersion). This division of labor, based upon dispersion, can occur when the *intrinsically diverse input and output variables* are operated upon by diverse control units and *intrinsically unitary input and output variables* are operated upon by unitary control units.

Multiple control units connected to a single sensor and actuator variable

Two or more predetermined control units can be connected to a single sensor and actuator variable if some means can be found to reconcile differences in the values sent to the actuator variable by the different controllers. One method of reconciliation is to make the actuator of one physically stronger (*dominant*) than another. Another method is to *sum the force or position* of the output of both control units in a mechanical *differential,* as shown in Figure 14.1.

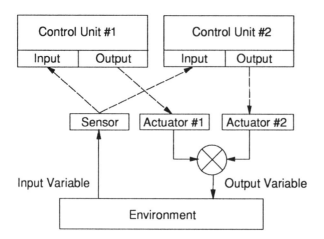

Figure 14.1. Multiple *control units connected to a single set of variables* can work together to produce successful behavior if their outputs are summed by some device such as a differential.

When the outputs of two or more units are connected to a differential as shown, each unit may provide some of the control behavior needed to produce successful behavior, or all units may act together to produce a greater force or displacement than is possible from one unit acting alone.

Value field selection on a single variable

In general, two or more control units connected to a set of sensor and actuator variables can produce different lines of behavior if the actuator *values* made active by one set of sensor values differ from the actuator *values* made active by another set of sensor values. This property of *value field selection* can be represented by the diagram in Figure 14.2.

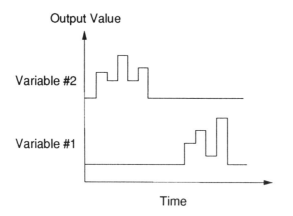

Figure 14.2. In a *control system with value field selection*, two or more units act upon different values of the same sensor variable.

Value field selection allows one unit to produce behavior at one moment and another unit connected to the same variable to produce behavior at another moment. The constant output values (null functions) required for different input values can be produced by a single intermediate value in a duplex controller without wasting memory cells on other intermediate values. Thus, *null functions* do not waste *matrix geography* (memory) in a duplex controller and allow behavior to be spread out over two or more smaller units.

A single predetermined control unit connected to multiple variables

We have seen how a single predetermined universal control unit or a predetermined coded unitary control unit can be connected to multiple sensor and actuator variables. For example, the direction error/rudder control system and the crosswind/bow thruster control system may be combined into one predetermined universal duplex ship direction control system, as shown earlier. This system can be represented generally by the *predetermined universal duplex control system* shown in Figure 14.3.

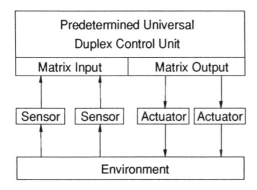

Figure 14.3. A predetermined universal duplex control unit can be programmed to form unitary or diverse relations among its sensor and actuator variables.

If a single coded unitary control unit were to be connected to these two input and two output variables, it would have to develop unitary relations among these variables such that a unique combination of responses would be produced for each combination of sensed values. As stated before, this may require a vast number of transitions. The universal duplex control unit can form as many *unitary relations* as required by the behavior task, or it can establish *asynchronous diverse relations* between specific sensor and actuator variables if necessary.

Two or more predetermined control units connected to multiple variables

In some cases, the system designer may not know if unitary or diverse relations are required in a particular control task. The designer may be required to connect two or more units to the same set of variables so that they can be programmed to act as separate units or a single unit. Thus, two or more universal controllers can be *superimposed* upon two or more input and output variables, as shown in Figure 14.4.

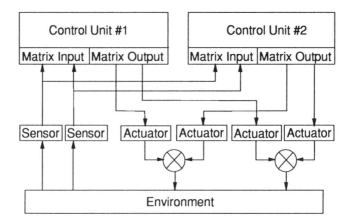

Figure 14.4. A system made up of *multiple control units connected to multiple sensor and actuator variables* can use one unit to form a unitary control system or use separate units to form a diverse control systems.

When the two universal controllers shown are connected to the two input variables such as a direction error and crosswind sensor, and two output variables such as a rudder and bow thruster, one controller can be programmed to produce rudder responses for direction error signals and to produce constant bow thruster values for all values of its two sensors. The other can be programmed to produce bow thruster responses for crosswind values and to produce constant rudder responses for all values of its two sensors. Thus, the rudder variable is made active by one unit and the bow thruster is made active by the other unit.

Dispersion in a system of multiple variables

When two or more universal controllers are connected to two or more input and output variables, one control unit can be programmed to produce one line of behavior, and another control unit can be programmed to produce constant responses for given input values (Fig. 14.5). In a system with *dispersion*, one variable is made active by one unit, and another variable is made active by the other unit. The active variables in one line of behavior are different from the active variables in the other line of behavior. Though the predetermined memory cells (P cells) involved in creating these null functions in each unit are in a sense wasted, many more P cells would be wasted in a single larger unit that can deal with all of the variables.

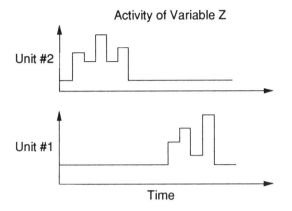

Figure 14.5. In a *control system with dispersion,* the variables made active by one unit differ from the variables made active by another Unit.

Value field selection in a system of multiple variables

When a unit is programmed to operate in a given sensor value field that is different from the sensor value field of another unit connected to the *same* set of variables, the division of labor among multiple control units is based upon *value field selection.* Each smaller unit may contain just the information needed to deal with a given set of input conditions belonging to a given line of behavior, rather than attempting to deal with all of the input conditions likely to be encountered in all of the lines of behavior. This reduces the number of P cells in each control unit. Though the total number of P cells may be the same in some cases, each unit is smaller and may be easier to program than a single larger control unit acting alone. In some cases, the units may act together to increase the range of action or the output force of the actuator variables.

Diverse predetermined system

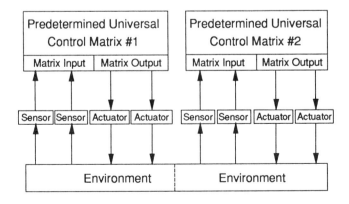

Figure 14.6. A *diverse predetermined system* is made up of a set of separate control units.

The simplest network is a *diverse predetermined system* consisting of separate control units connected only to the diverse variables, as shown in Figure 14.6. For example, a predetermined direction error/rudder control unit and a separate predetermined crosswind/bow thruster unit may form a successful diverse predetermined control system with each unit acting independently upon their variables.

Control units connected to a sigma actuator

Two or more control units also can be connected to a single actuator variable by a *sigma actuator,* as shown in Figure 14.7.

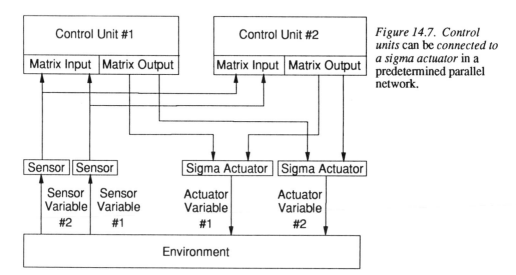

Figure 14.7. Control units *can be connected to a sigma actuator* in a predetermined parallel network.

The sigma actuator produces a force or position that is the sum of the signals from the multiple control units. The predetermined parallel network shown can produce unitary or diverse behavior based upon the behavior programmed into each control unit. Predetermined parallel networks provide alternative paths to create action from a given set of sensors to a given set of actuators.

Structured parallel networks

Connecting the control units in a system to all the variables in a system may be very wasteful of memory registers. Moreover, creating sets of connected units for every combination of sensor and actuator variables requires an astronomically large number of units. One practical method of creating a parallel network is to break the sensors and actuators into connected (coupled) pairs, divide these pairs into sets, and then create groupings of these sets, as shown in the *structured parallel network of predetermined control units* (Fig. 14.8).

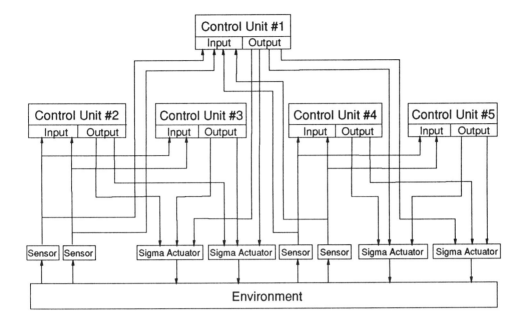

Figure 14.8. A *structured parallel network of predetermined control units* is made up of units that operate upon different variables.

Each unit can be programmed to produce a null function to any actuator or may be programmed to provide the primary source of behavior at any actuator. The structured parallel network can be extended to include any number of variables and any number of levels of organization.

Extended structured parallel network of predetermined controllers

Another controller at a higher level of organization can be connected to each sensor and actuator variable in the structured parallel network shown, forming the *extended structured parallel network* in Figure 14.9. Each sensor and actuator variable can be part of a system formed by a local unit, or be a part of a larger group of variables in much the same way that citizens of the United States belong to local, state, and Federal governments. The government service each citizen receives is the sum of the service from each government. Each citizen can then judge whether the local, state, or Federal government provides the best service value, and attempt to allocate the taxes paid to each level according to the quality of services provided by each level.

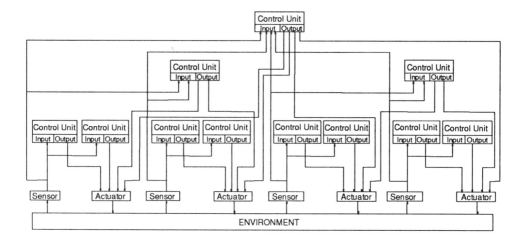

Figure 14.9. An *extended structured parallel predetermined network* shows how an actuator variable may be influenced by two or more control units at different levels of organization.

Uncoupled predetermined universal controlled machines

Until now, we have assumed that the actuators of a controlled machine are connected to its sensors through a given environment. For example, the furnace of a heating system influences its thermostat through an environment connected to both the furnace and thermostat. A rudder influences the direction error sensor of an auto-pilot by the action of the rudder upon the ship. Since each sensor is influenced by an actuator in the same environment, they are called *coupled control systems*. However, an actuator may influence variables in an environment other than the environment of its sensors, and a sensor may be influenced by an environment other than the environment of its actuator. For example, a computer may receive data from one location, operate upon it, and send the results to another location. Since the computer's sensors and actuators are not tied together by a given environment, it is called an *uncoupled control system*.

Coupled systems

If a computer *monitor* is considered the actuator of a *standalone computer*, and a *keyboard* as its sensor, then the *computer operator* is the environment of the computer, and action flows from the monitor through the operator to the keyboard, as shown in Figure 14.10.

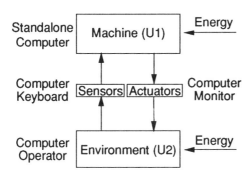

Figure 14.10. In a *coupled predetermined control system,* action flows from the actuator to the sensor of a controlled machine directly through a given environment.

In a coupled system, the sensors, environment, and actuators of a controlled machine are assumed to be a single operating system that is dependent upon the environment to provide the activity needed by the sensor variables to provoke an active line of behavior. These coupled machines can be connected in series and parallel, forming convergent, divergent, hierarchical, and nodal networks, as shown in Part I.

Uncoupled system

In some cases, it is possible to separate the variables in the environment that influence a machine's sensors from the variables in the environment that are influenced by a machine's actuators (Fig. 14.11).

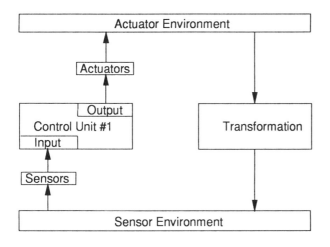

Figure 14.11. The environments in an *uncoupled predetermined control system* are classified as sensor or actuator environments.

The *sensor environment* may be particularly rich in useful sensor variables, and the *actuator environment* may be particularly rich in useful actuator variables. When the input and output variables of a unit are in different environments, a given input variable belonging to a given unit may not be influenced by a given output variable belonging to that unit.

Separating the sensor and actuator environments may be very useful. For example, a computer system can be divided into a *data collection network* that gathers information from a source environment, a central processing unit, and a *data presentation network* that displays the results of calculations on this data. This allows information to be collected from many sources, processed by a central unit, and the results distributed to many other locations.

Though the sensor and actuator environments are separated in an uncoupled system, the actuator variables may eventually influence the sensor variables through another environment in the manner of an *actuator/sensor machine*. Thus, action flows upward through the controlled machine, and downward through the environmental machine, creating the potential for the formation of a *closed line of behavior*.

Parallel predetermined uncoupled network

In a *parallel predetermined uncoupled network*, parallel units are connected to the same sensor and actuator variables, as shown in Figure 14.12.

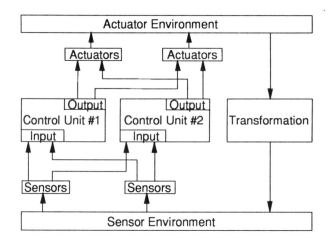

Figure 14.12. In a *parallel predetermined uncoupled network,* actions flow upward through the set of controlled machines and the results of these actions flow downward through the environments.

Parallel units in an uncoupled system can establish a division of labor to deal with sensor and actuator variables that are connected through the environment. For example, two or more units can control the rudder and propeller speed of a ship, which in turn determines the sensed values of the direction and velocity of the ship.

Direct networks of predetermined controllers

By uncoupling the input and output variables, the output of one controller can be connected directly to the input of another controller. In a *direct connection* between two controllers, there is no intervening environment. When the input and output terminals of two or more uncoupled controllers are connected directly, they can form a *direct network of predetermined controllers*.

Direct predetermined series network

In a *direct predetermined series network,* the output variables of one controller are connected directly to the input variables of a second controller without intervening sensors, environment, or actuators (Fig. 14.13). Thus, the sensors of one unit in a sensor environment influence the actuators of another unit in an actuator environment. The actions in the actuator environment must flow back to the sensor environment of the first controller to establish a closed line of behavior.

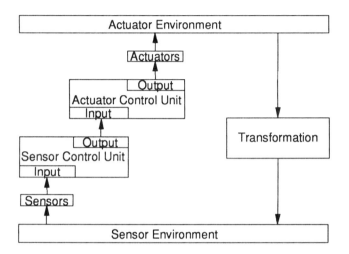

Figure 14.13. In a *direct predetermined series network,* the sensor variables of the first controller influence the actuator variables of the second controller, creating a machine with two controllers connected in series.

A *direct series connection* allows one controller to influence the input variables of another controller directly. For example, a motorcyclist does not provide the mechanical energy required to move the motorcycle along a roadway. A motorcyclist acts upon the throttle and brake controls of the motorcycle, which in turn act upon the speed and position of the rider through the environment. The operator does not "feel" the effects of hills in the environment through the throttle the way a bicyclist feels the effects of hills by the resistance of the foot pedals.

Direct predetermined convergent network

In a *direct predetermined convergent network,* the sensors of two or more responsive units influence the actuators of another responsive unit without a significant intermediary environment, as shown in Figure 14.14.

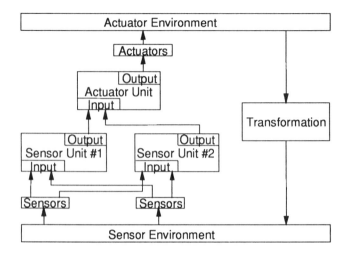

Figure 14.14. In a *direct predetermined convergent network,* the outputs of two or more units control the inputs of another unit directly.

For example, the actions of the pilot and copilot operate directly upon the control surfaces of their airplane at low force levels based upon the values of their instrument readings. These actions influence the velocity, direction, and position of the aircraft at very high force levels, which reflect back to the cockpit instrument readings through the aerodynamic environment. The pilot and copilot may establish a division of labor based upon value field selection and/or dispersion.

Direct predetermined divergent network

In a *direct predetermined divergent network,* the actuators of one unit directly control the sensors of two or more other units without an intermediate environment, as shown in Figure 14.15.

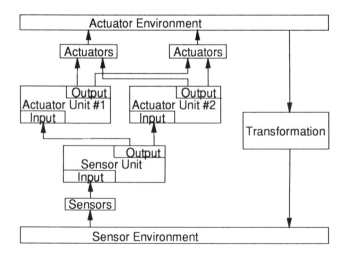

Figure 14.15. In a *direct predetermined divergent network,* the sensors of two or more controllers are directly influenced by the output of one other controller.

For example, an air traffic controller may attempt to influence the behavior of two or more aircraft at a given moment in time. The messages or signals sent by the controller are converted into the specific actions of the set of aircraft. The air traffic controller does not have to fly the aircraft or experience the environment of the actual pilots. Yet the instructions of the air traffic controller are usually carried out directly by the pilot of each aircraft.

Direct predetermined nodal network

Additional sensor and actuator units can be connected in parallel to the direct predetermined series network shown, forming the *direct predetermined nodal network* shown in Figure 14.16.

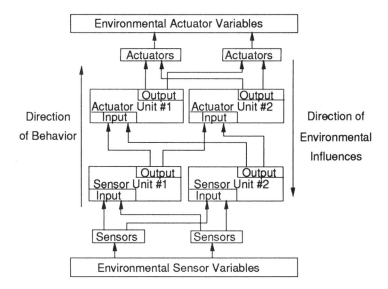

Figure 14.16. A *direct predetermined nodal network* can cause sensor information to converge upon or diverge to its actuator variables through two or more intervening controllers.

A direct predetermined nodal network can produce a unique combination of values of its actuator variables for each combination of values of its sensor variables (unitary behavior) or produce values of one set of sensor and actuator variables that are separate from the values of the other set of sensor and actuator variables (diverse behavior).

Direct predetermined hierarchical network

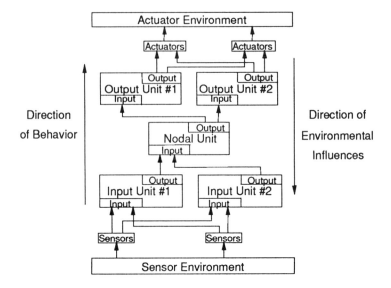

Figure 14.17. A *direct predetermined hierarchical network* is made up of a direct convergent network connected to a direct divergent network.

In a *direct predetermined hierarchical network*, each internal unit can deal with other internal units without dealing with an external environment (Fig. 14.17). Thus, external factors do not influence these internal transactions. For example, a direct hierarchical network is used in military organizations consisting of intelligence-gathering (converging) units that report directly through their chain of command to a command center (*nodal unit*), which then directs (diverges) military action to fighting units in their organization against enemy forces. Since the intelligence-gathering units are separate from the fighting units, the actions of the units in one set do not directly influence the action of units in the other set except through the single nodal unit. In this case, the influence of the nodal unit is not bypassed.

Direct predetermined multilevel nodal network

Like a direct hierarchical network, the units in a *direct predetermined multilevel nodal network* are connected directly. However, each unit can communicate with more than one nodal unit, as shown in Figure 14.18. Most tightly connected organizations with a defined set of sensor and actuator variables and with internal units that communicate only among themselves are examples of direct nodal networks. For example, computers in a system based upon parallel-distributed processing are usually connected in a direct nodal network.

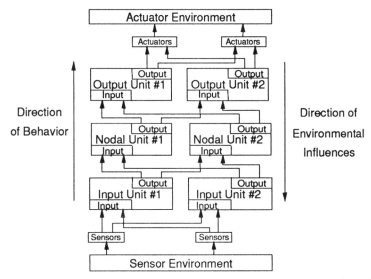

Figure 14.18. A *direct predetermined multilevel nodal network* allows more channels of communication to and from the sensor and actuator variables than a direct hierarchical network.

Superimposed network of direct nodal networks

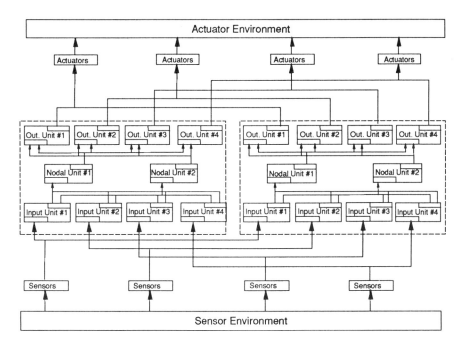

Figure 14.19. A *superimposed network of direct nodal networks* offers more alternative paths for behavior to flow from the sensor to actuator variables than a single network.

A two or more networks can be connected in parallel with a set of sensor and actuator variables, forming a *superimposed network of direct nodal networks,* as shown in Figure 14.19. Each set of superimposed networks can be viewed as assemblies that can act upon different parts of the behavioral task in much the same way that a single unit in a network can act upon different parts of the behavioral task.

Direct network of predetermined networks

Direct predetermined parallel, hierarchical and nodal networks may be considered single control units. These direct networks can be connected to form parallel, hierarchical, or nodal networks of networks, as shown in Figure 14.20. The *direct nodal network of predetermined networks* shown also can be considered a unit in a larger network of networks. This configuration of control units represents the design of a large complex machine controller made up of assemblies and subassemblies.

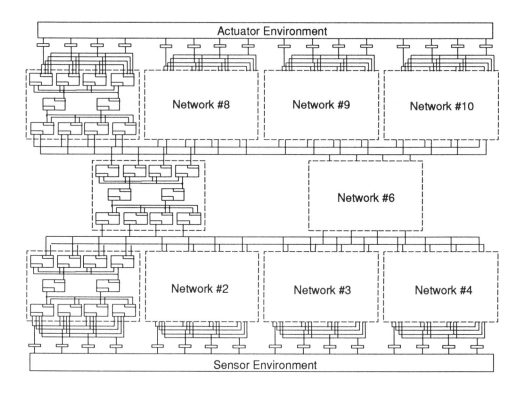

Figure 14.20. A *direct nodal network of predetermined nodal networks* is a single controlled machine made up of subassemblies of networks of controlled machines.

Indirect networks of predetermined controlled machines

In an uncoupled system, the actuators of one controlled machine can form an *indirect connection* to the sensors of another controlled machine through an intermediate environment. When two or more predetermined controlled machines are indirectly connected through environments, they can form an *indirect network of predetermined controlled machines*. While direct networks form a single *complex controlled machine*, indirect networks form an organized *group of machines*.

Indirect series connection

In an *indirect series connection,* the actuators of a machine are connected to a set of environmental variables that influence the sensors of a second machine. Then actuators of the second machine can influence another set of environmental variables that influence the sensors of the first machine (Fig. 14.21).

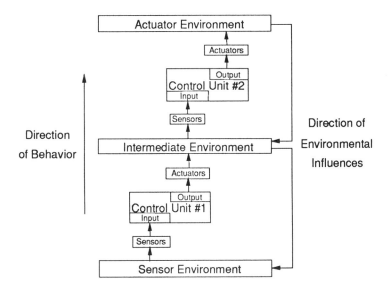

Figure 14.21. In an *indirect predetermined series network* involving two machines, the actuators of one are connected to the sensors of another through a set of environmental variables.

In an *indirect predetermined series network*, action flows from the sensors of the first machine to the actuators of the second machine through an intermediate environment. The effects of these actions must flow back through the actuator environmental variables to the sensor environment of the first machine to establish a closed line of behavior. For example, a person riding a bicycle must produce the mechanical energy required to go up a hill. Thus the bicyclist acts upon the bicycle environment, and the bicycle acts upon the roadway. The effect of hills is passed back to the actuators (legs) of the bicyclist, and the overall change in position of the bicycle and bicyclist is passed back to the sensors (eyes) of the bicyclist.

Indirect convergent network

Two or more machines can influence the sensor variables of another machine through their actuators and a set of environmental variables in an *indirect predetermined convergent network*, as shown in Figure 14.22. An obvious example of an indirect convergent connection is two people riding a tandem bicycle. Both act against and experience the bicycle environment and sense the effects of the bicycle upon the environment.

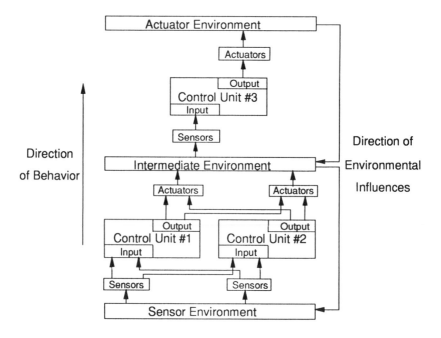

Figure 14.22. In an *indirect predetermined convergent network*, two or more machines influence the sensor variables of one other machine through an intervening environment.

Indirect divergent network

The actuators of one machine may influence the sensors of two or more other machines through an intermediate environment in an *indirect predetermined divergent network,* as shown in Figure 14.23. For example, the starting gate at a horse racetrack provides a physical barrier that can be removed to allow all the race horses to start simultaneously.

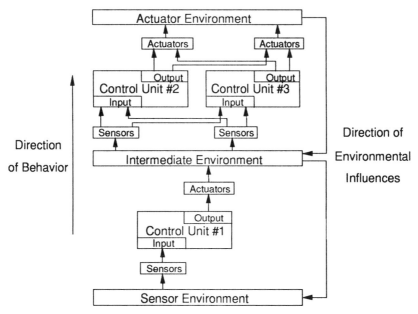

Figure 14.23. In an *indirect predetermined divergent network,* the actuator of one machine impinges upon the sensors of two or more other machines.

Indirect predetermined nodal network

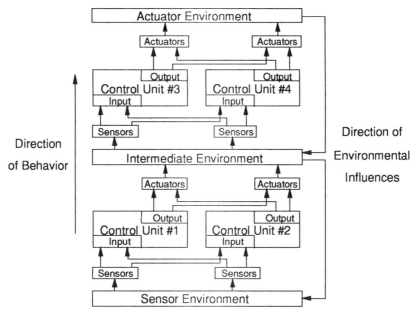

Figure 14.24. In an *indirect predetermined nodal network,* behavior flows to two or more actuator units from two or more sensor units via an intermediate environment.

Additional sensor and actuator units can be connected in parallel to the sensor and actuators units in the indirect predetermined series network shown earlier, forming the *indirect predetermined nodal network* in Figure 14.24. Changes in the actuator environment must flow to the sensor environment in order for the network to establish a closed line of behavior.

Indirect predetermined hierarchical network

A hierarchical network allows one unit to obtain information from one set of subordinate units and mediate the behavior of another set of subordinate units. An *indirect predetermined hierarchical network* is formed when the actions of two or more units converge upon a single unit through an intermediate environment, and when the action of this single unit then diverges to two or more other units through another intermediate environment, as shown in Figure 14.25.

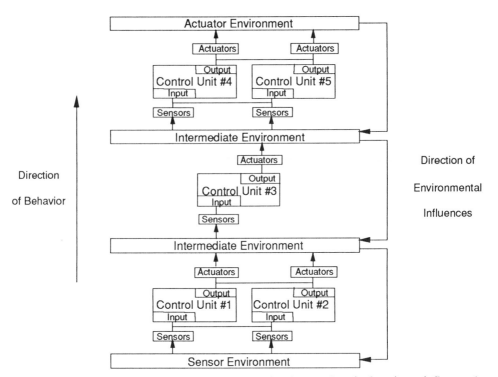

Figure 14.25. In an *indirect predetermined hierarchical network,* a single unit can influence the behavior of two or more units connected to an environment based upon the sensor values of two or more other units connected to another environment.

An indirect hierarchical network is made up of an indirect convergent network connected to an indirect divergent network by a single nodal unit. An example of an indirect hierarchical network is a system in which farm produce is collected through a network of local and regional produce buyers, who then form a separate network of regional and local farm product distributors. Each farming unit, local buyer, regional buyer, regional distributor, and local distributor must deal with a real supply and demand market environment.

Indirect predetermined multilevel nodal network

A nodal network is a multilevel hierarchical network with more than one nodal unit. Thus action can pass from one environment to the other through different pathways. All of the units in an *indirect predetermined multilevel nodal network* are connected through environments that may influence the communication between other units (Fig. 14.26). The internal nodes may establish a division of labor in which each node produces a different line of behavior according to conditions in the sensor environment. This may reduce the complexity of each internal unit and/or increase the variety of behavior of the network. Most indirect hierarchical networks can become indirect nodal networks if successful ways are found to bypass the single ultimate node.

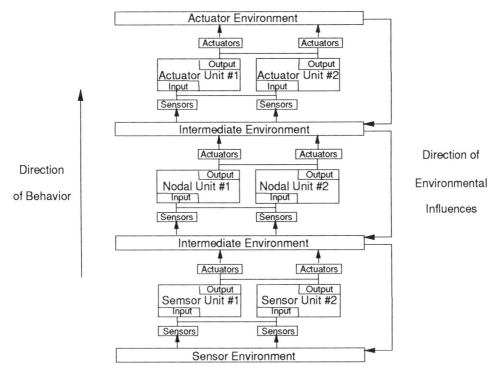

Figure 14.26. An indirect predetermined multilevel nodal network is similar to a social group of individuals acting together to produce some specific behavior.

Indirect network of predetermined direct networks

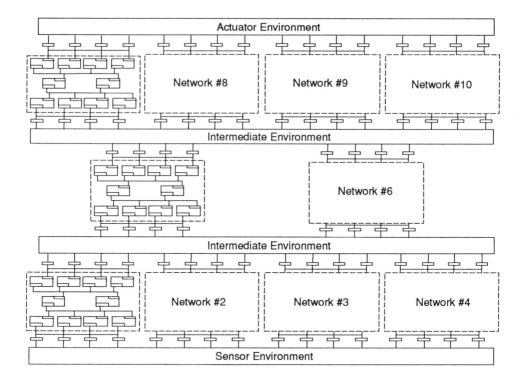

Figure 14.27. An *indirect network of predetermined direct networks* can be described as a set of network units acting in an organized group.

An *indirect network of predetermined direct networks* is similar to the indirect nodal network shown above, but replaces the individual units with direct nodal networks, as shown in Figure 14.27. Network units in an indirect network can operate upon different sensor and actuator variables to produce a wide variety of unitary and/or diverse behavior between the sensor and actuator environments.

Indirect network of predetermined indirect networks

An *indirect network of predetermined indirect networks* is made up entirely of separate units (Fig. 14.28). Each unit in an indirect network of predetermined indirect networks may be considered a separate unit or individual that has the opportunity to act in response to actions of other units or elicit action in other units through intervening environments.

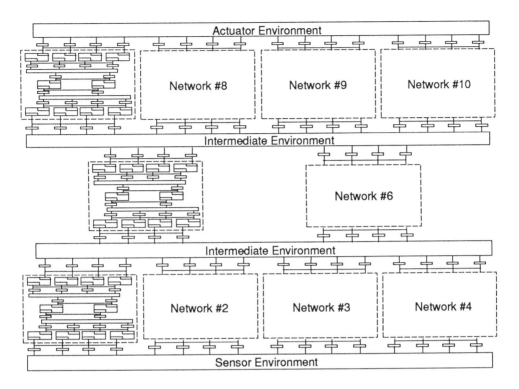

Figure 14.28. An *indirect network of predetermined indirect networks* is similar to a social group like a business or government organization made up of individuals who form departments and other subgroups within the organization.

Direct networks of predetermined bipartite controlled machines

Some controlled machines may have to deal with two types of environments, as shown in the following examples: A bank has to deal with its depositors and its borrowers; a business has to deal with its accounts payable and its accounts receivable; a sports team has to deal with offensive and defensive challenges. These units may be concerned with acquiring information in a *sentient environment* to which they must respond, and may have to produce an effect in a different *operant environment* over which they have some control. When the sensors and actuators of a machine are divided into two separate functions in different environments, it forms a *bipartite system*.

Predetermined bipartite system

The variables in the environment of a machine can be divided into two different categories considered useful in some way. For example, a bank may distinguish between its depositors and its loan customers, as shown in Figure 14.29.

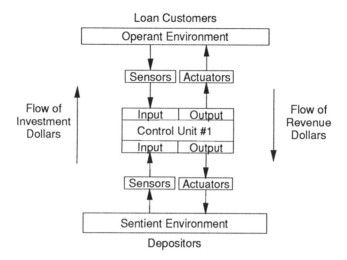

Figure 14.29. In a *predetermined bipartite system,* the input and output variables of a control unit are divided into sentient and operant categories. The sensor variables from both environments can be combined as the set of input variables of a universal duplex controller, and the actuator variables in both environments can be controlled by the set of output variables of a universal duplex controller.

Different bank personnel and different facilities may deal with its deposit customers and its loan customers. It may use different sensors and actuators to deal with both types of customers. The depositors may be considered sentient variables because the bank must respond to its depositors. The loan customers may be considered operant variables because the bank can determine its lending behavior. The performance of the bank is judged by its behavior in both environments since its net assets of cash and collateral on loans must exceed its liabilities to its depositors. Thus, action can flow from one set of environmental variables to the other through the machine itself. The determination of action in one environment may be determined by results in the other environment. For example, the amount of money lent and the interest rate paid to depositors can be based upon the total deposits and loan repayments.

Predetermined parallel bipartite network

Two or more units can establish a division of labor in a *predetermined parallel bipartite network* while dealing with environments that are essentially separate and different, as shown in Figure 14.30. When the sensor and actuator environment variables are not closely connected, two or more bipartite units may establish a division of labor and closed lines of behavior. For example, a commercial bank and a savings and loan association may divide the depositor community and borrowing markets into commercial and home mortgage segments.

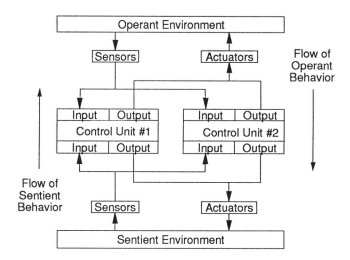

Figure 14.30. In a *parallel predetermined bipartite network*, action can flow directly from each set of environmental variables to each set of environmental variables through two or more intervening universal duplex controllers.

Direct predetermined series bipartite network

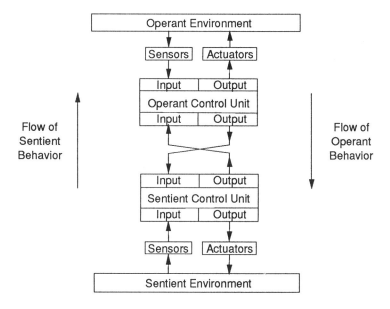

Figure 14.31. In a *direct predetermined series bipartite network*, some output variables of both units are directly connected to some input variables of other units.

A predetermined universal control unit with bipartite input and output variables can be connected direct to another predetermined bipartite universal control unit forming the *direct predetermined series bipartite network* shown in Figure 14.31. In a direct series bipartite network, influences can flow in both directions through two or more units. This configuration is similar to one bank (Control Unit #1) depositing money in another bank (Control Unit #2). Profits from the loans of the second bank (Control Unit #2) can flow to the first bank (Control Unit #1) in the form of the interest earned on its deposits. Thus, influences can flow in both directions in a series bipartite network.

Direct predetermined bipartite convergent network

A second sentient unit can be connected in parallel to the direct predetermined series bipartite network shown, forming the *direct predetermined bipartite convergent network* in Figure 14.32.

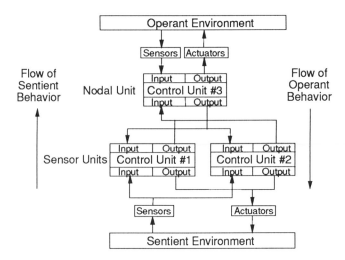

Figure 14.32. In a *direct predetermined bipartite convergent network,* some input and output variables of two or more units connected to a sentient environment are connected directly to some input and output variables of a unit connected to an operant environment.

The direct predetermined bipartite convergent network allows the behavior of two or more sentient units to be combined by an operant unit to produce unitary behavior in the operant environment.

Direct predetermined bipartite divergent network

A second operant unit can be connected in parallel to the direct predetermined series bipartite network shown earlier, forming the *direct predetermined bipartite divergent network* shown in Figure 14.33.

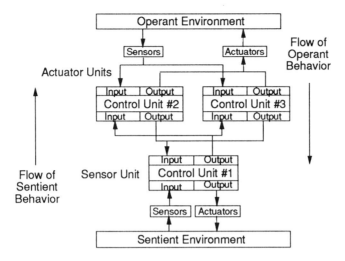

Figure 14.33. In a *direct predetermined bipartite divergent network,* one sentient unit can control the behavior of a set of operant variables through its influence upon two or more operant units.

The sentient unit can redefine the values of the sentient variables to influence the behavior of the operant units.

Direct predetermined bipartite nodal network

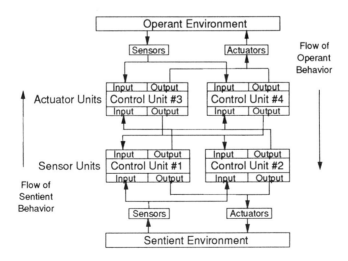

Figure 14.34. In a *direct predetermined bipartite nodal network,* any unit can influence the behavior of any other unit.

Additional sentient and operant units can be connected in parallel to the direct predetermined series bipartite network shown earlier, forming the *direct predetermined bipartite nodal network* shown in Figure 14.34. The direct predetermined bipartite nodal network can produce unitary or diverse behavior between any set of sentient and operant environmental variables.

Direct predetermined bipartite hierarchical network

The *direct predetermined bipartite hierarchical network* uses a single nodal unit to connect multiple control units connected to the sentient and operant environments, as shown in Figure 14.35.

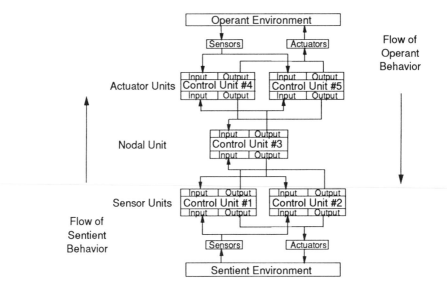

Figure 14.35. A *direct predetermined bipartite hierarchical network* allows influences to flow in both directions between the sentient and operant environments through a single bipartite control unit.

The single nodal control unit can collect and dispense influences to units connected to both environments. In general, a hierarchical network is much better at producing unitary behavior than producing diverse behavior.

Direct predetermined bipartite multilevel nodal network

In a bipartite nodal network, sensor and action information can flow in all directions. However, the bipartite nodal network is not an anarchic system in that the internal units can create unitary behavior while the peripheral units are more likely to produce diverse behavior. The *direct predetermined bipartite multilevel nodal network* is the most complex single machine shown so far. It is made up of a single network of bipartite control units, as shown in Figure 14.36.

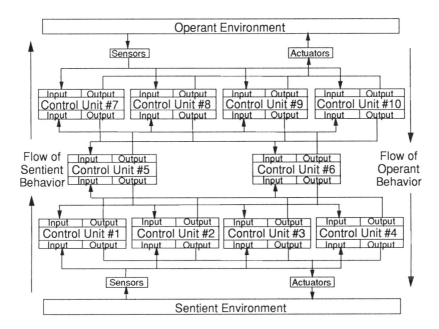

Figure 14.36. A *direct predetermined bipartite multilevel nodal network* can deal with more diverse behavior between sentient and operant environments than a hierarchical network.

The bipartite nodal network can pass influences in both directions like an industry that buys materials in a source environment, operates upon these materials to produce products that it sells in another environment. An industry of this type also can recycle these materials back through its organization to the source environment.

Direct network of parallel predetermined bipartite networks

A direct network may be considered a single machine unit made up of subassemblies. These units can be connected to a given set of sensor and actuator variables to form a *direct network of parallel predetermined bipartite networks* (Fig. 14.37).

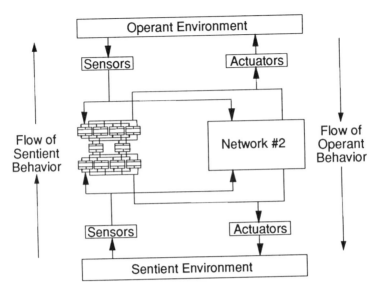

Figure 14.37. A *direct network of parallel predetermined bipartite networks* provides alternate pathways for the control of a given set of sensor and actuator variables.

A direct network of parallel bipartite networks can establish a unique method of dealing with a given set of sensor and actuator variables. A set of parallel networks may provide more effective control behavior by providing more alternative pathways for diverse behavior than a single, more complex, single network.

Direct nodal network of predetermined bipartite nodal networks

A machine can be made up of direct networks of networks of machines by directly connecting the output variables of networks to the input variables of other networks, as shown in Figure 14.38. *Bipartite predetermined universal duplex controllers* can be connected directly in series and in parallel to form direct networks of universal controllers. These networks create a control unit made up of parts that may provide the unitary and/or diverse relations required by the task environment, and may do so in a more effective manner than a single control unit acting alone. A business organization that is made up of divisions and departments within divisions operates as a direct network when its organizational units do not buy and sell products or services from and to each other. (If members of different departments within a company carry out extensive interactions, they would be part of a direct *tripartite nodal network of direct tripartite nodal networks.* If members of their departments carry out extensive interactions in special environments with members of the same department at their level, they would be part of a *quadrapartite network.*)

A direct network of networks provides many opportunities for the *unitization of behavior.* When each network embodies specific behavior, the behavior of the overall network can be changed by adding or subtracting networks, or modifying the behavior of specific networks. For example, computer programs are written as a set of *subroutines* within subroutines. A computer program made up of subroutines within subroutines is an example of a direct bipartite network of bipartite networks since subroutines within subroutines at a given level usually cannot interact directly with other subroutines at the same level in other subroutines.

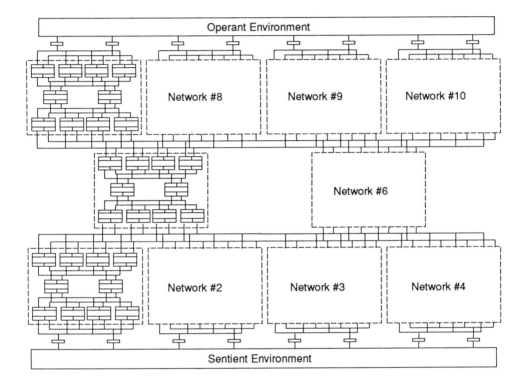

Figure 14.38. A *direct nodal network of predetermined bipartite direct nodal networks* forms the basis of the most complex single machine unit made up of assemblies and subassemblies.

Indirect bipartite networks

Bipartite universal duplex controllers can be connected indirectly in series through an environment and in parallel to the same environments to form *indirect bipartite networks* of predetermined universal controllers. These networks may provide the unitary and/or diverse relations required by both task environments, and may do so in a more effective manner than a single control unit acting alone.

Indirect predetermined series bipartite network

Two or more units can form an *indirect connection* through an environment and to unconnected environments in the *indirect predetermined series bipartite network* shown in Figure 14.39.

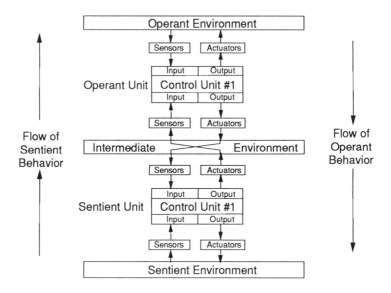

Figure 14.39. In an *indirect predetermined series bipartite network,* two or more universal control units can pass action in both directions between two environments through an intervening environment.

An indirect predetermined series bipartite network can establish closed lines of behavior between variables in separate environments through an intervening environment. For example, a savings bank may lend money to people who deposit money in a commercial bank. Some of the dividends on these people's savings accounts may be used to pay the interest on their loans from the savings bank.

Indirect predetermined bipartite nodal network

Two or more bipartite units can be connected in parallel, forming the *indirect predetermined bipartite nodal network* shown in Figure 14.40. Information can flow to and from the operant and sentient environments and to and from bipartite units connected to the same environments. This allows any unit in a bipartite nodal network to communicate with any other unit in the network.

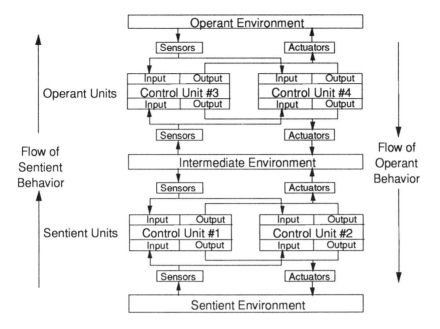

Figure 14.40. In an *indirect predetermined bipartite nodal network,* two or more bipartite units connected to a sentient environment can be connected to two or more bipartite units connected to an operant environment via an intermediate environment.

Indirect predetermined bipartite hierarchical network

In an *indirect predetermined bipartite hierarchical network,* the convergent and divergent units are combined into a single network. Some social and animal groups are based upon an indirect bipartite hierarchical network. Each unit can deal with subordinate units, pass information to a superior unit, and convert information from a superior unit into action in subordinate units through intervening environments, as shown in Figure 14.41. Each unit can influence other units on the same level through the sensor and actuator variables in the intermediate environments at each level of the organization. If units at a given level communicate with other units on the same level through specific designated environments, they form an *indirect tripartite hierarchical network.* A *planned economy* is usually an indirect predetermined tripartite hierarchical network.

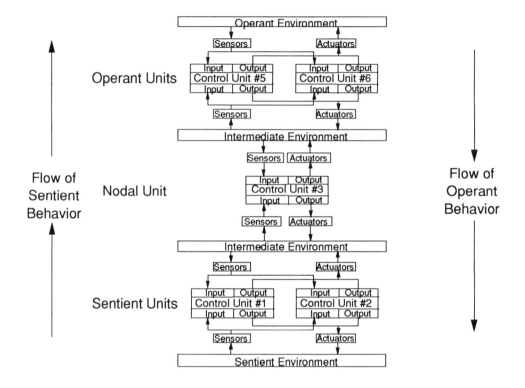

Figure 14.41. An *indirect predetermined bipartite hierarchical network* contains bipartite units that can both collect action for a nodal unit and distribute action from the nodal unit through intervening environments.

Indirect predetermined bipartite nodal network

Most activities within a society are based upon the design of an indirect bipartite nodal network in that each unit has more than just one subordinate and superior unit with which it must communicate (Fig. 14.42). In an *indirect predetermined bipartite multilevel nodal network*, sensor and action information can flow in all directions through the environments distributed throughout the network. A feature of a nodal network is that each unit can be connected to multiple superior units and multiple subordinate units. This may allow information and action to flow between the sentient and operant environments even if one nodal unit becomes inactive.

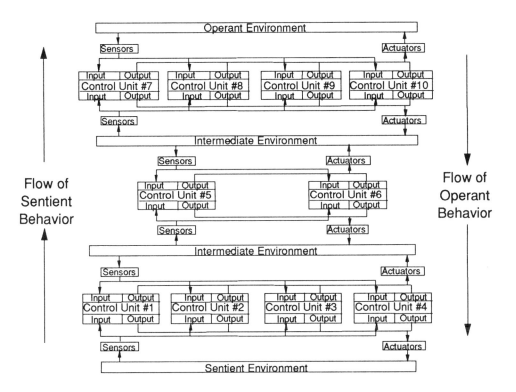

Figure 14.42. An *indirect predetermined bipartite multilevel nodal network* provides many channels of communication to all the units in the assembly through the intervening environments.

Indirect nodal network of predetermined bipartite direct nodal networks

Each unit in an indirect nodal network can be replaced by a direct nodal network, forming the *indirect nodal network of predetermined bipartite direct nodal networks,* as shown in Figure 14.43. A group of businesses that buy and sell goods and services to one another may be considered an indirect nodal network, with the intermediate environments forming the trading environments between each business. Materials flow from the sentient or source environment to the operant or customer (retail) environment, and used materials may be traded backward (recycled) to the source environment.

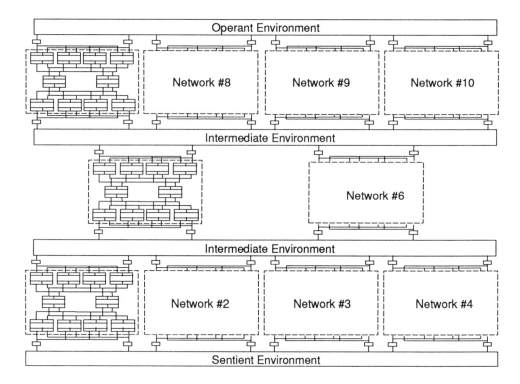

Figure 14.43. In an *indirect nodal network of predetermined bipartite direct nodal networks,* each subnetwork may be considered a single machine unit or business enterprise made up of subunits.

Indirect nodal network of indirect bipartite nodal networks

Indirect networks also may be connected through intermediary environments to form an *indirect network of predetermined indirect networks,* as shown in Figure 14.44. An indirect network of predetermined indirect networks is the general form of most human *economic structures.* Each internal network may represent the economic units within a given geographical area. Some of these organizations may be more hierarchical or more nodal depending upon how centralized or diverse is the flow of goods and services within each network. When units are organized into networks that themselves form units of other networks, many opportunities exist for modifying, optimizing, rearranging and redesigning the behavior of each unit and of reorganizing the system as the need arises.

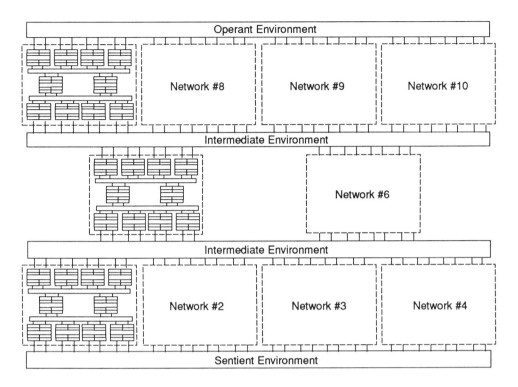

Figure 14.44. An *indirect nodal network of predetermined indirect bipartite nodal networks* acts like a group of businesses or other separate economic units that operate within an organized assembly having separate geographic areas.

Self-learning and self-organizing machines

When multiple control units deal with a control task, the task may be considered broken into subunits. This process of using subassemblies to deal with a task environment is called *unitization*. The size of each unit within the assembly may be reduced by connecting the units in a hierarchical or nodal network according to the *intrinsic organization* of the task environment. Then, the unitary variables can be connected into unitary control assemblies, and the diverse variables can be connected by separate pathways through diverse control units.

Unitization of predetermined universal machines

In many cases, it is difficult or impossible to determine the intrinsic organization of the task environment *a priori* (before a system is put into use). It may be necessary to attempt to operate a machine in a task environment *a posteriori* (after the machine is put into use) to determine the organization of the system variables. For example, it may be difficult to determine how an airplane flies until one attempts to fly it. So a *self-learning machine* that can quickly change its modulus of behavior according to the requirements of a given control task may be essential in dealing with a wide variety of control tasks. Part of the behavior established by dealing *a posteriori* with a task environment may reflect an understanding of the unitary or diverse organization of the task environment. Thus, a properly designed self-learning machine also may have to be *self-organizing*.

The purpose of the next part is to show the design of a standalone, standby, or cooperative controller that can establish its own behavior according to the needs of the task environment and the other machines to which it is connected. The next chapter will investigate the process of spontaneously creating self-organizing and self-sustaining behavior through a learning process that occurs *after a control system is built and put into use* (*a posteriori*) and that can even be carried out by the controller without outside help (autonomously) if the need arises.

PART III
EMPIRICAL MACHINES

Once a predetermined controlled machine is programmed and put into use, the operator may find that the modulus of behavior of the machine is not appropriate. Usually, the values in the control matrix of a predetermined controlled machine have to be modified after it is put into use.

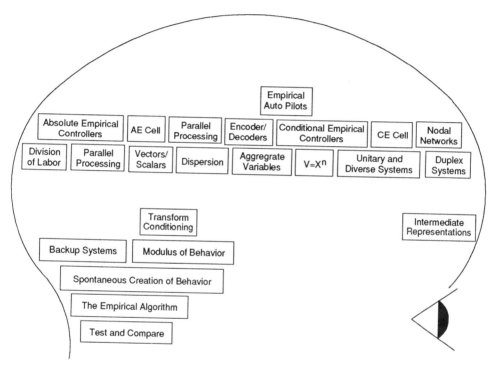

The helmsmen of a ship may have many opportunities to modify his or her control behavior through experience. In fact, the best ship auto-pilot control program acts like an experienced helmsman. The actions of this skilled helmsman may be recorded for each sensed condition *a posteriori* — after the auto-pilot has been installed in a given ship and while the helmsman attempts to control the ship. This can be done by having an *empirical controller* attempt to produce output states based upon its memory, allowing a more dominant helmsman to override those output states based upon the skill and judgment of the helmsman, and having the empirical controller record the actual output states that occur as each input state is encountered, according to the *empirical algorithm*. If the behavior of the empirical controller can be carried out without interference from the helmsman, the helmsman can withdraw from the control process and the empirical controller can produce the control behavior. The purpose of Part III is to show the circuit and logic design of the *natural computer* that can carry out this *process of empirical control*.

Chapter 15
Absolute Empirical Machines

The empirical algorithm discussed in Part I requires that the following three steps are carried out to produce successful empirical behavior:

1. An output state must be selected that has the highest *historical probability* of being carried out for a specific input state.
2. This *measure of probability* must be decreased after that output state is selected.
3. The measure of probability of the *actual output state* that occurs after that specific input state must be increased.

An *absolute empirical machine* can be designed that decreases the measure of probability of an output state to *zero* when it is selected and increases the measure of probability of the actual output state to *one* when it occurs. This all-or-nothing (absolute) empirical machine can reproduce the *last actual output state* produced after each input state. It does so by recording the actual output state produced after each input state. This output state can be produced by an operator after each input state, or the actual output state can be produced by the machine and allowed to occur by the task environment after each input state. The task environment may or may not include an operator.

Absolute empirical algorithm

The modulus of behavior of an empirical machine is established by the actual output values of an actuator variable that occur after specific values of an input variable. This machine can be created by replacing the predetermined connection matrix with an *absolute empirical memory matrix* made up of *absolute empirical memory cells* placed at the intersection of each input and output state. These absolute empirical memory cells *(AE cells)* can be made *sensitive* or *insensitive* like the AP cells at the intersections of a predetermined memory matrix. The sensitivity of an AE cell is determined by an *actuator feedback circuit* that measures the actual output value that occurs within a given period after a given input value. According to the *absolute empirical algorithm,* an AE cell is made insensitive when it produces an output. If the output state called for by an AE cell occurs within a specified period, it is restored to a sensitive state by the actuator feedback circuit. If the output does not occur within the specified period, it is left in an insensitive state, and the AE cell at the intersection of the actual output state and the original input state is placed in a sensitive state by the actuator feedback circuit.

Statement of the absolute empirical algorithm

The empirical algorithm says that the measure of probability of an output event occurring after a given input event must be changed *to some degree* by the number of times the output event occurs after the input event. One possibility is that the measure of probability is changed 100% by the occurrence of a single output event. This is the basis of the absolute empirical algorithm stated in Table 15.1.

Table 15.1. Statement of the absolute empirical algorithm

---

1. The controller selects the output state connected to the specific input state, and the connection between that output state and that input state is broken.
2. The controller attempts to produce that output state after a specific delay.
3. The controller compares the actual output state that occurs within a specific period with the prescribed output state.
4. If the selected output state can be carried out after the specified delay after an input state and within the specified period, the connection to that output from that input is remade so that output state will be selected when that input state occurs again.
5. If the output state selected for a specific input state cannot be carried out (because it is inhibited, restricted, or otherwise interfered with by the environment, an outside influence such as a trainer, or by other internal features such as its actuators or structure) and another output is carried out, a connection is made to the actual output state so that the actual output will be selected when the original input state occurs again.
6. If no outputs are carried out for a specific input (due to the previously mentioned influences and conditions), no connections are made to any output states with that input state, and no output will be selected when that input state occurs again.

This controller selects action based upon its internal records of the last action it experienced for each sensed condition. The absolute empirical algorithm says that the *confidence level* of an input/output relation is increased to unity each time it is successfully carried out and decreased to zero each time it is not carried out. This requirement can be carried out by the action of AE cells placed in a matrix at the intersection of each input and output terminal. In some applications, these (all-or-nothing) absolute empirical changes in confidence levels will provide adequate control behavior. For example, if a helmsman *always* produces the *same* rudder response for a given sensed condition, an absolute empirical memory matrix can record and produce these same rudder responses for each sensed condition, forming an absolute empirical auto-pilot. Many empirical control applications can operate successfully using the absolute empirical algorithm.

Absolute empirical memory cell

The active *absolute empirical memory cell* (AE cell), like the absolute predetermined memory cell (AP cell), uses a toggle switch and a relay to store and read an absolute value of "one" if its toggle switch is in the closed position, and a "zero" if its toggle switch is in the open position. A voltage appears at the output terminal only when the toggle switch is in the closed position, the output enable circuit is energized, and its input sense terminal is energized by a sensor switch. Thus the AE cell is sensitive only when its toggle switch is in the closed position and is insensitive when its toggle switch is in the open position. It is an active memory element because the input sense circuit is not connected directly to the output enable circuit. It is an empirical memory element because the value stored in the memory cell cannot be written in by manually placing the toggle switch in the "open" or "closed" position *before* the matrix is expected to produce behavior. A value can be written into the memory cell only through an actuator feedback circuit *after* an attempt has been made to produce action.

An electrically alterable absolute memory cell

The toggle switch of an *electrically alterable absolute memory element* can be placed in one position or the other using an electrical signal by making the armature of a toggle switch into a two-coil-relay, as shown in Figure 15.1.

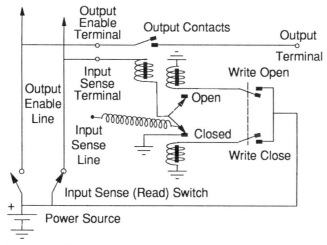

Figure 15.1. An *electrically alterable absolute memory element* uses a relay coil to flip its toggle switch to one position or the other.

The toggle can be moved to the closed position by applying a sufficient current to the *write closed coil* or can be moved to the open position by applying a sufficient current to the *write open coil*. The function and operation of the electrically alterable absolute memory element would be similar to a *binary memory register* in a digital computer if it had two output terminals. However, the electrically alterable absolute memory element used in a connection matrix has only one output terminal. Thus, it is an *absolute memory element* rather than a binary memory register. (See binary memory register in Appendix A.)

Toggle trip mechanism

As a practical matter, any *bistable device* may be difficult to switch from one position to the other. A mechanical toggle switch may be moved from one position to the other much more easily by using the *toggle trip mechanism* shown in Figure 15.2. The toggle trip mechanism provides a mechanical advantage that allows the toggle to be switched easily by an outside or remote electrical signal compared to a design that requires that the trip coils act upon the switch armature directly. The armature of the toggle trip mechanism is in the optimum position to the trip the toggle when the toggle is in its most difficult position.

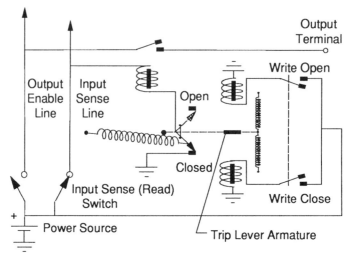

Figure 15.2. The *toggle trip mechanism* (TTM) keeps the armature of the trip mechanism equidistant from the trip coils and changes the position of the toggle switch by moving it past dead center.

Absolute empirical memory cell

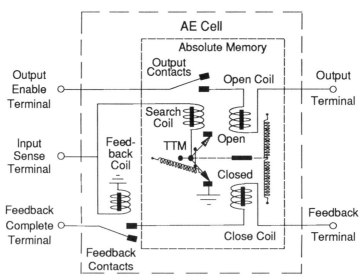

Figure 15.3. An *absolute empirical memory cell* can be set in a sensitive state if an output event occurs after an input event, and be left in an insensitive state if the output event does not occur after an input event.

The *absolute empirical memory cell*, shown in Figure 15.3, can be used to connect or disconnect a given input terminal to or from a given output terminal depending upon whether the output terminal was energized after the input terminal according to the absolute empirical algorithm. The absolute empirical memory cell consists of an input sense terminal, an output enable terminal, an output terminal, a feedback terminal, a feedback complete terminal, a search coil, a feedback coil, a bistable memory switch with a toggle trip mechanism, an open coil and a closed coil, and output and feedback contacts. The *state of sensitivity* of the AE cell is determined by the position of the toggle switch. The AE cell can produce a voltage at its output terminal only if its toggle switch is in the closed position, has a voltage on its output enable terminal, and receives a sufficient voltage to its input sense terminal to close its output contacts.

Operation of a single AE cell

The operation of a single AE cell is described in Table 15.2.

Table 15.2. Operation of a single AE cell

---

1. A voltage applied to the input sense terminal causes a current to flow in the feedback coil, causing the feedback contacts to close.
2. If the memory switch is in the open position, no output event is produced by this memory cell.
3. If the memory switch is in the closed position, the voltage at the input sense terminal also causes a current to flow in the search coil, causing the output contacts to close. If there is a voltage at the output enable terminal when the output contacts are closed, a voltage is produced at the output terminal. If the output terminal is grounded, a current can flow through the open coil.
4. The current in the open coil latches the output contacts closed and moves the memory switch to the open position by the action of the open coil on the armature of the toggle trip mechanism. (A current flow through the output terminal indicates that some output event has been attempted. This causes the voltage to the output enable terminal to be disconnected by means of an output disconnect circuit, not shown, after some delay, allowing the toggle trip mechanism to operate and return to its center position.)
5. If the output event associated with this AE cell occurs, even if it is not attempted by this AE cell, a voltage is created at its feedback terminal, causing a current to flow in its close coil if its feedback contacts are closed by a voltage applied to its input sense terminal.
6. The current in the close coil causes the memory switch to be moved to the closed position by the action of the close coil on the armature of the toggle trip mechanism. (A voltage at the feedback complete terminal can cause the voltage to the feedback terminal to be disconnected after some delay, allowing the toggle trip mechanism to return to its normal position.)

Thus, the toggle switch in an absolute empirical memory cell is switched to the open position whenever its input sense, output enable, and output terminals are energized simultaneously, and is switched to the closed position whenever its feedback terminal and its input sense terminal are energized simultaneously.

Summary of the operation of a single AE cell

A single AE cell operates according to the following logical statements when it receives a voltage at its input sense terminal and its output enable terminal:

1. If a signal is sent through its output terminal, and a signal comes back to its feedback terminal, the AE cell is restored to a sensitive state.
2. If a signal is not sent through its output terminal, and a signal comes back to its feedback terminal, the AE cell is placed in a sensitive state.
3. If a signal is sent through the output terminal, and no signal is sent back through the feedback terminal, the AE cell is left in an insensitive state.

An AE cell is similar to a *circuit breaker* in that it "trips off" when it selects an output (the output current is considered an overload). However, it differs from a circuit breaker in that it always trips off to a nonconductive state whenever an electrical load is applied above a certain level and is electrically "reset" to a conductive state by a remote signal rather than being manually "reset" to a conductive state at the breaker box.

Absolute empirical control matrix

An AE cell can be placed in a circuit that interrogates (reads) its state of sensitivity, provides the energy for it to change its state of sensitivity and to produce an output, and places (writes) it into a state of high sensitivity after the appropriate delays required by the discontinuous state ontology. This circuit is called an *absolute empirical control matrix.*

AE cell actuator feedback circuit

The AP cells in the absolute predetermined control matrix shown in Part II can be replaced with AE cells. Once an output latch is selected by an AE cell, the electromechanical actuator attempts to position itself according to the position of the energized actuator stator coil. A feedback signal must be sent from the feedback contacts at the location of the actual position of the actuator to the feedback close coil in the AE cell connected to the latched input terminal, as shown in Figure 15.4. Even though all of the AE cells connected to the actual feedback contacts receive a feedback signal, current can flow in the open coil only in the AE cell connected to the latched input terminal.

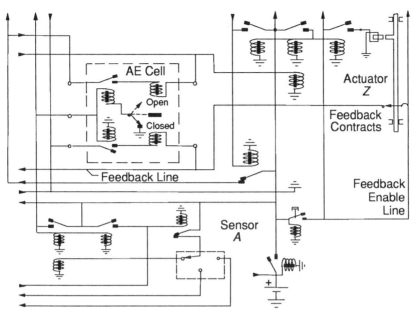

Figure 15.4. The *AE cell actuator feedback circuit* sends a signal to all of the AE cells connected to the actual position of the actuator.

AE cell actuator feedback delay circuit

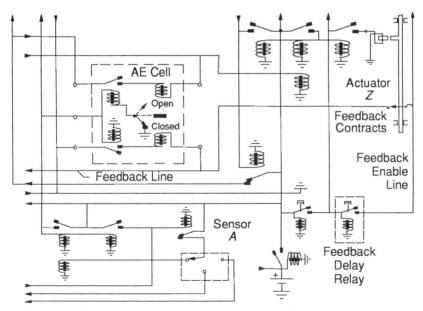

Figure 15.5. The *AE cell actuator feedback delay circuit* provides a period during which the actuator can move to its selected position without sending a feedback signal.

Many feedback contacts may be closed as the actuator moves to its new position. However, no feedback signal should be sent until the actuator has been given time to moved to its selected position or an operator can place the actuator in some desired position. After some predetermined delay, the *feedback delay relay* (Fig. 15.5) closes. This energizes the feedback enable line, allowing a feedback signal to be sent from the feedback contacts at the actual location of the actuator armature at that moment.

AE cell feedback complete disconnect

Only one position feedback signal can be sent in a given matrix in a given transition cycle according the empirical algorithm. Therefore, all of the feedback contacts must be disconnected after a feedback signal has been sent from one position. This is accomplished by the *feedback complete disconnect circuit* shown in Figure 15.6. Once a feedback signal has passed through the close coil of one AE cell, it flows to ground through the feedback complete latch. This closes the contacts of the latching coil and causes a current to flow in the feedback disconnect coil while a voltage is present on the actuator enable line. This also opens the contacts of the feedback disconnect coil and removes the voltage from all of the feedback contract for the remainder of the transition cycle. Because the voltage is removed from the feedback circuit, the toggle trip mechanism in the selected AE cell can return to its normal center position.

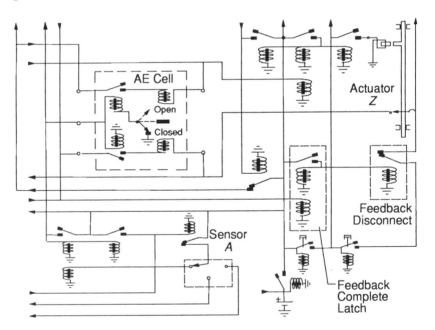

Figure 15.6. The *AE cell feedback latched disconnect circuit* allows a feedback signal to be sent from only one output device during each transition cycle.

AE cell end-of-cycle feedback disconnect circuit

The empirical algorithm also places a limit on how long a feedback signal can be sent after a transition cycle begins. The time of this *action period* is governed by the *end-of-cycle feedback disconnect circuit* shown in Figure 15.7.

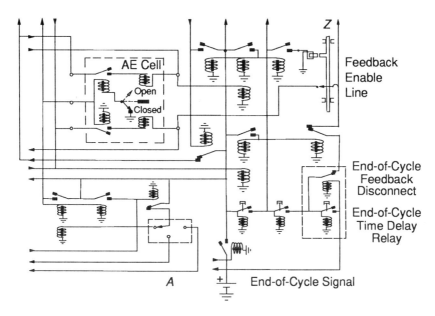

Figure 15.7. The *AE cell end-of-cycle feedback disconnect circuit* is controlled by a relay that disconnects the feedback enable line after a given delay.

The end-of-cycle delay relay energizes a second end-of-cycle feedback disconnect after a fixed period after the start of the action period. This disconnects the feedback enable line and prohibits a feedback signal from being sent if no feedback has occurred within the action period. If this condition occurs, all AE cells connected to the active sensor terminal may be left in an insensitive state. The end-of-cycle relay also sends an end-of-cycle signal to the outside timing source indicating that the *action period* of that matrix in that transition cycle is over. When the cycle switch is opened by the outside timing source, all disconnects and latching relays are returned to their normal positions in preparation for the start of the next transition cycle.

AE cell actuator brake circuit

While the transition cycle switch is in the open position, there is no power applied to the AE cell matrix or any actuator coil. However, the actuator must still be held firmly in some position until an actuator stator coil is energized during a transition cycle. The position of the actuator can be maintained by means of the *AE cell actuator brake circuit* shown in Figure 15.8.

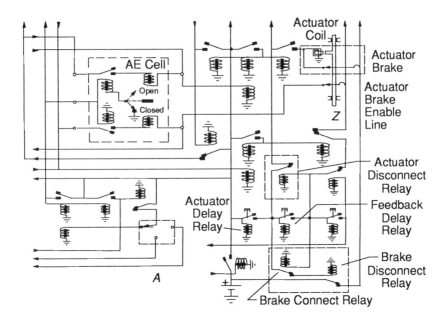

Figure 15.8. The *AE cell actuator brake circuit* holds the actuator in a fixed position between transition cycles.

The contacts of the brake disconnect relay are normally closed, allowing current to flow from the power source to the actuator brake despite the position of the cycle timing switch. This causes the actuator to be held at whatever position it is in by the current from the actuator brake circuit flowing through the stator coil at the actual location of the actuator armature except when the actuator must be positioned by the control matrix.

Operation of the AE cell actuator brake circuit

The actuator brake circuit maintains current to the actuator coil at the position of the closed actuator brake switch corresponding to the actual armature position and attempts to hold the actuator armature in this position until some time in the next transition cycle when the actuator delay relay times out again. When the feedback signal is sent from this actuator position, the feedback enable line is disconnected as described earlier. If the actuator brake is overpowered and the actuator armature is repositioned any time during the remainder of this period, it will be held by the energized brake circuit in whatever position it is placed and no other feedback signals will be sent. The AE cell actuator brake circuit operators according to the steps in Table 15.3.

Table 15.3. Operation of the AE cell actuator brake

---

1. After the actuator delay relay times out at the start of the action period when an AE cell is supposed to energize the selected actuator coil, the brake disconnect relay is opened by the current flowing through the contacts of the brake connect relay. This interrupts the voltage on the actuator brake enable line and releases the actuator brake. This allows an actuator coil or an outside agent to position the armature of the actuator.
2. The feedback delay relay times out at the beginning of the feedback period. This causes a feedback signal to be sent from the position of the armature at that moment, causes the actuator brake to be re-engaged, holding the actuator in this position even if it is not the position selected by the actuator coil, and causes the actuator coil to be disconnected for the remainder of the transition cycle.

Thus, the actuator brake is released to allow the matrix to position the actuator during its action period, and the actuator brake is engaged to maintain the actuator in a fixed position the rest of the time.

Application of a single AE cell System

The single AE cell system shown can be used to test the *reliability* of an output device. If the output device produces the action called for by its electromagnet within the time limits set by the cycle switch for every transition cycle and sends a feedback signal to the AE cell that caused it to operate, then that AE cell will be returned to a sensitive state. If the output device fails *once* to produce the action called for within the time limits in any cycle, that AE cell will be left in its insensitive state, and will be incapable of operating its output device. Examining the sensitivity of the AE cell will reveal whether the output device has ever failed to operate.

Absolute empirical control matrix

The *absolute empirical control matrix* shown in Figure 15.9 deals with a single input variable (A) having two possible values, (0) and (1). The sensor switch is connected to input terminal (1). It also has a single output variable (Z) consisting of two output values, (0) and (1). Because its variables have two values, it can be called a *binary memory matrix*. This matrix contains four AE cells, which are connected in parallel to the input and output terminals of the matrix.

The sensor switch is connected to value (1), and the memory switch in AE cell (1,0) is shown in the conductive state. This causes the output contacts of that AE cell to close at the start of a transition cycle. This causes the actuator to move to the (0) output position and the memory switch of the selecting AE cell to move to the nonconducting position (not shown as such) after the actuator delay period. The actuator feedback switch is therefore connected to the actuator feedback circuit connected to AE cells (0,0) and (1,0). Since the feedback contacts are closed in AE cells (1,0) and (1,1) only, the memory switch in only AE cell (1,0) is restored to the conductive state.

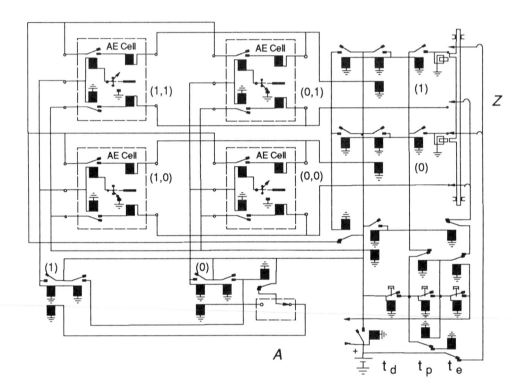

Figure 15.9. An *absolute empirical binary control matrix* records the last output value of its binary actuator, which occurs after a specific delay and within a specific period for each value of its binary sensor, as measured by the actuator feedback circuit.

Operation of an absolute empirical control matrix

The absolute empirical control matrix shown operates according to the sequence of steps summarized in Table 15.4 according to the absolute empirical algorithm. After a period of experiencing different input conditions and selecting different outputs, the absolute empirical control matrix will select the last output state that has been allowed to occur for each input state within the cycle operating times and delays in that AE cell matrix driver circuit.

Table 15.4. Operation of an absolute empirical control matrix

---

1. At the beginning of a transition cycle, a voltage is applied to the input sense terminal corresponding to a given input state at that moment.
2. The AE cell on this input column with a conducting memory switch causes a voltage to appear at the output terminal connected to this AE cell, and the memory switch in this AE cell is changed from conducting to nonconducting by the toggle trip mechanism.
3. After a specified delay (t_d), the actuator attempts to move to the selected output terminal. After another specific period of time (t_p), a voltage is applied by the actuator feedback circuit to whatever feedback terminal the actuator is connect to at this moment.
4. If the actuator arrives at the output terminal selected by the original AE cell within the specified time frame $(t_d) + (t_p)$, the original AE cell is restored to its original conducting state and the same output will be selected in the future for that input.
5. If the actuator does not arrive at the selected output terminal in the time frame called for by the controller, but arrives at another output terminal in that time frame, the AE cell at the intersection of the original input terminal and the actual output terminal is made conducting, and the controller will select that output in the future for that input.
6. If the actuator does not arrive at any terminal in the time frame called for by the controller, there will be no conducting AE cell connected to the original input terminal, and the controller will no longer attempt to select an output for that input.

Absolute empirical trinary control matrix

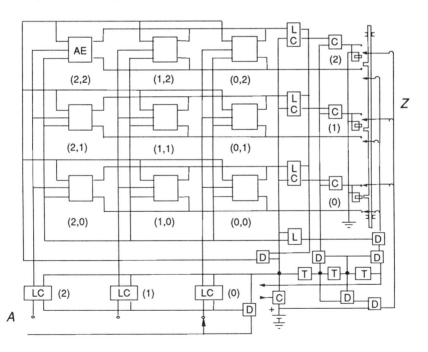

Figure 15.10. An *absolute empirical trinary control matrix* has three values of its input and output variables *A* and Z.

The number of values of the input and output variables of an AE cell memory matrix can be increased indefinitely. If the number of values of the binary matrix shown above is increased to three, as shown in Figure 15.10, the matrix can be called a *absolute empirical trinary control matrix*. The *maximum variety* of behavior in a scalar matrix for a given number of AE cells is produced if the number of output values matches the number of input values, as postulated by the theorem on *matched input/output variety* in Chapter 6. If a matrix has more input values than output values, it can be called an *input matrix*. If a matrix has more output values than input values, it can be called an *output Matrix*.

Application of an AE cell control matrix

An absolute empirical control matrix can be used effectively as an empirical controller for any application having one input variable and one output variable when a given output value always occurs with a given input value and when there are a relatively few values of these variables. If the resolutions of these input and output variables are 5%, as in the ship auto-pilot shown earlier, the input and output variables both require 20 terminals. This matrix would contain 400 AE cells. Of these 400 AE cells, 20 are placed in the conductive state to specify an output state for each possible input state. The rules of a connection matrix apply to an AE cell memory matrix in that only one high-sensitivity AE cell can exist on a given input sense terminal, but any number of high-sensitivity AE cells (limited only by the number of input terminals) can be connected to a given output terminal.

Transform conditioning

The absolute empirical control matrix requires one more circuit than the absolute predetermined control matrix shown in Part II. Besides connecting a specific input terminal to a given output terminal, the AE cell matrix must establish a connection at the intersection of the actual output terminal and the input terminal that is active at that time. This is accomplished by an *actuator feedback signal,* which carries out the empirical function of *transform conditioning*.

Actuator feedback signal

Any control system that records the actual output states for each input state encountered while the system is in use is called an empirical controller and forms the basis of the absolute empirical control system shown in Figure 15.11.

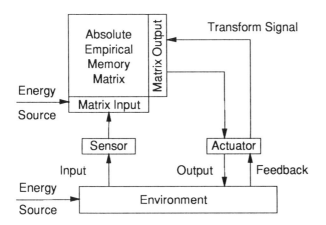

Figure 15.11. An *absolute empirical controller with transform conditioning* records the actual output state (transform) that occurs after some delay from each input state (operand) as each input state occurs.

If the environment allows some actions to be carried out under given conditions, and prohibits other actions from being carried out, the absolute empirical controller will develop a *repertoire of behavior* that can be carried out within the constraints of the environment.

Interferences imposed by the limitations of the machine

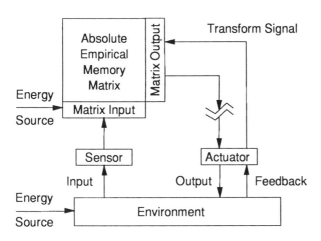

Figure 15.12. The *internal interferences* imposed by the physical limitations of the machine may affect the behavior of an empirical system.

Many factors may prohibit the actions called for by the empirical controller from being carried out. For example, the controller may call for an actuator position that is beyond the physical limits of its actuator (Fig. 15.12). Since this action cannot be carried out, the controller will learn to select a position that is within the range of its actuator. In an empirical system with more than one actuator, one actuator may strike into another, not allowing the behavior of one or both actuators to be carried out. The behavior of this empirical system will change automatically until its actuators no longer interfere.

Interferences imposed by the environment

A mobile robot controlled by an absolute empirical matrix may try to move through obstacles such as walls (Fig. 15.13). Since this behavior cannot be carried out, the behavior of the robot will change until it finds behavior that allows it to move around without hitting any walls.

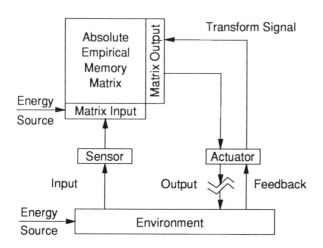

Figure 15.13. Environmental interferences cause the behavior of an empirical machine to change until it reflects the constraints imposed by the environment.

Thus, an empirical machine will learn to adjust to the limitations imposed by an operator, the environment, and its own physical abilities.

Dominance of the operator

The absolute empirical controller shown may select an action that does not coincide with the wishes of the operator (Fig. 15.14). In this case, the operator may have to overpower the empirical machine to make sure that the desired actuator state occurs.

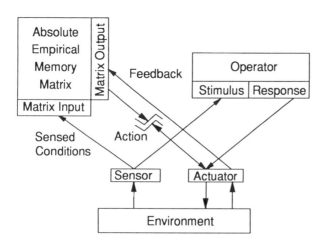

Figure 15.14. The operator must be the domi-nant controller in an empirical control system.

In some cases, the operator may have to be physically stronger than the actuators of the empirical machine to prohibit its actuators from carrying out behavior that is not considered desired. In the example given, an operator may move an actuator to some desired position while the actuator brake is applied, or the operator may prohibit the actuator from reaching the position called for by the empirical memory matrix. In any case, the position recorded for a given input state is determined by the actual position of the actuator at the beginning of the feedback period. By recording the positions produced by an operator, an empirical machine can develop a repertoire of successful behavior that reflects the skill of the operator in that application.

Application for an absolute empirical controller

The absolute empirical controller can be used for any application previously carried out by an absolute predetermined control matrix. For example, the absolute empirical controller can be used to determine the behavior of a ship auto-pilot, which attempts to steer the ship according to a direction error signal by remembering and producing the input/output relations produced by a helmsman (Fig. 15.15). The helmsman may not be aware that he or she is programming an empirical auto-pilot when he or she steers the ship. After a period of learning, during which time the controller and the helmsman attempt to produce a given rudder position for each direction error encountered, the absolute empirical controller will produce the last rudder position produced by the helmsman for each error signal encountered.

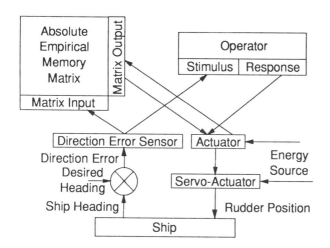

Figure 15.15. An absolute empirical auto-pilot uses an absolute empirical matrix with an actuator feedback circuit, position sensors, position actuators, and a skilled operator to connect specific sensor states to specific actuator states.

An absolute empirical control matrix with 400 AE cells will provide the same quality of behavior as a predetermined control matrix with 400 AP cells. In the example of the ship auto-pilot given, the helmsman may dominate the control system by manipulating the *servo-valve*, which operates at a low force level, rather than the rudder itself, which operates at an amplified force level.

Coded unitary AE cell controllers

The AE cell matrix shown consists of a single input variable and a single output variable. The matrix can deal with two or more unitary input variables if they are decoded into a single matrix input variable and can deal with two or more unitary output variables if they are encoded from the single matrix output variable. The actuator feedback terminals must also be decoded to provide a single variable transform signal to the empirical matrix.

AE cell coded unitary controller

A *coded unitary absolute empirical controller* is formed when a set of input variables is decoded into an input value to a single variable AE cell matrix, which transforms it into a single variable matrix output state, which in turn is converted into a set of values of the output variables by an output encoder in the same manner described in the coded unitary DS predetermined controlled machine shown in Part II. These output variables must then be decoded to produce a single variable actuator feedback signal to the empirical matrix, as shown in Figure 15.16.

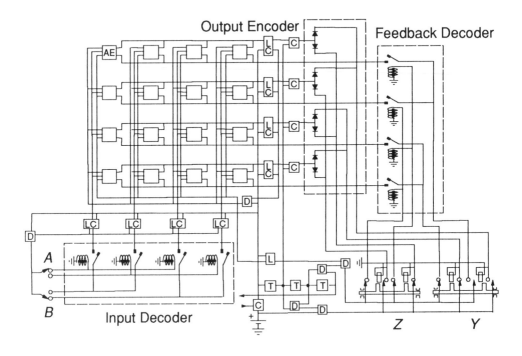

Figure 15.16. An *AE cell coded unitary controller* can deal effectively with two or more sensor and actuator variables in a unitary task environment where a given output always occurs after a given input.

The AE cell matrix deals with the single input and output terminals created by the input decoder and output encoder. To feed back the actual output that occurs in each transition cycle, a feedback decoder must be connected to the actuator feedback terminals. It then converts each combination of output values into a value of a single feedback variable, which is then used by the AE cell matrix to set the AE cell at the intersection of the actual input and output state into a conductive state. To prevent more than one feedback signal from being sent in a given transition cycle, the feedback complete terminals of all of the AE cells in the matrix are connected to a feedback complete disconnect.

AE cell coded unitary control system

The *AE cell coded unitary control system* can be used as a self-programming ship direction control system that has two or more input and output variables, as shown in Figure 15.17.

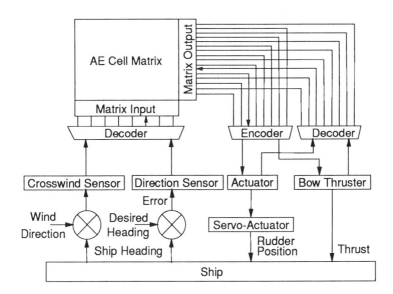

Figure 15.17. An *AE cell coded unitary ship direction control system* uses an AE cell control matrix and requires a feedback decoder to measure and decode the actual actuator and bow thruster positions in each transition cycle.

This absolute empirical system can be used effectively in a unitary application where the operator always produces the same responses for each input condition. However, the AE cell coded unitary controller suffers the same deficiencies as a coded unitary predetermined controller in that its matrix input and output terminals are committed to combinations of values that may never be needed, and the system is committed to a unitary organization.

Absolute empirical multivariable matrices

Since AE cells are active memory elements, they can be used to create absolute empirical multivariable input and output matrices similar to the absolute predetermined multivariable matrices shown in Part II. Like a predetermined matrix, the input or the output side of an empirical matrix can have multiple variables. These input and output matrices can be combined into a single absolute empirical monolithic control matrix.

AE cell binary multivariable input matrix

Two or more AE cell submatrices can be connected in parallel to form an *AE cell binary multivariable input matrix,* which can deal with multiple input variables and a single output variable, as shown in Figure 15.18.

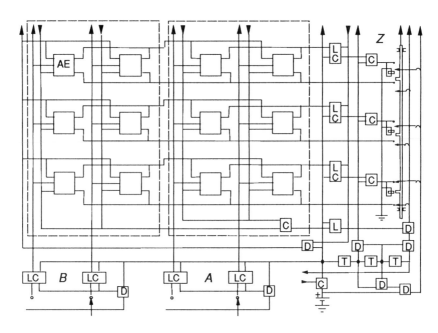

Figure 15.18. An *AE cell binary multivariable input matrix* can produce a different output value for different combinations of values of the multiple input variables.

The AE cell multivariable input matrix is the same as an AP cell multivariable input matrix, shown in Part II, except that it contains AE cells instead of AP cells, and contains a transform conditioning circuit that measures the actual output state at given moments in time. It uses this information to change the conductive state of the matrix. The input matrix shown is called a *binary* multivariable input matrix because each of its input variables have *two values.*

AE cell trinary multivariable input matrix

The number of values of each variable in an AE cell multivariable input matrix can be increased indefinitely to match the variety of its sensors and actuators if each submatrix is connected as shown in Figure 15.19.

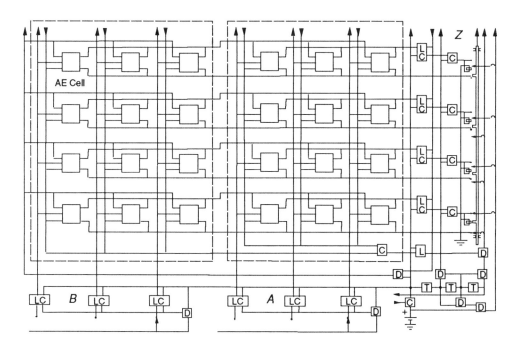

Figure 15.19. In an *AE cell trinary multivariable input matrix,* each input variable has three possible values.

However, the *AE cell trinary multivariable input matrix* can produce a unique output from the maximum number of combinations of input values with a minimum number of AE cells. For example, an AE cell *binary* multivariable input matrix with three input variables uses $2 \times 3 = 6$ AE cells per transition and can produce $2^3 = 8$ unique transitions. An AE cell *trinary* multivariable input matrix with two input variables also has 6 AE cells per transition, but can produce $3^2 = 9$ transitions. The improvement of a trinary multivariable input matrix over a binary multivariable input matrix increases as the number of input variables increases, as shown in Part II.

AE cell multivariable input matrix control system

An AE cell multivariable input matrix can be used in any absolute application having two or more input variables and one output variable. For example, an AE cell multivariable input matrix can be used as a *ship direction control system* with two input variables, such as direction error and crosswind velocity, and a single output rudder position variable, as shown in Figure 15.20.

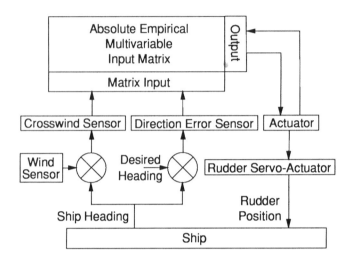

Figure 15.20. An *AE cell multivariable input matrix control system* can use two or more input variables to control one output variable.

The number of values of each input variable must be increased to match the variety of each sensor. For example, a system with a 5% resolution would require twenty terminals for each sensor and actuator variable. This system can learn to produce twenty different rudder positions for any twenty of 400 different combinations of values of direction error and crosswind if the helmsman always produces the same rudder response for each of these twenty combinations of values of the crosswind and direction error sensors.

AE cell multivariable output matrix

Two or more AE cell matrices can also be connected in parallel to form an *AE cell multivariable output matrix* that can deal with a single input variable and multiple output variables (Fig. 15.21). The AE cell multivariable output matrix is the same as a AP cell multivariable output matrix shown in Part II except that it contains a transform conditioning circuit that measures the actual output state of each actuator variable at given moments in time and uses this information to change the distribution of sensitivities of the AE cells in the matrix.

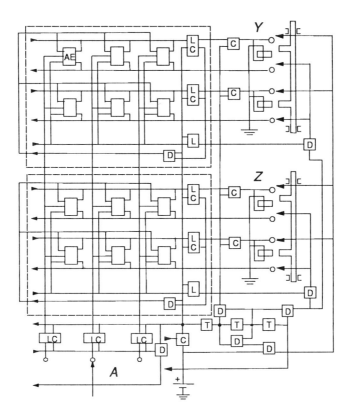

Figure 15.21. An *AE cell multivariable output matrix* can produce a different combination of values of its multiple output variables for different values of its single input variable.

Rotated AE cell multivariable output matrix

The AE cell multivariable output matrix shown can be rotated 90° to show it in the position that output matrices are used in duplex controllers (Fig. 15.22). With an output matrix shown in this position, the forward selection process occurs across the page from the input variable *A*, and the delays and action proceed downward toward the actuator variables *Y* and *Z*. The transform condition process occurs upwards from the actuator variables.

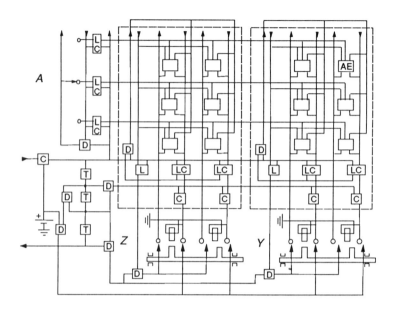

Figure 15.22. A *rotated AE cell multivariable output matrix* shows the more common orientation of an output matrix.

AE cell trinary multivariable output matrix

The number of values of each output variable in an AE cell output matrix can be increased indefinitely if connected as shown in the *AE cell trinary multivariable output matrix* (Fig. 15.23). The AE cell trinary multivariable output matrix makes the most efficient use of its AE cells. For example, an absolute empirical *binary* multivariable output matrix with 3 output variables requires $2 \times 3 = 6$ AE cells per transition and can produce any one of $2^3 = 8$ combinations of output values. An AE cell *trinary* multivariable output matrix with 2 output variables also requires $3 \times 2 = 6$ AE cells per transition but can produce $3^2 = 9$ combinations of output values. The advantage of the trinary matrix over the binary matrix increases as the number of variables increases as shown in Part II. The trinary matrix can produce more combinations of output variables for a given number of AE cells than a matrix based upon any other number system.

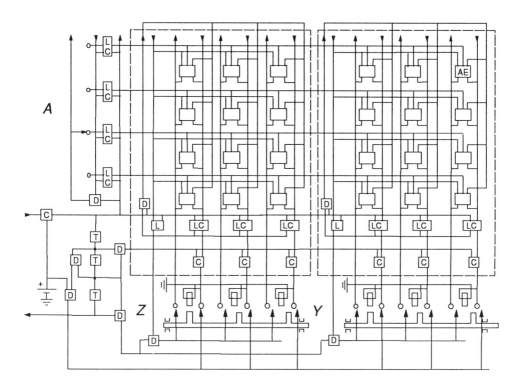

Figure 15.23. An *AE cell trinary multivariable output matrix* uses three values in each of its output variables.

AE cell multivariable output matrix control system

An AE cell multivariable output matrix can be used in any absolute application having a single input variable and two or more output variables. For example, an AE cell multivariable output matrix can be used as a *ship direction control system* with one input variable, such as direction error, and two output variables, such as rudder position and bow thruster, as shown in Figure 15.24. The number of values of each output variable must be increases to match the variety of each actuator. For example, if each actuator variable has twenty terminals, each actuator can be positioned with a resolution of 5%. If the input variable has a resolution of 5%, the control matrix would require $20 \times 40 = 800$ AE cells. This system can learn to produce different combinations of rudder correction and bow thruster action for each sum of the direction error and crosswind signal if the helmsman always produces the same rudder response for each different combination of crosswind and direction error.

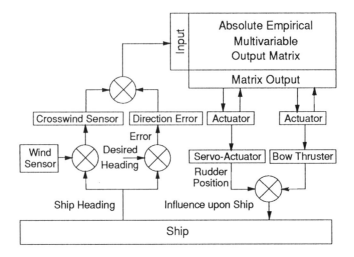

Figure 15.24. An *AE cell multivariable output matrix control system* can control two or more actuator variables from a single sensor variable.

AE cell binary monolithic control matrix

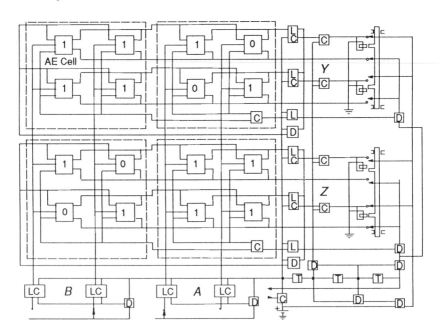

Figure 15.25. An *AE cell binary monolithic control matrix* can learn to connect different combinations of values of its input variables with different combinations of values of its output variables in a unitary manner, or can learn to produce one or more lines of behavior that occur at different times in a synchronous diverse manner.

The design features of the AE cell multivariable input matrix can be combined with the design features of the AE cell multivariable output matrix, forming the *AE cell binary monolithic control matrix* shown in Figure 15.25. An AE cell monolithic control matrix cannot learn asynchronous diverse behavior because it cannot create a null submatrix, which requires two or more memory cells on a given input terminal in a given submatrix to be in a conductive state. The AE cell monolithic control matrix works best when its input variety matches its output variety.

AE cell trinary monolithic control matrix

Any number of additional input and output variables can be added to the AE cell monolithic control matrix shown, and any number of values can be added to each of these variables. If one value is added to each sensor and actuator variable, the *AE cell trinary monolithic control matrix* shown in Figure 15.26 is formed.

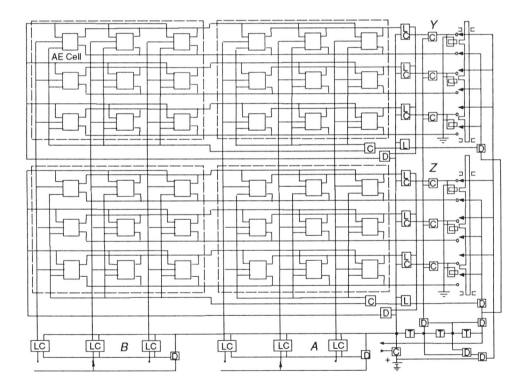

Figure 15.26. An *AE cell trinary monolithic control matrix* can produce more unique combinations of values of its actuator variables for more unique combinations of values of its sensor variables than a binary monolithic control matrix with the same number of AE cells.

The advantage of the trinary monolithic control matrix over the binary monolithic control matrix increases as the number of sensor and actuator variables increases. For example, a binary monolithic control matrix with six sensor and actuator variables requires $12 \times 12 = 144$ AE cells, and can connect any one of $2^6 = 64$ sensor states to any one of $2^6 = 64$ actuator states. A trinary monolithic control matrix with four sensor variables and four actuator variables also requires 144 AE cells, but can connect any one of $3^4 = 81$ sensor states to any one of $3^4 = 81$ actuator states.

Application of an absolute empirical monolithic control system

The multiple input and output variables of the *ship direction control system* shown in Figure 15.27 can be connected by an absolute empirical monolithic control matrix.

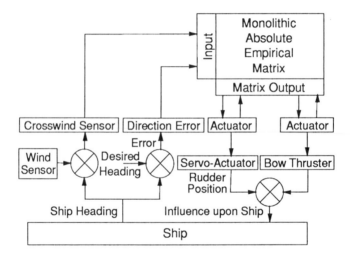

Figure 15.27. An *application of an absolute empirical monolithic control system* is a multivariable ship direction control system that requires a unique combination of values of its output variables for each combination of values of its input variables.

The AE cell monolithic control matrix can learn the last combination of output values produced by a helmsman for each combination of input values. The resolution of the sensor and actuator variables is limited to about 5% because it is impractical to make multiposition sensor switches with more than twenty or thirty contacts, and vernier actuators with more than twenty to thirty coils. For greater resolution, it is necessary to digitize the sensor and actuator variables as shown in the next chapter.

Chapter 16
Digitized Absolute Empirical Controlled Machines

If the number of values of the sensor and actuator variables is more than about eight, consideration should be given to *digitizing* these variable into binary or trinary aggregate variables in the manner shown in the digitized predetermined matrices in Part II. The maximum resolution for the minimum number of AE cells can be achieved using a trinary digitized system.

Binary digitized AE cell matrices

The binary digitized system is the simplest possible digitizing system although it is not the most efficient in terms of the number of memory cells required to provide a given resolution of the sensor and actuator variables.

AE cell binary digitized input matrix

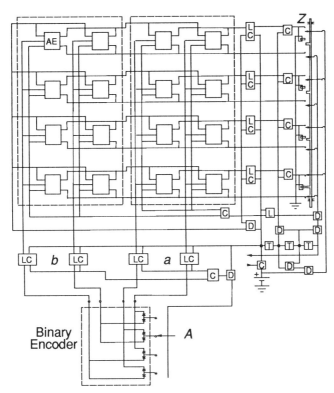

Figure 16.1. An *AE cell binary digitized input matrix* can use a set of aggregate variables to represent a value of an input variable.

An input variable can be broken into a set of *aggregate variables* using a binary encoder. The encoder can be connected to an AE cell multivariable input matrix, as shown in Figure 16.1. An *AE cell binary digitized input matrix* can be used effectively when the sensor variables have many different possible values. It can recognize many more input values using fewer input terminals than an undigitized matrix, and can learn to produce a specific output based upon all of its sensor aggregate variables, or can produce a general output based values of its most significant sensor aggregate variables.

AE cell binary digitized input matrix with three aggregate variables

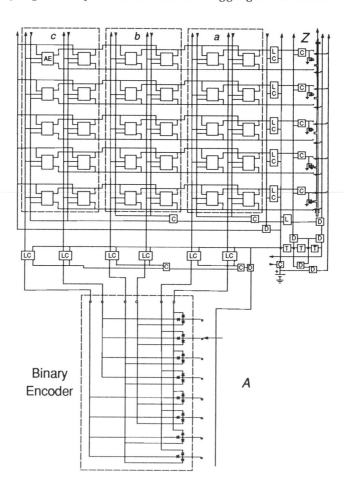

Figure 16.2. An *AE cell binary digitized input matrix with three aggregate variables* can identify more values of its input variable than a system with only two aggregate variables.

The number of aggregate variable in a digitized input matrix can be increased indefinitely if the memory cells are connected as shown in the *AE cell binary digitized input matrix with three aggregate variables* in Figure 16.2. The number of distinctly different input states *(variety)* of a digitized input variable increases according to the equation $V = x^n$, where V is the variety, x is the number of values of the aggregate variables, and n is the number of aggregate variables. For example, an AE cell binary digitized input matrix with 8 binary aggregate variables can identify any one of 2^8 = 256 different input values, providing a sensor resolution of 0.4% using only 2 * 8 = 16 AE cells per identification. An undigitized input matrix requires 256 AE cells to identify any one of 256 different input values.

Application of an AE cell binary digitized input matrix

An *engine temperature alarm system* may require a high-variety sensor, since an engine is usually designed to operate over a wide temperature range including a temperature close to the boiling point. However, only a few different output signals may need to be sent, such as: "engine temperature below normal," "engine temperature normal," or "warning, engine too hot." This control system is can be dealt with using an AE cell binary digitized input matrix, as shown in Figure 16.3.

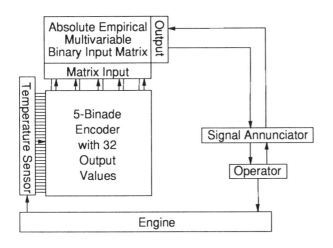

Figure 16.3. An application of an AE cell binary digitized input matrix is an engine temperature alarm system.

The AE cell binary digitized input matrix can be programmed by an operator specifying a value of the output *annunciator* for each temperature encountered. Any one of 32 different sensor values can be identified by the AE cell binary digitized input matrix with 5 aggregate variables, allowing an alarm to sound if the engine reaches some fairly precise critical temperature.

AE cell binary digitized multivariable input matrix

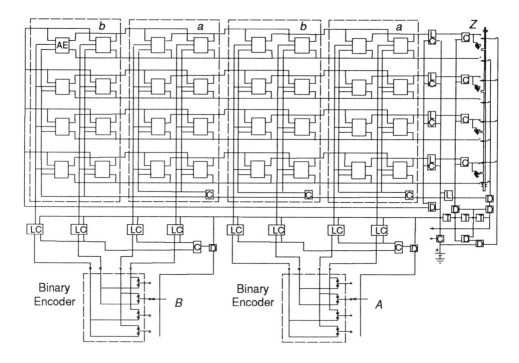

Figure 16.4. An *AE cell binary digitized multivariable input matrix* can handle two or more medium-resolution sensor variables.

The *AE cell binary digitized multivariable input matrix* shown in Figure 16.4 can deal with two or more separately digitized sensor variables. It can learn to produce a specific output based upon all of its sensor variables in a unitary manner, or can produce one (synchronous) diverse line of behavior at a time based upon less than all of its sensor variables.

Application of an AE cell binary digitized multivariable input matrix

The resolution of the crosswind and direction error sensors in a ship direction control system can be greatly increased by digitizing the input variables of an AE cell binary digitized multivariable input matrix without greatly increasing the number of AE cells. The output variable of the matrix can be used to control a single actuator like the rudder servo-actuator in the *absolute empirical ship direction control system* shown in Figure 16.5. The absolute empirical binary digitized multivariable input matrix can learn to keep a ship on a straight course if the helmsman always produces the correct specific rudder position for each combination of values of the crosswind and direction error sensors.

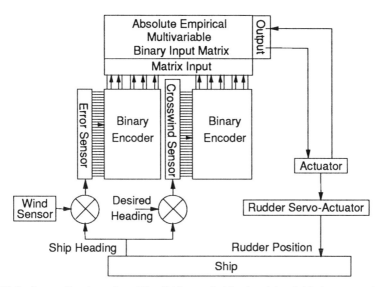

Figure 16.5. An *application of an AE cell binary digitized multivariable input matrix* is an empirical ship direction control system that has multiple input variables and one output variable.

AE cell binary digitized output matrix

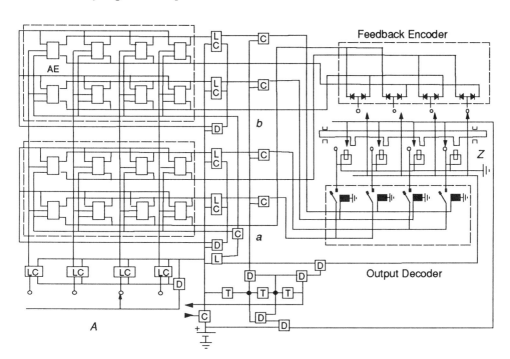

Figure 16.6. An *AE cell binary digitized output matrix* can produce many different output states using a relatively few output terminals.

The actuator variable of an AE cell multivariable output matrix can also be broken into a set of *binary aggregate variables*, as shown in Figure 16.6. Each combination of activation of the output terminals produces a unique output state. This greatly expands the total number of possible output states. For example, the three binary aggregate variables can produce eight output states. This feature is very desirable when a relatively few medium-resolution output states are required in the control task.

Rotated AE cell digitized output matrix

To show the aggregate variables of the binary digitized output matrix in the orientation used in a duplex system, it is desirable to rotate the digitized output matrix by 90°, as shown in Figure 16.7.

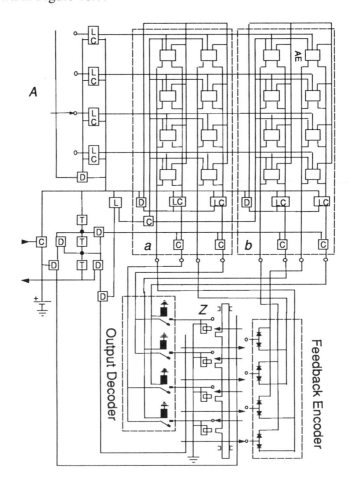

Figure 16.7. A *rotated AE cell binary digitized output matrix* shows the orientation of an output matrix in a duplex controller.

In the rotated AE cell digitized output matrix, the *forward selection process* proceeds horizontally, and the *action and transform conditioning processes* proceed vertically.

AE cell digitized output matrix with three binary aggregate variables

The number of aggregate variable can be increased indefinite if the AE cells are connected as shown in the *AE cell digitized output matrix with three binary aggregate variables* in Figure 16.8.

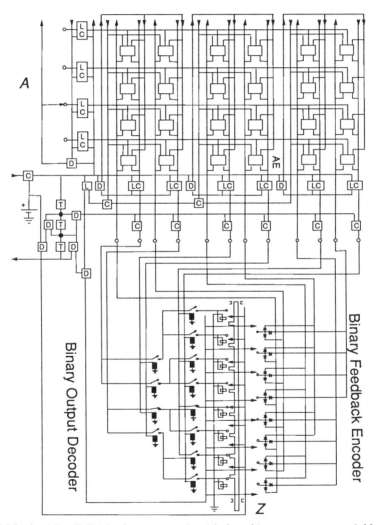

Figure 16.8. An *AE cell digitized output matrix with three binary aggregate variables* can produce twice as many output values using only two more AE cells per transition than a digitized output matrix with two binary aggregate variables.

The digitized output matrix with three binary aggregate variables can produce 2^3 = 8 output values using only 6 AE cells per output value compared to the digitized output matrix shown earlier, which can produce 2^2 = 4 output values using 4 AE cells. If the number of binary aggregate variables is increased to 8, the binary digitized output matrix can produce 2^8 = 256 different output values using only $2 \times 8 = 16$ AE cells per output value. An undigitized output matrix would require 256 AE cells to produce an output value with the same 0.4% resolution.

An application of an AE cell binary digitized output matrix

A linear actuator in a controlled machine system may have to move to only a few very precise locations based upon information from a single input variable. An AE cell binary digitized output matrix with a *medium-resolution linear actuator* can be taught to move to very precise locations (Fig 16.9).

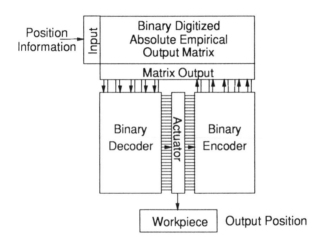

Figure 16.9. An *application for an AE cell binary digitized output matrix* is a medium-resolution linear actuator that must produce a relatively few fairly precise positions according to the values of a single input variable.

The input variable may consist of only two or three values such as "home location," "out position," and "in position." An AE cell binary digitized output matrix with 5 aggregate output variables can produce any one of $2^5 = 32$ different actuator positions. Many single-axis medium-resolution motion control machines, such as material sorting or palletizing devices, fall into this category.

AE cell binary digitized multivariable output matrix

An AE cell multivariable output matrix can be used with multiple digitized actuator variables, forming the *AE cell binary digitized multivariable output matrix* shown in Figure 16.10.

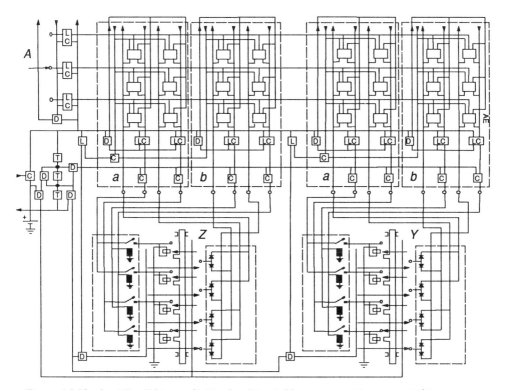

Figure 16.10. An *AE cell binary digitized multivariable output matrix* can control two or more high-resolution output variables.

The AE cell binary digitized multivariable output matrix can be taught to produce unitary behavior that positions both actuators into unique combinations of values for each value of the input variable, or it can position one actuator and hold the position of the other constant for each value of the input variable in a synchronous diverse manner.

Application of an AE cell binary digitized multivariable output matrix

The AE cell binary digitized multivariable output matrix can be used in a *ship direction control system* with the crosswind sensor and direction error sensor summed by a differential and used as a single input variable to control the two output variables consisting of rudder position actuator and a bow thruster, as shown in Figure 16.11. The digitized output variables can provide far more resolution for a given number of memory cells than an undigitized system. For example, the output matrix shown with two binary digitized actuator variables, each with four aggregate variables, can select any one of $2^8 = 256$ different combinations of values using only $2 \times 8 = 16$ AE cells per selection.

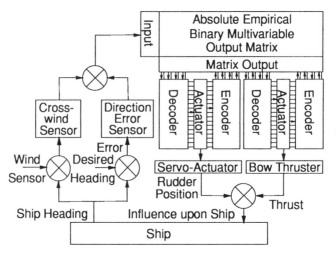

Figure 16.11. An *application for an AE cell binary digitized multivariable output matrix* is in a ship direction control system with two or more medium-resolution output devices.

AE cell binary digitized monolithic control matrix

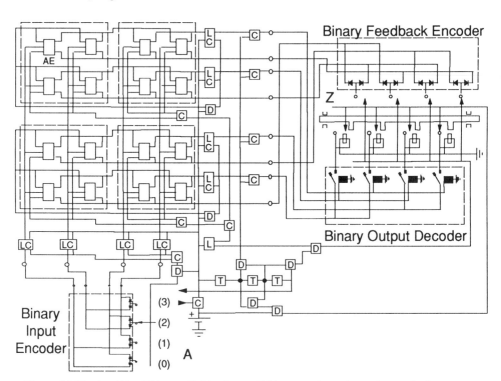

Figure 16.12. An *AE cell binary digitized monolithic control matrix* can be used when many values of the input variable are used to produce many different values of the output variable.

A sensor encoder, an actuator decoder, and a feedback encoder can be added to the AP cell monolithic control matrix shown earlier, forming the *AE cell binary digitized monolithic control matrix* shown in Figure 16.12. The AE cell binary digitized monolithic control matrix can learn to produce a specific output based upon all of its sensor and actuator variables in a unitary manner, or it can produce one or more (synchronous) diverse lines of behavior based on less than all of its sensor variables. (An AE cell monolithic controllers cannot learn to produce asynchronous diverse behavior because it cannot set all of the AE cells in a submatrix into the conductive state.) The AE cell binary digitized monolithic control matrix is shown because it is the simplest digitized system. However, an AE cell trinary digitized monolithic control matrix can learn to produce more unique transitions for a given number of memory cells than the AE cell binary digitized monolithic control matrix shown.

Application of an AE cell binary digitized monolithic control matrix

The multiple variables of an AE cell monolithic control matrix can be used as aggregate variables to digitize a single input and single output variable, as shown in the *AE cell binary digitized monolithic temperature control system* in Figure 16.13.

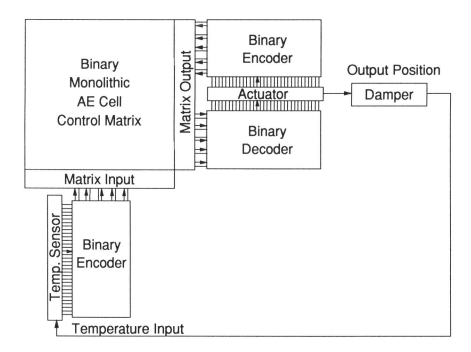

Figure 16.13. An AE cell binary digitized monolithic control matrix can be used to position a damper in a *temperature control system*.

An AE cell binary digitized monolithic control matrix with 5 sensor and 5 actuator aggregate variables can learn to produce any one of 32 damper positions whenever any one of 32 different temperature values occurs, and it uses only 100 AE cells. An undigitized scalar matrix would require 1024 AE cells to do the same job.

AE cell binary digitized multivariable monolithic control matrix

The aggregate variables of an AE cell monolithic control matrix can be used to digitize multiple input and multiple output variables, as shown in the *AE cell binary digitized multivariable monolithic control matrix* (Fig. 16.14).

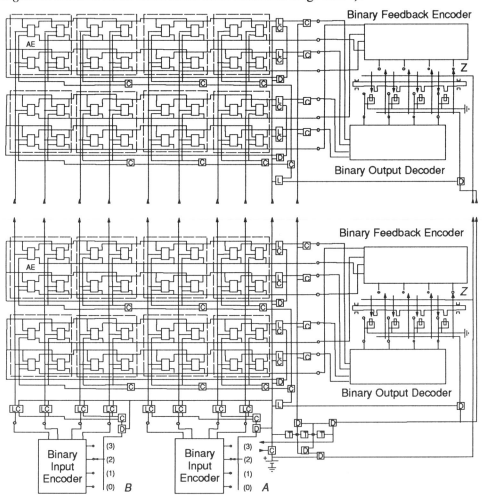

Figure 16.14. An *AE cell multivariable binary digitized monolithic control matrix* can learn to produce outputs based upon all of its sensor and actuator variables in a unitary manner, or it can produce one or more (synchronous) diverse lines of behavior based on less than all of its sensor and actuator variables.

Any number of digitized sensor and actuator variables can be added to the AE cell binary digitized multivariable monolithic control matrix shown. Yet, the number of AE cells in this controller increases according to the product of the number of sensor and actuator variables. This type of controller may be the most useful and effective empirical controller for most absolute applications having multiple medium-resolution sensor and actuator variables.

AE cell binary digitized multivariable monolithic control system

The number of sensor and actuator variables in the AE cell binary digitized monolithic control matrix can be increased indefinitely, forming an AE cell binary digitized multivariable monolithic control system, as shown in Figure 16.15.

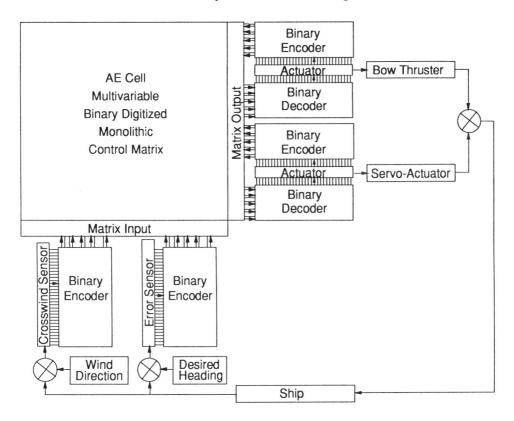

Figure 16.15. An AE cell binary digitized multivariable monolithic control system can be used to control two or more sensor variables and two or more actuator variables in a *ship direction control system.*

The AE cell binary digitized multivariable monolithic controller is a useful and versatile absolute empirical multivariable controller since it can learn to produce unitary or synchronous diverse relations between any number of sensor and actuator variables.

Trinary digitized AE cell control matrices

All digital computers and most control systems are based upon the binary system. Many control system designers believe that the binary system is the best system to deal with high-resolution variables. However, a quadrary (four-value) system has the same efficiency as a binary system, while the trinary digitized system, shown in this section, is an even better choice in almost all applications.

AE cell trinary digitized input matrix

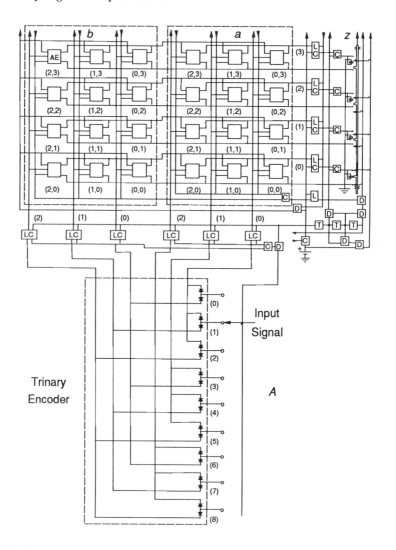

Figure 16.16. An *AE cell trinary digitized input matrix* can provide a higher resolution of its input variable for a given number of AE cells than an AE cell binary digitized input matrix.

A multivariable input matrix can be digitized with a *trinary encoder* to form the *AE cell trinary digitized input matrix* shown in Figure 16.16. A binary digitized input matrix with 12 aggregate variables requires $2 \times 12 = 24$ AE cells per output value to identify any one of $2^{12} = 4096$ input values. A trinary digitized input matrix with 8 aggregate variables also requires $3 \times 8 = 24$ AE cells but can identify any one of $3^8 = 6561$ input values. Thus, the trinary digitized input matrix can provide a higher resolution for a given number of AE cells than the AE cell binary digitized input matrix shown earlier and any other type of digitized system.

Application of an AE cell trinary digitized input matrix

An *engine temperature warning system* can also be digitized using an AE cell trinary digitized input matrix, as shown in Figure 16.17.

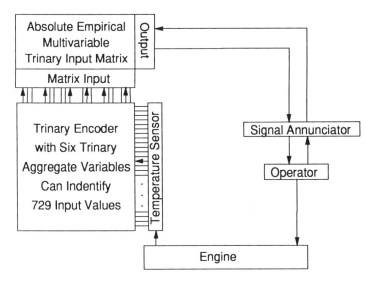

Figure 16.17. An *application of an AE cell trinary digitized input matrix* is an engine temperature warning system used to identify a critical temperature and to sound a warning. It provides a higher resolution than a binary digitized system with the same number of memory cells.

If 9 binary aggregate variables are used to digitize the temperature sensor, providing a variety of $2^9 = 512$ input values, then $2 \times 9 = 18$ AE cells are required to produce each output value. If 6 trinary aggregate variables are used to digitize the temperature sensor, providing a variety of $3^6 = 729$ input values, then $3 \times 6 = 18$ AE cells also are required to produce each output value. If an undigitized input matrix were used with a sensor having this resolution, then 729 AE cells would be required to produce each output.

AE cell trinary digitized multivariable input matrix

Two or more sensor variables can be digitized using trinary encoders, forming the
AE cell trinary digitized multivariable input matrix shown in Figure 16.18. A trinary
digitized system is usually the preferred embodiment for any digitized system.

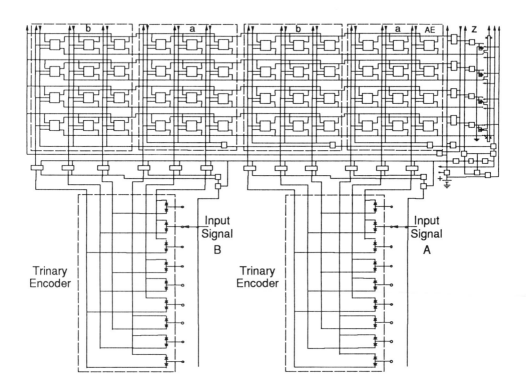

Figure 16.18. An *AE cell trinary digitized multivariable input matrix* can identify any possible
combination of values of its digitized input variables using fewer AE cells than any other system.

Application of an AE cell trinary digitized multivariable input matrix

The *multivariable ship direction control system* shown previously can also be digitized using trinary encoders (Fig. 16.19), resulting in a higher-resolution system
using the same number of memory cells as a binary system. If 6 trinary aggregate
variables are used for each of 2 input variables, providing a resolution of 1 part in
729, then $3 \times 12 = 36$ AE cells can be used to identify any one of $729 \times 729 = 531,441$
different combinations of input values. If 9 binary aggregate variables are used for

each input variable, providing a sensor resolution of 1 part in 512, then $2 \times 18 = 36$ AE cells are also required to produce each output value, and any one of $512 \times 512 = 262,144$ different combinations of input values can be identified. This AE cell trinary digitized multivariable input matrix can identify more than twice as many combinations of values of multiple input variables, using the same number of AE cells as the binary digitized system. If a trinary decoder were used to combine these input variables, then the coded input matrix would require 531,441 input terminals and would require 531,441 AE cells to make each transition instead of the 36 AE cells required in the trinary digitized input matrix with the same sensor resolution.

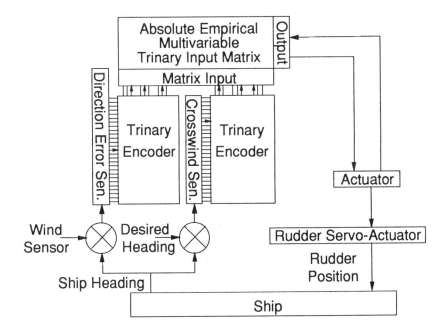

Figure 16.19. An *application of an AE cell trinary digitized multivariable input matrix* is in a multivariable ship direction control system that can learn a unique rudder position for specific combinations of values of its crosswind and error direction sensors.

AE cell trinary digitized output matrix

An AE cell trinary multivariable output matrix can be connected to a trinary decoder, forming the *AE cell trinary digitized output matrix* shown in Figure 16.20. The AE cell trinary digitized output matrix is the preferred embodiment of a digitized absolute empirical output matrix.

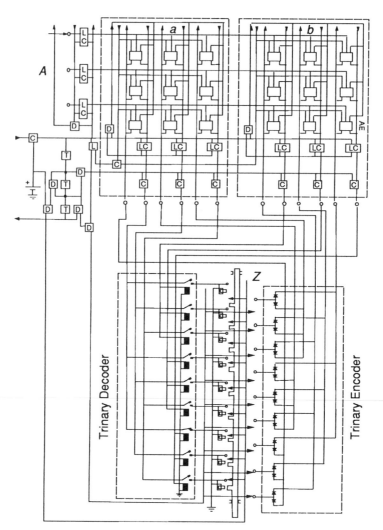

Figure 16.20. An *AE cell trinary digitized output matrix* can produce any one of more output values than the AE cell binary digitized output matrix shown earlier.

Application of an AE cell trinary digitized output matrix

An AE cell trinary digitized output matrix (Fig. 16.21) can be used to position a *linear actuator* to a higher-resolution than a binary digitized system with the same number of memory cells. A trinary digitized output matrix with 8 aggregate variables requires $3 \times 8 = 24$ AE cells, and can produce any one of $3^8 = 6561$ output values. A binary digitized output matrix with 12 aggregate variables also requires $2 \times 12 = 24$ AE cells per output value, and can produce any one of only $2^{12} = 4096$ output values.

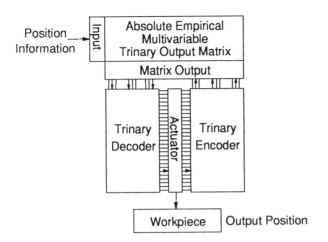

Figure 16.21. An *application of an AE cell trinary digitized output matrix* is as a medium-resolution linear actuator system, which can be taught to position a workpiece in any one of more positions than an AE cell binary digitized output matrix with the same number of AE cells per position.

AE cell trinary digitized multivariable output matrix

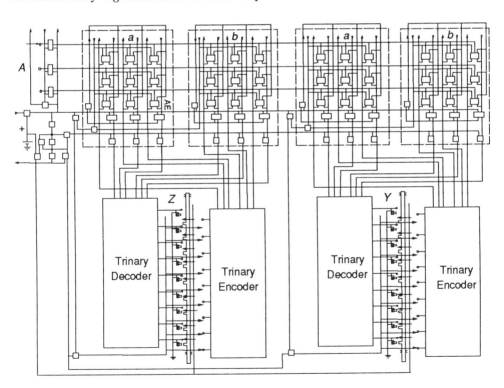

Figure 16.22. An *AE cell trinary digitized multivariable output matrix* can produce more unique combinations of output values for a given number of AE cells than any other type of output matrix.

The number of variables in the output matrix shown can be increased, forming the *AE cell trinary digitized multivariable output matrix* shown in Figure 16.22. An AE cell trinary digitized output matrix with two output variables, each digitized with 8 trinary aggregate variables requires $16 \times 3 = 48$ AE cells per selection and can produce any one of $3^{16} = 43,046,721$ combinations of values per selection. The AE cell binary digitized output matrix shown earlier with two output variables, each digitized with 12 binary aggregate variables, also requires 48 AE cells per selection, but can produce any one of only $2^{24} = 16,777,216$ different combinations of output values. Thus, the advantage of the trinary system over the binary system increases as the number of aggregate variables increases.

Application of an AE cell trinary digitized multivariable output matrix

The values of the crosswind sensor and the direction error sensor of the ship direction control system shown earlier can be summed in a differential and presented to the input variable of an AE cell trinary digitized output matrix connected to a rudder servo-actuator and a bow thruster, as shown in Figure 16.23.

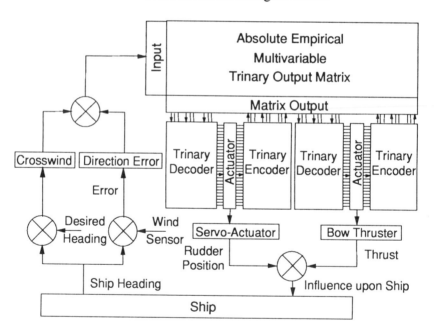

Figure 16.23. An *application of an AE cell trinary digitized multivariable output matrix* is a ship direction control system with multiple output variables.

The AE cell trinary digitized multivariable output matrix can learn to produce some combination of rudder and bow thruster values for each value of its input variable. If a limited number of input values can provide adequate control and the last actuator values produced by the helmsman are acceptable, the *ship direction control system* shown can be a practical control system.

AE cell trinary digitized monolithic controller

The AE cell trinary digitized input matrix can be combined, with an AE cell trinary digitized output matrix, forming the AE cell trinary digitized monolithic controller shown in Figure 16.24.

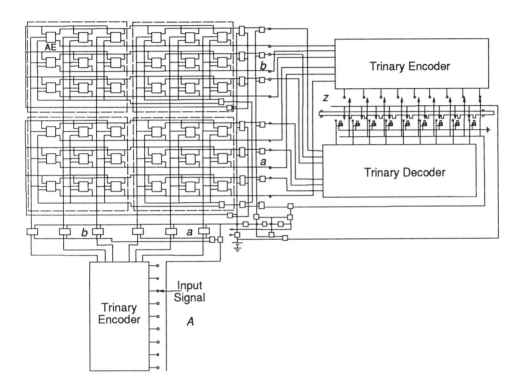

Figure 16.24. An *AE cell trinary digitized monolithic controller* can provide higher resolution of its sensor and actuator variables using fewer AE cell than a binary digitized monolithic controller.

The AE cell trinary digitized monolithic control shown, with two sensor and actuator aggregate variables can produce nine different output positions from nine different input values using only 36 AE cells. A binary digitized monolithic controller with 36 AE cells can produce only eight different output positions from eight input values.

Application of an AE cell trinary digitized monolithic control system

A temperature sensor can be connected to a trinary input encoder, and a heat duct damper can be connected to a trinary output decoder and feedback encoder of an AE cell trinary monolithic control matrix, forming the *temperature control system* shown in Figure 16.25.

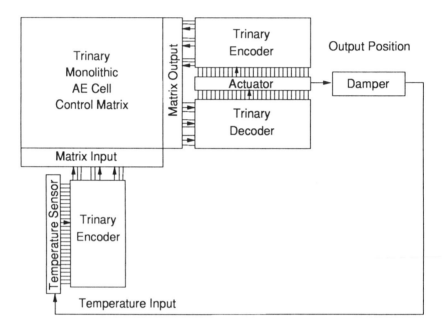

Figure 16.25. An *application of an AE cell trinary digitized monolithic control matrix* is that of a temperature control system, which uses many values of its input variable to produce different values of its output variable.

A trinary digitized monolithic system with input and output resolutions of 243 values requires $3 \times 5 = 15$ matrix input and output terminals and uses $15 \times 15 = 225$ memory cells. However, if only 5 transitions are required by the control task, a trinary digitized duplex system with the same resolution would require only $15 \times 5 = 75$ memory cells in each matrix, for a total of 150 memory cells compared to the 225 memory cells in a monolithic matrix with the same resolution.

AE cell trinary digitized multivariable monolithic control system

The number of sensor and actuator variables in the AE cell trinary digitized monolithic control matrix can be increased indefinitely, forming an AE cell trinary digitized multivariable monolithic control system, as shown in Figure 16.26. The AE cell trinary digitized multivariable monolithic controller is a useful and versatile absolute empirical multivariable controller since it can learn to produce unitary or synchronous diverse relations between any number of sensor and actuator variables.

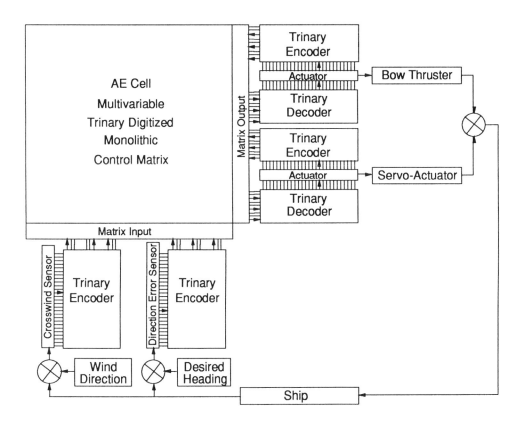

Figure 16.26. An AE cell trinary digitized multivariable monolithic control system can be used to control two or more sensor variables and two or more actuator variables in a *ship direction control system.*

Mechanically digitized sensors and actuators

As shown in Part II, x/y tables and gear trains can be used to create aggregate variables that can represent the components of a vector variable. These digitizing mechanisms can be connected to AE cell matrices to form high-resolution absolute empirical machines.

AE cell multivariable input matrix with vector sensor

Sensors can be used as components of a vector, as shown in Part II. These sensors can be connected to an *AE cell multivariable input matrix with a vector sensor*, as shown in Figure 16.27.

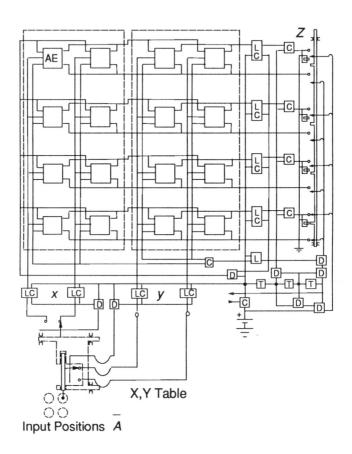

Figure 16.27. An *AE cell multivariable input matrix with vector sensor* converts specific states of an n-dimensional vector into values of a single output variable.

Each input position is converted into a unique combination of values of the component variables x and y by the sensors mounted on an x/y table. The AE cell multivariable input matrix can then be taught to produce specific behavior by an operator producing a specific value of the single output variable for each different position of the x/y table.

Actuators as components of a vector

A set of actuators can be connected in a rational coordinate system such as an *x/y* table. They can act as components of a position vector in an empirical system if some means is provided to sense the actual position of each actuator, as shown in Figure 16.28.

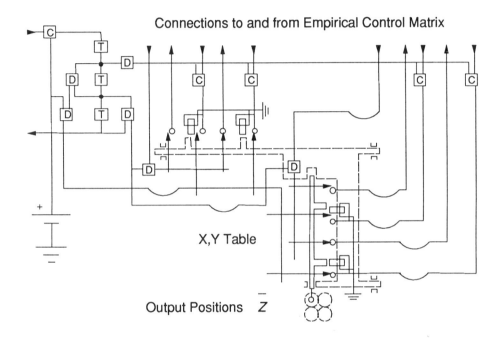

Figure 16.28. A *vector actuator with position feedback* can produce an *n*-dimensional vector motion using *n* actuators.

The actuators act like decoders that compose a vector motion, and the feedback sensors act like encoders that naturally decompose the space vector into a set of transform signals to the aggregate variables of the output matrix.

AE cell multivariable output matrix with a vector actuator

The vector actuator with position feedback shown can be connected to an AE cell multivariable output matrix, forming the *AE cell multivariable output matrix with a vector actuator* shown in Figure 16.29. This system can be empirically programmed to position an *x/y table* at specific locations according to values of the input variable by an operator manually placing the *x/y* table in the desired position as each input state is encountered.

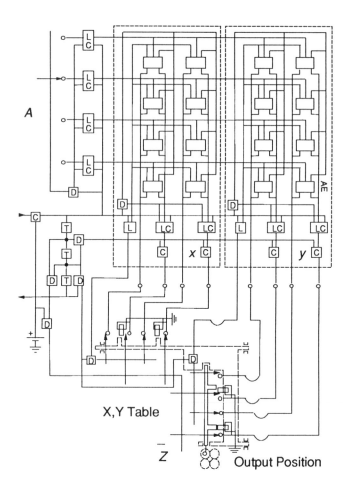

Figure 16.29. An *AE cell multivariable output matrix with a vector actuator* can convert a specific value of a single input variable into an *n*-dimensional vector state.

AE cell monolithic vector controller

The AE cell multivariable input matrix with vector sensor can be combined with the AE cell multivariable output matrix with vector actuator, forming the *AE cell monolithic vector controller* shown in Figure 16.30. The number of values of each component can be increased by adding more sensor and actuator terminals, and the number of components of the sensor and actuator vectors can be increased by adding more axes to the sensor and actuator tables.

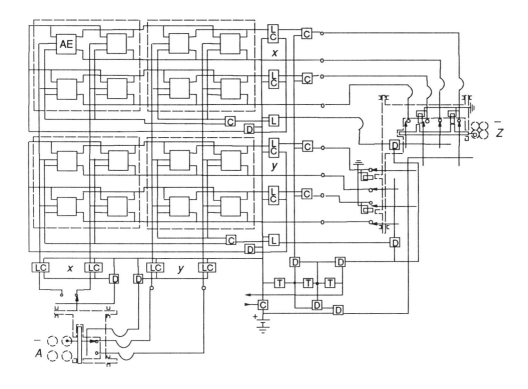

Figure 16.30. An *AE cell monolithic vector controller* can learn to convert a given set of input vectors into a given set of output vectors.

AE cell binary multivariable input matrix with a mechanical encoder

The mechanical encoder shown in Part II can also be connected to an AE cell binary multivariable input matrix, forming the *AE cell binary multivariable input matrix with a mechanical encoder* shown in Figure 16.31. An AE cell multivariable input matrix with a mechanically encoded sensor can be programmed empirically to respond in a unique way to an input position of extremely high resolution. For example, a mechanical encoder with ten binary component input variables (binades) can produce 1024 different input states. Any one of the states can be programmed to produce a unique output by an operator producing that output value when a particular input state occurs.

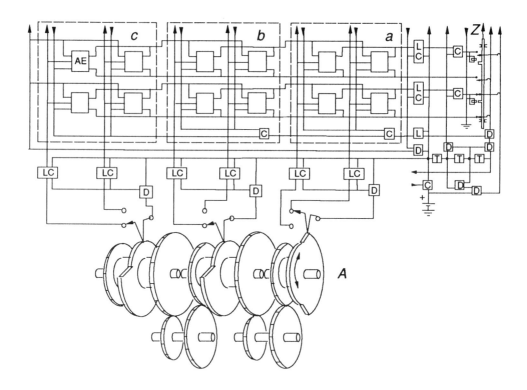

Figure 16.31. An *AE cell binary multivariable input matrix with a mechanical encoder* can produce a unique value of its output variable for one of many possible input states.

Binary pi actuator with position feedback

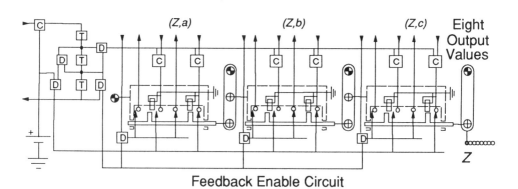

Figure 16.32. A *binary pi actuator with position feedback* can be used as the output device of any empirical system with multiple output variables.

Feedback contacts can be added to a binary pi actuator shown in Part II, forming a mechanical decoder that can produce a unique output position for each combination of values of its aggregate variables and can make contact with the feedback co-terminals representing its actual position. This *binary pi actuator* with position feedback is shown in Figure 16.32. The feedback sensors on each aggregate actuator act as natural mechanical encoders that provide a unique combination of transform values to aggregate variable co-terminals for each output position.

AE cell binary multivariable output matrix with a binary pi actuator

The binary pi actuator with position feedback shown can be connected to the aggregate output variables of an AE cell multivariable output matrix, as shown in Figure 16.33.

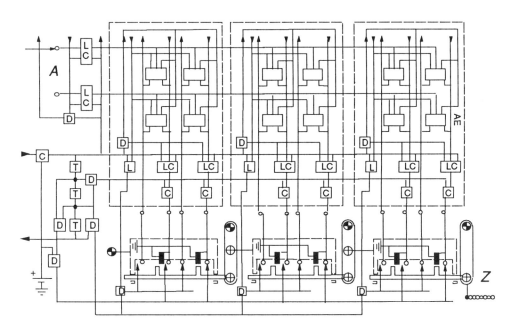

Figure 16.33. An *AE cell binary multivariable output matrix with a binary pi actuator with position feedback* can learn to produce any one of many different output positions for each value of its input variable.

An *AE cell multivariable output matrix with a binary pi actuator with position feedback* can be taught to produce any one of many output positions by an operator placing the actuator in the desired position for a given value of its input variable. This system can be used as a high-resolution linear actuator. For example, a pi actuator with ten binary actuators can be taught to produce any one of 1024 different output positions for a given value of the matrix input variable.

AE cell binary monolithic control matrix with mechanically digitized sensor and actuator

The AE cell binary multivariable input matrix with a mechanical encoder can be combined with the AE cell binary multivariable output matrix with a pi actuator to form the *AE cell binary monolithic control matrix with mechanically digitized sensor and actuator* shown in Figure 16.34.

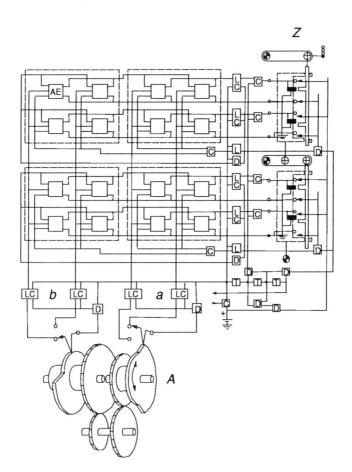

Figure 16.34. An *AE cell binary monolithic control matrix with mechanically digitized sensor and actuator* can learn to produce many different output positions for many different input positions.

The AE cell binary monolithic control matrix with mechanically digitized sensor and actuator can be used in any control application requiring many high-resolution transitions that occur in a consistent manner.

AE cell trinary multivariable input matrix with a mechanical encoder

The mechanical encoder shown in Part II can be connected to an AE cell multivariable input matrix with trinary aggregate variables, as shown in Figure 16.35.

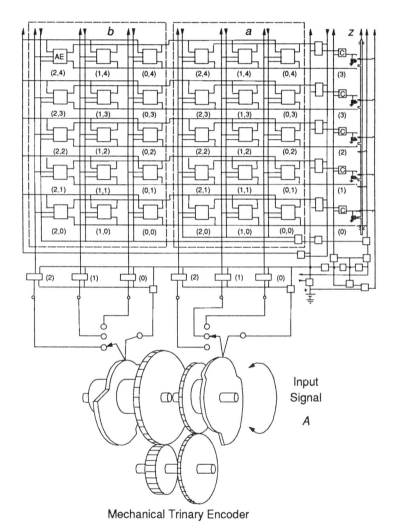

Mechanical Trinary Encoder

Figure 16.35. An *AE cell multivariable trinary input matrix with a mechanical encoder* can resolve more positions than a binary digitized sensor with the same number of contacts.

A *trinary mechanical encoder* with 6 contacts can resolve 9 positions as compared to a binary mechanical encoder with 6 contracts that can only resolve 8 positions. A trinary digitized sensor is preferred to the binary digitized sensor particularly when the number of aggregate variables is large.

Trinary pi actuator with position feedback

An additional stator pole and feedback co-terminal can be added to each aggregate variable of the binary pi actuator shown previously, forming the *trinary pi actuator with position feedback* shown in Figure 16.36. If 4 poles are used for each aggregate variable, the quadrary pi actuator would produce the same number of output positions for a given number of poles as a binary pi actuator. A pi actuator with more than 4 poles for each aggregate variable would produce fewer positions for a given number of poles than the binary, trinary or quadrary pi actuator.

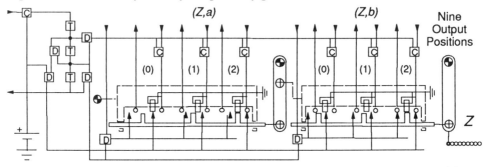

Figure 16.36. A *trinary pi actuator with position feedback* can produce more output positions for a given number of stator coils than a binary pi actuator.

AE cell multivariable trinary output matrix with a pi actuator

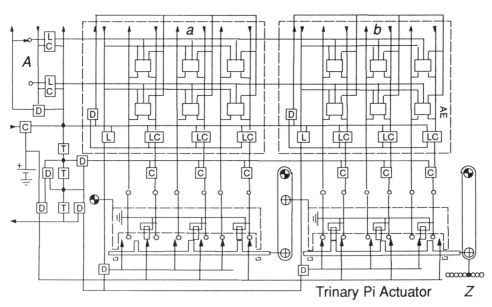

Figure 16.37. An *AE cell trinary multivariable output matrix with a pi actuator* can produce more unique output positions than a binary digitized AE cell output matrix having an actuator with the same number of components.

The trinary pi actuator shown can be connected to an AE cell trinary multivariable output matrix, forming the *AE cell trinary multivariable output matrix with a pi actuator* shown in Figure 16.37. The trinary pi actuator is preferred to a binary pi actuator particularly when many aggregate variables are required to obtain very high-resolution positions. Thus, trinary digitized systems are the preferred embodiment of most responsive control systems.

AE cell trinary monolithic control matrix with mechanically digitized sensor and actuator

The AE cell trinary multivariable input matrix with a mechanical encoder can be combined with the AE cell trinary multivariable output matrix with a pi actuator to form the AE cell trinary monolithic control matrix with mechanically digitized sensor and actuator shown in Figure 16.38.

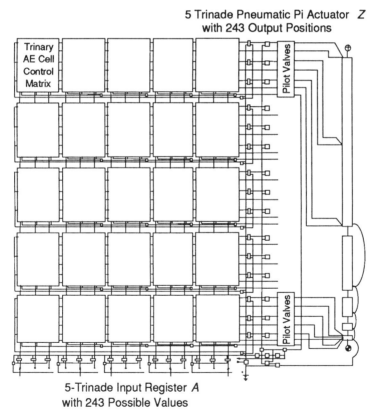

Figure 16.38. An *AE cell monolithic trinary control matrix with mechanically digitized sensor and actuator* can produce more unique transitions for a given number of memory cells than any other type of control matrix.

The AE cell trinary monolithic control matrix with a five-trinade input register and a five-trinade pi actuator requires $5 \times 3 = 15$ input terminals and 15 output terminals for a total of 225 AE cells. Thus, it can produce $3^5 = 243$ different output positions from 243 different input states using only 225 AE cells. The AE cell binary monolithic control matrix with an eight-binade input register and an eight-binade pi actuator can produce 256 output positions from 256 different input states using 256 AE cells since the ratio of output variety to number of memory cells is higher in the trinary system than in the binary system. Thus, the AE cell trinary monolithic control matrix is the preferred design for all high-resolution control applications with many input/output relations in which a given output is always expected to occur for a given input on an absolute basis.

Problems with AE cell controllers

A predetermined input matrix can be connected to a predetermined output matrix, forming a predetermined duplex controller that can deal with multiple input and output variables, as shown in Part II. This controller can be programmed by manually setting the memory switches in both its input and output matrices. However, the values in the memory switches in an empirical system are determined by the actual positions of the actuators in each transition cycle in contrast to a predetermined system, where the internal memory switches are directly influenced by the programmer. Therefore, an AE cell output matrix cannot be connected to an AE cell input matrix to form an absolute empirical duplex system because the AE cell output matrix cannot *back select* a transform signal to the input matrix. This is because the output matrix sets its forward selecting AE cells to zero when it forward selects an output. Thus it cannot find the forward selecting AE cell when its action is carried out. (See duplex empirical controller.)

Also, the AE cell monolithic matrix cannot create a null matrix. So it cannot learn to create asynchronous diverse behavior. However, the AE cell coded multivariable input and output matrices and the AE cell monolithic control matrices can deal with some absolute applications of multiple sensor and actuator variables, as shown previously. But the unique features of the *conditional empirical matrices,* shown in the next chapter, are needed to make a monolithic and duplex empirical controller that can handle multiple input and output variables in a unitary and asynchronous diverse manner, and can learn to deal with sensor/actuator relations that occur on a probabilistic basis.

Chapter 17
Conditional Empirical Machines

In some cases, a helmsman may not provide the same rudder response for each sensed condition, and the last response for a given input may not be the most appropriate. In these cases, the behavior of the trainer or the environment appears to be *probabilistic* rather than state determined. A *conditional empirical controller* can deal with an operator or an environment that does not always provide the same responses to given conditions, where an absolute empirical controller cannot. A conditional empirical controller attempts to produce behavior based upon its record of the relative success of its different attempts to produce action and the actual responses that occur in these attempts. It then selects the responses that are most likely to be carried out, according to the *empirical algorithm*. Thus, a conditional empirical controller provides a more effective method of determining what can and cannot be done within the confines of its environment and the restrictions of its operators than an absolute empirical controller, which requires a state-determined system.

A conditional empirical controller consists of a *conditional empirical memory matrix* that selects the most probable output state for a given input state. It is made up of *conditional empirical memory cells* that change their propensity to select action according to their *success rate*. Their success rate is equal to the logarithmic average of the ratio of the number of times their output occurs divided by the number of times they attempt to cause an output plus the number of times their output occurs.

The empirical process

Empirical behavior is established through experience in an *empirical machine* by means of the *empirical process*. The empirical process leads to the creation of *successful behavior* that can be carried out in a given environment. This behavior can be made *acceptable* to those people who can prohibit the behavior of an empirical machine that is considered undesirable. This successful empirical behavior is *doable* because it does not violate the laws of nature and is acceptable because it may provide some *useful* function to its operators. Because the behavior is doable and acceptable, it may be allowed to continue and thus may be *ongoing*. The *empirical algorithm* is a description of the empirical process that empirical machines and other *natural empirical systems* such as the *animal brain* use to establish useful and ongoing control behavior. This process is similar to the *test and compare process* used in science. This procedure can also be called the process of "trial and incorporation of successful behavior."

The empirical process in a machine

For an empirical machine to operate successfully within a given environment, it must predict and produce behavior that can be carried out without *interference* from the environment or the people in this environment. It must discover some laws of nature within that environment, learn the geographic features of its surroundings, determine its own physical limitations, and ascertain the wants and needs of the people with whom it is involved to predict what behavior can be carried out. For example, if an *empirical ship auto-pilot* can accurately predict the rudder position that would be produced by a helmsman for a given direction error, then it can steer the ship without receiving conflicting commands from the helmsman. It can then act as if it knows the response characteristic of the ship and will appear to possess the skill of the helmsman. After that, it may not have to change the way it predicts behavior. Thus its behavior may continue unchanged.

The scientific method

The *scientific method* has evolved into a successful procedure of discovering the laws of nature. It can be stated using the four steps listed in Table 17.1.

Table 17.1. Steps involved in the scientific method

---

1. Make a prediction as to the occurrence of an event according to some possible basis for its cause.
2. Test the prediction by attempting to produce the event under specific known conditions.
3. Compare the actual results with the predicted results.
4. Alter the basis of the prediction so that new predicted results match the actual results.

If an empirical machine can produce events based upon some prediction, measure whether the events occurs as predicted, and change the basis of the predictions based upon these results, it can select events in the future that have the highest likelihood of happening as predicted.

Test and compare

The *empirical process* is summarized in Table 17.2.

Table 17.2. The empirical process

---

1. Predict (test) an output event based upon the *confidence level* established for each possible output in relation to each input event.
2. Attempt to produce the predicted results.
3. Compare the actual output event to the predicted output event.
4. If the prediction is correct, continue to use that prediction for those input events.
5. If no consistent results can be found, abandon that prediction and try a prediction that leads to the actual result.

These steps can be condensed into one statement: Discard predictions that lead to unworkable results and create new predictions that lead to the actual results.

Rules of an empirical machine

This process can be carried out in the control matrix of an empirical machine if it acts according to the rules listed in Table 17.3.

Table 17.3. Rules of an empirical machine

---

1. Predict the output having the highest confidence level for given input conditions.
2. If the prediction is correct, increase the confidence level of that prediction.
3. Conversely, if the prediction is incorrect, decrease the confidence level of that prediction.
4. If another result occurs, incorporate the new results by increasing the confidence level of the actual results that occur for a given input.

If consistent sensor/actuator relations are possible, the predictions of the machine's controller will begin to reflect the relations that can be carried out, and after a number of trials the machine will appear intelligent, skillful, or mindful of the opportunities for behavior in its environment.

The empirical algorithm

The test and compare process and the requirements of a real machine are combined in the *empirical algorithm* shown in Table 17.4, which describes the method that living or nonliving empirical machines use to develop skillful behavior based upon the discontinuous state ontology described in Part I.

Table 17.4. Statement of the Empirical Algorithm

---

1. At a given moment, the empirical controller selects the output state having the highest confidence level according to its memory for a specific input state, and decreases the confidence level for selecting that transition.
2. It attempts to produce that output state after a specific delay.
3. The controller measures the actual output state that occurs within a specific period of the attempted output.
4. If the selected output state can be carried out after the specified delay after an input state and within the specified period, the confidence level stored in its memory of producing that output for that input is increased more than its was decreased.
5. If the selected output state cannot be carried out after the specified delay after an input state and within the specified period (because it is inhibited, restricted, or otherwise interfered with by the environment, an outside influence such as a trainer, or its internal signals, actuators, or structure), the confidence level stored in memory of the selected output is decreased, and the probability of the controller later selecting the selected output for that input is decreased.
6. If another output state is carried out after the previously mentioned delays, the confidence level of the actual output is increased, and the probability of selecting that output later is increased relative to other possible outputs.
7. If no outputs are carried out for a specific input (due to the previously mentioned influences and conditions), the confidence level stored in memory for the selected output remains decreased, and the probability of the controller later selecting that output for that input is decreased relative to all of the other possible outputs.

Thus, an empirical controller selects action based upon its internal records of the success or failure of its attempts to carry out specific actions under specific conditions.

Diagram of the empirical algorithm

This algorithm can be shown by the flow chart of events within an empirical control system shown in Figure 17.1.

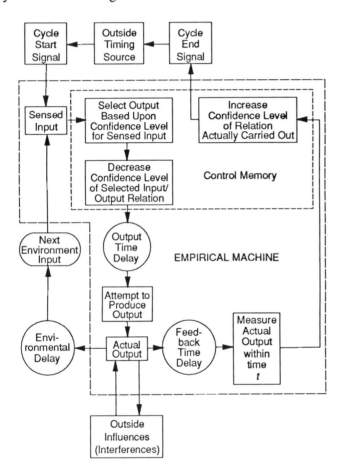

Figure 17.1. The *diagram of the empirical algorithm* shows the sequence of steps needed to establish successful behavior.

The best place to enter the flow diagram of the empirical algorithm is at the *sensed input* block. Some specific input state is present at the beginning of a transition cycle. At this point, the empirical controller determines an output state based upon the confidence level that it has established for all of the possible outputs states for that input state, and it decreases the confidence level of the select output state for that input state. After a given delay (t_d), the controller attempts to produce that output state. The actuator may need some time to arrive at the prescribed position. So after another delay (t_p), the controller measures the actual output state, and the empirical controller increases the confidence level of the measured output state for the input state in that transition.

The environment and/or other influences may or may not allow the prescribed output state to be carried out within the action period. If the measured output is the same as the prescribed output, the empirical controller increases the confidence level it has for that relation more than it had been decreased, as explained by the operation of the logarithmic subtraction mechanism. If another output occurs for that input, the confidence level of that relation is increased. If no output appears within the prescribed period, the confidence level for the selected output is decreased, and no confidence levels are increased.

The environment or another outside influence establishes the *intrinsic delays* within each transition and the next input state to the controller. The controller measures the next input state at the time determined by its *transition sampling rate*, and carries out the empirical algorithm according to its *internal delays*.

Timing

The empirical algorithm is similar to the basic *Hebb rule*[7] of D. O. Hebb (1949), which is based upon the reinforcement of input/output relations that occur together. However, the empirical algorithm requires that the organism attempt to produce an output based upon the input conditions, and that this attempt first creates a negative change in the confidence level of the organism. It also requires a specific delay between the measurement of the input state and the initiation of an output state. It also requires that the output be measured only after another delay to allow the actuator time to arrive at its selected position and that the measurement can be made only for a predetermined period. According to the empirical algorithm, the output relation is reinforced only if it occurs within a specific delay from the input (t_d) and within a specific period (t_p) after the initiation of the action. These time relations are needed to define the behavior of machines that are expected to operate in the real world where objects having mass act according to *intrinsic relations* between force, position, and time.

Conditional empirical memory cells

As discussed previously, an absolute empirical memory matrix selects the output terminal connected to the only conductive AE cell connected to the latched input terminal. The operation of the absolute empirical memory matrix requires that the remaining AE cells connected to that input terminal be nonconductive. In a conditional memory matrix, all of the memory cells connected to a given input terminal may have a varying propensity to conduct. Since each conditional memory cell contains some value of an infinitely varying propensity to conduct, *relative* to the other memory cells connected to a given input terminal, they are considered *conditional memory cells,* in contrast to absolute memory cells, which have a propensity to conduct based entirely upon their own sensitivity value. When these conditional memory cells are arranged in a *conditional memory matrix*, they can be used to select the output terminal connected to the conditional memory cell with the highest propensity to conduct *relative* to all of the other cells connected to the working input terminal rather than the value of a single memory cell. The basic memory device in a conditional empirical controller is the conditional empirical memory cell (CE cell). It can calculate and store an *analog* (nondiscrete) *value* representing the *logarithmic average of the success rate* of its output terminal carrying out action in relation the number of times it attempts to carry out action.

Changing confidence level of a conditional memory remotely

Statement 4 of the empirical algorithm requires that the confidence level of an association between a given input state and output state be increased if the output called for actually occurs. Statement 5 requires that the confidence level be decreased if the output called for does not occur. This process can be carried out by an *adjustable sensitivity relay,* provided it has the following support mechanisms:

1. A way of determining whether a given output actually occurs; and
2. A way of actively and remotely changing the wiper arm of the adjustable sensitivity relay based upon these results.

The sensitivities of the conditional predetermined memory cells (CP cells) shown in Part II were changed manually by an outside programmer. The following section shows how a CE cell with an adjustable sensitivity relay can change its sensitivity by itself according to the outcome of its attempt.

Remote controlled logarithmic subtracting mechanism

Obviously, the conductive state of a conditional memory cannot keep getting more or less sensitive by an equal amount every time it is changed. A solution to this requirement is to move the wiper arm of the adjustable sensitivity relay some fraction of the distance remaining toward its maximum or minimum position, as shown in the conditional predetermined memory cell (CP cell) in Part II. These changes can also be carried out according to the results of each selection by a conditional empirical memory cell using the *remote-controlled logarithmic subtracting mechanism* shown in Figure 17.2.

The bail B is moved in the minimum direction when the write minimum contacts are closed, causing current to flow in the min coil. This causes the bail to pick up the minimum lever L2 at the position determined by the moveable stop and push it to the minimum position. This causes a movement of the minimum ratchet R1 proportional to the distance of the moveable stop from the minimum position. This motion is transmitted to the minimum subtracting member of the differential D1, causing the wiper arm to move toward the minimum position in a proportion determined by the gear ratios between the ratchet and the differential and the gear ratio between the differential output shaft and the wiper arm shaft. (The differential also divides the motion of the subtracting member D1 by one half.)

A motion of the bail toward the maximum position is created when the maximum coil is energized by the closing of the write maximum contacts. This causes a movement in the wiper arm toward the maximum position in proportion to its original distance to the maximum position through the action of the maximum ratchet R2 through the differential side member D2 to the differential output shaft and the wiper arm shaft in the same manner described above. Thus, the wiper arm is moved toward the maximum or minimum positions by an amount proportional to the position of the wiper from either position each time one or the other coil is energized.

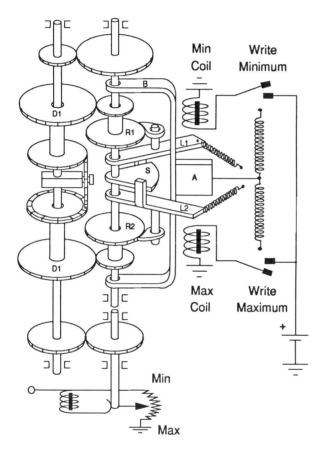

Figure 17.2. A remote-controlled logarithmic subtracting mechanism uses two solenoids, one to move the bail in the maximum direction and the other to move the bail in the minimum direction.

Logarithmic response curve

The action of the logarithmic subtracting mechanism (LSM) produces the *logarithmic response curves* shown in Figure 17.3. If more write minimum movements occur than write maximum movements, the sensitivity of the conditional memory will reach a value closer to the minimum sensitivity. If more write maximum movements occur than write minimum movements, the sensitivity of the conditional memory will reach a position closer to the maximum sensitivity. If an equal number of write maximum and write minimum movements occur, for a sufficient number of times, the conditional memory will approach the mid position.

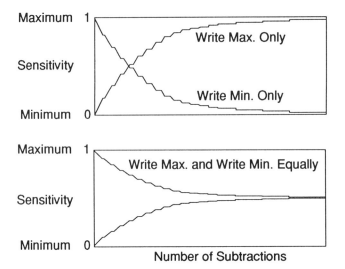

Figure 17.3. The *logarithmic response curves* of a conditional memory show how its sensitivity approaches its maximum or minimum value asymptotically.

Calculation of success rate

The shooting probability (P) of a basketball player (shooting percentage/100) is usually given by Equation 17.1.

$$P = N_1/N_2, \qquad (17.1)$$

where N_1 is the number of baskets made, and N_2 is the number of attempts (over some given period, number of games, or number of attempts). This equation assumes that a player cannot make a basket without making an attempt; so P is always equal to or less than one. However, the logarithmic subtracting mechanism may experience more write maximum events than write minimum events. These windfall conditions occur when a transform signal indicates that the output of a memory cell occurs without it having been attempted by that memory cell. To keep the value of the logarithmic subtracting mechanism from exceeding one, a *success rate (R)* is calculated by the logarithmic subtracting mechanism according to Equation 17.2.

$$R = N_1/(N_1 + N_2). \qquad (17.2)$$

The success rate (R) of the conditional memory never exceeds one even when many successes occur without any attempts. For example, if 100 write maximum events occur without any attempts being made, the value of the logarithmic subtracting mechanism (R) will approach 1. If 100 successes occur for every 100 attempts, then $R = 100/(100 + 100) = 1/2$. If 50 successes occur in 100 attempts, then $R = 50/(50 + 100) = 1/3$. This success rate is made equal to the sensitivity of a conditional memory cell by connecting the output of the logarithmic subtracting mechanism to the wiper arm of its adjustable sensitivity relay.

A conditional empirical memory with a logarithmic subtracting mechanism

Since the wiper arm of a conditional memory can be placed in an almost infinite number of different positions, it can store an almost infinite number of different values compared to an absolute memory, which can store only an all-or-nothing value. If energizing the minimum coil of a logarithmic subtracting mechanism (LSM) represents one event and energizing the maximum coil represents the other event, a LSM can be used to compute and record the *logarithmic average* of the number of times one event happens versus the number of times another event happens.

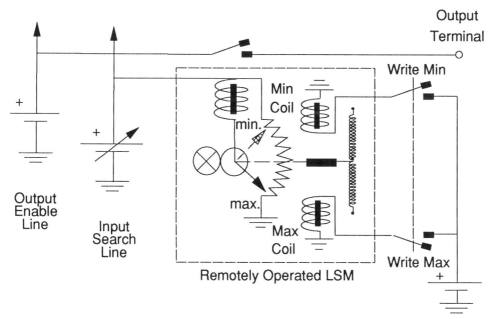

Figure 17.4. A conditional memory with a remotely operated logarithmic subtracting mechanism can be used to compute and store a quantity related to the number of times an event happens versus the number of times it does not happen.

The conditional memory in Figure 17.4 also can be used to measure how many times an event is attempted versus the number of times it is carried out successfully. Energizing the minimum coil can indicate that an event has been attempted, and the energizing of the maximum coil can represent that the event has been carried out successfully. For example, if the minimum coil is energized each time a basketball player shoots the ball at the hoop, and the maximum coil is energized each time that player makes the basket, the position of the wiper arm will represents the player's *shooting success rate*. The player's shooting success rate approaches the player's shooting average as the shooting average goes to zero, and approaches 1/2 the player's shooting average as the shooting average approaches one.

Operation of a conditional memory cell with a LSM

A conditional memory cell with a remotely operated logarithmic subtracting mechanism operates according to the operations listed in Table 17.5.

Table 17.5. Operation of a conditional memory with a remotely operated logarithmic subtracting mechanism

---

1. The sensitivity of a conditional memory with a logarithmic subtracting mechanism is interrogated when an increasing voltage ramp is applied to its input search terminal.
2. The sensitivity of the memory is measured by the voltage level required to close its output contacts. This produces a voltage at its output terminal and indicates the position of its wiper arm.
3. The sensitivity of the conditional memory is decreased if current is made to flow through the minimum coil, causing the wiper arm to move some fraction of the distance toward the minimum position.
4. The sensitivity of the conditional memory is increased if current is made to flow through the maximum coil, causing the wiper arm to move some fraction of the distance toward the maximum position.

Thus, the sensitivity of the conditional memory can be changed by moving the wiper arm to the minimum position, where no level of input voltage will cause the relay to close, or by moving the wiper arm to the maximum position, where the maximum current will flow through the relay coil when an input voltage is applied. In this high-sensitivity state, even a very low voltage will cause the relay contacts to close.

CE cell

A *conditional empirical memory cell (CE cell)* can be used to maintain a record of the number of times its output occurs in relation to the number of times it attempts to produce an output. The CE cell is similar to the absolute empirical memory cell (AE cell) except that the CE cell uses a logarithmic subtracting mechanism (LSM) to change its sensitivity incrementally (Fig. 17.5). Like the conditional predetermined memory cell (CP cell), the sensitivity of the CE cell is determined by the input voltage required to cause the search relay contacts to close. The sensitivity of the search relay is changed when its logarithmic subtracting mechanism moves the wiper arm toward the less sensitive position when the CE cell selects an output, and moves the wiper arm toward the more sensitive position when its output is carried out.

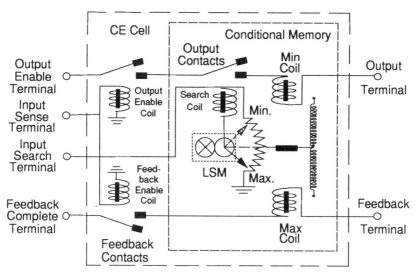

Figure 17.5. A *CE cell* is a conditional memory element having a sensitivity that is determined by the number of times it attempts to produce an output versus the number of times the output is successfully carried out.

Operation of a CE cell

A CE cell operates according to the sequence of steps listed in Table 17.6.

Table 17.6. Operation of a CE cell

1. A voltage is applied to the output enable and input sense terminals at the beginning of a transition cycle.
2. To test the sensitivity of the CE cell, an increasing voltage ramp is applied to its input search terminal. At some voltage level, determined by the position of the wiper arm, a sufficient current flows through the search relay coil to cause its output contacts to close.
3. When the output contacts close, the voltage from the output enable terminal causes a voltage to appear at the output terminal. If the output terminal is connected to ground, a current can flow through the minimum coil.
4. The current through the minimum coil latches the output contracts closed and causes the logarithmic subtracting mechanism to move the wiper arm a specific percentage of the distance remaining toward the minimum sensitivity position. (A current through the output terminal indicates that an output device has been energized. This causes the voltage to the output enable terminal to be disconnected by the matrix circuit driver to be shown, which unlatches the output contacts and allows the armature of the logarithmic subtracting mechanism to return to its normal center position.)
5. If a voltage is applied to the feedback terminal while the input sense terminal is still energized, a current can flow to ground through the maximum coil through the feedback complete terminal. This causes the logarithmic subtracting mechanism to move the wiper arm incrementally toward the maximum sensitivity position. (A current through the feedback complete terminal indicates that a feedback signal has been received by a CE cell. This signal is used by the matrix circuit driver, to be shown, to disconnect the voltage to the feedback circuit. This prohibits any more feedback signals during that transition cycle and allows the armature of the LSM to return to its normal center position.)

If an equal number of input and output events occur at a given CE cell, the wiper arm of that CE cell will approach the midpoint. If more input events occur than output events, the wiper arm will reach a position closer to the minimum sensitivity position. If more outputs events occur than input events, the sensitivity will reach a position closer to the maximum sensitivity position.

Propensity to increase sensitivity

After a CE cell experiences many consistent input/output relations, in which its output event always occurs for its input event, its wiper arm arrives at its midpoint, where it moves an equal distance in both directions as long as its output event always occurs with every input event. Since most relations are less than perfectly consistent, the wiper arm of a CE cell is normally farther from the maximum position than the minimum position. This causes the logarithmic subtracting mechanism to move the wiper arm a greater distance toward the maximum position than it moves the wiper arm toward the minimum position in a given transition cycle. Thus, a single output/feedback cycle of a CE cell usually leaves its wiper arm in a more sensitivity position than before the cycle.

If the wiper arm gets above the midpoint, due to the occurrence of many windfall conditions in which output events occur without input events, the CE cell will gain such a high sensitivity that it will always select its output when its input occurs. If these input/output relations always occur, the wiper arm will then move back toward the midpoint.

Confidence level

After attempting to cause outputs and measuring their outcomes, the wiper arm of a CE cell will be in a position that approximates the ratio of successful outputs to attempts to produce outputs while its input sense terminal is held closed. Thus, the position of the wiper arm is a measure of the *confidence level* of a given CE cell that its output and feedback terminals will be energized while its output and feedback contacts are closed. The value of the *ramp voltage* that causes the contacts of the CE cell to close is a measure of the position of its wiper arm. Thus, the value of the voltage ramp that causes a CE cell to select an output is a measure of the confidence level of that selection. The higher the voltage level required to close the output contacts of a CE cell, the lower is its confidence level. The lower the voltage required for a given CE cell to produce an output, the higher the confidence level of that CE cell.

Other implementations of the CE cell

There are many other devices that can carry out the empirical algorithm. The CE cell designed around the adjustable sensitivity relay and the logarithmic subtracting mechanism was selected because it demonstrates the essential operation of the empirical algorithm in a way that many people can understand. The individual operations carried out by a CE cell can also be carried out by other special-purpose devices using other types of analog memories. For example, *electrochemical devices* such as small batteries can be charged and discharged to different levels. The level of charge can be used as an analog memory value. Also, an electrical charge can be stored within the junction of a *metal nitride oxide semiconductor (MNOS)* transistor gate that changes its operating characteristics. These changes can be used as an analog record of information stored at that gate. A single *microcomputer* can be programmed to duplicate the function of a single CE cell or even a matrix of CE cells. The most likely state-of-the-art hardware candidate for the CE cell is an *EE-PROM (electrically erasable—programmable read-only memory)*. A CE cell is interrogated by the ramp signal at a very low frequency compared to the read/write cycle of memories in a digital computer. So the slowness of an EE-PROM poses no problem when used in an empirical matrix. However, the adjustable sensitivity relay with a logarithmic subtracting mechanism will continue to be used in this discussion because it does not require an understanding of solid state electrical technology.

Truth tables

The CE cell can be used by itself in some applications such as recording the probability of an event occurring or not occurring. However, the primary use of the CE cell is to select the highest relative probability option among a set of possible choices represented by a set of CE cells. When a matrix of CE cells is used to select a unique output value for a given input value, the confidence levels of the matrix of CE cells can be represented by a *truth table of success rates.*

The empirical algorithm summarized

An empirical controller operates according to the empirical algorithm. The empirical algorithm is summarized in Table 17.7.

Table 17.7. Summary of the empirical algorithm

---

1. An empirical controller selects the output (after a given delay) having the highest confidence level, according to its memory, for a specific given input.
2. If the selected output can be carried out (not inhibited, restricted, or otherwise interfered with by the environment, an outside influence such as a trainer, or by other internal signals, actuators or structure of its own), the confidence level stored in its memory of that output for that input is increased so that the probability of later selecting that output for that input is increased.
3. If the output selected for a specific given input cannot be carried out (due to the above mentioned influences), the confidence level stored in memory is decreased and the probability of the controller later selecting that output for that input is decreased.

The first rule requires a circuit that can select an output for a given input based upon whether the output occurs with that input consistently. This implies that three types of components are needed to carry out this function:

1. Input elements.
2. Output elements.
3. Selector elements.

These three types of components can be put together to form a memory matrix. This matrix can be represented by a *truth table* made up of these three elements. The truth table contains the information needed to connect specific input elements with specific output elements, and is used to define the modulus of behavior of a machine.

Truth table of probabilities

A truth table defines the modulus of behavior of a controlled machine since it shows the probability of an actuator value being selected for each sensor value. In an absolute empirical machine a "1" indicates that the output element in that row is always selected for the input element in that column. A "0" indicates that the output element in that row is never selected for the input element in that column. Values between 0 and 1 in a conditional matrix indicate the historical probabilities of each output element occurring with each input element, as shown in the *stochastic truth table* in Figure 17.6.

Output State

0.1	0.3	0.2	0.4	1.0	(4)
0.2	0.4	0.3	0.9	0.7	(3)
0	0.1	0.9	0.2	0.1	(2)
0.3	0.2	0.4	0.1	0.5	(1)
0.9	0.8	0.1	0.2	0.3	(0)
(4)	(3)	(2)	(1)	(0)	

Input State

Figure 17.6. The values in a *stochastic truth table* show the historical probabilities of each output element occurring after each input element.

The truth table shown can be called a *stochastic matrix* because it contains the *historical probability* of each output element occurring with each input element. If some output state always occurs after an input state, the sum of the probabilities in each row is 1.

Truth table of success rates

The empirical algorithm says that an output has to be selected that has the highest confidence level associated with the input value identified at a given moment. Confidence levels ranging from 0 to 1 must be based upon the success rate rather than the historical probability of the experienced events because output events can occur without input events in an empirical matrix. In general, the values of success rate approach 1/2 the values of historical probability as the values of probability approach 1, and the values of success rate approach the values of the historical probability as the values of the probability approach 0, as shown in the *truth table of success rate* in Figure 17.7.

Output State

0.091	0.231	0.167	0.286	0.500	(4)
0.167	0.286	0.231	0.474	0.412	(3)
0	0.091	0.474	0.167	0.091	(2)
0.231	0.167	0.286	0.091	0.333	(1)
0.474	0.444	0.091	0.167	0.231	(0)
(4)	(3)	(2)	(1)	(0)	

Input State

Figure 17.7. The values in a *truth table of success rate* can represent the confidence levels of each possible input/output relation.

The *confidence level* of the CE cells at the intersection of each input and output terminal in a conditional matrix can be represented by a specific number from 0 to 1. The truth table of success rate represents the confidence levels of a matrix of conditional memories. A value of success rate greater than 0.5 indicates a windfall condition in which more successes have occurred in the *near-term* than attempts have been made in the near-term.

CE cell control matrices

To carry out the requirements of the discontinuous state ontology, the CE cell control matrix must respond to a transition cycle timing signal that indicates when a transition cycle starts. It must also select the output state most likely to occur within a given period of an input state, must provide the energy source and information to adjust the CE cells in the matrix according to the behavior of its sensors and actuators, must indicate when a transition is complete, and must hold the actuators in a constant position between transition cycles. These functions are carried out by the *CE cell matrix driver circuit* shown in this section.

CE cell driver

To determine which output state has the highest probability of being carried out with each input state, the CE cells representing each input/output relation must be connected to each input and output terminal, forming a CE cell matrix. The set of CE cells connected to the latched input terminal must be interrogated by the *CE cell driver* shown in Figure 17.8.

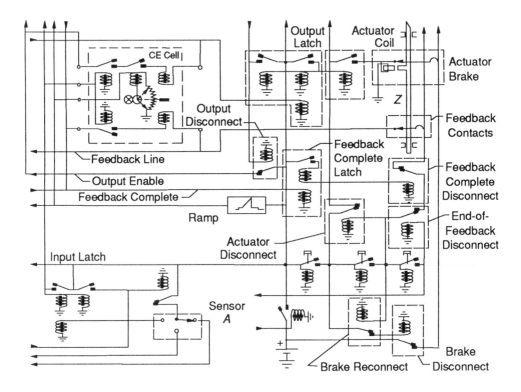

Figure 17.8. The *CE cell driver* provides the electrical power sources and circuitry needed to operate a matrix of CE cells.

The CE cell driver supplies a power source, a cycle switch to initiate the transition cycle, a sensor switch, output latch, and conductors to the output enable, input search, input sense, and feedback complete terminals for all of the CE cells in the matrix. The CE cell driver uses all of the delays and disconnects in the AE cell matrix, and adds a *ramp generator* like that used in the conditional predetermined (CP cell) matrix to interrogate its CE cells. Only one input terminal, one CE cell, and one actuator coil are shown connected to the driver circuit. In most cases, multiple input terminals, multiple actuator coils, and a matrix of CE cells are connected to the driver circuit.

Operation of a CE cell control matrix

The CE cell control matrix operates in the steps given in Table 17.8 during a single transition cycle. These steps are based upon the empirical algorithm.

Table 17.8. Operation of the CE cell control matrix

---

1. The cycle switch is closed at the beginning of a transition cycle by an outside timing source. This causes a voltage to appear at the sensor switch, ramp generator, output latch line, and output device feedback line.
2. The position of the armature of the sensor switch energizes one matrix input terminal and causes the input sense and feedback complete relays to close in all of the CE cells connected to this input terminal.
3. The ramp generator initiates a voltage increase, which is conducted to the search coils of all the CE cells in the matrix.
4. The output contacts close at some voltage level in the CE cell in the selected input column with the highest sensitivity. This latches its output contacts closed and energizes its output latch.
5. The current through the minimum coil of the selecting CE cell also produces an incremental change in the minimum direction in the position of its wiper arm.
6. The output latch also sends a voltage to the output disconnect coil, which disconnects the output enable circuit of that matrix for the remainder of the transition cycle. This prohibits any other CE cell from producing an output for the remainder of the cycle and allows the armature of the selecting LSM to return to its normal center position.
7. After a delay created by the damper on the armature of the actuator delay relay, the actuator coil of the latched output terminal is energized. At the same time, the actuator brake is released by current flowing through the brake reconnect contacts, and armature of the actuator attempts to move to its selected position.
8. After another delay to allow the actuator to move to the position selected by the control matrix, the feedback delay relay times out. This disconnects the actuator coil, re-engages the actuator brake, and allows a voltage to appear on the feedback enable line. This holds the armature in a fixed position and causes a feedback signal to appear on the feedback line connected to the actual position of the actuator. This causes current to flow through the maximum coil in the CE cell connected to this feedback line and the latched input terminal, which creates an incremental change in the position of the wiper arm of that CE cell in the maximum direction.
9. The feedback complete signal through this restored CE cell also causes the feedback complete disconnect to latch in and disconnect the feedback enable line so that no more feedback signals can be sent for the remainder of the transition cycle. This terminates the feedback signal and allows the armature of the LSM in the restored CE cell to return to its normal center position.
10. After another delay caused by the end-of-feedback delay relay, current to the feedback enable line is disconnected so that no feedback can be sent in that transition cycle even if none had been sent, and a signal is sent to the outside timing source, indicating that the transition cycle is over. When the cycle switch is opened by the outside timing source, all of the CE cell and circuit driver switches can return to their normal positions.

The CE cell matrix driver provides the input and output logic needed for a matrix of CE cells to learn behavior that can be carried out according to the steps involved in the empirical algorithm.

Application of a single CE cell

A driver circuit with a single CE cell can be used to keep track of the success rate of a device based upon the number of times an event is attempted by the device versus the number of times it is carried out. For example, a single CE cell can compute and record the approximate percentage of successful operations of an output device based upon the total number of times the device operates successfully divided by the sum of the total number of attempts to operate the device and the total number of times the device operates successfully.

If an equal number of output and feedback events occur for a given CE cell, the wiper arm will lie close to the midpoint of its range. This corresponds to a CE cell calling for an output that always occurs, indicating that the output device is *perfectly* reliable. In most cases, a CE cell will call for an output that may not occur each time. Thus, the normal operating region of a wiper arm is at or below its midpoint. This causes the logarithmic subtracting mechanism to move the wiper arm toward the minimum position in a smaller incremental move whenever it attempts to produce an output than when it receives a feedback signal. Thus, a single output/feedback cycle results in a net increase in sensitivity unless the wiper arm is at or above its midpoint.

If the output device is caused to operate by an outside source without the action of the CE cell, the wiper arm can move above the midpoint. This unusual but legitimate "windfall" condition is similar to what happens when an investor obtains income without having made an investment. If the device operates successfully only 50% of the time, the wiper arm will approach a point that is 1/3 of the way between the minimum and maximum position. If the device never operates successfully, the wiper arm will approach the minimum position.

Advantages of the empirical algorithm

This procedure of "charging" the CE cell when it selects an output and "paying" it more than it was charged if the output is carried out simplifies the internal *book-keeping* of the empirical controller because all of the CE cells in a matrix are always "paid up" after each transition cycle. This procedure is like paying cash in the real world, which requires no credit accounts.

A logarithmic response is created by reducing or increasing the sensitivity of a CE cell by some fraction of its position to or from its minimum or maximum position. This reduces the likelihood that two or more CE cells will have the same value. (See logarithmic response curves.)

The logarithmic response also allows CE cells to *learn new responses* with the minimum number of trials. Those CE cells that have fallen into a low-sensitivity state can quickly catch up to the other CE cells in a matrix, while the high-sensitivity CE cells never get hopelessly far ahead of the other CE cells in the matrix.

The logarithmic response also causes the CE cell with a higher initial sensitivity value to remain higher than a CE cell with a lower initial sensitivity value even if both subsequently receive the same ratio of attempts to successes. This preserves *long-term memory* data.

Long-term memory also is preserved in another way. For example, if a given CE cell attempts to produce an output but fails to succeed too often, its sensitivity level will drop below the sensitivity of another CE cell in the matrix. Then the first (now *quiescent*) *CE cell* will no longer attempt to produce an output, and will remain inactive. The other *working CE cell* will attempt to select action. However, the last probability value in the quiescent CE cell will remain undisturbed as *latent information* that may have some bearing upon later changes in behavior. For example, if the working CE cell also fails to produce enough successful outputs, the first (quiescent) CE cell may reestablish itself as the working CE cell.

If capacitor-type conditional memories such as EE-PROMS are used, the logarithmic decay of their charge will allow the *relative levels of charge* in each memory cell to remain the same, thus preserving the original state of the conditional memory. However, new learning will more rapidly predominate over old learning if the average sensitivity level of a relative memory decays significantly with time. This is because the *rate of increase in sensitivity* in a given transition increases as the value of sensitivity approaches zero.

CE cell binary control matrix

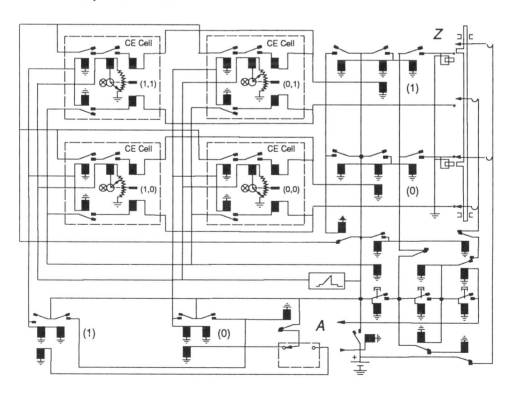

Figure 17.9. A *CE cell binary control matrix* can be used to find the most probable output state for each input state encountered when the actual behavior is probabilistic.

The binary control matrix shown in Figure 17.9 has a single input variable A with two possible values, (0) and (1). It also has a single output variable Z with two output values (0) and (1). The sensor switch is connected to input terminal $A(0)$, which is connected to CE cells (0,0) and (0,1). The CE cell (0,0) is shown in the most sensitive state, which causes its output contacts to close, causing the actuator to move to the $Z(0)$ output position, and its memory switch to move to a less sensitive position as has been described. The actuator feedback switch is therefore connected to the actuator feedback circuit connected to CE cells (0,0) and (1,0). Since the contacts are closed in CE cells (0,0) and (0,1) only, the memory switch in (0,0) is restored to a more sensitive state as has been described.

CE cell trinary control matrix

The number of values of the input variable and output variables in a scalar control matrix can be increased indefinitely using the method of connection shown in the *CE cell trinary control matrix* in Figure 17.10.

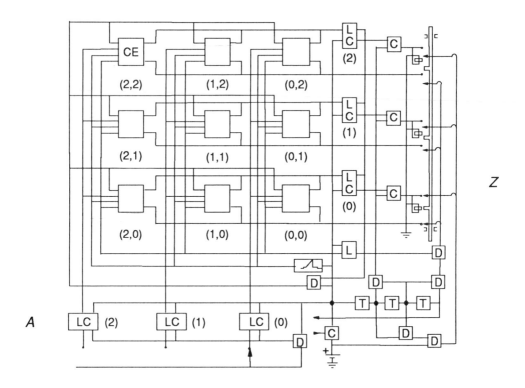

Figure 17.10. A *CE cell trinary control matrix* has three values of its input and output variables.

A CE cell quadrary control matrix has four values of its input and output variables, and a pentary control matrix has five values of its input and output variables, etc. The number of values can be expanded to match the variety of the sensor and actuator variables.

Conditional empirical control system

A CE cell matrix can be incorporated into a *conditional empirical control system* represented by the diagram in Figure 17.11.

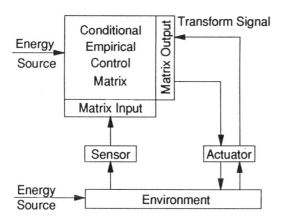

Figure 17.11. A conditional empirical control system is made up of a CE cell memory matrix, an environment, a sensor, an actuator, and a transform signal representing the actual value of the actuator at specific moments.

The behavior of a conditional empirical control system is established by information carried by the input, output, and transform signals.

Application of a CE cell control matrix

The CE cell control matrix can be used effectively as an empirical controller for any application having one input variable and one output variable, each with a limited number of values. For example, the CE cell control matrix can be connected to a direction error sensor and a rudder servo amplifier to form the *conditional empirical ship auto-pilot* shown in Figure 17.12. Once the system is installed and put into use, the empirical controller will attempt to produce specific rudder positions for each error signal. The helmsman must prohibit inappropriate rudder responses and supply the appropriate steering correction value for each direction error encountered. After a period of learning, the conditional empirical auto-pilot can supply the most probable relation between input and output states produced by the helmsman, and the helmsman may relinquish his or her task to the auto-pilot. If the resolutions of the input and output variables are 5%, the number of input and output states is equal to 20 states. Then the matrix would require $20 \times 20 = 400$ CE cells. Of these 400 CE cells, 20 must be placed in a most sensitive state in order for the matrix to specify an output for each input state.

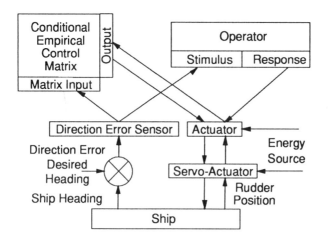

Figure 17.12. A conditional empirical ship auto-pilot can produce the specific rudder position most often used by the helmsman for each specific direction error state encountered.

Uncertainty level indicator

The ramp voltage required to close the contacts of the CE cell representing given input conditions can also be used as an *uncertainty level indicator,* as shown in the ramp generator for the conditional predetermined control matrix in Part II. If a very high ramp voltage is required to close the contact of a CE cell in a set of CE cells, it indicates that those CE cells have a high degree of uncertainty. Thus, a high ramp voltage value may be used to trigger a warning signal to an operator that none of the interrogated CE cells for those input conditions has a high confidence level and that the operator may wish to take over the control task. The condition/action relations supplied by the operator under these circumstances may then become incorporated in the behavior of the empirical controller and increase the confidence level of the controller for those conditions.

Averaging learning events

A conditional empirically controlled machine establishes its behavior not by a predetermined program or by a single recording event, but by *averaging many learning events* and selecting those events that occur most consistently. This feature makes it much easier to train a conditional empirical machine, compared to an absolute empirical machine, since the operator can make a few mistakes while training the machine without these mistakes become a part of the actual behavior of the machine system.

Preconditioned empirical matrices

The sensitivity of the cells in an empirical control matrix can be set initially to specific predetermined values to produce specific initial behavior. This *preconditioned behavior* can then be reinforced or discouraged by experience.

Parallel search

In a large control matrix, many CE cells have to be interrogated to select an output. The use of a CE cell matrix allows these operations to be carried out in parallel. The actual selection is based upon the relative values of many CE cells connected in parallel to a given input terminal. Regardless of the number of memory cells that have to be interrogated, the time to select an output is based only on the rate of rise of the ramp voltage and the sensitivity of the selecting memory cell.

A digital computer can be programmed to simulate a set of CE cells. Also, individual "cells" in *spreadsheet software* applications can be programmed to behave like a CE cell. However, a single digital computer has to query and update each cell in sequence. If a system has many cells, it may take too much time for a serial processor to operate a real machine in a real environment.

A CE cell is a more complex device than an absolute empirical memory cell (AE cell). It is more like a complete computer than a memory element since it can determine the logarithmic average of one event occurring in relation to another event. In view of this requirement, an individual *microcomputer* may be needed to carry out the function of an individual CE cell. These microcomputers can be connected in the matrix arrangements described in this volume, providing a unique configuration for a parallel and distributed microprocessor system.

Mixed memory matrices

It is possible to mix absolute and conditional empirical memory cells in a given matrix if the ramp voltage required to activate the absolute memory cells is high enough to allow high-sensitivity conditional cells to be activated. The mixture of cells can be arranged so that preferred behavior is carried out on the absolute cells, and other contingency behavior can be carried out on the conditional cells. Also, absolute and conditional predetermined cells can be mixed with the empirical cells to provide predetermined (instinctive) behavior that can be overridden by the empirical cells if the ramp voltage needed to activate the predetermined cells is low enough to allow the empirical cells to work. The absolute predetermined cell (AP cell), conditional predetermined cell (CP cell), and the absolute empirical cell (AE cell) must be modified slightly for them to work in a conditional empirical (CE cell) matrix, as shown in this section.

Mixed absolute predetermined memory cell

The absolute predetermined memory cell (AP cell) can be modified to operate in a mixed empirical matrix by adding an output enable relay, feedback complete relay, and an input sense terminal, forming the *mixed absolute predetermined memory cell (MAP cell)* shown in Figure 17.13. The set of mixed absolute predetermined memory cells acts like a persistent *instinctive memory* in that they can be preset to cause certain behavior, but can be overridden by empirical cells if the empirical cells acquire a higher sensitivity than the intrinsic sensitivity of the predetermined cells.

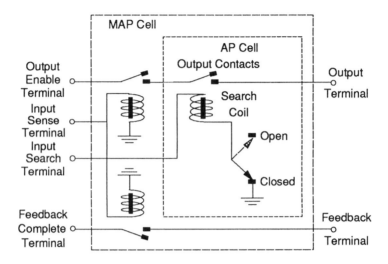

Figure 17.13. A *mixed absolute predetermined memory cell* (MAP cell) can operate in a conditional empirical matrix.

Mixed conditional predetermined memory cell

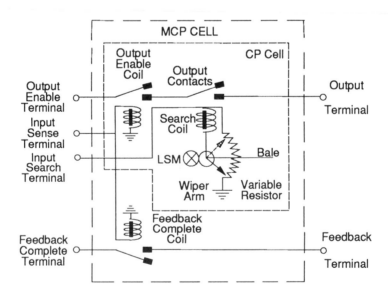

Figure 17.14. A *mixed conditional predetermined memory cell (MCP cell)* can produce predetermined behavior in a conditional empirical matrix.

The conditional predetermined memory cell (CP cell) also can be modified to operate in a mixed empirical matrix by adding a feedback complete relay, as shown in Figure 17.14. A *mixed conditional predetermined memory cell* (MCP cell) has a fixed sensitivity that can be set manually by a programmer or established by design before the system is put into use. The addition of a feedback complete relay allows it to carry the feedback signal to the feedback disconnect though it does not use the feedback signal to change its sensitivity. It acts like a CE cell that does not change its sensitivity (learn) through experience.

Mixed absolute empirical memory cell

An output enable relay can be added to an absolute empirical memory cell (AE cell), forming the *mixed absolute empirical memory cell (MAE cell)* shown in Figure 17.15.

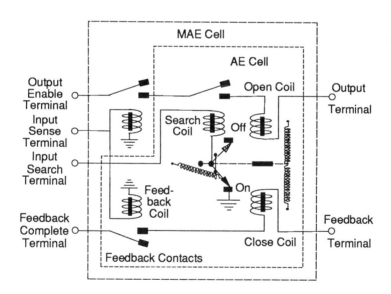

Figure 17.15. A *mixed absolute empirical memory cell* (MAE cell) can produce absolute responses in a conditional empirical matrix.

The addition of the output enable relay allows the MAE cell to use the separate output enable and input sense lines needed in a conditional empirical memory matrix. If the ramp voltage required to close the contacts of the MAE cell is high enough to give other CE cells a chance to produce behavior, the MAE cell can work as an absolute empirical memory cell in a conditional empirical memory matrix.

Mixed conditional control matrix

The mixed absolute predetermined cell, mixed conditional predetermined cell, the mixed absolute empirical cell, and the conditional empirical cell can be connected in a *mixed conditional control matrix*, as shown in Figure 17.16.

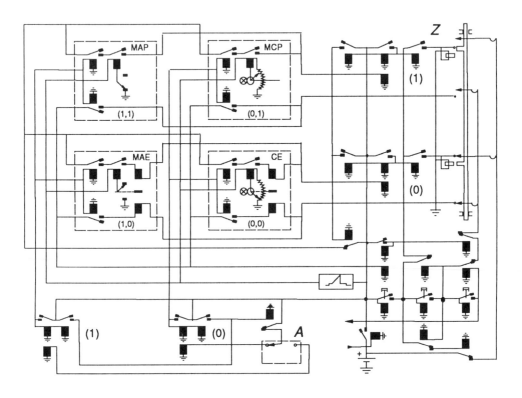

Figure 17.16. A *mixed conditional control matrix* may contain all four types of memory cells.

A mixed conditional control matrix can be designed and programmed before the matrix is put into use to produce specific behavior or its behavior can be changed after it is put into use. The predetermined cells can be used to provide *fallback behavior* that will appear if the confidence levels of the empirical cells fall below the intrinsic sensitivities of the predetermined cells. The animal brain may be made up of mixed memory cells.

Chapter 18
Conditional Empirical Multivariable Machines

All of the conditional empirical control matrices discussed so far are made up of a single input and output variable. Special matrix designs are required to create control matrices with two or more input and/or output variables.

Conditional empirical coded unitary controllers

The conditional matrices shown previously are single input/output variable (scalar) control matrices. One way of dealing with two or more input and/or two or more output variables is to decode the input variables into a single variable, encode the matrix output variable into a set of actuator variables, and then decode the transform signal into the single matrix feedback variable, forming the *CE cell coded unitary controller* shown in this section.

CE cell coded unitary controller

A scalar conditional empirical memory matrix can be used to form the *CE cell coded unitary controller* shown in Figure 18.1.

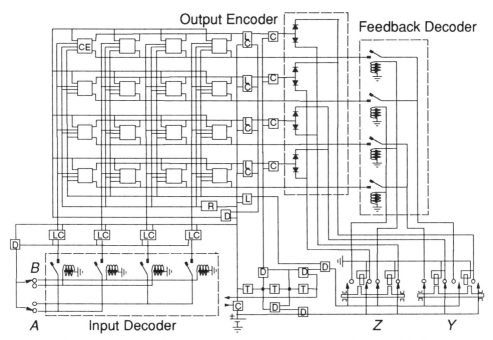

Figure 18.1. A *CE cell coded unitary controller* uses a scalar conditional empirical memory matrix to produce the most probable relations between multiple input and output variables.

This system also uses an sensor decoder to transform the multiple sensor variables into a single matrix input variable, an actuator encoder to transform single matrix output variable into a set of actuator variables, and a feedback decoder to transform the actual values of the actuator variables into the single transform signal to the CE cell matrix to increase the sensitivity of the CE cell at the intersection of the actual input and output states in each transition cycle. The CE cell unitary controller can deal effective with a multivariable unitary environment where there are *consistent input/output relations* (some probability greater than zero that a given output state can occur after a given input state) and where most combinations of values of the sensor variable produce unique combinations of values of the actuator variables.

Coded unitary conditional empirical control system

The *coded unitary conditional empirical control system* (Fig. 18.2) can include a limited number of sensor and actuator variables each with a limited number of values, and can provide adequate control behavior when each combination of sensor values must produce a specific combination of actuator values.

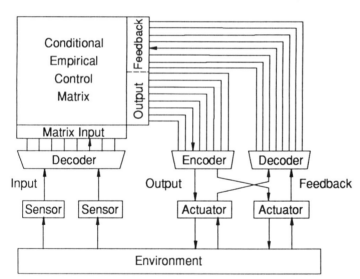

Figure 18.2. The *coded unitary conditional empirical control system* uses a scalar conditional empirical control matrix, an input decoder, an output encoder, and a feedback decoder to discover and produce the control behavior that can be carried out.

The coded unitary conditional empirical control system can be used as a practical control system if each output state occurs according to some distribution of probabilities for each input state. The empirical duplex control matrix to be shown may be required if only a relatively few input/output relations are required in relation to the number of possible relations.

Application of a coded unitary conditional empirical control system

The two input and two output variables of the ship auto-pilot shown earlier can be connected to a coded unitary conditional empirical controller to form a successful multivariable conditional empirical control system, as shown in Figure 18.3.

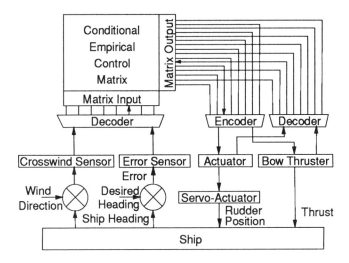

Figure 18.3. A *CE cell coded unitary ship direction control system* produces the most consistently produced combination of actuator values for each combination of sensor values.

The wind and error sensors can be connected to the input decoder, which produces a unique output value of each combination of sensor values. The decoder output value is used by the scalar empirical matrix to select the most consistent output value, which is then converted to a set of rudder and bow thruster values by the output encoder. The actual actuator values are measured by the sensor of the feedback decoder, which produces a value that is used to increase the sensitivity of the CE cell at the intersection of the feedback decoder value and the current input decoder value.

If the resolutions of the sensors and actuators are 5%, requiring 20 terminals for each variable, then 160,000 CE cells are required in this application. This number is quite impractical, and can be reduced to 8000 or less by using the coded duplex empirical controller to be described later, and can be reduced to 1600 using a monolithic control matrix to be described in this chapter. Digitizing these variables can reduce the number of required memory cells even more. Each of these improvements comes about by reducing the commitment of each matrix terminal to specific sensor and actuator values that may never be used.

Superimposed coded unitary control units

Two or more coded unitary control units can be connected to a specific set of sensors and sigma actuators, forming a *superimposed system of CE cell coded unitary controllers*, as shown in Figure 18.4.

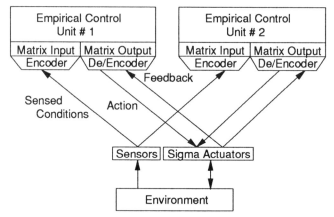

Figure 18.4. A *superimposed system of CE cell coded unitary controllers* can learn to produce unitary or diverse behavior according to the needs of the task environment.

One unit can learn to produce diverse behavior of one set of sensor and sigma actuators and the other unit can learn to produce diverse behavior of the other set through dispersion, or both units can work together to produce unitary behavior of all of the sensors and sigma actuators through the process of value field selection.

Assorted empirical control system

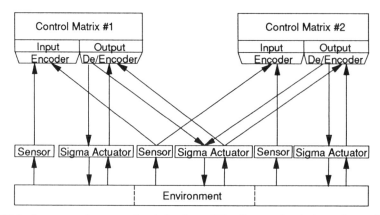

Figure 18.5. An *assorted system of coded unitary CE cell control units* is made up of two or more diverse systems and at least one superimposed system.

Two or more coded unitary control units can also be connected to an assorted set of sensors and sigma actuators, forming an *assorted superimposed system of CE cell coded unitary controllers*, as shown in Figure 18.5. Coded unitary control units in an assorted system can establish a unitary relation with the sensor and actuator variables to which they are connected exclusively, and learn to work together upon the sensor and actuator variables upon which they are superimposed.

Combined system of coded unitary control units

A coded unitary controller can be connected to produce unitary relations at its sensor and actuator variables. If some of these sensor and actuator variables can act independently from other sets, separate controllers must be connected to these variables in parallel with the coded unitary controller, forming the *combined system of CE cell coded unitary controllers* shown in Figure 18.6.

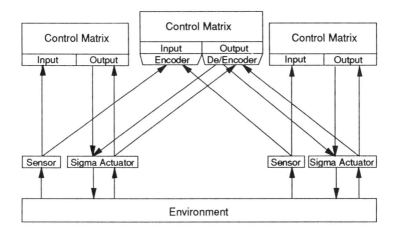

Figure 18.6. A coded unitary control unit in a *combined system of CE cell controllers* can learn to produce unitary behavior in a set of sensor and sigma actuators, or allow other more diverse control units to determine the behavior of these sensors and sigma actuators.

The coded unitary unit can produce consistent results if the task environment is unitary. However, the diverse units can produce more consistent results if the task environments are diverse. The units that produce the most consistent results will dominate the behavior of the sensor and actuator variables.

Composite empirical control system

Different levels of coded unitary controllers can be connected to different sets of sensor and actuator variables in the manner of a structured superimposed network, as shown in Figure 18.7, creating a composite empirical system that can learn to produce unitary or diverse behavior among more variables than a combined system.

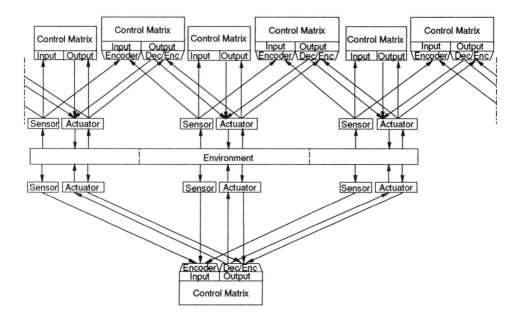

Figure 18.7. A *CE cell composite control system* can learn to produce unitary or diverse behavior according to the needs of the task environment.

The number of control units and the number of CE cells in a composite system increases astronomically with even small increases in the number of variables and the number of values of these variables. For example, the number of CE cells in a composite system with n binary variables increases by 10^n. Obviously, another method must be found to provide a control system that can learn to deal with unitary and/or diverse variables.

Conditional empirical multivariable input controllers

We have shown the need for a conditional empirical controller that can deal with multiple input and output variables that does not commit memory cells and circuitry to specific combinations of values of input and output variables that may never be used. This need can be met by a multivariable conditional empirical input controller that provides a unitary and/or diverse organization of its multiple input or output variables and does so with fewer CE cells than a coded unitary or composite design with the same variety.

The absolute empirical multivariable input matrix shown earlier can learn to produce unitary or diverse relations between two or more input variables and one output variable if a given input state always produces a given output state. This matrix can be converted into a conditional empirical matrix by substituting conditional empirical cells for the absolute empirical cells and providing the appropriate CE cell matrix driver circuit.

The CE cell multivariable input matrix

Two or more single variable empirical submatrices can be combined to create a matrix with multiple input variables and a single output variable, as shown in the circuit diagram in Figure 18.8. Each single-variable submatrix represents a given input variable, and the input terminals of each submatrix represent values of each input variable. Since each input variable has two values, the multivariable input matrix shown is called a *CE cell binary multivariable input matrix.*

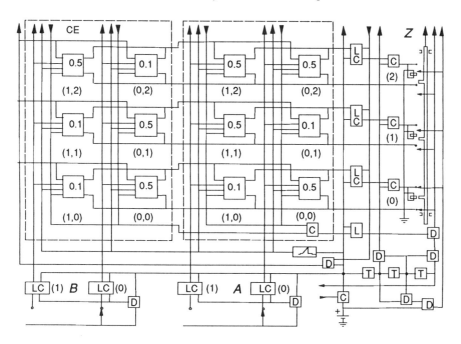

Figure 18.8. A *CE cell binary multivariable input matrix* consists of a set of single-variable binary conditional empirical submatrices connected in series such that they energize a single output terminal when a CE cell in a given row in each matrix becomes conductive.

An output is produced when a conductive path is found along a row of CE cells between the output enable line and the output terminal on that row. To be conductive, these CE cells must be connected to the latched input terminals in each matrix and have their output contacts closed by the action of the input ramp upon their search coils. For example, CE cells $B(0,0)$ and $A(0,0)$ are the highest-sensitivity CE cells connected to the input terminals shown to be active. These cells produce the output $Z(0)$. Once an output is selected, no other outputs may be produced, due to the action of the output enable disconnect. The sensitivities of these selecting CE cells are reduced according to their logarithmic subtracting mechanisms. If the selected output value can be executed as called for, the sensitivity of the selecting CE cells is increased according to their logarithmic subtracting mechanisms by an amount greater than they were decreased, leaving them more likely to select that output in the future.

Operation of the CE cell multivariable input matrix

Figure 18.1. Operation of the multivariable CE cell input matrix

---

1. The cycle switch is closed at the start of a new transition cycle.
2. One input value terminal is energized in each matrix according to the positions of the sensor switches at the beginning of the transition cycle. These input terminals are latched "on," and the sensor switches are disconnected to prohibit any additional input terminals from being energized during that transition cycle.
3. The sensitivity of all of the CE cells is interrogated by the ramp signal, but only the CE cells in the columns of the latched input terminals can produce an output.
4. When a CE cell on a given row in each matrix becomes conductive, a voltage from the output enable line on the left-hand side of the multivariable matrix can flow through the set of matrices to the right-hand output terminal.
5. This output signal disconnects the output enable line so that no more CE cells can become conductive and the sensitivities of the selecting CE cells are decreased according to their logarithmic subtracting mechanisms.
6. After the time delay specified by the actuator time delay relay in the matrix driver, the actuator brake is released, and the actuator coil connected to the selected output terminal is energized and attempts to align the armature pole of the actuator in that position.
7. After another time delay created by the feedback delay relay, to allow time for the actuator to position itself, the actuator coils are disconnected and the actuator brake is re-engaged. This holds the actuator at the actual position it is in when the brake is re-engaged. A feedback signal is also sent along the row of CE cells corresponding to the actual position of the actuator. This signal increases the sensitivity of the CE cells on this row that are connected to the latched input terminals, and it operates the feedback disconnect to prevent another feedback signal from being sent if the actuator moves to a new position during this period.
8. After another time delay created by the end-of-feedback delay relay to give the actuator time to produce a feedback signal, the feedback line is disconnected. This ends the period during which a feedback signal can occur, and a signal is sent to the outside timing source indicating that the transition cycle is over. (This signal could be sent directly to the cycle switch to end the transition cycle locally.)
9. The cycle switch is opened, causing power to be shut off to all elements in the matrix except the actuator brake circuit, and all of the latches and disconnects are restored to their normal positions.

The multivariable input matrix can be used in any application having multiple input variables and a single output variable. The steps given in Table 18.1 allow the multivariable conditional empirical input matrix to operate in the discontinuous state ontology.

The coattail effect in a multivariable conditional empirical input matrix

The multivariable input matrix selects the output value on the row of CE cells having the *highest set of lowest sensitivity values* for the set of current (latched) values of the input variables, and it changes the sensitivities of the CE cells in the matrix according to the actual output position of the actuator. The multivariable conditional empirical input matrix *forward selects* the output terminal connected to the row of CE cells in which *the least sensitive CE cell in this row connected to the active input terminals is higher than the least sensitive CE cell connected to the active input* terminals in any *other row*. This *coattail effect* is similar to the way a weak candidate for political office may get elected when other members of his or her party do well in an election. Their election victories may then strengthen the weak candidate. The coattail effect may create a set of very high sensitivity CE cells on a given intermediate terminal, causing that intermediate terminal to be strongly selected when a given input state occurs.

Unitary conformation of a CE cell multivariable input matrix

A multivariable input matrix can act like a unitary controller without using a sensor decoder when its memory cells contain the distribution of sensitivity values shown in Figure 18.9.

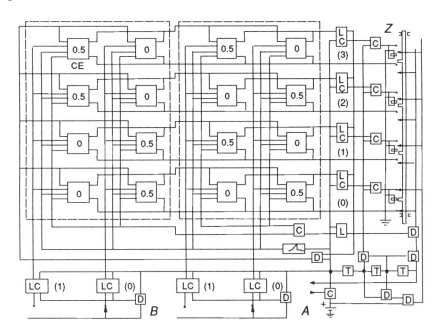

Figure 18.9. A *unitary conformation of a CE cell multivariable input matrix* can produce a unique output for each combination of sensor values in a unitary manner.

In the example of the multivariable input matrix shown, the output value $Z(0)$ would be selected since it contains the row of CE cells with the *highest of the lowest sensitivities* for the positions of the sensor switches shown. If both sensor switch contacts were moved to the left, the output value $Z(3)$ would be selected. If there are x^n output terminals, every combination of x values of the n input variables can select an output in a unitary manner. However, if only a few combinations of values of the input variables are expected to produce unique outputs, then only this number of output terminals are needed, and the matrix can learn to produce these outputs when those combinations of input values occur.

Synchronous diverse conformation of a CE cell multivariable input matrix

A multivariable input matrix can produce a set of values of its output variable based upon the combined values of a set of its input variables in a unitary manner, as shown, or can produce specific values of its output variable based upon the values of a specific input variable, regardless of the values of the other input variables, in a synchronous diverse manner, as shown in Figure 18.10.

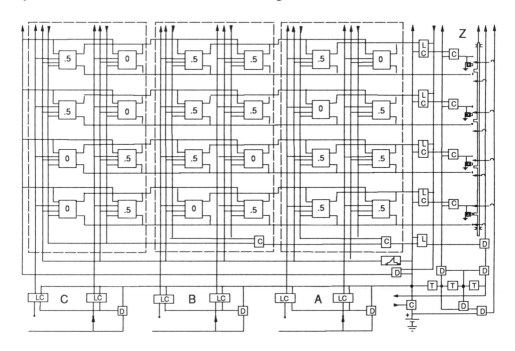

Figure 18.10. In a *synchronous diverse conformation of a CE cell multivariable input matrix,* the distribution of sensitivities of the CE cells is such that the values of one input variable alone determine the value of the output variable.

With the distribution of sensitivities given and with the input variable *C* in the position shown, each value of *B* produces a different output value regardless of the values of *A*. When *C* is in the other position, *A* produces a different output value despite the value of *B*. Thus, *A* can act independently of *B*, or *B* can act independently of *A*, depending upon the value of *C*. Since the two diverse lines of behavior cannot occur at the same time, and when they occur is determined by the value of *C*, they are not asynchronous.

CE cell trinary multivariable input matrix

The number of values of each variable can also be increased to three, forming the *CE cell trinary multivariable input matrix* shown in Figure 18.11.

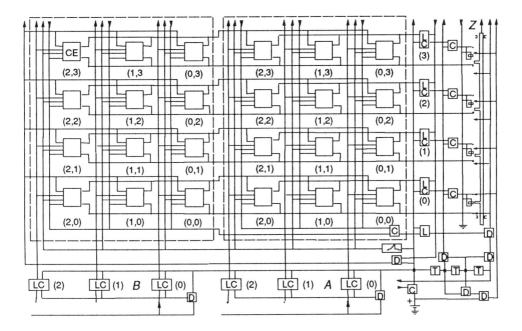

Figure 18.11. A *CE cell trinary multivariable input matrix* can identify any one of more unique combinations of values of its input variables than a CE cell binary multivariable input matrix with the same number of input terminals.

The number of input variables can be increased indefinitely, and the number of values of each input variable can be increased to match the variety of the sensor variables. However, the maximum variety of behavior can be produced for a given number of CE cells when each input variable has only three values.

Self-organization in a multivariable input matrix

If each output state of a multivariable input matrix is determined by different combination of values of its input variables, then the variables of the matrix form a unitary system. If all of the CE cells representing the values of a given variable are set in a high sensitivity value, then individual values of that variable have no meaning in that they cannot produce a unique output state. That variable becomes inactive and is essentially disconnected from the matrix. It no longer is a member of a unitary system. If all of the CE cells representing values of all of the variables except the values of one variable are set in a high sensitivity state, then that one variable is independent of the other input variables, and it can form a diverse relation with its output variable. If all of the CE cells representing values of a given variable are set in a low sensitivity state, the whole matrix is essentially shut down in a superimposed system with other units in which one unit can dominate behavior by being the first to act. (See Part IV) That means that none of the variables in the matrix can produce an output state, and all of those variables become inactive.

Thus, depending upon the sensitivities established by the CE cells in the multivariable input matrix, the matrix forms a unitary relation among its input/output variables, forms a diverse relation of one or more input variables and its output variable, or becomes inactive, forming no relations between its input and output variables. The multivariable input matrix can be used as a unitary or diverse empirical controller in any system having multiple input variables and a single output variable, or it can eliminate the influence of one or more input variables to the point of forming a null or inactive matrix. These conformations can be established by the CE cells changing their sensitivities according to the behavior of their actuators, after the system is put into use. The actual conformation selected in an empirical matrix will be based upon whatever conformation produces the most consistent results.

Application of a CE cell multivariable input matrix

The CE cell multivariable input matrix can be used to connect two or more input variables to a single output variable in a unitary or diverse fashion. For example, a multivariable input matrix can be used to connect the direction error sensor and the crosswind sensor in the *ship direction control system* shown earlier to a single rudder position actuator as shown in Figure 18.12.

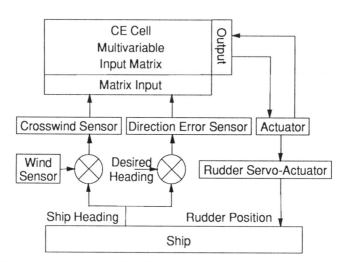

Figure 18.12. An *application of a CE cell multivariable input matrix* is a ship direction control system, which can use one or all of its input variables to control one output variable.

If both the crosswind and direction error signals are needed to maintain the course of a ship, unique rudder actuator values can be established empirically for each combination of values of crosswind and direction error. If the crosswind has no influence upon the amount of rudder correction required to keep the ship on course, high sensitivity can be established in the crosswind matrix for all values of crosswind, thus eliminating it as a control variable.

Conditional empirical multivariable output controllers

We have seen how an absolute empirical multivariable output matrix can learn to produce unitary or diverse relations between its input variable and two or more output variables. This matrix can be converted into a conditional empirical multivariable output matrix by substituting conditional empirical cells for the absolute empirical cells and providing the appropriate CE cell driver circuit.

Conditional empirical binary multivariable output matrix

Two or more single-variable conditional empirical matrices can be combined to produce a *CE cell binary multivariable output matrix* having a single input variable and multiple binary output variables, as shown in Figure 18.13.

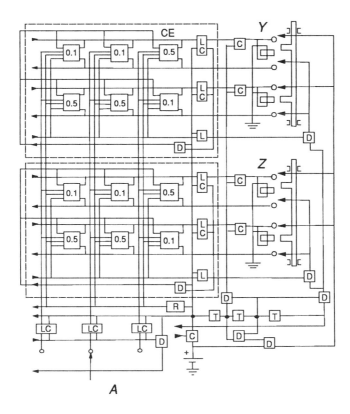

Figure 18.13. A *CE cell binary multivariable output matrix* is made up of two or more binary single-variable empirical conditional submatrices connected to a single input variable.

The most sensitive CE cell in the matrix of each output variable connected to the current sensor switch position select an output value, and the sensitivity of these CE cells are reduced. Then the sensitivities are increased in the cells at the intersections of the actual actuator positions and the current sensor switch position. If the contacts of the sensor switch were moved to the right, both actuators would move to the other position. If there are x^n sensor switch positions terminals, where x is the number of values of each actuator and n is the number of actuators, the output matrix can select a unique combination of actuator positions for every sensor switch position in a unitary manner.

Unitary conformation of a CE cell multivariable output matrix

The CE cell binary multivariable output matrix shown can be rotated 90° to show the position it assumes in a duplex controller. Depending upon the distribution of sensitivities of the CE cells in a multivariable output matrix, the matrix can produce a unique combination of values of its output variables for each input value in a unitary fashion, as shown in Figure 18.14.

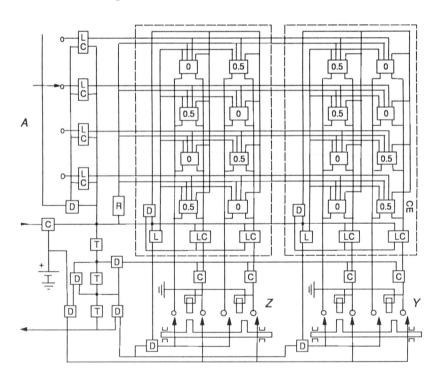

Figure 18.14. A unitary conformation in a CE cell multivariable output matrix produces a unique set of output values for each value of the input variable.

In the example given, all four possible combinations of output values are produced by the four values of the input variable. It can produce every possible combination of values of n output variables having x values because it has x^n input terminals.

Operation of the conditional empirical multivariable output matrix

Table 18.2. Operation of the CE cell multivariable output matrix

---

1. The cycle switch is closed, starting a new transition cycle.
2. One input terminal is energized according to the initial position of the sensor switch representing the input variable. This input terminal is latched "on," and the sensor switch is disconnected so that no additional input terminals can be energized during this transition cycle.
3. The sensitivities of all of the CE cells in both submatrices are interrogated by the ramp signal.
4. When a CE cell in a submatrix representing an output variable becomes conductive, a voltage from the output enable power source can flow through this CE cell to the output terminal in that column.
5. This output signal disengages the output enable power source to that submatrix so that no more CE cells in that submatrix can produce an output. The logarithmic subtracting mechanism in the conducting CE cell decreases the sensitivity of that CE cell according to its sensitivity value. The ramp continues to rise so an output can be selected in both submatrices. Though the output contacts close in the other CE cells in a submatrix that has selected an output, no other outputs are produced and no other subtractions occur because the output enable power source to these CE cells has been disconnected.
6. After the delay specified by the output time delay relay, the actuator brake is released, the actuator coils connected to the selected output terminals are energized, and the actuators attempt to align their armature poles to the positions of the energized actuator coils.
7. After another delay, to allow the actuators to position themselves, the actuator brake is re-engaged, holding the actuators in their current positions and feedback signals are sent to the CE cells connected to the actual positions of the actuators. These signals increase the sensitivities of the CE cells that are connected to the latched input terminal and the actual output co-terminals. When the sensitivity of the selected CE cell in a given submatrix has been influenced by the feedback signal, a signal is sent to the feedback disconnect in that submatrix to prevent another feedback signal from being sent if the actuator in that submatrix moves to a new position during this period.
8. After another delay, to give all of the actuators time to produce a feedback signal, the feedback power source is disconnected, signifying the end of the period in which feedback signals can occur, and a signal is sent to the outside timing source indicating that the transition cycle is over. (This signal could be sent directly to the cycle switch to end the transition cycle locally.)
9. The cycle switch is opened causing power to be shut off to all components except the actuator brake circuit, and all of the latches and disconnects are restored to their normal positions.

The multivariable output matrix selects the set of output values in each output variable connected to the CE cell with the highest sensitivity in each matrix and changes the sensitivities of the CE cells in each matrix, according to the actual output position of the actuator. This process is carried out in the steps given in Table 18.2. The multivariable output matrix can be used in any application having a single input variable and multiple output variables. The sequence of actions described allows the multivariable output matrix to operate in the discontinuous state ontology. Since the actuator variables have two values, the conditional empirical multivariable output matrix shown is called a *binary* output matrix.

Synchronous diverse conformation of a CE cell multivariable output matrix

A multivariable output matrix can produce a set of values in some of its output variables based upon the values of its input terminals, and can hold a constant value in the other output variables in a synchronous diverse manner, as shown in Figure 18.15.

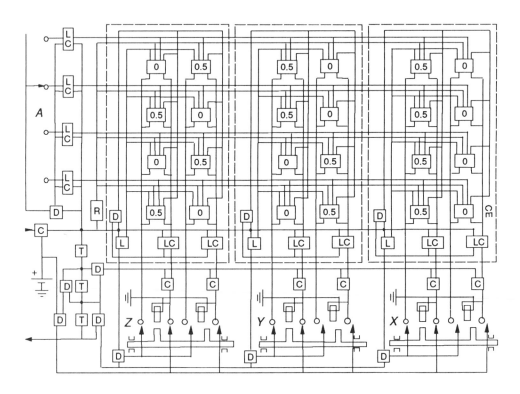

Figure 18.15. In a *synchronous diverse conformation in a CE cell multivariable output matrix,* the value of the input variable does not attempt to change the values of all of the output variables.

In the example of a synchronous diverse conformation of a multivariable output matrix shown, the values of actuators Z and Y are determined by the values of the input variable A and the distribution of high-sensitivity values of the CE cells in their submatrices. Since the sensitivities are high in the CE cells connected to only one value of the actuator X, its submatrix does not attempt to change the position of its actuator. When this immobile actuator is summed with another actuator, it will produce no change in behavior. Thus, the multivariable matrix attempts to create activity in only two of its three actuators in a diverse manner. The multivariable output matrix can be used as a unitary or synchronous diverse empirical controller in any application having a single input variable and multiple output variables.

CE cell trinary multivariable output matrix

The number of values of each output variable in a CE cell multivariable output matrix can be increased to three if each CE cell is connected as shown in Figure 18.16.

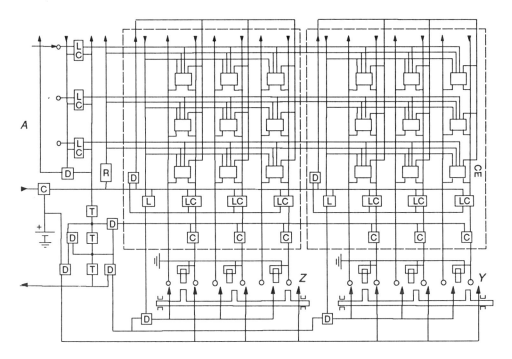

Figure 18.16. A *CE cell trinary multivariable output matrix* can produce more combinations of values of its output variables than a CE cell binary multivariable output matrix with the same number of CE cells per transition.

If the number of values of each output variable is increased to 3 as shown, a multivariable output matrix with 2 output variables can produce any one of $3^2 = 9$ different output states using only 6 CE cells per transition, as compared to a binary matrix with 3 output variables shown earlier, which can produce only $2^3 = 8$ different output states using the same number of 6 CE cells per transition. The number of values of the output variables can be increased to match the variety of its output actuators, although the maximum variety of behavior for a given number of CE cells is produced with the trinary system shown.

Application of a CE cell multivariable output matrix

A multivariable CE cell output matrix can be used to control two or more output variables from two or more input variables when the input variables are summed into a single input variable. Thus, the value of the crosswind sensor and direction error sensor in the *ship direction control system* shown earlier can be summed in a differential and the output of the differential used as the single input variable of a multivariable output matrix. Then the output variables of the matrix can be used to control the rudder servo-actuator and the bow thruster, as shown in Figure 18.17.

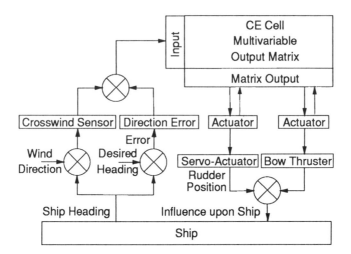

Figure 18.17. An *application of a CE cell multivariable output matrix* is that of a ship direction control system that can produce multiple outputs, using a differential to sum multiple inputs into a single input variable.

The output variables can be controlled by the helmsman such that unique combinations of values of both variables are produced for each value of the input variable in a unitary manner, or one output variable may be held constant for all values of the input variable so that the other output variable can form a synchronous diverse relation with the input variable. The ability of an input variable to change behavior is eliminated if it is held constant.

Conditional empirical monolithic controllers

The features of a conditional empirical multivariable input matrix can be combined with the features of a conditional empirical multivariable output matrix to form a single conditional empirical multivariable monolithic control matrix.

CE cell monolithic control matrix

The design of the CE cell input and output matrices shown can be combined into a single multivariable *CE cell binary monolithic control matrix,* as shown in Figure 18.18.

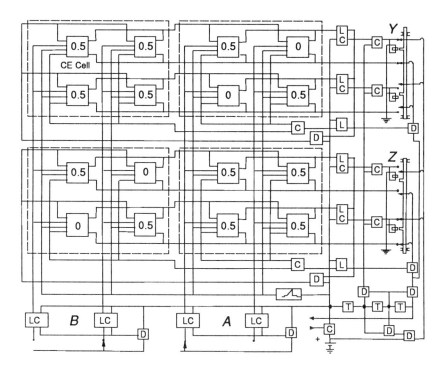

Figure 18.18. The *CE cell binary monolithic control matrix* can discover and incorporate those consistent sensor/actuator relations that occur between multiple input and multiple output variables.

The CE cell binary monolithic control matrix shown with two binary input variables and two binary output variables is made up of four scalar submatrices, each connecting a particular input variable to a particular output variable. It can operate in a *unitary manner* in which a particular combination of values of the input variables produces a unique combination of values of the output variables, or it can act in an *asynchronous diverse manner* in which any particular input variable can influence a particular output variable at any time without regard for the values of the other variables. This is accomplished by setting the sensitivity values of the CE cells connected to a diverse input and output variables in a high sensitivity state as shown.

CE cell monolithic control matrix is particularly effective when the input variety matches the output variety and every combination of values of the input variables produces a different combination of values of the output variables. This is partly because all of the memory cells in a monolithic control matrix are used to produce behavior. The CE cell monolithic control matrix is not particularly efficient when only a few input/output relations are required to produce a desired line of behavior. In this case, a CE cell duplex controller, to be shown, may provide a more effective control matrix.

Operation of a CE cell monolithic control matrix

The CE cell monolithic control matrix operates like a CE cell input matrix and a CE cell output matrix, as shown in Table 18.3.

Table 18.3. Operation of a CE cell monolithic control matrix

---

1. The cycle switch is closed at the start of a new transition cycle.
2. One input terminal of each input variable is energized according to the initial position of their sensor switches. These input terminals are latched "on" and the sensor switches are disconnected so that no additional input terminals can be energized during this transition cycle.
3. The sensitivities of all of CE cells in all of the matrices are interrogated by the ramp signal.
4. When a CE cell in a given row in each matrix becomes conductive, a voltage from the output enable power source can flow through these CE cells to the output terminal on that row.
5. This output signal disengages the output enable power source from those matrices so that no more CE cells in those matrices can produce an output. The logarithmic subtracting mechanisms in the conducting CE cells decrease the sensitivities of these CE cells according to their individual sensitivity values. The ramp continues to rise so an output can be selected in the matrices connected to both output variables. Though the output contacts close in the other CE cells in the matrices, no other outputs are produced and no other subtractions occur because the output enable power source to these CE cells has been disconnected.
6. After the time delay specified by the output time delay relay, the actuator brake is released, the actuator coils connected to the selected output terminals are energized, and the actuators attempt to align their armature poles to the positions of the energized actuator coils.
7. After another time delay, to allow the actuators to position themselves, the actuator brake is re-engaged, holding the actuators at their current positions, and feedback signals are sent along the rows of CE cells corresponding to the actual positions of the actuators. These signals increase the sensitivities of the CE cells that are connected to the latched input terminals and the actual output terminal. When the sensitivity of a CE cell has been influenced by the feedback signal, a signal is sent to its feedback disconnect to prevent another feedback signal from being sent if its actuator moves to a new position during this period.
8. After another time delay, to give all of the actuators time to produce a feedback signal, the feedback power source is disconnected, signifying the end of the period in which feedback signals can occur, and a signal is sent to the outside timing source indicating that the transition cycle is over. (This signal could be sent directly to the cycle switch to end the transition cycle.)
9. The cycle switch is opened causing power to be shut off to all components except the actuator brake circuit, and all of the latches and disconnects are restored to their normal positions.

The CE cell monolithic controller can learn to produce a unique combination of actuator values for each combination of sensor values in a unitary manner, or can learn to produce a given value of a specific actuator variable in response to a specific value of a given sensor variable regardless of the values of the other sensor variables in an asynchronous diverse manner.

CE cell binary monolithic control matrix with three binary sensor and actuator variables

The number of input and output variables in a CE cell monolithic control matrix can be increased to three if they are connected as shown in the *CE cell monolithic control matrix with three binary sensor and actuator variables* in Figure 18.19.

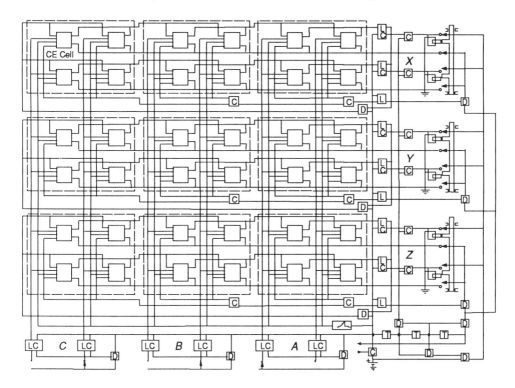

Figure 18.19. A CE cell monolithic controller with three binary sensor and actuator variables can learn to produce eight combinations of values of its actuator variables in response to eight combinations of values of its sensor variables in a unitary manner while using only 36 CE cells. A coded unitary controller would require $8 \times 8 = 64$ CE cells to produce the same behavior.

A CE cell monolithic controller can also learn to produce specific values of any one of its actuator variables in response to specific values of any one of its sensor variables regardless of the values of the other sensor variables. Thus, it can learn to produce asynchronous diverse behavior or unitary behavior according to the requirements of the task environment.

CE cell trinary monolithic controller

The number of values of the variables in a monolithic control matrix can be increased to three, as shown in the *CE cell trinary monolithic controller* in Figure 18.20.

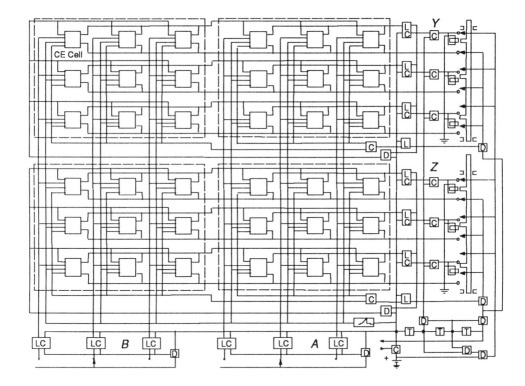

Figure 18.20. A *CE cell trinary monolithic controller* can learn to produce more different combinations of values of its actuator variables in response to different combinations of values of its sensor variables using fewer CE cells than a binary monolithic controller or a monolithic controller based upon any other number system.

The CE cell trinary monolithic controller shown with two sensor and actuator variables can produce nine different combinations of values of its actuator variables while using only 36 CE cells. The CE cell binary monolithic controller with three sensor and actuator variables shown earlier also requires 36 CE cells, but can produce only eight different combinations of values of its actuator variables.

Application of a CE cell monolithic controller

The number of values of the variables in a monolithic control matrix can be increased to match the variety of its sensor and actuator variables, although the greatest variety of behavior for a given number of CE cells can be achieved when each variable has only three values. A CE cell monolithic control matrix can learn to produce different combinations of actions of the rudder position and bow thruster actuators of the *ship direction control system* shown in Figure 18.21 for different combinations of values of its crosswind and direction error sensors in a unitary manner.

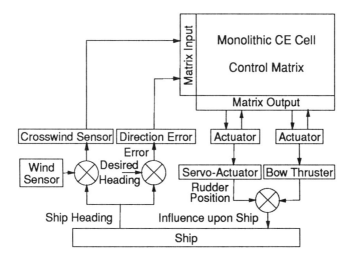

Figure 18.21. An *application of a CE cell monolithic controller* is a multivariable ship direction control system in which each combination of values of the input variables requires a different combination of values of the output variables.

It can also learn to use the direction error sensor to control the rudder position, and the crosswind sensor to control the bow thruster in an asynchronous diverse manner, according to which type of organization produces the most consistent results. In applications where the number of values of the sensor and actuator variables exceeds about eight, consideration should be given to digitizing the sensor and actuator variables as shown in the next chapter.

Chapter 19
Digitized Conditional Empirical Matrices

In many applications, the number of values of sensor and actuator variables far exceeds the number of values made active by the line of behavior of the control system. For example, an automatic damper control system may require a temperature sensor with many input values and a damper actuator with many output values. However, in practice, the control system may have to operate over a very narrow temperature range, requiring very small changes in damper position. This kind of application is not handled efficiently by an undigitized matrix, because it contains far more CE cells than are required to do the job. In these cases, the CE cells connected to the unused sensor and actuator values do not participate in the selection of behavior and are therefore wasted. Perhaps worse, these CE cells must be trained to support the behavior of the active terminals. This increases the learning time of the system and increases its cost. The number of CE cells can be reduced by reducing the number of input and output terminals. However, this reduces the resolution of the sensors and actuators. What can be done to reduce the number of *nonselecting* CE cells without reducing the resolution of the system?

The answer to this question is to design a control matrix in which specific CE cells are not committed to specific input or output states. This can be done by representing the values of a sensor or actuator variable by the values of another set of aggregate variables, as shown in Part II. For example, the decimal number 11 can be represented by four binary variables with values of 1011. The process of converting the values of one kind of variable into the values of another kind of variables is called *digitization*. If a variable is represented by n aggregate variables, each having x values, then x^n values of the variable can be represented using only $x \times n$ aggregate variable terminals, as shown in Part II.

CE cell digitized input matrices

As shown earlier, a multivariable input matrix can have two or more input variables and one output variable. Therefore, a sensor variable can be broken into two or more aggregate input variables in which each combination of values of the aggregate variables represents a given value of the sensor variable.

Digitized sensor variable

Each value of a sensor variable can be encoded to produce a unique combination of values of a set of *aggregate input variables*, as shown in the *CE cell binary digitized input matrix* in Figure 19.1.

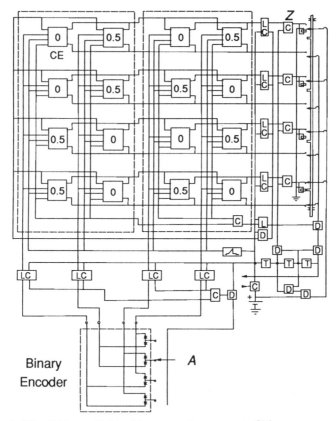

Figure 19.1. A *CE cell binary digitized input matrix* uses a set of binary aggregate variables to represent its input variable rather than the values of the input variable itself.

A sensor variable with a variety V requires n aggregate variables with x values to represent it according to the simple equation of variety of a unitary system, which is $V = x^n$. The number of output terminals of this encoder and thus the number of input terminals in an input matrix digitized by this encoder is equal to $x \times n$. For example, a input variable with four values requires two aggregate variables when a *binary encoder* is used, and the digitized input matrix requires four input terminals, as has been shown. The *specific conformation* of CE cell sensitivities shown allows the binary digitized input matrix to produce a specific output for each different value of the input variable in a *unitary* manner.

Characteristics of an input variable

While encoding an input variable having the four values shown, one value of the *most significant aggregate variable* represents a generally low set of values of the input variable, and the other value represents a high set of values of the input variable. Each value of the *least significant aggregate variable* represents the lower half of the low or high values, and the other represents the upper half of the low or high values of the input variable. Each *characteristic* of an input variable can be represented by an aggregate variable or a combination of aggregate variables.

Generalized input values

If the value of the most significant aggregate variable adequately represents an input state of a variable, it can be used to select a matrix output value by itself if the sensitivities in the remaining CE cells of that output value are set in a high-sensitivity state, as shown in Figure 19.2.

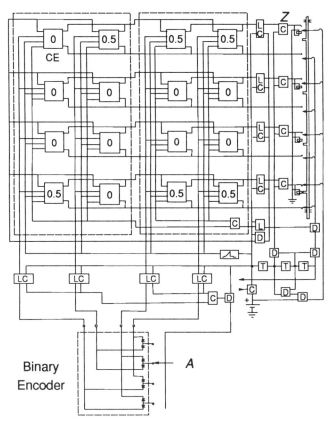

Figure 19.2. In a *generalized conformation of CE cells in a binary digitized input matrix,* some category of values adequately defines an input state.

That is to say, if the high or low *generalized values* of the input variable are sufficient to specify one or another output state, the value of the most significant aggregate variable can be used to select the output of the matrix in a diverse manner. This condition is likely to occur when the resolution of the input variable is greater than what is required by the task environment. The remaining CE cells connected to these outputs must be set in a high-sensitivity state, as shown, to produce this *generalized conformation*. On the other hand, some *specific value* of an input variable may be required to provide successful control. In this category of values, the complete set of aggregate variables must be used to select a matrix output terminal in the unitary manner shown previously, where each combination of values of the aggregate variables produces a unique output state.

CE cell binary digitized input matrix with three aggregate variables

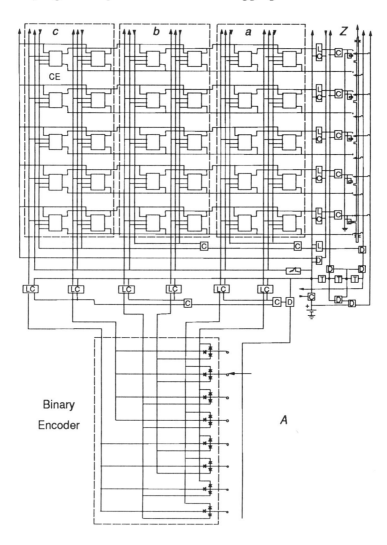

Figure 19.3. A *CE cell binary digitized input matrix with three aggregate variables* requires far fewer CE cells than an undigitized matrix with the same sensor variety.

The digitized input matrix with a single input variable shown produces the same result and uses the same number of CE cells as a scalar matrix with four input terminals, except the digitized input matrix uses two CE cells instead of one to select an output. This is because the digitized input matrix uses two aggregate variables to represent the input variable. However, if the number of values of the input variable were increased to 8, the undigitized input matrix would require 8 input terminals, and the digitized input matrix would require only $x \times n = 6$ input terminals, where $x^n = 8$ input values, $n = 3$ aggregate variables, and $x = 2$ values of each aggregate variable, as shown in Figure 19.3. If 8 matrix output terminals are required, then this digitized input matrix requires $6 \times 8 = 48$ CE cells. An undigitized scalar matrix would require $8^2 = 64$ CE cells. The advantage of the digitized input matrix increases greatly over the single-variable matrix as the number of values of the input variable increases and the number of aggregate variables increases.

Application of a CE cell binary digitized input matrix

By increasing the number of aggregate variables to match the variety of a sensor variable, a digitized CE cell input matrix can respond to any one of a large number of possible input values. This feature allows a digitized CE cell input matrix to be used as a *medium-resolution temperature warning system,* as shown in Figure 19.4.

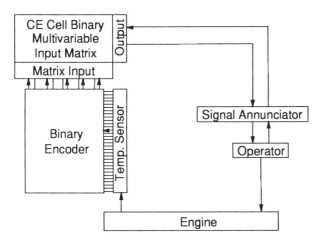

Figure 19.4. An application of a CE cell binary digitized input matrix is a temperature warning system that can learn to respond to some specific temperature value.

An operator only needs to sound an alarm at or above a given temperature. The feedback signal from the annunciator changes the sensitivities of the CE cells in the input matrix, causing the matrix to sound the alarm whenever those temperatures occur again. The five binary aggregate variables provide a sensor variety of $2^5 = 32$ values for a resolution of 3%, and the matrix uses only $2 \times 5 = 10$ CE cells for each output selected.

CE cell trinary digitized input matrix

The number of values of the binary aggregate variables in the CE cell binary digitized input matrix shown can be increased to three, forming the *CE cell trinary digitized input matrix* shown in Figure 19.5.

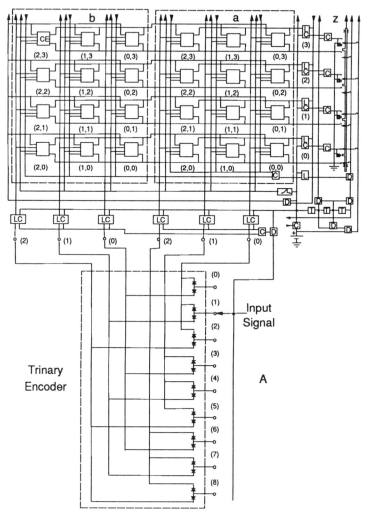

Figure 19.5. A *CE cell trinary digitized input matrix* can identify more values of its input variable using the same number of CE cells as a binary digitized input matrix.

The CE cell trinary digitized input matrix shown can select a unique output value for any one of nine different input values using only six CE cells per selection. A CE cell binary digitized input matrix, using six CE cells to select an output, can identify any one of only eight different input values. If the number of trinary aggregate variables is increased to 6, the trinary digitized input matrix could identify any one of $3^6 = 729$ input values using $3 \times 6 = 18$ CE cells per selection. A binary digitized input matrix, with the same number of 18 CE cells to select an output, could identify any one of only $2^9 = 512$ different input values.

Application of a CE cell trinary digitized input matrix

A CE cell trinary digitized input matrix is the preferred design in any application involving an input variable with many possible values and a single output variable having only a few possible values, such as the *engine temperature warning system* shown in Figure 19.6.

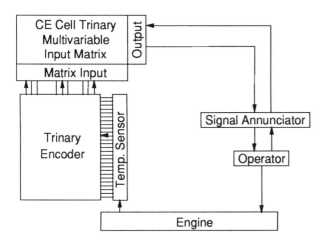

Figure 19.6. An application for a CE cell trinary digitized input matrix is an engine temperature warning system that uses fewer memory cells than a binary digitized system with the same resolution.

The engine temperature warning system can be taught to produce a warning signal at some specific temperature by an operator turning on a warning signal at that critical temperature. The 3 trinary aggregate variables provide a sensor variety of $3^3 = 27$ values for a resolution of approximately 3%, and the matrix uses only $3 \times 3 = 9$ CE cells for each output selected in contrast to the binary digitized sensor with a variety of 32 values using 5 aggregate variables, which requires 10 CE cells per selection.

CE cell binary digitized multivariable input matrix

Each of two or more sensor variables can be encoded into a set of aggregate variables in a multivariable input matrix. The set of values of these aggregate variables can be used to represent a given multivariable input state, as shown in Figure 19.7.

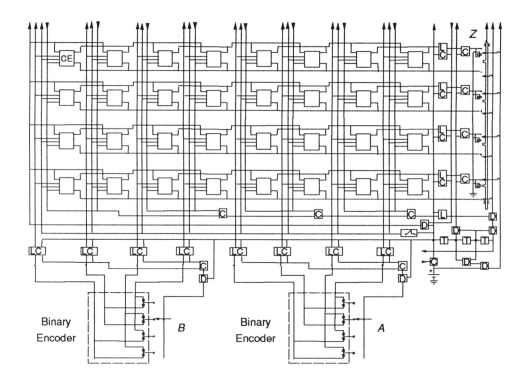

Figure 19.7. The *CE cell binary digitized multivariable input matrix* is made up of multiple sets of aggregate variables that are used to represent the state of a set of input variables.

The number of CE cells in a binary digitized matrix is greatly reduced over an undigitized or coded unitary multivariable controller, particularly when each variable has many values. For example, a multivariable output matrix requires a row of 8 CE cells to select an output from the 4 binary aggregate input variables used to represent the two sensor variables with 4 values each. However, a coded unitary input controller would require 16 CE cells for each output state. If the number of values of each input variable were increased to 16, the binary digitized multivariable controller would require $2 \times 8 = 16$ CE cells per output state, while the coded unitary input controller would require $16^2 = 256$ CE cells to make a single selection.

Application of a CE cell binary digitized multivariable input matrix

The crosswind sensor and the direction error sensors of a *ship direction control system* can be connected to binary encoders that are connected to a CE cell multivariable input matrix. The single output variable of the matrix can then be connected to a rudder servo-actuator, as shown in Figure 19.8.

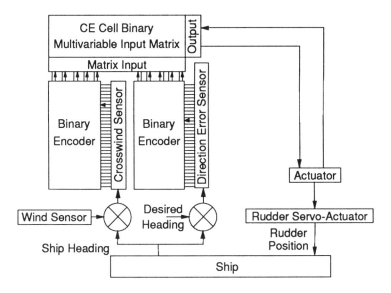

Figure 19.8. An *application of a CE cell binary digitized multivariable input matrix* is a ship direction control system with two input variables and one output variable.

As each rudder position is established by the helmsman for each combination of crosswind and direction error, the CE cell multivariable input matrix will change its sensitivities until it establishes the same modulus of behavior as the helmsman. The crosswind sensor may or may not have any influence upon behavior depending upon whether the conditional empirical multivariable input matrix assigns sensitivities in a unitary or diverse manner.

CE cell trinary digitized multivariable input matrix

Two or more trinary encoders can be connected to a CE cell multivariable input matrix, forming the *CE cell trinary digitized multivariable input matrix* shown in Figure 19.9. Each row of CE cells can be taught to identify a particular combination of values of any number of the digitized input variables. In some cases, many combinations of values of the input variables may be used to produce the same output value in a general manner. In other cases, each combination of input values may be used to produce a different output value in a specific manner.

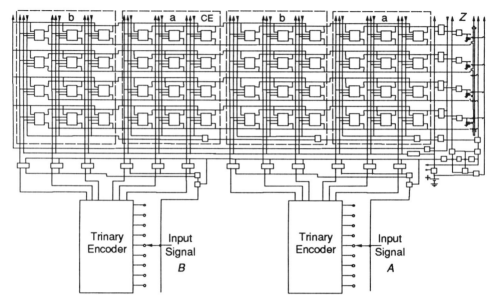

Figure 19.9. A *CE cell trinary digitized multivariable input matrix* can deal more effectively with multiple medium-resolution input variables and a single low-resolution output variable than any other type of digitizing system.

Application of a CE cell trinary digitized multivariable input matrix

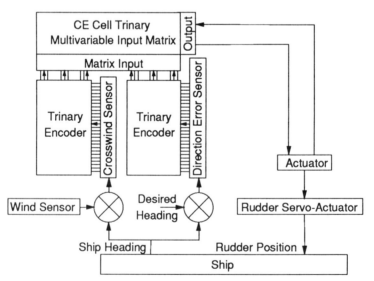

Figure 19.10. An *application of a CE cell trinary digitized multivariable input matrix* is a ship direction control system with two medium-resolution sensor variables and a single actuator variable, which can produce higher quality control with fewer memory cells than any other type of digitizing system.

The CE cell trinary digitized multivariable input matrix shown can be used to operate the *ship direction control system* shown in Figure 19.10. A unique rudder servo-actuator position can be established by the CE cell trinary digitized multivariable input matrix for each combination of values of the crosswind sensor and direction error sensor produced by a skilled helmsman after the empirical control system is installed and put to use in a ship.

CE cell digitized output matrices

As shown earlier, a multivariable output matrix can have two or more actuator variables and one sensor variable. Each actuator variable can be broken into a set of aggregate output variables that in combination determine the value of the actuator variable they represent.

CE cell binary digitized output matrix

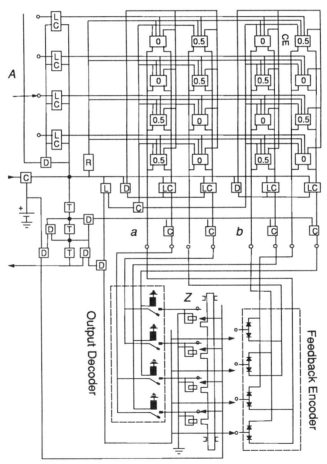

Figure 19.11. In a *CE cell binary digitized output matrix*, the actuator variable is connected to a set of binary aggregate matrix variables through an output decoder and a feedback encoder.

The output variables of a multivariable output matrix can be used as aggregate variables if they are connected to an output decoder. The actual actuator value is then encoded and used to activate the set of aggregate feedback co-terminals corresponding to the actual value of the output variable.

The CE cell digitized output matrix requires that contacts close in a CE cell connected to the latched input terminal in each submatrix connected to each aggregate variable to select a value of the actuator variable.

Specific representation

When all the aggregate variables are used to represent a value of the actuator variables in a unitary manner, the output is a *specific representation* of the input variable. In the specific conformation shown in Figure 19.11, each combination of values of the aggregate output variables is used to produce a unique value of the single actuator variable. ⁻

Generalized representation

The position of an output variable can be *characterized* as generally high, low, or in the middle. Each of these categories can be described by values of the most significant, least significant, and intermediate aggregate variables in the same way that values of an input variable can be characterized. Thus, the digitized output matrix also can select an output state that represents a *generalized value* rather than a specific output value.

Generalized conformation of an output state

The performance of the digitized output matrix with two aggregate variables shown is similar to the scalar matrix shown earlier except that it uses two CE cells instead of only one to connect a value of an input variable to an output variable. In the example given in Figure 19.12, the most significant values of both aggregate variables determine the position of the actuator. Either of the two low values is an acceptable output for one input state, and either of the two high values is an acceptable output for the other input state. In this case, the feedback encoder increases the sensitivies of the CE cells that represent this field of output values. This condition occurs when the resolution of the actuator is greater than that required by the task environment.

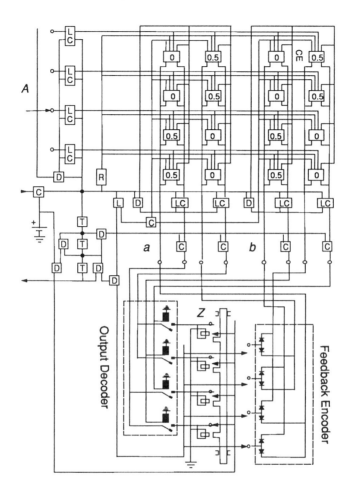

Figure 19.12. A *generalized conformation of CE cells in a binary digitized output matrix* is made when a generally low or high value of the output variable is sufficient to satisfy the needs of the task environment.

CE cell binary digitized output matrix with three aggregate variables

The number of aggregate variables in a digitized output matrix can be increased indefinitely using the method of connection shown in Figure 19.13, resulting in a great increase in the resolution of the input variable with only a small increase in the number of aggregate variables. For example, by including one additional binary aggregate variable, requiring two more CE cells per input terminal, the resolution of the actuator variable is doubled from four values to eight values while the number of CE cells is increased from four to only six. The CE cell binary digitized output matrix with three aggregate variables requires that the contacts close in three CE cells connected to the latched input terminal to select an output value.

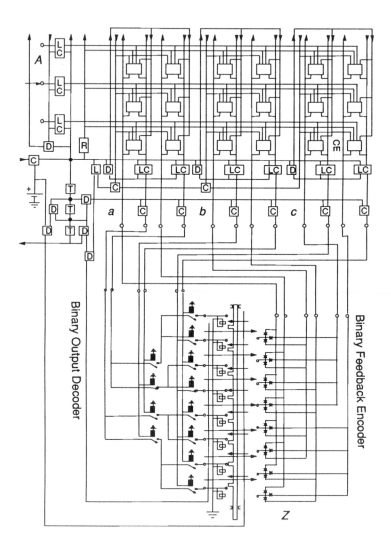

Figure 19.13. A *CE cell binary digitized output matrix with three aggregate variables* can select twice as many values of an output variable and uses only two more CE cells per selection than a digitized output matrix with only two aggregate variables.

Application of a CE cell binary digitized output matrix

The number of binary aggregate variables can be increased to five, creating a *medium-resolution positioning system* (Fig. 19.14), which can be trained to produce any one of 32 different output positions using only ten CE cells per selection.

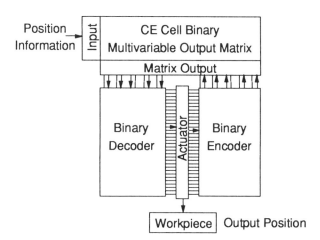

Figure 19.14. An *application of a CE cell binary digitized output matrix* is a medium-resolution machine actuator system.

The actuator can be positioned by an operator for each value of the matrix input variable, causing the CE cell multivariable matrix to produce specific positions when specific input values are produced later. These input values can be produced by a higher-level machine controller, while the machine actuator system may be just a subsystem within a larger machine system.

CE cell trinary digitized output matrix

The CE cell binary digitized output matrix shown previously can be converted into a *CE cell trinary digitized output matrix* by adding another value element to each aggregate variable, as shown in Figure 19.15. The CE cell trinary digitized output matrix can produce more output positions using a given number of CE cells per position than any other method of digitization.

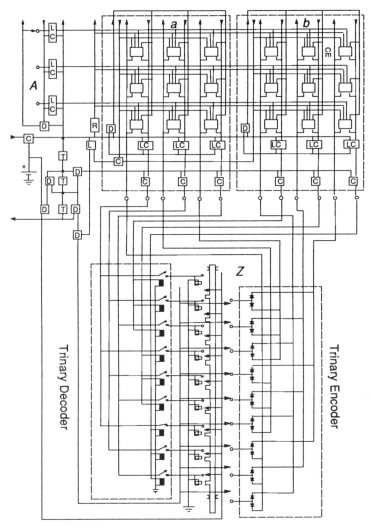

Figure 19.15. A *CE cell trinary digitized output matrix* is more efficient than the CE cell binary digitized output matrix shown earlier or a digitized system of any other number base.

Application of a CE cell trinary digitized output matrix

A CE cell trinary digitized output matrix can be used to operate the *medium-resolution linear actuator system* shown in Figure 19.16. The medium-resolution linear actuator system can be programmed empirically by placing the actuator in the desired position of each input value. This machine can be used as a damper control system, medium-resolution auto-pilot, or as a component in a medium-resolution actuator in a production machine system requiring resolutions not greater than about 3%.

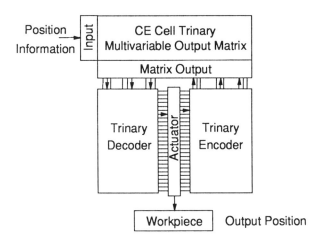

Figure 19.16. A good *application of a CE cell trinary digitized output matrix* is a medium-resolution linear actuator where a relatively few input values of a single input variable are used to produce output positions with a resolution to 3%.

CE cell binary digitized multivariable output matrix

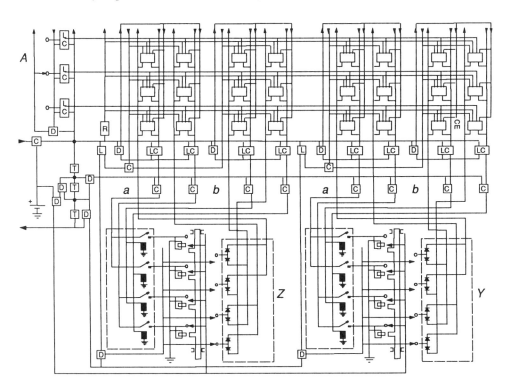

Figure 19.17. The *CE cell binary digitized multivariable output matrix* can be used to produce a unique combination of values of its output variables for each value of its input variable, or it can hold the value of one output variable constant and produce specific values of the other output variables for each value of its input variable.

Each input state of a multivariable digitized output matrix can produce a unique combination of values of a set of aggregate output variables. Each set of aggregate variables can be used to drive an output decoder, which in turn can determine the position of an actuator, as shown in Figure 19.17. Thus, each actuator position is determined by a set of values of its aggregate variables.

Each actuator in a CE cell binary digitized multivariable output matrix can learn either a general high or low position for a given input state, or it can learn one of the four specific positions for a given input state. The binary digitized multivariable output matrix greatly reduces the number of required CE cells to produce a given variety of behavior in the same manner that the binary digitized multivariable input matrix greatly reduces the number of CE cells required to identify a given variety of input conditions.

Application of a CE cell binary digitized multivariable output matrix

When a CE cell binary digitized multivariable output matrix is used to control the servo-actuator and bow thruster of a *ship direction control system* (Fig. 19.18), the matrix may determine that one or the other actuator has the greatest influence upon the direction of the ship. For example, the helmsman may use the bow thruster more than the rudder to control the direction of the ship without being aware of doing so.

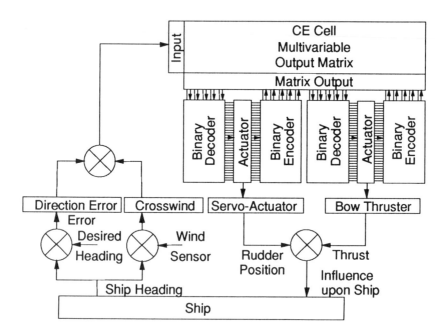

Figure 19.18. An *application of a CE cell binary digitized multivariable output matrix* is a ship direction control system with a single input variable and multiple output variables.

An empirical system acquires the actual behavior of an operator rather than the theoretical preconceptions of a control system designer or programmer.

CE cell trinary digitized multivariable output matrix

Two or more sets of trinary digitized aggregate output variables can be connected, forming the *CE cell trinary digitized multivariable output matrix* shown in Figure 19.19. The CE cell trinary digitized multivariable output matrix can be used as a pick-and-place robot, an industrial robot with more than two axes, or any other actuator system having two or more output variables with resolutions not greater than about 3% and a single input variable with a relatively few values.

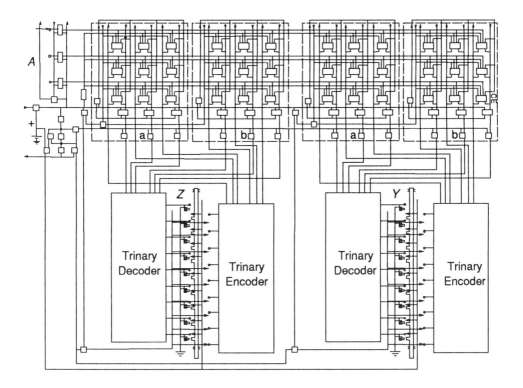

Figure 19.19. A *CE cell trinary digitized multivariable output matrix* can produce any one of more combinations of values of a set of output variables using fewer memory cells than any other digitized system.

Application of a CE cell trinary digitized multivariable output matrix

A CE cell trinary digitized output matrix also can be used to operate a rudder position servo-actuator and a bow thruster in the *ship direction control system* with a single input variable such as direction error or the sum of the direction error and a crosswind sensor, as shown in Figure 19.20.

The empirical control system can be taught to produce a unique combination of values of its output variables for each value of its input variable, or it may discover that a generally high or low value of one of its actuators produces the most consistent results.

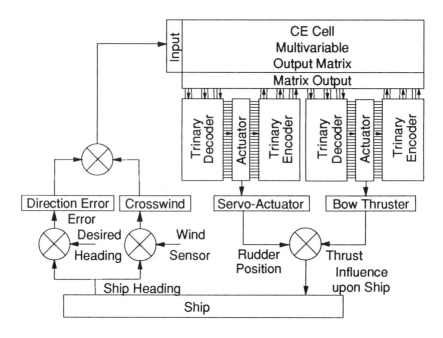

Figure 19.20. An *application of a CE cell trinary digitized multivariable output matrix* is as a ship direction control system with a single input variable with a limited number of values and two or more output variables having resolutions not greater than about 3%.

CE cell digitized monolithic control systems

Since a monolithic matrix can deal with multiple input and output variables, it can connect the aggregate variables of digitized sensor and actuator variables, forming an efficient medium-resolution control system.

CE cell binary digitized monolithic control matrix

The CE cell binary digitized input matrix can be combined with a CE cell binary digitized output matrix, forming the CE cell binary digitized monolithic controller shown in Figure 19.21. If many different values of a digitized input variable must produce many different values of a digitized output variable, then the CE cell binary digitized monolithic control matrix can use far fewer CE cells than the digitized duplex control matrix, to be shown, because each combination of values of the aggregate variables is never represented by a value of a single intermediate variable.

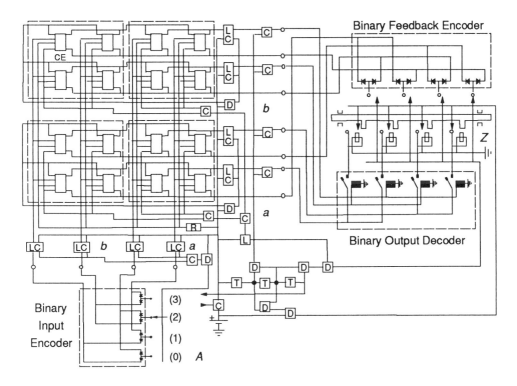

Figure 19.21. A *CE cell binary digitized monolithic control matrix* can learn to produce a unique value of its output variable for a unique value of its input variable using far fewer CE cells than an undigitized monolithic control matrix.

Application of CE cell binary digitized monolithic control matrix

A CE cell binary digitized monolithic control matrix can be used in a temperature control system (Fig. 19.22) where each measured value of temperature is used to control the position of a damper. The binary digitized monolithic control matrix with 5 aggregate variables for each sensor and actuator variable requires $2 \times 5 = 10$ input and output CE cells for a total of $10 \times 10 = 100$ CE cells, and can sense 32 different temperature values and producing 32 damper positions. An undigitized monolithic control matrix would require $32 \times 32 = 1024$ CE cells to do the same job.

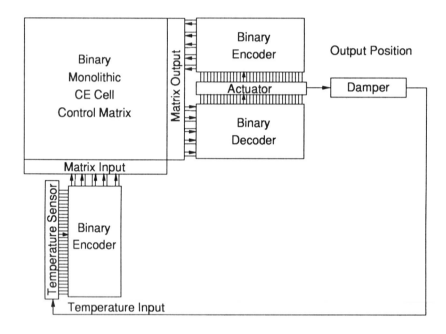

Figure 19.22. An *application of a CE cell binary digitized monolithic control matrix* is as an empirical temperature control system, using a medium-resolution temperature sensor and a medium-resolution heat duct damper.

CE cell trinary digitized monolithic control matrix

The number of values of each aggregate input and output variable in the binary digitized monolithic control matrix shown can be increased to three from two, forming a *CE cell trinary digitized monolithic control matrix* that is more efficient than a binary digitized monolithic control system. For example, a binary digitized monolithic control matrix with 3 aggregate input and output variables requires $6 \times 6 = 36$ CE cells, and has a sensor and actuator variety of 8 values. The trinary digitized monolithic control matrix with 2 aggregate input and output variables also requires 36 CE cells, but has a sensor and actuator variety of 9 values. The advantage of a trinary digitized monolithic control matrix over a binary digitized control matrix increases as the number of aggregate variables increases.

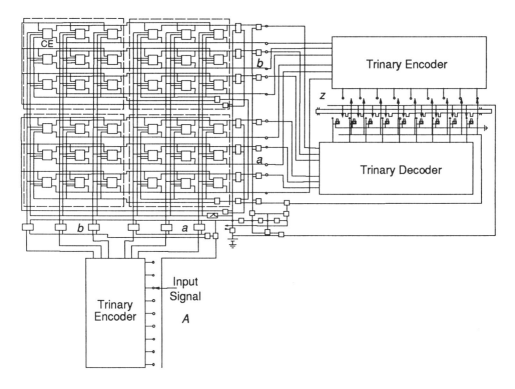

Figure 19.23. A *CE cell trinary digitized monolithic control matrix* can learn to produce more unique values of its output variable for more unique values of its input variable than an binary digitized monolithic control matrix with the same number of CE cells.

Application of a CE cell trinary digitized monolithic control matrix

A temperature sensor and a heat duct damper can be added to a CE cell trinary digitized monolithic control matrix, forming the *empirical temperature control system* shown in Figure 19.23. A trinary digitized monolithic control system with 27 values of its sensor and actuator variables requires 3 trinary aggregate for each sensor and actuator variables. This means that the trinary control matrix requires $3 \times 3 = 9$ input terminals and 9 output terminals for a total of $9 \times 9 = 81$ memory cells. A binary digitized monolithic controller with approximately the same resolution requires 100 CE cells. An undigitized scalar matrix would require $27 \times 27 = 729$ memory cells and a trinary digitized duplex controller, to be shown, would require $9 \times 27 = 243$ memory cells in each matrix for a total of 486 memory cells to do the same job. If only one of the many possible transitions are used, as in a temperature warning system that must sound and alarm at some critical temperature, then the trinary digitized duplex control system would do the job with only $9 \times 2 = 18$ memory cells, which is far fewer than even the digitized monolithic system.

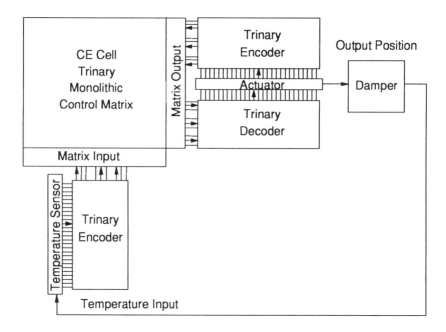

Figure 19.24. An *application of a CE cell trinary digitized monolithic control matrix* is as an empirical temperature control system that uses fewer CE cells for a given resolution of the sensor and actuator variables than a binary digitized monolithic controller.

CE cell binary digitized multivariable monolithic control matrix

The multiple input and multiple output variables of a CE cell monolithic control matrix can be used as aggregate variables in the *CE cell multivariable binary digitized monolithic control matrix* shown in Figure 19.24. Any number of digitized sensor and actuator variables can be added to the CE cell binary digitized multivariable monolithic control matrix shown. The number of CE cells in this controller increases according to the product of the number of sensor and actuator variables. The complete circuit for the CE cell multivariable binary digitized monolithic control matrix is shown because the simplest version requires only $8 \times 8 = 64$ CE cells compared to a trinary digitized multivariable monolithic control matrix, which requires $12 \times 12 = 144$ CE cells. However, a trinary digitized monolithic controller, shown in the next section, is more effective than the binary digitized monolithic empirical controller.

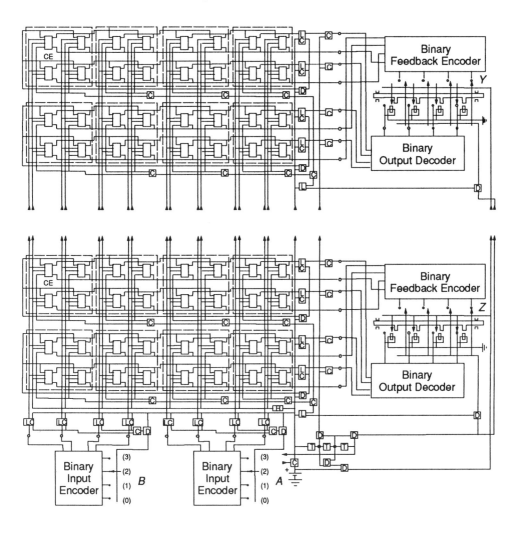

Figure 19.25. A *CE cell binary digitized multivariable monolithic control matrix* can connect general or specific values of two or more sensor variables to general or specific values of two or more actuator variables.

CE cell trinary digitized multivariable monolithic control system

The number of sensor and actuator variables in the CE cell trinary digitized monolithic control matrix can be increased indefinitely, forming a CE cell multivariable trinary digitized monolithic control system, as shown in Figure 19.25.

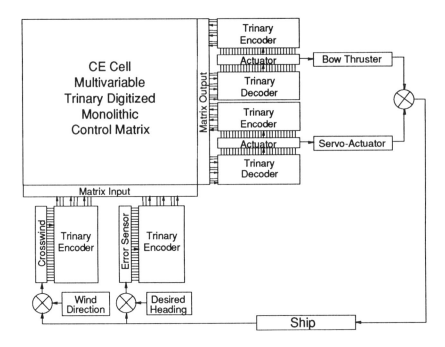

Figure 19.26. A *CE cell trinary digitized multivariable monolithic control system* can be used to control two or more sensor variables and two or more actuator variables, as in a ship direction control system.

The CE cell multivariable trinary digitized monolithic controller is a useful and versatile multivariable empirical controllers since it can be used to produce unitary or asynchronous diverse relations between any number of sensor and actuator variables, and can produce many transitions in its line of behavior. However, the resolution of a single vernier actuator is limited to less than about one part in thirty. Thus, a mechanically digitized sensors and pi actuators, to be shown, may be needed in applications requiring resolutions higher than about 3%.

CE cell monolithic control matrix operating upon a vector sensor and actuator

A CE cell monolithic control matrix can be connected to vector sensors and actuators such as the *x/y* tables shown in Figure 19.26. The CE cell monolithic control matrix connected to a vector sensor and actuator may use each combination of values of the components of the sensor vector to produce a unique combination of values of the components of the actuator variable in a unitary manner, or it may learn to use specific values of some components of the sensor vector to produce specific values of some components of the actuator variable in a diverse manner.

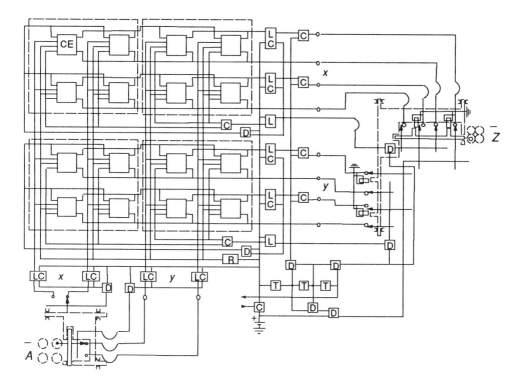

Figure 19.27. A *CE cell monolithic control matrix operating upon a vector sensor and actuator* can transform one vector state into another vector state.

CE cell binary monolithic control matrix with mechanically digitized sensor and actuator

A CE cell binary monolithic control matrix can also be connected to a mechanically digitized binary sensor and a binary pi actuator, as shown in Figure 19.27. The CE cell binary monolithic control matrix with mechanically digitized sensor and actuator, with two aggregate sensor and actuator variables, can produce four output positions for four input positions using sixteen CE cells. If the number of binary sensor and binary actuator aggregate variables is increased to eight, any one of 256 output positions could be produced by any one of 256 different input positions using only 256 CE cells. An undigitized matrix would require $256 \times 256 = 65,536$ memory cells to do the same job.

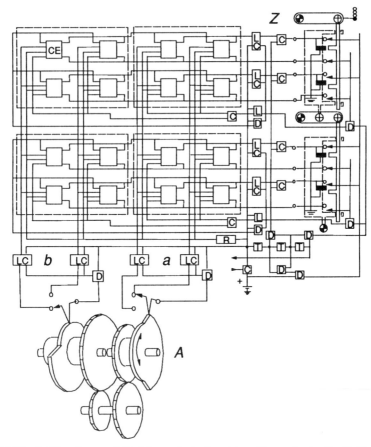

Figure 19.28. A CE cell monolithic binary control matrix with mechanically digitized sensor and actuator can produce many high-resolution output positions for many high-resolution input values.

CE cell trinary monolithic control matrix with mechanically digitized sensor and actuator

A trinary digitized sensor register and a trinary digitized pi actuator can be connected to a CE cell trinary monolithic control matrix, forming the high-resolution control system shown in Figure 19.28. The CE cell trinary monolithic control matrix shown, with a five-trinade input register and a five-trinade pi actuator, can learn to create transitions between 243 different sensor values and 243 different actuator states using only $15 \times 15 = 225$ CE cells. A binary monolithic control matrix with an eight-binade input register and an eight-binade pi actuator can learn to create transitions between 256 different sensor states and 256 different actuator states using 256 CE cells. An undigitized scalar matrix would require $243 \times 243 = 59,049$ memory cells to do the same job as the trinary digitized monolithic controller with five sensor and five actuator aggregate variables. If only one of the possible transitions is used, as in a temperature warning system that must sound and alarm at some critical temperature, then a trinary digitized duplex control system, shown in the next chapters, can do the job with only $15 \times 2 = 30$ memory cells, which are far fewer than even the digitized monolithic system.

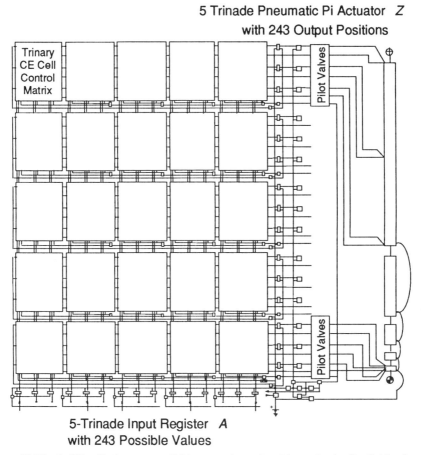

Figure 19.29. A CE cell trinary monolithic control matrix with mechanically digitized sensor and actuator can produce more transitions for a given number of CE cells than a monolithic controller with a binary digitized sensor and actuator.

The CE cell multivariable monolithic controller, with mechanically digitized trinary sensor and actuator variables, is a useful and versatile multivariable empirical controller since it can be used to produce many unitary or asynchronous diverse relations between many sensor and actuator variables that have any level of resolution. However, the number of CE cells in this controller increases exponentially, according to the number of sensor and actuator variables. If a relatively few transitions are required in the line of behavior of a system with a large number of variables, the CE cell duplex controllers, shown in the following chapters, might accomplish the task using fewer CE cells a monolithic system.

Chapter 20
Duplex Empirical Controllers

Some applications may require high-resolution sensor and actuator variables. However, only a few input and output relations (transitions) may be needed between their many possible values to produce the required line of behavior. This kind of application can be dealt more effectively by using two nonsquare matrices connected in the manner of the predetermined duplex controller shown in Part II rather than a monolithic matrix. CE cell matrices can be joined to form a duplex controller if some method can be found to change the sensitivities of the CE cells in the input and output matrices in the duplex controller so that they will select the most probable output state for a given input state. This can be accomplished in at least three different ways:

1. By using Type 1 bidirectional memory cells (BD1 cells) in the Type 1 output matrix based upon the standard CE cell.
2. By using Type 2 bidirectional memory cells (BD2 cells) in a Type 2 bidirectional output matrix to provide a scalar duplex controller.
3. By using Type 3 bidirectional memory cells (BD3 cells) in a Type 3 bidirectional output matrix to provide a multivariable duplex controller.

The *forward selection process* in all duplex empirical controllers operates according to the empirical algorithm in which the sensitivity of the CE cells in the input matrix are decreased when they *forward select* a *forward-selected intermediate state* in the manner of the CE cell input matrices shown previously. Then, a bidirectional output matrix *forward selects* an output state by finding the most sensitive CE cells connected to each intermediate state in the manner of the CE cell output matrices shown previously. A bidirectional output matrices must then *back select* the intermediate state most likely to select the actual output state, and this *back-selected intermediate state* is used to increase the sensitivity of the CE cells in the input matrix at the intersection of the back-selected intermediate state and the latched input state. Three different procedures for carrying out the *back selection process* will be presented in this chapter based upon the type of bidirectional CE cell used.

The *transform conditioning process* in the *Type 1 bidirectional output matrix* is almost identical to the transform conditioning process in the input matrix. The Type 1 bidirectional output matrix decreases the sensitivity of the BD1 cell that forward selects the output and then increases the sensitivity of the BD1 cell at the intersection of the actual output and the forward-selecting intermediate state in the usual way. It then back selects the intermediate state connected to the BD1 cell with the highest sensitivity connected to the actual output state. It then increases the sensitivity of the CE cell in the input matrix at the intersection of the latched input state and the back-selected intermediate state.

The Type 1 bidirectional output matrix favors the back selection of the intermediate state selected in the forward-selection process because it increases the sensitivity of the BD1 cell on the forward-selected intermediate state *before* it back selects. A duplex controller with a Type 1 bidirectional output matrix tends to put more behavior on fewer intermediate states. This can cause new learning to interfere with old learning in some cases.

However, the transform conditioning process of the *Type 2 bidirectional output matrix* works in a slightly different way. When a BD2 cell in the Type 2 bidirectional output matrix forward selects an output, the sensitivities are decreased in *all* of the BD2 cells connected to the forward-selecting intermediate terminal *and all* of the BD2 cells connected to the forward-selected output terminal. After the actual output occurs, the bidirectional output matrix back selects the intermediate state connected to the BD2 cell with the highest sensitivity connected to the actual output terminal and *then* increases the sensitivity of this selecting BD2 cell. It then increases the sensitivity of the CE cell in the input matrix at the intersection of the latched input state and the back-selected intermediate state, as described previously.

The Type 2 bidirectional output matrix tends to shun the back selection of the intermediate states selected in the forward-selection process because it back selects and increases the sensitivity of the back-selected BD2 cell *after* it decreases the sensitivity of all of the BD2 cells connected to the forward-selecting intermediate state. Since it decreases the sensitivity of *all* of the BD2 cells on the forward-selected output terminal, it does not change the relative sensitivities of these BD2 cells. So the probability of back selecting the forward-selecting BD2 cell is not increased prior to the back-selection process. This procedure tends to allow only one high-sensitivity BD2 cell on each intermediate and output terminal. This feature is very desirable in a single-variable (scalar) output matrix because it creates a distribution of BD2 cell values that conforms to the rules of a connection matrix and overcomes the problem with the Type 1 matrix. This makes the Type 2 bidirectional output matrix very desirable in a coded unitary duplex controller.

However, to create a duplex empirical controller that can deal with multiple input *and* multiple output variables without using unitary decoding and encoding, it is necessary to create a Type 3 bidirectional output matrix that can have a set of high-sensitivity bidirectional memory cells on one value of an output variable if that variable is to remain constant for different values of the other output variables. This conformation cannot be achieved using the Type 1 or Type 2 bidirectional output matrices because they attempt to create just one-high sensitivity bidirectional memory cell on each output terminal. The Type 3 bidirectional cell (BD3 cell) and the Type 3 bidirectional output matrix can create many high-sensitivity conditional memory cells on a given output terminal and is therefore required in the output matrix of a conditional empirical multivariable duplex controller and in any other multi-variable bidirectional control matrix.

Type 1 empirical duplex controllers

The simplest type of duplex empirical controller consists of a CE cell input matrix connected to a CE cell output matrix with additional search coils located in collateral cells that can back select the intermediate state most likely to produce the actual output state. The CE cells in the output matrix can be combined with the collateral cells, forming a Type 1 bidirectional output matrix that can be connected to a CE cell input matrix to form a Type 1 empirical duplex controller.

Expanded Type 1 bidirectional output matrices

To understand the forward- and back-selection processes in a duplex empirical controller, it is desirable to expand the output matrix into a bidirectional output matrix with a *forward-selecting matrix* and *back-selecting matrix,* as shown in Figure 20.1

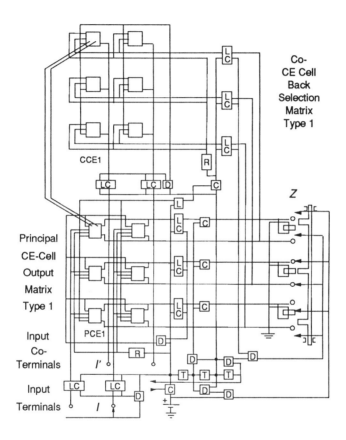

Figure 20.1. An *expanded Type 1 bidirectional output matrix* uses a standard CE cell output matrix to forward select an output state and a Type 1 back-selecting matrix to back select the input co-terminal associated with the input terminal most likely to forward-select the actual output state.

The forward-selecting CE cell matrix selects an output terminal and decreases the sensitivity of the selecting CE cell in the usual manner. After the appropriate delays, it increases the sensitivity of the CE cell at the intersection of the latched input terminal and the actual output terminal in the manner of the CE cell matrices shown so far. However, the expanded bidirectional output matrix uses an additional back-selecting matrix of collateral Type 1 CE cells to back select the input co-terminal associated with the input terminal most likely to forward-select the *actual* output terminal based upon the values the variable resistors stored in the CE cells in the forward-selecting matrix.

Type 1 principal and collateral CE cells

A *Type 1 principal and collateral CE cell* is made up of a standard CE cell and a collateral CE cell that contains a *back-select search coil* and a set of contacts. The back-select search coil in the collateral CE cell is connected to the variable resistor of its principal CE cell, as shown in Figure 20.2.

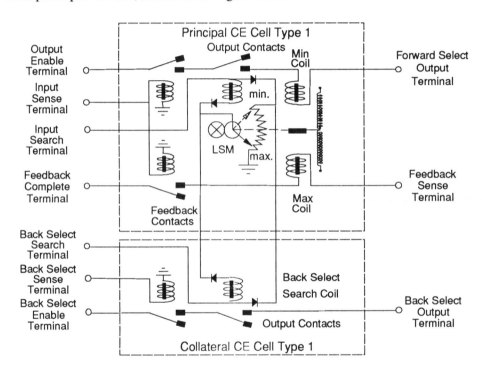

Figure 20.2. A *Type 1 principal and collateral CE cell* uses the same variable resistor to forward select an actuator value and back select the input value most likely to produce the actual output value.

The Type 1 principal CE cell is identical to a standard CE cell except that its variable resistor is connected to a back-select search coil in a collateral CE cell located in a back select matrix, and both cells contain *diodes* that isolate the search coils from each other. When the principal CE cell forward selects an actuator value, its minimum coil decreases its sensitivity value in the usual manner. Then, the most sensitive Type 1 collateral CE cell connected to the actual output terminal back selects an input co-terminal. The sensitivity of the collateral CE cell is determined by the value of the variable resistor of its principal CE cell located in the forward-selecting matrix. However, the Type 1 collateral CE cell cannot change the sensitivity of its Type 1 principal CE cell.

Expanded Type 1 duplex controller

The input terminals of an expanded bidirectional output matrix can be connected to the output terminals of a standard CE cell input matrix by means of the intermediate terminals of the intermediate variable (I) of an *expanded Type 1 duplex controller,* as shown in Figure 20.3. One of the back-selected intermediate co-terminals of the intermediate co-variable (I') can then be used to increase the sensitivity of the CE cell connected to the latched input terminal in the input matrix that is most likely to produce the actual output state when that input terminal is energized in the future.

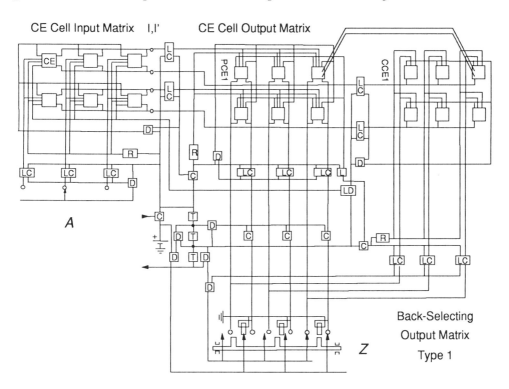

Figure 20.3. An *expanded Type 1 duplex controller* consists of a standard CE cell input matrix, a standard CE cell output matrix, and a back-selecting output matrix with Type 1 collateral CE cells, all connected to a common set of intermediate terminals (I) and co-terminals (I').

An expanded Type 1 duplex controller uses a separate back-selecting output matrix to back select the intermediate state most likely to produce the actual output value encountered. This intermediate state is then be used to reinforce the CE cell in the input matrix connected to the latched input state. This *expanded duplex system* requires only as many intermediate terminals as there are desired input/output relations. When the number of input/output relations is small in relation to the number of possible input and output values, the duplex system is more efficient in its use of CE cells than a single control matrix.

Transform Conditioning in an expanded Type 1 duplex system

The process of selection and learning in a duplex system is similar to the sequence of steps required by the scalar CE cell control matrix shown previously. These steps are shown in Table 20.1.

Table 20.1. Operation of an expanded Type 1 duplex controller

---

1. The CE cells along the column connected to the latched input terminal of the input matrix are interrogated by a *forward-selecting ramp generator* in the input matrix to find the CE cell with the highest confidence level.
2. The intermediate terminal on the row of the selected CE cell is energized, and the confidence level of the selecting CE cell in the input matrix is reduced by an amount proportional to its confidence level.
3. The CE cells in the output matrix connected to the selected intermediate terminal are interrogated by a *forward-selecting ramp generator* in the output matrix to find the CE cell with the highest confidence level.
4. The output terminal of the column connected to the CE cell in the forward-selecting output matrix with the highest confidence level is energized, and the confidence level of this second selecting CE cell is also reduced by an amount proportional to its confidence level.
5. If an output can be carried out within a given period, then the confidence level of the CE cell in the forward-selecting output matrix at the intersection of the actual output terminal and the selected intermediate terminal is increased.
6. Then the back-selecting matrix is interrogated by a *back-selecting ramp generator* to find the collateral CE cell with the highest confidence level connected to the actual output co-terminal. A signal is sent back to the intermediate co-terminal connected to this collateral CE cell, which is used to increase the confidence level of the CE cell in the input matrix connected to the latched input terminal and the energized intermediate co-terminal. This increases the probability that the *actual output* will be selected by the *original input* at a later trial.
7. If no output is completed within the given period, then both selecting CE cells are left at their lower confidence values, decreasing the probability that they will select action in the future.

The number of intermediate terminals can be chosen to match the number of input/output relations required to produce a closed line of behavior. This makes it possible to separate the resolution design parameter from the number of relations design parameter. This duplex system can provide an empirical ship auto-pilot having fewer CE cells than the single empirical matrix shown before, having the same resolution of its input and output variables.

Type 1 bidirectional memory cell

A Type 1 principal CE cell and its Type 1 collateral CE cell can be combined into one *Type 1 bidirectional CE cell (BD1 cell)* with two search coils, as shown in Figure 20.4.

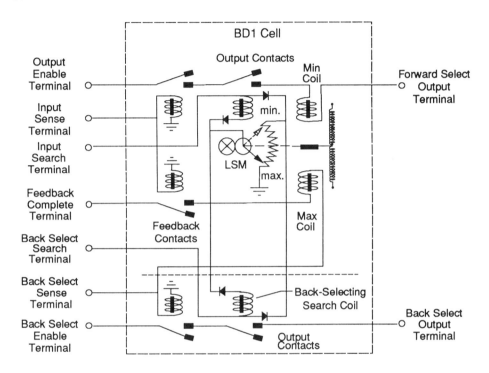

Figure 20.4. A *Type 1 bidirectional memory cell* (BD1 cell) acts like a typical CE cell but can also be used to back select the most probable intermediate terminal for a given actual actuator value.

The Type 1 conditional empirical bidirectional memory cell (BD1 cell) uses nine terminals to select an actuator value and decrease its sensitivity, to increase its sensitivity if its actuator value occurs, and to produce a feedback signal for the input matrix if it is the most sensitive cell for a given actual actuator value. Joining the principal and collateral cells greatly reduces the length and number of conductors required if both separate cells are used in separate expanded matrices.

Type 1 bidirectional output matrix

Joining each principal cell and collateral cell into a single Type 1 bidirectional empirical memory cell allows the actuator and feedback matrices to be combined into one *Type 1 bidirectional output matrix*, as shown in Figure 20.5. The Type 1 bidirectional output matrix selects the Type 1 bidirectional memory cell (BD1 cell) having the highest sensitivity for a given input value and attempts to produce the output value connected to this cell. An actuator feedback signal is sent back along the row of the actual output value. This signal is used to increase the sensitivity of the cell at the intersection of the actual output and the current input state in the manner of a typical conditional matrix. The *back-selecting ramp generator* then interrogates all of the back-select search coils. However, an output can be produced only from cells along the row of the actual output value that have their feedback sense terminals energized, causing the most sensitive BD1 cell on this row to produce a signal to the input co-terminal connected to this cell.

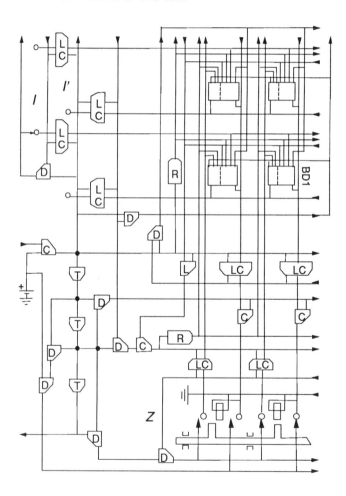

Figure 20 5. A *Type 1 bidirectional output matrix* combines the function of an actuator matrix and a back-selecting matrix by using a set of Type 1 bidirectional empirical memory cells (BD1 cell).

Type 1 duplex empirical controller

As shown before, a control system can be split into two matrices (Fig. 20.6) that are connected by a set of intermediate terminals that represent the values of a single intermediate variable. Each intermediate value can then be used to represent a unique input/output relation. A scalar duplex empirical controller uses two memory cells to produce a given input/output relation. One cell in the input matrix specifies a value of the intermediate variable, and a cell in the output matrix specifies a given output state. The Type 1 duplex empirical controller works well when each input state produces a different output state under most conditions. But if a given output state is specified by two or more input states, the Type 1 duplex controller cannot configure itself so that all of those inputs will produce a single given intermediate value.

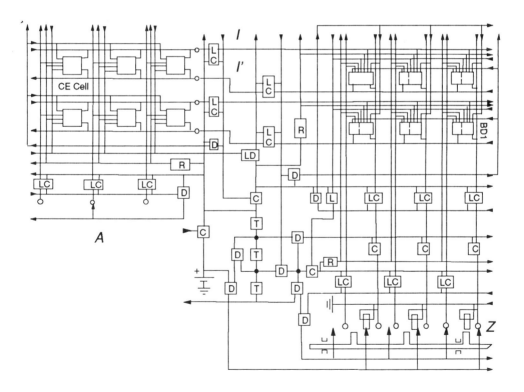

Figure 20.6. A *Type 1 duplex empirical controller* can reduce the number of memory cells required to produce an effective control system when two or more input states produce the same output states.

Operation of the Type 1 duplex empirical controller

The process of selection and learning in a Type 1 duplex controller is similar to the sequence of steps required by the expanded Type 1 duplex system except that the forward- and back-selection process is carried out in one bidirectional output matrix, as described in Table 20.2. The sensitivity of the BD1 cell at the intersection of the actual output co-terminal and the selected intermediate terminal is increased *before* the back-selection process.

Table 20.2. Operation of a Type 1 empirical duplex controller

---

1. The CE cells along the column connected to the latched input terminal of the input matrix are interrogated by its forward-selecting ramp generator to find the CE cell with the highest confidence level.
2. The intermediate terminal on the row of the selected CE cell is energized, and the confidence level of the selecting CE cell in the input matrix is reduced by an amount proportional to its confidence level.
3. The BD1 cells in the output matrix connected to the selected intermediate terminal are interrogated by a forward-selecting ramp generator in the output matrix to find the BD1 cell with the highest confidence level.
4. The output terminal in the column connected to the BD1 cell in the output matrix with the highest confidence level is energized, and the confidence level of this selecting BD1 cell is also reduced by an amount proportional to its confidence level.
5. If an output can be carried out within a given period, then the confidence level of the BD1 cell in the output matrix at the intersection of the actual output co-terminal and the selected intermediate terminal is increased.
6. Then the BD1 cells in the output matrix are interrogated by a back-selecting ramp generator to find the BD1 cell with the highest confidence level connected to the actual output co-terminal. A signal is sent back to the intermediate co-terminal connected to this BD1 cell, which is used to increase the confidence level of the CE cell in the input matrix connected to the original latched input terminal and the energized intermediate co-terminal. This increases the probability that the *actual output* will be selected by the *latched input* at a later trial.
7. If no output is completed within the given period, then both selecting cells are left at their lower confidence values decreasing the probability that they will select action in the future.

Problems with the Type 1 bidirectional output matrix

Since the sensitivity of a BD1 cell is increased before the back-selecting process, the Type 1 bidirectional matrix is likely to back select the intermediate co-terminal belonging to the forward-selected intermediate terminal even when the actual output is different from the selected output. This tends to create two or more high-sensitivity cells on a given intermediate value, which violates the rules of a connection matrix, which says that only one output may appear on a given input line. (See Part II). This tendency may cause new learning to interfere with previously learned relations in that two or more output states may try to appear on the same intermediate value. This problem can be eliminated if the sensitivity of the bidirectional cell in the output matrix is increased *after* the back-selection process, as shown in the next sections.

Type 2 scalar empirical duplex controllers

The Type 1 empirical duplex controller may disrupt old learning with new learning in some cases because it reinforces the BD1 cell on the forward-selecting intermediate state and the actual output state before it back selects the intermediate co-terminal used to reinforce the CE cell in the input matrix connected to the latched input state. This potential problem can be overcome by using a Type 2 bidirectional conditional empirical memory cell (BD2 cell) in a Type 2 bidirectional output matrix that decreases the sensitivities of *all* of the bidirectional cells on the *row and column* of the selecting BD2 cell, and reinforces the most sensitive BD2 cell connected to the actual output state *after* it back selects an intermediate state.

Expanded Type 2 bidirectional output matrix

To understand the forward- and back-selection processes in a Type 2 duplex empirical controller, it is desirable to expand the bidirectional output matrix into a forward-selecting and back-selecting matrix, as shown in Figure 20.7.

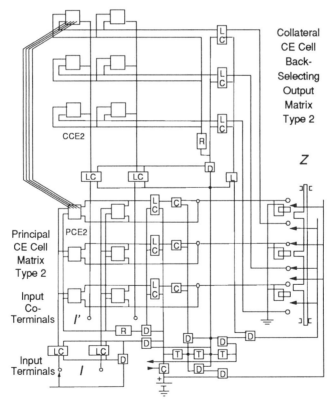

Figure 20.7. An *expanded Type 2 bidirectional output matrix* uses a matrix of Type 2 principal CE cells to forward select an output state and a matrix of Type 2 collateral CE cells to back select the input co-terminal whose input terminal is most likely to select the actual output state.

The forward-selecting Type 2 matrix selects the output terminal connected to the most sensitive Type 2 principal CE cell on the latched input terminal. It decreases the sensitivity of *all* of the CE cells connected to the latched input terminal *and* the selected output terminal. After the appropriate delays, it interrogates the back-selecting matrix to find the Type 2 collateral CE cell with the highest sensitivity connected to the *actual* output co-terminal. It then increases the sensitivity of the Type 2 principal CE cell connected to this collateral CE cell and produces an output at the input co-terminal connected to this collateral CE cell. This increases the probability that the input terminal belonging to the back-selected co-terminal will forward select the actual output terminal, based upon the values of the variable resistors stored in the forward-selecting matrix, when that input terminal is energized in the future.

Type 2 principal and collateral CE cells

The forward-selecting and back-selecting functions of a Type 2 bidirectional CE cell can be carried out by separate *Type 2 principal and collateral CE cells,* as shown in Figure 20.8.

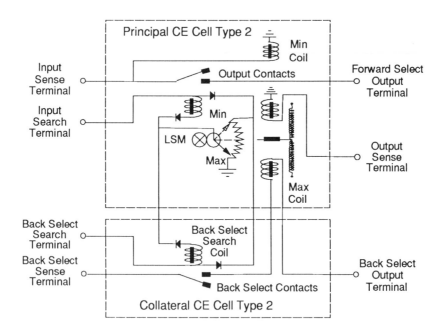

Figure 20.8. Type 2 principal and collateral CE cells use the same variable resistor to forward select an actuator value and back select the input value most likely to produce the actual output value.

The Type 2 principal CE cell is similar to a standard CE cell, except that its variable resistor is connected to a back-select search coil in a collateral CE cell in another matrix, and contains diodes that isolate its search coil from the collateral back-select search terminal in the other matrix. It also contains two minimum coils: One is energized whenever its input sense terminal is energized, and the other is energized whenever its output sense terminal is energized. Its maximum coil is energized whenever its collateral CE cell back selects an input co-terminal.

Duplex controller with expanded Type 2 output matrix

The input terminals of the input variable *A* of an expanded Type 2 bidirectional output matrix can be connected to the output terminals of a standard CE cell input matrix, creating the intermediate terminals of the intermediate variable *I* of a *duplex empirical controller with an expanded Type 2 output matrix*, as shown in Figure 20.9.

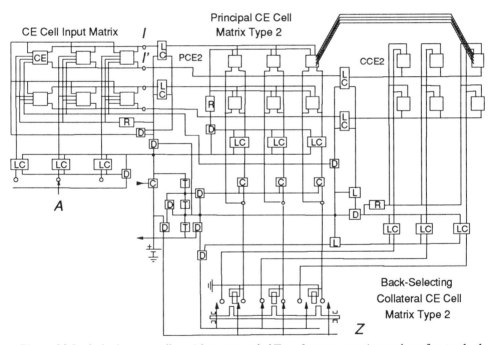

Figure 20.9. A *duplex controller with an expanded Type 2 output matrix* consists of a standard CE cell input matrix, a forward-selecting output matrix of Type 2 principal CE cells, and a back-selecting output matrix of Type 2 collateral CE cells, all connected to a common set of intermediate terminals and co-terminals.

A duplex controller with an expanded Type 2 output matrix uses a separate back-selecting matrix to select the intermediate state most likely to produce the actual output value encountered. This intermediate state is then used to reinforce the CE cell in the input matrix connected to the original input state.

Operation of a duplex controller with an expanded Type 2 output matrix

One of the intermediate terminals *I* of a duplex controller with an expanded Type 2
output matrix is selected by the most sensitive CE cell in the input matrix connected
to the latched input terminal. The most sensitive Type 2 principal CE cell connected
to this intermediate terminal selects an actuator terminal. The sensitivities of all of
the cells on this row and column in the output matrix are decreased. The most
sensitive Type 2 collateral cell connected to the actual actuator co-terminal then back
selects an intermediate co-terminals of the intermediate co-variable *I'* and the sensitivity of this cell is increased. The selected intermediate co-terminal is then used
to increase the sensitivity of the CE cell in the input matrix most likely to energize
the actual output co-terminal of the output variable *Z* when that latched input terminal
is energized in the future.

Type 2 bidirectional CE cell

The Type 2 principal and collateral CE cells can be combined into one Type 2
bidirectional CE cell with two search coils and two minimum coils, as shown in
Figure 20.10.

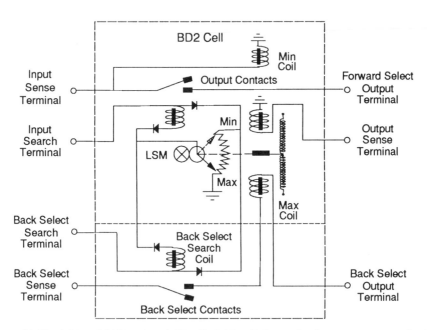

Figure 20.10. A *Type 2 bidirectional CE cell (BD2 cell)* forward selects an actuator value like a
typical CE cell but can also back selects the most probable intermediate co-terminal for a given
actual actuator co-terminal based upon the value of its variable resistor.

The Type 2 bidirectional memory cell (BD2 cell) uses only seven terminals to select an actuator value and decrease its sensitivity, and then increase its sensitivity if it back selects an intermediate co-terminal. Joining the principal and collateral CE cells greatly reduces the length and number of conductors required when both cells are used in separate expanded matrices.

Type 2 bidirectional output matrix

Joining each Type 2 principal and collateral cell into a single Type 2 bidirectional CE cell (BD2 cell) allows the actuator and feedback matrices to be combined into one *Type 2 bidirectional output matrix,* as shown in Figure 20.11.

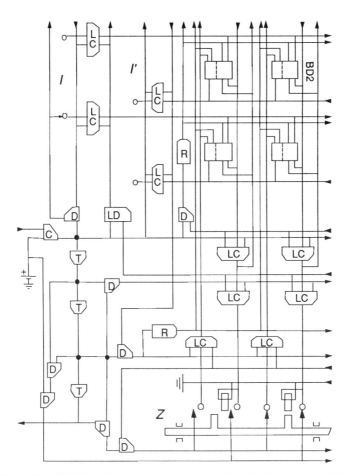

Figure 20.11. A *Type 2 bidirectional output matrix* combines the function of a forward-selecting output matrix and a back-selecting output matrix into one bidirectional matrix by using a set of Type 2 bidirectional memory cells (BD2 cells).

The Type 2 bidirectional matrix first decreases the sensitivity of all of the BD2 cells in the row of the latched input terminal and then forward selects the Type 2 bidirectional memory cell (BD2 cell) connected to the latched input terminal and having the highest sensitivity. (The relative sensitivities of all of the cells in this row are not changed by decreasing their sensitivity by the LSM before they are interrogated.) The selecting BD2 cell energizes the output latch connected to this cell. An output sense signal is immediately sent back along the column of the selected output terminal, which decreases the sensitivity of all of the BD2 cells on this column. (The relative sensitivities of all of the cells in this column are not changed by logarithmically decreasing all of their sensitivities.) After the appropriate delays, the back-selecting ramp generator then interrogates the BD2 cells along the column of the actual output co-terminal. The most sensitive BD2 cell in this column back selects the input co-terminal connected to this BD2 cell. This increases the sensitivity of the back-selecting BD2 cell.

Operation of the Type 2 bidirectional output matrix

Table 20.3. Operation of a Type 2 bidirectional output matrix

---

1. The cycle switch is closed starting a new transition cycle.
2. One input value terminal is energized according to the initial position of the sensor switch representing the input variable. This input terminal is latched "on" and the sensor switch is disconnected so that no additional input terminals can be energized.
3. The sensitivities of all of the BD2 cells in the row connected to the latched input terminal are decreased and are then interrogated by the forward-selecting ramp signal to find the BD2 cell with the highest sensitivity value. (Because the sensitivities of all of the BD2 cells on the row are decreased logarithmically, their relative sensitivities are not changed.)
4. When a BD2 cell becomes conductive, a voltage from the latched input terminal can flow through this BD2 cell to the latched connect on that column. This latched connect disconnects the forward-selecting ramp so that no other BD2 cells can produce an output. The output of this latched connect also decreases the sensitivity of all of the BD2 cells in the column to which it is connected. (The sensitivity of the selecting BD2 cell is not decreased a second time in this process because its minimum coil has not been released by the latched input terminal.)
5. After the delay specified by the output delay relay, the second selected latched connect is energized, and the actuator brake is released. The actuator coil connected to the selected latched connect is energized and the actuator attempts to align an armature pole to the position of the energized actuator coil. The selected latched connect also disconnects the latched input terminal, allowing the armature of the LSM to return to its normal position.
6. After another delay, to allow the actuator to position itself, the actuator brake is re-engaged, holding the actuator in whatever position it is. The column of BD2 cells connected to the actual output co-terminal is then interrogated by the back-selecting ramp to find the BD2 cell in this column with the highest sensitivity. The selected BD2 cell sends a signal to its input co-terminal, which increases the sensitivity of this selecting BD2 cell according to its logarithmic subtracting mechanism. The input co-terminal latch disconnects the feedback ramp generator so that no other BD2 cells can energize another input co-terminal during this transition cycle.
7. The cycle switch is opened, causing power to be shut off to all elements except the actuator brake circuit, and all of the latches and disconnects are restored to their normal positions.

The process of selection and learning in a Type 2 bidirectional output matrix is similar to the sequence of steps required by the expanded Type 2 output matrices except that the forward- and back- selection process is carried out in one bidirectional output matrix, as described in Table 20.3. The input co-terminal made active in a given transition represents the input terminal that is most likely to produce the actual output state that occurred in that transition.

Type 2 empirical duplex controller

As shown before, the output variable of an input matrix can be connected to the input variable of a bidirectional output matrix by a set of intermediate terminals that represent the values of a single intermediate variable (I). Each intermediate value can be used to represent a unique input/output relation, as shown in Figure 20.12.

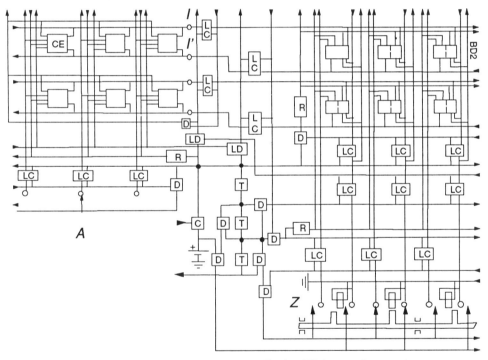

Figure 20.12. A *Type 2 duplex controller* can be effective if it has one input and one output variable, if there are a relatively few required transitions compared to the variety of the input and output variables, and if two or more input states produce the same output states.

The input matrix of the Type 2 duplex controller selects a value of the intermediate variable based upon the value of the latched input variable, and decreases the sensitivity of the selecting CE cell. The output matrix uses the value of the intermediate variable to decrease the sensitivity of all of the BD2 cells connected to this value of the intermediate variable and then selects a value of the output variable. It then decreases the sensitivity of all of the BD2 cells in the output matrix connected to this output variable. It then back selects the highest-sensitivity BD2 cell on the actual output co-terminal, increases the sensitivity of this BD2 cell, and uses it to energize its intermediate co-terminal. The input matrix increases the sensitivity of the CE cell connected to this intermediate co-terminal and the latched input terminal so that the duplex controller is more likely to select the actual output state that occurred in that transition when that input value occurs again.

Operation of the Type 2 duplex controller

Table 20.4. Operation of a Type 2 duplex empirical controller

---

1. The input latch energizes and sustains a voltage on the input column connected to the input value made active at the beginning of the transition cycle and excludes any other input column in the input matrix from being energized.
2. The CE cells along the column connected to the active input terminal of the input matrix are interrogated to find the CE cell with the highest confidence level.
3. The intermediate terminal on the row of the selected CE cell is energized, and the confidence level of the selecting CE cell in the input matrix is reduced by an amount proportional to its confidence level.
4. The sensitivity of all of the BD2 cells in the output matrix connected to the selected intermediate terminal are decreased logarithmically.
5. Then the BD2 cells in the output matrix connected to the selected intermediate terminal are interrogated by a ramp voltage to find the BD2 cell with the highest confidence level. The BD2 cell with the highest confidence level energizes its output terminal, which latches on the connect in this column causing the ramp voltage to be terminated to prohibit any other BD2 cell from selecting an output, and the sensitivities of all of the BD2 cells in this column are decreased logarithmically.
6. After the delay required by the discontinuous state ontology for this machine, a second set of latched connects are enabled and the coil of the selected actuator terminal is energized, causing the actuator to attempt to move to this position. The selected latched connect also disconnects the latched intermediate terminal, releasing the logarithmic subtracting mechanisms in the BD2 cells energized by the intermediate variable.
7. After a delay to allow the actuator to move to its selected position, the feedback latches are enabled and the actual position of the actuator at this time is latched in, causing the feedback enable circuit to be disconnected so that no other actuator feedback terminals can be energized.
8. The output matrix is interrogated by the back-selecting ramp to find the BD2 cell with the highest confidence level connected to the actual output co-terminal. A signal is sent back to the intermediate co-terminal connected to this BD2 cell, and the confidence of this BD2 cell is increased. When an intermediate co-terminal is energized, its latch disconnects the back-selecting ramp and feedback latches so that no other co-terminal can be selected in this transition.
9. The confidence level is increased in the CE cell in the input matrix connected to the back selected intermediate co-terminal and the latched input terminal. This increases the probability that the *actual output* will be selected by the *latched input* at a later trial.
10. The cycle switch is opened at the end of the transition cycle, allowing all of the latches, timers, and disconnects to return to their normal positions.

The process of selection and learning in a Type 2 duplex system is similar to the sequence of steps required by the expanded Type 2 duplex controller except that the forward- and back-selection process is carried out in one bidirectional output matrix, as described in Table 20.4. Since the sensitivity of all of the BD2 cells on the row of the selecting intermediate terminal and the column of the selected output terminal are decreased logarithmically before the back-selection process, two important features appear:

1. The relative sensitivities of the BD2 cells in the selected output *column* are decreased equally when a BD2 cell in this column selects an output. Thus, no BD2 cell on this column is given an advantage or disadvantage in the back-selecting process. So the Type 2 bidirectional matrix is not more or less likely to back select the forward-selecting BD2 cell.
2. The relative sensitivities of all of the BD2 cells on the *row* connected to the selecting intermediate terminal are decreased equally after a BD2 cell selects an output. Thus, the output matrix is less likely to back select a BD2 cell on the forward-selecting intermediate value row if the actual output occurs on a *different* output terminal.

Thus, the Type 2 bidirectional matrix is less likely to back select the intermediate co-terminal belonging to the forward-selecting intermediate terminal if a different output occurs, and it is still likely to select the same intermediate co-terminal if the selected output does occur. This is exactly the way a scalar matrix is supposed to work.

Application of the Type 2 empirical duplex controller

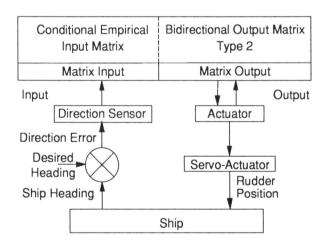

Figure 20.13. The *diagram of an empirical duplex control system* shows that the size of a duplex controller is determined by the resolution of the sensor and actuator variables and the number of expected input/output relations.

As stated before, the *empirical ship auto-pilot* (Fig. 20.13) may have many input states that produce the same output state. For example, a small deviation from set point may require a 1/2 rudder or even a full rudder displacement, and any greater deviation may require a continued full rudder displacement in the opposite direction of the error. In this case, only 3 or 5 values of the intermediate variable are required, and the size of the duplex system can be reduced in comparison to a single empirical control matrix.

In the example given above, if the number of different input/output relations is 5, then each matrix can contain $20 \times 5 = 100$ cells, resulting in a total of 200 cells instead of the 400 cells in the single matrix. Each matrix in this duplex controller must still establish 20 connections, for a total of 40 connections. Each value of the intermediate variable represents an active output state. If only 3 output states are required, then only 3 of the intermediate values have to be used in a given line of behavior. Since this application has only one input and one output variable, a duplex controller used in this application should use a Type 2 bidirectional output matrix.

Type 2 coded unitary duplex empirical controllers

A coded unitary duplex empirical controller can deal with multiple input and output variables in a unitary fashion using a Type 2 bidirectional output matrix. The coded unitary duplex controller is useful only when a few input and output variables with a limited number of values require only a limited number of input/output relations to establish effective control behavior.

Coded unitary Type 2 duplex empirical controller

A coded duplex empirical controller consists of a coded scalar input matrix and a coded scalar Type 2 bidirectional output matrix, as shown in Figure 20.14. The number of intermediate states designed into the system can be determined beforehand to match the expected number of required input/output relations. The empirical duplex controller can determine and record the active transitions in a line of behavior and use each value of the intermediate variable to represent each active output state. The number of memory cells in the unitary duplex controller is equal to the input variety (x^n) times the number of intermediate values (x_I) plus the output variety (x^n) times the number of intermediate values (x_I). In the examples shown, where the resolution of the sensors and actuators are 50%, then $x = 2$, $n = 2$, and $x_I = 4$, the number of cells in the unitary duplex controller is $2(4 \times 4) = 32$ cells, as compared to the 16 cells required in the single matrix coded unitary controller. Though this example favors the single coded matrix, the coded duplex controller would be more favorable when the sensor and actuator variables have more values, and where there are relatively few required transitions.

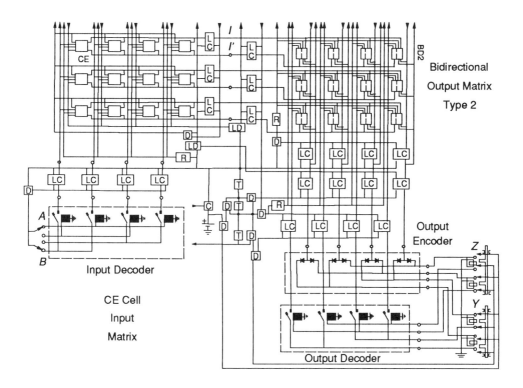

Figure 20.14. A *coded Type 2 duplex empirical controller* can use a scalar Type 2 duplex controller, an input decoder, an output encoder, and a feedback decoder to deal effectively with multiple input and output variables in a unitary environment.

Application of a coded unitary Type 2 duplex controller

The two input and two output variables of the *ship direction control system* shown earlier can be connected to a coded unitary duplex controller to form a successful multivariable empirical control system as shown in Figure 20.15. In this application, the resolutions of the sensors and actuators could be 5%, and it might get by with 10 different transitions. Then, $x = 20$, $n = 2$, and $x_I = 10$, and the number of cells in the unitary duplex controller = $2(400 \times 10) = 8000$ cells, as compared to 160,000 cells in the single matrix coded unitary controller. The Type 2 bidirectional output matrix can be used to determine which intermediate states produce the most consistent input/output relations in this multivariable ship direction control system.

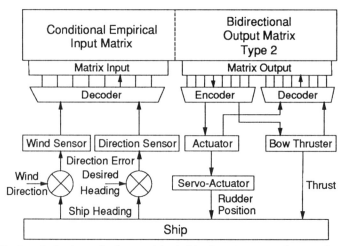

Figure 20.15. A *coded unitary duplex empirical ship direction control system* is a far more efficient multivariable controller than a single matrix coded unitary controller if only a limited
number of specific combinations of sensor and actuator values can produce successful control
behavior.

Operation of the coded unitary Type 2 duplex controller

Again, the coded duplex system operates almost exactly like the single input and
output variable Type 2 duplex controller shown earlier (see Operation of a Type 2
duplex controller). The only difference is that the input variables are decoded into
the single input matrix input variable, the value of the output matrix is encoded into
a set of values of the output variables, and the actual output values are decoded into
a single actuator feedback signal in the output matrix. The two matrices operate as
if they are connected to a single input and a single output variable. However, only
the most consistent input/output relations will become represented by the limited
number of intermediate states as a result of the transform conditioning process used
in the duplex controller shown earlier.

Problems with the coded Type 2 duplex controller

The design of the coded duplex controller allows for a reduction of the number of
memory cells that are wasted when not all of the possible input/output relations are
needed. However, like the coded scalar matrix, the coded duplex controller is still
very inefficient in its use of cells because too many input and output terminals are
committed to combinations of values of the input and output variables that will not
be used. This problem is made worse when the following conditions occur:

1. If the resolutions of the input or output variables are greater than what is
   needed to provide adequate control behavior,
2. If there are too many impertinent variables in the system,
3. If there are one or more random variables in the system.

A coded empirical matrix, which is a unitary system, cannot deal effectively with the scalar relations between input and output variables of a diverse system. Attempting to use a coded unitary empirical controller in a diverse situation creates two problems:

1. It creates the need for far more memory cells than is required to learn the control task,
2. It makes it necessary for the unitary controller to learn more than is needed and thus it takes too long to learn the control task.

These problems can be overcome by the multivariable Type 3 duplex empirical controller shown in the next section.

Type 3 multivariable duplex empirical controllers

A successful duplex controller with multiple input and output variables cannot be obtained using the Type 1 or Type 2 duplex controllers because they attempt to put output values on a relatively few bidirectional cells. A Type 3 bidirectional memory cell and a Type 3 bidirectional control matrix described in this section are required to make a multivariable duplex empirical controller since they tend to spread out the distribution of high sensitivity bidirectional memory cells over many intermediate value terminals. This makes the Type 3 bidirectional output matrix ideal for use in the multivariable output matrix of duplex controllers where multiple bidirectional memory cells are used to forward select output values and back select the intermediate value needed to carry the transform conditioning process to the input matrix.

Expanded Type 3 bidirectional output matrix

A Type 3 bidirectional output matrix can be separated into a forward-selecting matrix and a back-selecting matrix, as shown in Figure 20.16. The forward-selecting matrix decreases the sensitivity of the most sensitive Type 3 principal CE cell connected to the latched input terminal when it forward selects an output terminal, and increases the sensitivity of the Type 3 principal CE cell belonging to the most sensitive Type 3 collateral CE cell connected to the actual output co-terminal when it back selects the input co-terminal belonging to the input terminal most likely to select the actual output value.

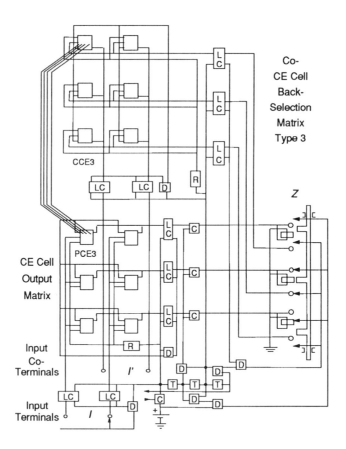

Figure 20.16. In an *expanded Type 3 bidirectional output matrix,* a matrix of Type 3 principal CE cells forward select an output value and another matrix of Type 3 collateral CE cells back select the input value most likely to select the actual output value.

Type 3 principal and collateral CE cells

The Type 3 principal CE cell contains all of the elements of a standard CE cell except that its feedback enable coil and feedback contacts reside in its collateral cell, as shown in Figure 20.17. The Type 3 collateral CE cell consists of a back-select sense terminal, back-select search terminal, back-select search coil, back-select contacts, and a back-select output terminal. The Type 3 principal CE cell decreases the sensitivity of both search relays when it forward selects, and the Type 3 collateral CE cell increases the sensitivity of both search relays when it back selects.

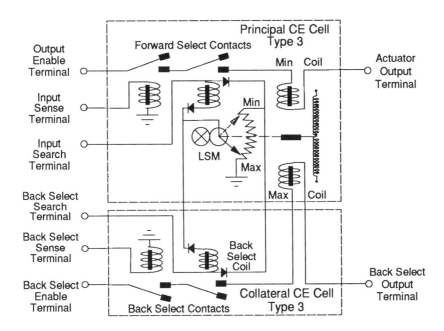

Figure 20.17. In *Type 3 principal and collateral CE cells,* the principal cell decreases the sensitivity of both search relays, and the collateral cell increases the sensitivity of both search relays.

Type 3 expanded duplex controller

The Type 3 expanded bidirectional output matrix can be connected to a standard CE cell input matrix, forming the Type 3 expanded duplex controller shown in Figure 20.18. The input matrix selects an intermediate terminal of the intermediate variable *I* based upon the latched value of the input variable *A* and the sensitivities of the CE cells in this column. The forward-selecting output matrix then selects an output terminal of the actuator variable *Z* based upon the value of the intermediate terminal selected by the input matrix and the sensitivities of the Type 3 principal CE cells in this row. The sensitivities of both the principal and collateral cells of the selecting cell are decreased according to the LSM in the principal cell. The back-selecting matrix back selects the intermediate co-terminal of the intermediate co-variable *I'* based upon the value of the actual latched output and the sensitivities of the Type 3 collateral cells in that column and increases the sensitivity of both this collateral cell and its principal cell. The selected intermediate co-terminal then provides the transform conditioning signal needed to increase the sensitivity of the CE cell in the input matrix connected to the original latched input state so that it is more likely to select the back-selected intermediate terminal whenever the original input state occurs again.

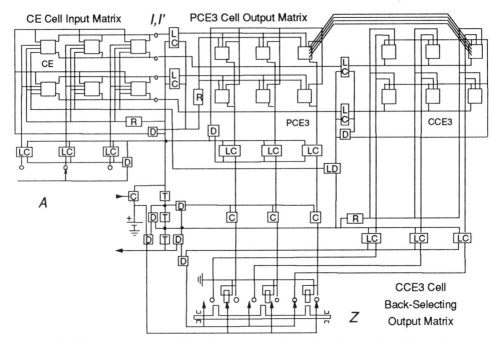

Figure 20.18. A *Type 3 expanded duplex controller* shows how the output matrix can be divided into forward- and back-selecting matrices.

Type 3 bidirectional CE cell

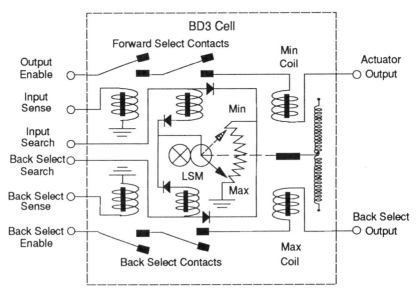

Figure 20.19. The *Type 3 bidirectional memory cell (BD3 cell)* decreases its sensitivity when it forward selects and increases its sensitivity when it back selects.

The Type 3 principal and collateral CE cells can be combined in the single *Type 3 bidirectional memory cell (BD3 cell)* shown in Figure 20.19. The Type 3 bidirectional memory cell (BD3 cell) is the basic bidirectional memory cell that can be used in all multivariable bidirectional matrices. It is also the simplest type of bidirectional memory cell because it decreases its sensitivity when it forward selects and increases its sensitivity when it back selects. It is easy to see how it can *evolve* from the standard CE cell by the feedback enable relay in the CE cell changing into the back-selecting search relay in the BD3 cell.

Type 3 bidirectional binary output matrix

The *Type 3 bidirectional binary output matrix* is made up of Type 3 bidirectional CE cells, as shown in Figure 20.20.

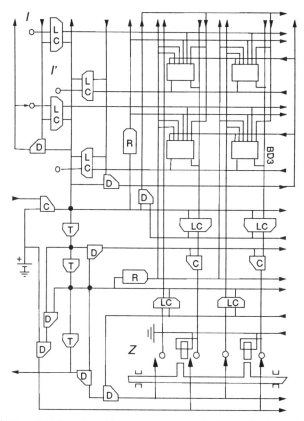

Figure 20.20. A *Type 3 bidirectional binary output matrix* forward selects the output having the bidirectional memory cell (BD3 cell) with the highest sensitivity on the latched input terminal and decreases the sensitivity of that BD3 cell. It then back selects the input co-terminal having the BD3 cell with the highest sensitivity on the latched actual output terminal and increases the sensitivity of that BD3 cell.

The Type 3 bidirectional matrix tends to shun the back-selection of the input co-terminal belonging to the input terminals used in the forward-selection process because it back selects *after* it decreases the sensitivity of the forward selecting BD3 cell. This might cause the Type 3 bidirectional output matrix to create more than one high sensitivity BD3 cell on a given output terminal. This could cause new learning to interfere with old learning in a single variable (scalar) matrix. However, the Type 2 matrix shown previously is intended for single variable (scalar) applications. The Type 3 bidirectional output matrix tends to spread out the same behavior onto two or more input terminals, unlike the Type 1 matrix, which tends to concentrate behavior on fewer input terminals. This feature makes the Type 3 matrix work well with multiple output variables (vectors) such as the multivariable bidirectional output matrix, to be discussed, where multiple BD3 cells are used to back select an input co-terminal.

Type 3 empirical duplex controller

The Type 3 bidirectional output matrix can be connected to a standard CE cell input matrix by means of a common *intermediate variable I* and intermediate co-variable *I'*, forming the *Type 3 duplex empirical controller* shown in Figure 20.21.

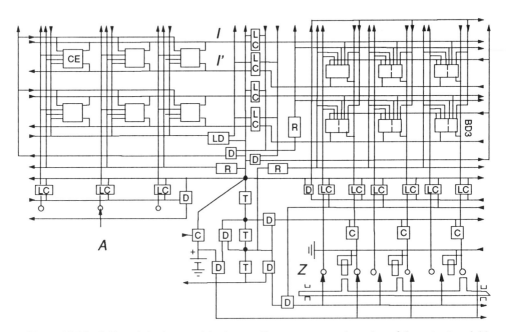

Figure 20.21. A *Type 3 duplex empirical controller* can connect the value of the output variable *Z* that occurs most often with a given value of the input variable *A* by means of one value of its intermediate variable *I*.

The number of rows of CE cells and thus the number of values of the intermediate variable is determined by the number of required input/output relations. (The Type 3 bidirectional output matrix is intended for use in the multivariable output matrix of a multivariable duplex empirical controller. The scalar duplex empirical controller shown here should use the Type 2 bidirectional output matrix shown earlier. The Type 3 bidirectional output matrix with a single output variable is shown only for the sake of simplicity.)

Bidirectional multivariable output matrix

Additional Type 3 bidirectional output matrices can be connected to form the *Type 3 bidirectional multivariable output matrix* in Figure 20.22.

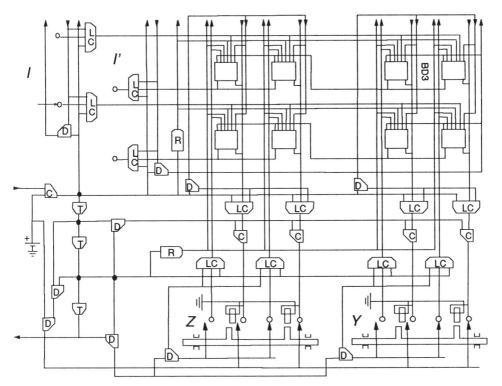

Figure 20.22. A *Type 3 bidirectional multivariable output matrix* forward selects the output value in each matrix connected to the BD3 cell in each matrix with the highest sensitivity for the current input value, and back selects an input co-terminal that corresponds to the input value that will most likely select the set of actual output values the next time that input value occurs again.

A bidirectional multivariable output matrix is not used by itself unless there is some reason to back select an input value that represents the set of actual output values, as required by a multivariable duplex controller or a bidirectional multivariable matrix in some other kind of directly connected network of empirical matrices. (See Part IV.)

Operation of the Type 3 bidirectional multivariable output matrix

The bidirectional multivariable output matrix operates according to the steps listed in Table 20.5.

Table 20.5. Operation of the Type 3 bidirectional multivariable output matrix

---

1. The cycle switch is closed at the start of a new transition cycle.
2. One input terminal is energized according to the initial position of the sensor switch, representing the current value of the input variable. This input terminal is latched "on," and the sensor switch is disconnected so that no additional input terminals can be energized in that transition cycle.
3. The sensitivities of the BD3 cells in the row connected to the latched input terminal are interrogated by the forward-selecting ramp signal.
4. When a BD3 cell in this row becomes conductive in each submatrix, a current can flow from the output enable power source through these cells to the output terminal in that column.
5. This output signal disengages the output enable power source in that submatrix so that no more BD3 cells in that submatrix can produce an output. The ramp must continue to rise until an output has been selected in the submatrix of each output variable. The logarithmic subtracting mechanisms in the conducting BD3 cells decrease the sensitivities of these cells according to their sensitivity values. (Though the ramp may cause output contacts close in many other cells in a matrix that has selected an output, no other outputs can be produced, and no other subtractions can occur because the power source to these cells has been disconnected.)
6. After the delay specified by the output delay relay, the actuator brake is released, the actuator coils connected to the selected output terminals are energized, and the actuators attempt to align their armature poles to the positions of the energized actuator coils.
7. After another delay, to allow the actuators to position themselves, the actuator brake is re-engaged, and a signal is sent to latch "on" the *actual* output co-terminals. This signal is also sent to the actual output co-terminal disconnect to prevent different co-terminals from being energized if the actuator in that matrix moves to a new position in this period.
8. The back-selecting ramp generator then interrogates the BD3 cells connected to the latched output co-terminals corresponding to the actual positions of the actuators. As the contacts of each BD3 cell close, conductive paths are created in the column from the back-select enable power source toward the input co-terminals.
9. When a *conductive path* is completed to one input co-terminal, the sensitivities of all of the BD3 cells on this conductive path are increased, and the back-select enable power source is disconnected so that no more input co-terminals can be energized.
10. After another delay, the actual output circuit is shut off, and a signal is sent to the outside timing source, indicating that this transition cycle is over. (This signal could be sent directly to the cycle switch to end the transition cycle locally.)
11. The cycle switch is opened, causing power to be shut off to all elements except the actuator brake circuit. The actuator brake holds the actuators at whatever position they are until sometime in the next transition cycle, and all of the latches, delay relays, ramps, and disconnects are restored to their normal positions.

The bidirectional multivariable output matrix uses the same kind of Type 3 bidirectional memory cells used in the single variable Type 3 bidirectional output matrix shown above. However, the multivariable output matrix requires that multiple BD3 cells become conductive in the row of a given intermediate variable in order for its intermediate co-terminal to be back-selected.

The BD3 cells on the back-selected input co-terminal in the bidirectional multivariable output matrix may not be the highest confidence level BD3 cells connected the actual output terminals. The back selected co-terminal represents the *set of BD3 cells in which the lowest sensitivity cell is higher than the lowest sensitivity in any other row of cells* connected to the actual output terminals. Being part of the set of back-selecting cells increases the sensitivity of the low-sensitivity BD3 cells in the set and tends to bring them up to the level of the higher-sensitivity BD3 cells in the set. If one of the output variables is *inactive* (remains in a constant position), the higher-confidence BD3 cells in the selecting set are likely to force up the confidence level of the BD3 cell connected to this constant value. If one of the output variables is *impertinent* (has no specific relation to the input variable), the higher-confidence BD3 cells in the selecting set are likely to force up the confidence levels of all of the BD3 cells connected to this variable and the back-selected input value. This decreases the probability that the impertinent variable will interfere with the back-selection process.

The coattail effect in a bidirectional output matrix

The likelihood of a given BD3 cell on a latched actual output co-terminal back-selecting its intermediate co-terminal in a multivariable bidirectional output matrix is determined in part by its sensitivity relative to the sensitivities of the other BD3 cells on this output terminals *and* in part by the sensitivity of the other BD3 cells in other variables connected to this intermediate co-terminal. Since the sensitivity of all of the BD3 cells on the row of the selected intermediate co-terminal and the columns of the selected output terminals are increased logarithmically after the back-selection process, even the low-confidence BD3 cells are brought up toward the level of the high-confidence BD3 cells on that row. This *coattail effect* also occurs in the forward-selection process in a multivariable CE cell input matrix. The coattail effect in a bidirectional output matrix may create a set of very high sensitivity BD3 cells on a given intermediate terminal, causing that intermediate terminal to strongly select a given output state, and causing that output state to strongly back select that intermediate co-terminal.

Multivariable duplex controller

To create an empirical duplex controller that can deal with multiple input *and* output variables, it is necessary to create a duplex system having a multivariable input matrix and a Type 3 multivariable bidirectional output matrix, as shown in Figure 20.23.

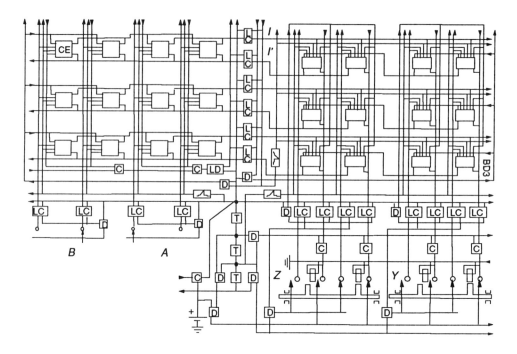

Figure 20.23. The *multivariable duplex empirical controller* consists of a multivariable input matrix using CE cells connected by a single intermediate variable to a Type 3 multivariable bidirectional output matrix using Type 3 bidirectional memory cells (BD3 cells).

The multivariable duplex controller can be used for any empirical control application having multiple input and output variables where each variable has relatively few values and where each value is potentially meaningful to the control application. The Type 3 multivariable duplex controller can also be used in a scalar application by digitizing the sensor and actuator variables as shown in the next chapter. Since most applications required digitized sensor and actuator variables, the scalar Type 2 duplex controller may not have to be used.

Operation of the multivariable duplex controller

The multivariable duplex empirical controller uses a multivariable input matrix and a multivariable bidirectional output matrix Type 3 based upon the BD3 cell. It operates according to the steps in Table 20.6.

The intermediate co-terminal made active by the back-selection process denotes the intermediate terminal that is most likely to be forward-selected by a given input state and to forward-select a given output state. Thus, a given intermediate terminal learns to represent a given input/output relation.

Table 20.6. Operation of a Type 3 multivariable duplex controller

---

1. At the beginning of the transition cycle, the input latch in the input matrix energizes and sustains a voltage on the input columns connected to the input values of each input variable made active at the beginning of the transition cycle, and excludes any other input column in the input matrix from being energized.
2. The CE cells along the columns connected to the latched input terminals in each submatrix of the input matrix are interrogated to find the set of CE cells in which the lowest-confidence CE cell in the set is higher than the lowest-confidence CE cell in any other set.
3. The intermediate terminal is energized on the first row of CE cells to form a *conductive path* from the output enable line to that intermediate terminal, and the confidence levels of all of the selecting CE cells in the that row are reduced by their respective LSMs by an amount proportional to their confidence levels.
4. Then the BD3 cells in the output matrices connected to the selected intermediate terminal are interrogated to find the BD3 cell in each submatrix with the highest confidence level. The BD3 cell in each submatrix with the highest confidence level is used to energize its output terminal, and the confidence levels of the BD3 cells connected to the selected output terminals are reduced by their LSMs by an amount proportional to their individual confidence levels.
5. After a delay mandated by the discontinuous state ontology, the actuator delay relay times out. This disengages the actuator brake and enables the set of output connects, which allows the output terminals selected by the output matrix to be energized. This causes the actuators to attempt to move to the selected positions.
6. After a delay to allow the actuators to move to their selected positions, the actuator brakes are restored, and the output matrix is interrogated by a voltage ramp to find the BD3 cell in each submatrix with the highest-confidence level connected to the actual output terminal in each submatrix. As the contacts of the higher-confidence BD3 cells close, *conductive paths* begin to form between the back-select enable bus and the intermediate co-terminals. When a conductive path is completed between the back-select enable bus and a single intermediate co-terminal, that co-terminal is energized, the confidence levels of the BD3 cells on this conductive path are increased, and the back-select enable bus is disconnected to keep any other conductive paths from being completed.
7. The confidence level is increased in the CE cell in the input matrix connected to the back-selected intermediate co-terminal and the latched input terminal. This increases the probability that the *actual combination of output values* will be selected by the *original combination of input values* of this transition at a later trial.
8. After another delay, to allow the last step to be completed, a signal is sent to the outside timing source indicating that this transition is complete. Then the cycle switch is opened and all of the latches, timers, disconnects, and ramps are restored to their initial conditions.

Summary of the operation of the multivariable duplex empirical controller

The multivariable duplex empirical controller operates in much the same way as a scalar duplex controller, but can deal with multiple input and output variables rather than a single sensor and actuator variable. The operation of the multivariable duplex empirical controller is summarized in Table 20.7.

The forward selecting and back selecting process allows a specific combination of values of multiple sensor variables to forward select the specific combination of values of multiple actuator variables that is most likely to occur with that specific combination of values of the input variables. These relations are established through empirical trials in which the controller attempts to produce the specific output states that can occur most often for each input state.

Table 20.7. Summary of the operation of the multivariable duplex empirical controller

---

1. At the beginning of the transition cycle, the input latches capture a set of values of the sensor variables.
2. The input matrix then selects a single intermediate value that best fits this set of latched input values and *deducts* from the confidence levels of the CE cells connected to this intermediate value and the latched input values.
3. This intermediate value activates the most sensitive BD3 cell in each output variable in the output matrix and *deducts* from the confidence levels of those BD3 cells that energize an output value.
4. The set of actual actuator values then back selects the intermediate co-terminal that best represents the actual output co-terminal and more than *restores* the confidence levels of the back-selecting BD3 cells.
5. This back-selected intermediate value is then used to more than *restore* the confidence levels of the CE cells in the input matrix connected to the latched values of the sensor variables.

Parallel and series processing in a multivariable duplex empirical controller

The selection of an output in a duplex controller requires a selection in its input matrix *and* a selection in its output matrix. Thus, two selections must be carried out in *series* to produce an output. However, the input and output matrices interrogate their CE and BD3 cells in *parallel* to make each selection. Thus, a duplex empirical controller produces action in a *series and parallel process*.

Distributed processing in a duplex empirical controller

Thus, in a large multivariable duplex empirical controller, many conditional empirical memory cells have to be interrogated to select an output. The actual selection is based upon the relative values of many conditional empirical cells distributed over a large portion of the memory. Thus, the selection of the values of an output state in a multivariable duplex controller is determined by the *distribution of sensitivities* in all of the conditional empirical cells connected to the actual input values and the possible output values. In most cases, the sensitivities of many empirical memory cells in a multivariable duplex controller may have to be altered in each transition. Since many transactions have to be made in each transition in a duplex empirical control, its operation is based upon the *distributed processing* of sensitivity information.

Self-organization in multivariable empirical duplex control systems

A multivariable duplex controller can combine values of a set of input variables into a set of intermediate variables in a unitary manner, or can use its intermediate values to represent specific values of a specific input variable regardless of the values of the other input variables in a diverse manner. These intermediate values can then select combinations of values of output variables or select specific values of only one output variable in a unitary or diverse manner. The selections are based upon which active conditional memory cells have the highest of the lowest sensitivity values.

Unitary organization in a multivariable duplex empirical controller

The sensitivity values of the conditional memory cells in a multivariable duplex empirical controller shown in Figure 20.24 can establish a unitary relation between the input and output variables such that every combination of input values produces a unique combination of output values.

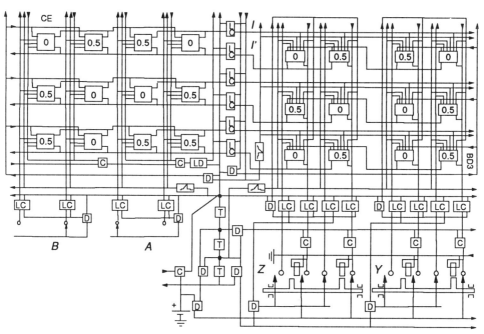

Figure 20.24. In the *unitary conformation of a binary multivariable duplex empirical controller*, any combination of values of the input variables can produce a unique combination of values of the output variables.

Each selection of an output state in a binary multivariable duplex empirical controller requires that two CE cells in the input matrix and two BD3 cells in the bidirectional output matrix must be set in a high-sensitivity state. If every combination of values is important, the controller must have x^n intermediate values, assuming x is the number of values of the input and output variables and n is the number of input or output variables.

Diverse organization in a binary multivariable duplex empirical controller

A diverse relation can be formed between specific input and output variables when specific values of the input variable produce specific values of the output variable regardless of the values of the other variables. A diverse relation has been formed between the left-hand sensor and the left-hand actuator in the multivariable duplex controller shown in Figure 20.25.

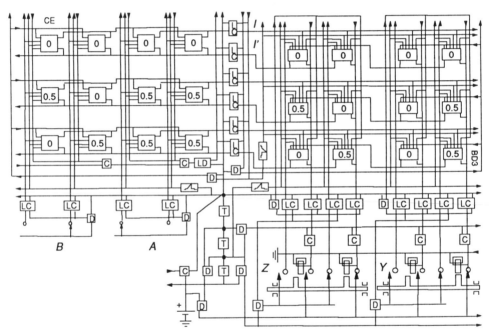

Figure 20.25. In a *diverse conformation in a binary multivariable duplex empirical controller,* the values of some input variables are ignored and the values of some output variables are held constant.

The left-hand sensor controls the left-hand actuator as if they were in a separate system. If the right-hand actuator is allowed to assume position it chooses to assume, it will tend to remain at some constant value due to the dominance of small changes (see Part V). Since the Type 3 bidirectional output matrix can replicate many high sensitivity BD3 cells on a given output terminal, the sensitivities of the BD3 cells connected to this value of the right-hand actuator can be set close to their normal maximum (0.5) as this variable remains constant while the matrix learns the values of the active output variable. There must be enough intermediate values to specify each diverse transition involving the active output variable. The remaining intermediate values may be used to form *synchronous diverse relations* among the remaining variables if another input variable is available to determine which diverse relation is made active at a given time, as discussed in Part II.

Application of a multivariable empirical duplex control system

The multivariable duplex controller can be used to deal effectively with the multiple input and output variables of the ship auto-pilot discussed earlier (Fig. 20.26).

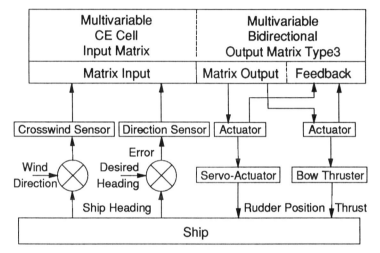

Figure 20.26. A *multivariable duplex empirical control system* can be used as an empirical ship direction control system that can use each combination of sensor values and actuator values to specify the input/output relations, or it can connect a specific sensor variable to a specific actuator variable while holding the value of the other actuator constant, depending upon the success of both methods of organization.

It can use each combination of values of the position error sensor and the crosswind sensor to create unique combinations of rudder position and bow thruster action if these relations satisfy the control requirements of the ship. Or it can use values of one or the other sensor to produce the action of one or both actuators. For example, successful control behavior may be achieved by having the bow thruster operate in the opposite direction and in proportion to the crosswind while the rudder is held constant, *or* the rudder can act in the opposite direction of the direction error sensor while the bow thruster remains constant. One of these diverse control systems may produce more consistent control behavior than a unitary organization of these variables. Whether the system is organized in a diverse or unitary manner is determined empirically by the multivariable duplex controller according to which produces the most consistent results. Consistent results can be produced only when the behavior of the controller corresponds to the wishes of its operator, conforms to the laws of nature, and meets the requirements of its mechanical structure.

Self-organization in a universal duplex empirical controller

The multivariable duplex empirical controllers can produce unitary relations and synchronous diverse relations among multiple sensor and actuator variables. However, it cannot produce asynchronous diverse relations, which can occur together any time regardless of the state of the other diverse relations. These asynchronous diverse relations can be produced by a monolithic empirical control matrix made up of CE cells, as shown earlier, and can be produced in a *universal* duplex empirical controller made up of a monolithic empirical input matrix made up of CE cells and a monolithic bidirectional empirical output matrix made up of BD3 cells.

Monolithic bidirectional output matrix

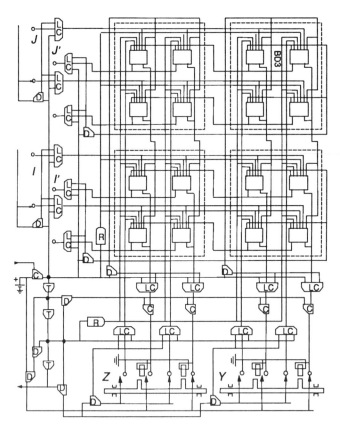

Figure 20.27. A *monolithic bidirectional output matrix* can learn to find the combination of values of its output variables that is most likely to be produced by a given combination of values of its input variables, and then it can find the combination of values of its input co-terminals most likely to produce the actual combination of output values.

The single input variable of the multivariable bidirectional output matrix can be broken into two or more input variables, forming the *monolithic bidirectional output matrix* shown in Figure 20.27. The monolithic bidirectional control matrix would not normally be used as a standalone controller because there is no need to back select the input co-terminals that represent the input state most like to select the actual output state in this system. However, it must be used in a direct network of empirical controllers, as shown in Part IV, or as the output matrix in a universal duplex empirical controller.

Universal duplex empirical controller

The duplex empirical controller shown earlier cannot deal effectively with diverse behavior that has to occur in an asynchronous manner. This problem can be overcome by connecting a CE cell monolithic input matrix to the monolithic bidirectional output control matrix, forming the *universal duplex empirical controller* (UDEC) shown in Figure 20.28, which can create as many asynchronous diverse lines of behavior as it has intermediate variables $I, J, \ldots$.

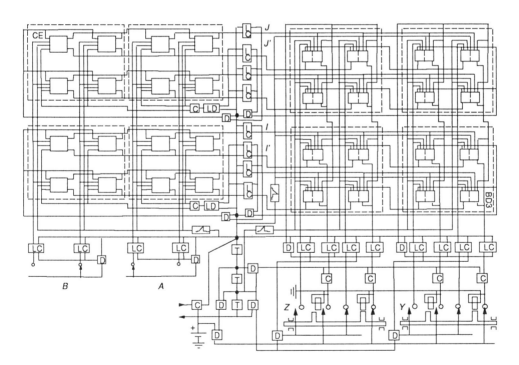

Figure 20.28. The *universal duplex empirical controller* can learn to connect the most consistently occurring combinations of values of its output variables with each combination of values of its input variables in a unitary manner, or can form asynchronous diverse lines of behavior through each of its intermediate variables.

The universal duplex empirical controller can be enlarged to deal with any number of unitary and/or diverse variables having any number of values and can deal with two or more *asynchronous diverse lines of behavior*. However, trinary variables make better use of memory cells, and the resolution of most sensors and actuators are limited to about thirty values.

Operation of the universal duplex empirical controller

A universal duplex empirical controller with multiple input and multiple output variables operates according to the steps listed in Table 20.8 that occur during a single transition. The operation of a universal duplex controller is quite similar to the operation of a multivariable duplex empirical controller except that the universal duplex controller uses multiple intermediate variables to represent each asynchronous line of behavior.

Table 20.8. Operation of a universal duplex empirical controller

---

1. A set of sensor values is latched in at the beginning of a transition cycle.
2. The input matrix then selects one intermediate value in each intermediate variable for the particular combination of latched input values.
3. The sensitivities of all of the CE cells in the input matrix connected to these intermediate values and the latched input values are diminished according to the sensitivity level of each of these CE cells.
4. The output submatrices use the values of the intermediate variables to select a set of values of the actuator variables. (The value of each actuator variable can be selected by only one intermediate variable, although one intermediate variable can select the values of two or more actuator. In a unitary relation, parts of the output state can be selected by different intermediate variables. However, each asynchronous diverse relation must be selected by a different intermediate variable.)
5. The sensitivities of all the selecting BD3 cells on the latched intermediate terminals are diminished according the sensitivity values of each cell as they energize their output terminals.
6. After a delay, the actuators attempt to move to their prescribed positions.
7. After another delay, the actual actuator values are latched into the actuator co-terminals. (These values may or may not be the same as those selected by the actuator matrix, depending upon whether the selected output actually occurs.)
8. The bidirectional output matrix back selects an intermediate co-terminal for each intermediate variable that best represents the set of actual actuator values.
9. The sensitivity is increased in all the memory cells connected to these intermediate co-terminals, the latched input terminals in the input matrix, and the latched co-terminals in the output matrix.
10. After another delay, the transition cycle switch is opened, and all of the latches, disconnects, timers, and ramps are returned to their initial conditions.

Application of a universal duplex empirical controller

The multiple input and output variables of a *ship direction control system* can be connected by means of a universal duplex empirical controller, as shown in Figure

20.29. The universal duplex empirical controller is particularly desirable when two or more asynchronous diverse lines of behavior must occur between specific sensors and actuators. For example, successful control behavior may be achieved when the bow thruster operates in the opposite direction and in proportion to the crosswind at any time regardless of the values of the direction error sensor. Also, the best results may be achieved when the rudder acts in the opposite direction of the direction error sensor at any time regardless of the values of the crosswind sensor. In this case, the auto-pilot and crosswind compensation systems have emerged as separate systems.

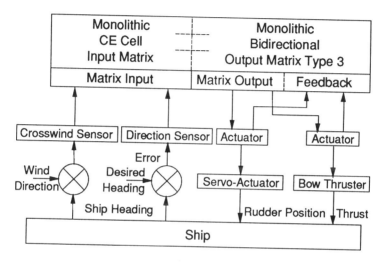

Figure 20.29. An *application of a universal duplex empirical control system* is a multivariable ship direction control system that can learn to produce any combination of output values for each combination of input values in a unitary manner, or it can establish as many asynchronous diverse lines of behavior as it has intermediate variables, whichever produces the most consistent results.

The universal empirical controller

The universal duplex empirical controller, made up of a monolithic input matrix and a monolithic bidirectional output matrix, is the most general and useful empirical controller. It can deal effectively with any number of unitary or/diverse variables, and can be used to develop the more specialized control systems shown in the following sections. Also, it can be used to create networks of control units that can deal more effectively with many unitary and/or diverse variables than a single control unit acting alone. For these reasons, it is called a *universal empirical controller*. However, it still commits input and output terminals to specific values of its input and output variables. If a particular value or set of values is not needed in the control process, all of the memory cells and circuitry connected to those terminals are wasted. This problem can be largely overcome by the digitized universal empirical controllers shown in the next chapter.

Chapter 21
Digitized Duplex Empirical Controllers

Though the multivariable duplex controller can deal effectively with multiple input and output variables organized in a unitary system, and can connect a particular input variable to a particular output variable at a given moment in time in a synchronous diverse manner, it is still very "literal" about each value of each variable. It cannot generalize about "high" or "low" values of its variables. This means that the controller may have to establish transitions for the complete set of values of a given variable although many of these values may represent a single *general state* defined by a specific value field. This problem becomes acute when the resolutions of the sensor and/or actuator variables greatly exceed the requirements of the task environment. This limitation can be overcome by *digitizing* the input and output variables into a set of aggregate variables that represent each input and output variable, and connecting these digitized variables by means of multivariable or universal duplex empirical controllers that can establish a limited number of general or specific input/output relations.

Binary digitized duplex empirical controllers

We have seen how a sensor variable can be digitized into a set of aggregate variables by an encoder. These aggregate variables are then used as the input variables of the input matrix of a duplex controller that can select a value of an intermediate variable, as shown previously in digitized predetermined and absolute duplex empirical controllers. This intermediate variable can select a set of values of the output variables of a bidirectional output matrix. These variables also can be defined as aggregate variables that can be converted back to a single actuator variable by means of a decoder, as shown before. The actual output of an actuator can then be converted back to a set of aggregate co-variables by means of an output encoder, and these values can be used to back select a value of the intermediate co-variable through the bidirectional output matrix to change the sensitivity of the empirical memory cells in the input matrix.

Scalar binary digitized bidirectional output matrix

The bidirectional output matrix, shown in the *scalar binary digitized bidirectional output matrix* in Figure 21.1, is a Type 3 multivariable bidirectional output matrix. Its multiple output variables are used as aggregate variables. These aggregate variables are decoded into a single (scalar) output variable. The actual value of the output variable is encoded into a set of aggregate feedback variables that are then used to back select the value of the input variable that is most likely to produce the actual output state.

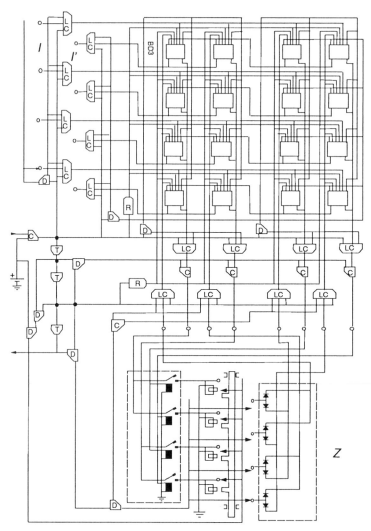

Figure 21.1. The *scalar binary digitized bidirectional output matrix* uses a set of Type 3 bidirectional output submatrices to control the set of aggregate variables of a single output variable.

The aggregate variables determine the position of a single actuator through the decoder, and the actuator feedback encoder transmits the actual output position to the appropriate values of the aggregate variables. The digitized bidirectional matrix is not expected to be used by itself because back selecting is used only when the matrix is the output matrix of a duplex controller or when the matrix is used in a direct series network (see Chapter 22). The digitized Type 3 bidirectional output matrix shown can be used as the output matrix of a digitized duplex controller with only one actuator in place of an undigitized Type 2 bidirectional output matrix with only one actuator.

Scalar binary digitized duplex empirical controller

The *scalar binary digitized duplex controller* is formed by connecting a scalar binary digitized input matrix to a scalar binary digitized bidirectional output matrix, as shown in Figure 21.2. Like the output matrix in a multivariable duplex controller, a scalar digitized output matrix must select the output state that is most likely to occur for a given value of the intermediate variable and send back a signal to the input matrix as to the value of the intermediate variable that is most likely to produce the output state that actually occurs.

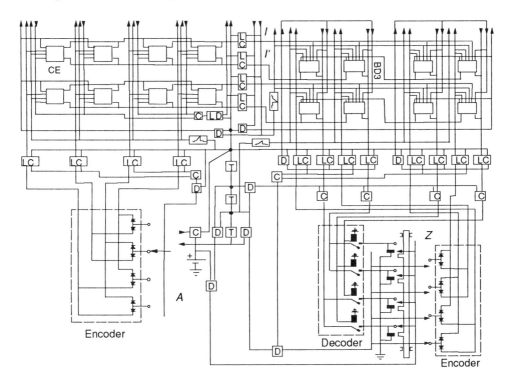

Figure 21.2. The *scalar binary digitized duplex empirical controller* can provide general or specific representations of its input and output states, can be tailored to provide the number of intermediate values equal to the number of expected transitions, and can do so with a minimum number of memory cells and a minimum learning time.

It can identify a value of a sensor or actuator variable as the value of the most significant place or it can identify more specific sensor or actuator states as values of the most and lesser significant places if need be. The digitized multivariable duplex controller is the basic building block of far more complex empirical control systems. (A scalar digitized controller, like the one shown, uses the bidirectional Type 3 output matrix because the digitization process converts the scalar actuator variable into a set of variables. Thus, any scalar system can use the Type 3 duplex controller instead of the Type 2 duplex controller by digitizing its scalar variables.)

Operation of a scalar binary digitized duplex empirical controller

A digitized duplex empirical controller with a single set of input and output variables operates according to the steps listed in Table 21.1 during a single transition cycle.

Table 21.1. Operation of a scalar digitized duplex empirical controller

---

1. Some value of the sensor exists at the start of a transition cycle.
2. This sensor value is digitized into a set of values of the aggregate input variables, according to the code of the sensor encoder.
3. These values are captured by the input latches of the input matrix.
4. The input matrix then selects one intermediate value for the particular combination of latched values of the aggregate input variables.
5. The sensitivities of all of the CE cells in the input matrix connected to this intermediate value line and the latched input values are diminished according to the sensitivity level of each of these CE cells.
6. The output matrix interrogates the BD3 cells on this intermediate value terminal to find the BD3 cell belonging to each aggregate output variable having the highest sensitivity. Each selected BD3 cell energizes its aggregate value terminal and the sensitivity of each selected BD3 cell is diminished according its sensitivity value.
7. After a time delay required by the DS ontology, the aggregate actuator variables are decoded to produce the final actuator output value, and the actuator attempts to move to this decoded position.
8. After another time delay to allow the actuator to arrive at the decoded position, the actual actuator value is digitized by the actuator encoder, causing a set of co-terminals of the aggregate variables to be latched into the back-selection circuit. (These co-terminals may or may not coincide to those selected by the actuator matrix, depending upon whether the output selected is actually carried out.)
9. The bidirectional output matrix back selects the intermediate co-terminal that best represents the set of latched aggregate co-terminals.
10. The sensitivities are increased proportionally in all the memory cells connected to this intermediate co-terminal, the latched input terminals of the aggregate variables in the input matrix, and the latched co-terminals of the aggregate variables in the output matrix.
11. After another time delay to complete these transactions, the cycle switch is opened and all of the latches, timers, ramps, and disconnects are restored to their normal positions.

The operation of a scalar digitized duplex controller is quite similar to a multivariable duplex empirical controller.

Application of a scalar binary digitized duplex empirical controller

The scalar binary digitized duplex empirical controller can connect a digitized input variable, such as a direction error sensor, with a digitized output variable, such as a rudder servo-actuator (Fig. 21.3), using far fewer conditional memory cells than an undigitized system.

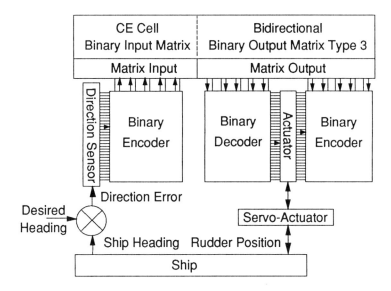

Figure 21.3. An *application of a scalar binary digitized duplex empirical controller* is as a medium-resolution empirical ship auto-pilot.

For example, a binary digitized duplex empirical controller with five sensor and five actuator aggregate variables can identify any one of $2^5 = 32$ different sensed conditions and produce any one of 32 different responses using only 20 memory cells per transition. An undigitized nonduplex scalar matrix would require 32×32 = 1024 memory cells to establish just one specific transition, but can then establish any 32 out of 1024 different transitions.

Binary digitized multivariable bidirectional output matrix

More actuator variables can be added to the binary digitized bidirectional output matrix, forming the *binary digitized multivariable bidirectional output matrix* shown in Figure 21.4. The binary digitized multivariable bidirectional output matrix is not expected to be used by itself because back selecting is used only when the matrix is the output matrix of a duplex controller or when the matrix is used in a direct series network (see Chapter 22). The digitized multivariable bidirectional output matrix is used as the output matrix of a digitized multivariable duplex controller, which can position an actuator in a general or specific manner according to the requirements of the control system and can do so with far fewer memory cells than a undigitized multivariable controller.

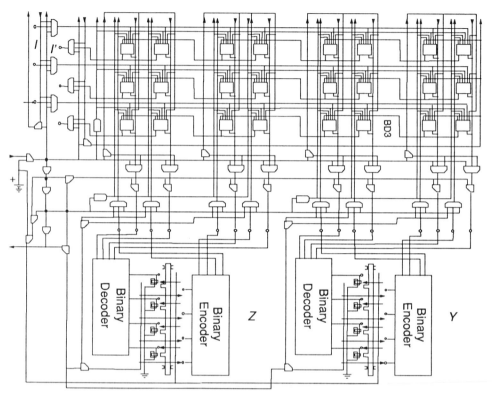

Figure 21.4. A *binary digitized multivariable bidirectional output matrix* can provide general or specific representations of multiple output variables.

Binary digitized multivariable duplex empirical controller

More variables can be added to a digitized duplex controller by adding more sub-matrices as shown in a *binary digitized multivariable duplex controller* (Fig. 21.5). The sensor and actuator matrices treat the additional aggregate variables the same way they treat any additional variables. The operation of this system is the same as the digitized duplex empirical controller with a single set of sensor and actuator variables. The multivariable input matrix forward selects an intermediate terminal connected to the CE cell with the lowest sensitivity on a latched input terminal that is higher than any other CE cell with the lowest sensitivity connected to a latched input terminal and any other intermediate terminal. The bidirectional output matrix forward selects the set of output terminals connected to the BD3 cells with the highest sensitivities in each aggregate variable connected to this intermediate terminal. It then back selects the intermediate co-terminal connected to the BD3 cell with the lowest sensitivity connected to the latched actual output co-terminal that is higher than any other BD3 cell with the lowest sensitivity connected to a latched actual output co-terminal and any other intermediate co-terminal. This intermediate co-terminal increases the sensitivity of its CE cells in the input matrix connected to the latched input terminals.

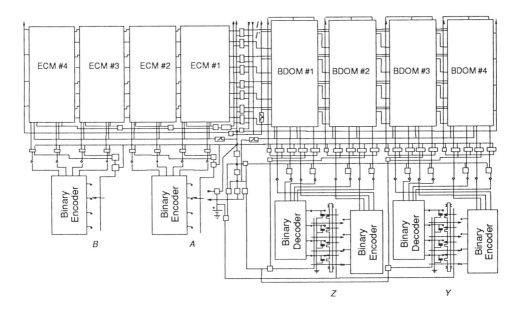

Figure 21.5. A *binary digitized multivariable duplex empirical controller* can connect generalized or specific values of multiple sensor variables to generalized or specific values of multiple actuator variables.

Application of a binary digitized multivariable duplex empirical controller

A crosswind sensor and a direction error sensor can be connected to binary digitized multivariable input matrix, and a rudder servo-actuator and a bow thruster can be connected to a binary digitized multivariable bidirectional output matrix, forming the *ship direction control system* in Figure 21.6, using a binary digitized multivariable duplex empirical controller. The binary digitized multivariable duplex empirical controller provides a far superior ship direction control system than the undigitized nonduplex control matrix. For example, if 5 binary aggregate variables are used for each sensor and actuator variable, the system can identify and produce any one of 32 sensor and actuator values for a resolution of better than 3%, and can produce any one of 1024 possible combinations of values of the two actuator variables from any one of 1024 possible combinations of sensor values using only 20 memory cells per transition. An undigitized, nonduplex coded unitary CE cell matrix would require $1024 \times 1024 = 1,048,576$ CE cells to do the same unitary task. Moreover, the binary digitized multivariable duplex empirical controller can establish *synchronous diverse relations* between specific sensors and actuators if the need arises.

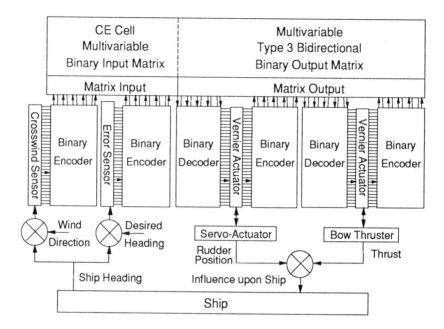

Figure 21.6. An *application of a binary digitized multivariable duplex empirical controller* is as a ship direction control system with two sensor and two actuator variables.

Comparison of the coded unitary system with a digitized duplex system

The digitized control systems shown in this chapter are different from the coded unitary empirical duplex controller that uses a Type 2 bidirectional output matrix. In a coded unitary duplex empirical controller, the multiple sensor variables are *decoded* into a single input variable. The single output variable of the Type 2 bidirectional output matrix is then *encoded* to recreate a set of output variables, and the actual output states are *decoded* to provide the transform feedback signal. The digitized duplex controller has the opposite arrangement. It *encodes* each sensor variable into a set of aggregate variables, *decodes* the output aggregate variables into each actuator variable, and *encodes* the actual output state of each actuator variable into a set of aggregate co-variables to provide the transform conditioning signal. The coded unitary system can only by a *unitary system,* while the digitized duplex system can be either a *unitary or synchronous diverse system.*

Trinary digitized duplex empirical controllers

The digitized duplex empirical controller is the basic building block for most empirically controlled systems. As shown earlier, the trinary digitized system is more effective than a binary digitized system and is the most effective method of digitization.

Scalar trinary digitized bidirectional output matrix

Another value can be added to each aggregate variable in the binary digitized bidirectional output matrix shown earlier, forming the *scalar trinary digitized bidirectional output matrix* in Figure 21.7.

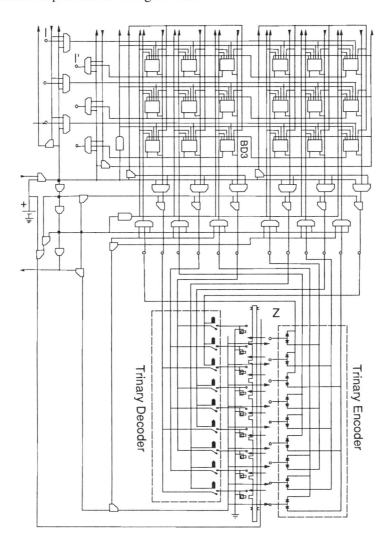

Figure 21.7. A *scalar trinary digitized bidirectional output matrix* is the most desirable design for an empirical bidirectional output matrix.

The scalar trinary digitized bidirectional output matrix is not used by itself unless it is a part of a duplex controller or a direct series network of empirical units (see Chapter 22). This is because there is no need to back select input co-terminals in a standalone controller. The trinary digitized output matrix shown earlier, which does not select in both directions, can be used as a standalone empirical controller with a single input variable and digitized output variables.

Trinary digitized scalar duplex empirical controller

A trinary digitized CE cell input matrix can be connected to a trinary digitized bidirectional output matrix, forming the *scalar trinary digitized duplex empirical controller* shown in Figure 21.8.

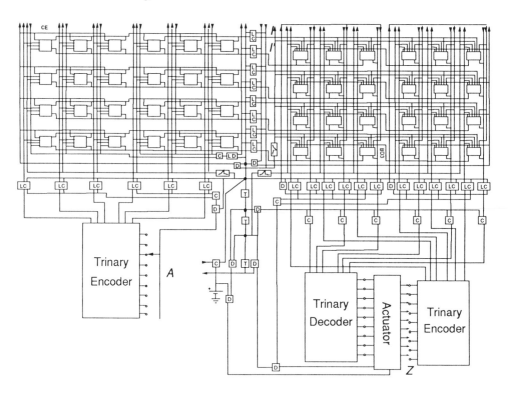

Figure 21.8. The *scalar trinary digitized duplex empirical controller* is the preferred design for any scalar digitized duplex empirical controller.

The scalar trinary digitized duplex empirical controller can learn to respond in more different ways to more different sensed conditions using fewer memory cells than any other system, including the binary system shown earlier. Though the actual input and output variables are scalar quantities, the digitizing process requires multivariable input and output matrices. This requires that a Type 3 bidirectional output matrix be used rather than the Type 2 bidirectional output matrix designed specifically for a single undigitized actuator variable.

Application of a scalar trinary digitized duplex empirical controller

The most desirable design for scalar applications involving a single input variable and a single output variable with more than a few values is the trinary digitized duplex empirical controller. For example, the best design for the *empirical ship auto-pilot* discussed earlier is the trinary digitized duplex empirical control system shown in Figure 21.9.

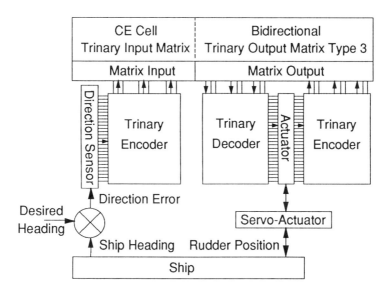

Figure 21.9. An *application of a scalar trinary digitized duplex empirical controller* is as an empirical ship auto-pilot.

The scalar trinary digitized duplex empirical controller with three aggregate sensor and actuator variables can be taught to respond to any one of 27 different values of a direction error sensor and produce any one of 27 different rudder servo-actuator positions using only 18 memory cells for each sensor/actuator relation. The number of these sensor/actuator relations that can be produced by the duplex empirical controller is determined by the number of values of the intermediate variable connecting the input and output matrices. This number can be estimated fairly accurately for any given control task, and the duplex controller designed accordingly.

Trinary digitized multivariable duplex empirical controller

Any number of additional trinary digitized sensor and actuator variables can be added to the trinary digitized duplex empirical controller shown, forming the *trinary digitized multivariable duplex empirical controller* shown in Figure 21.10.

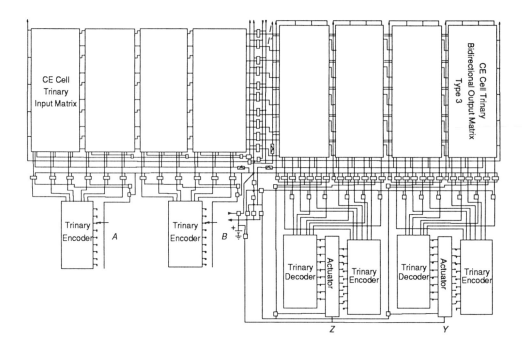

Figure 21.10. A *trinary digitized multivariable duplex empirical controller* can learn to produce any specific combination of values of a set of digitized actuator variables upon the occurrence of any specific combination of values of a set of digitized sensor variables.

The trinary digitized multivariable duplex empirical controller is the most desirable design for any unitary application having multiple medium-resolution sensor and actuator variables that requires a relatively few transitions in its line of behavior.

Application of a trinary digitized multivariable duplex empirical controller

The trinary digitized multivariable duplex controller shown can be used in the multivariable *ship direction control system* shown in Figure 21.11. The trinary digitized multivariable duplex empirical controller can establish some predetermined number of unitary transitions between multiple sensor and actuator variables while using a minimum number of memory cells. It can also establish *synchronous diverse relations* wherein some of its sensors can control some of its actuators at one time, and other sensors can control other actuators at another time. However, if multiple diverse relations between individual sensors and actuators must be carried out asynchronously, additional intermediate variables must be added to the duplex empirical controller, forming the digitized universal duplex empirical controllers (UDEC) shown in the next section.

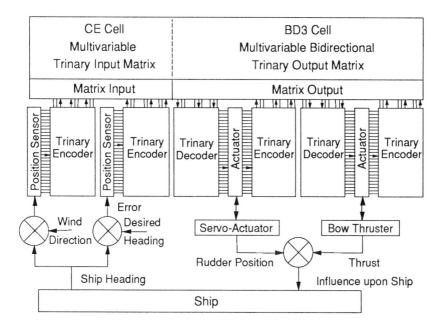

Figure 21.11. An *application for a trinary digitized multivariable duplex empirical controller* is as a ship direction control system with two or more digitized sensor variables and two or more digitized actuator variables.

Digitized universal duplex empirical controllers

A control system with multiple sensors and actuators may have to connect one or more sensors to one or more actuators at any time and without regard for the values of the other sensors. These asynchronous diverse relations can be produced by the digitized universal duplex controllers shown in this section.

CE cell binary digitized monolithic input matrix

We have seen how a CE cell monolithic control matrix can connect multiple input variables with multiple output variables. We have also seen how these multiple input and output variables can be used as aggregate input and aggregate output variables in a CE cell digitized monolithic control matrix. The digitized sensor variable can also be used to control two or more actuator variables, as shown in the *CE cell binary digitized monolithic input matrix* in Figure 21.12.

Empirical Machines

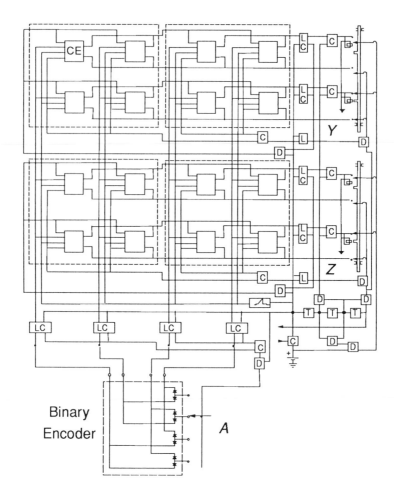

Figure 21.12. A *CE cell binary digitized monolithic input matrix* uses a single (scalar) digitized sensor variable to control two or more actuator variables.

Certain conformations of the CE cell binary digitized monolithic input matrix allow each input state to produce a unique combination of positions of the actuator variables. Other conformations can cause the position of one actuator to remain constant for all of the input states while the position of the other actuator is determined by a limited number of the input states.

CE cell binary digitized multivariable monolithic input matrix

More sensor variables can be added to the CE cell binary digitized monolithic input matrix shown, forming the *CE cell binary digitized multivariable monolithic input matrix* shown in Figure 21.13.

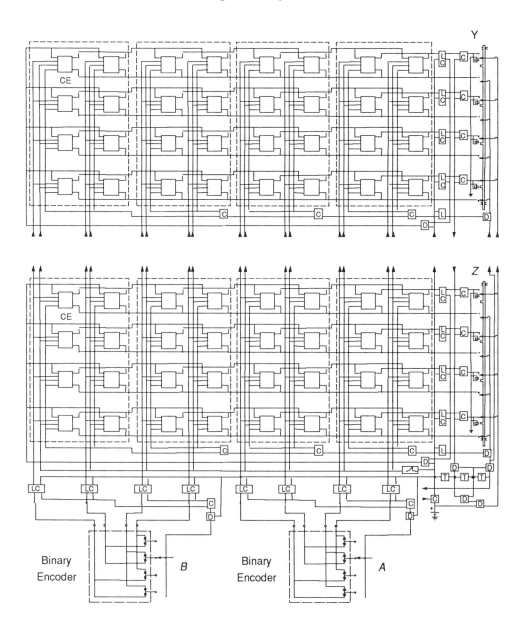

Figure 21.13. A *CE cell binary digitized multivariable monolithic input matrix* uses two or more digitized sensor variables to determine the position of two or more actuator variables.

The number of input states in the active line of behavior of the machine shown must be limited to the number of possible output states of each actuator.

BD3 cell binary digitized monolithic bidirectional output matrix

More input variables can be added to the BD3 cell binary digitized multivariable bidirectional output matrix shown earlier, forming the *BD3 cell binary digitized monolithic bidirectional output matrix* shown in Figure 21.14.

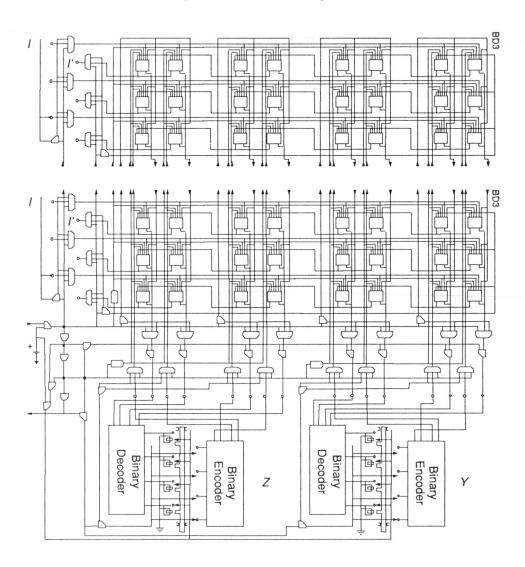

Figure 21.14. A *BD3 cell binary digitized monolithic bidirectional output matrix* can use any one of two or more input variables to select values of one or more binary digitized actuator variables.

In some cases, each input variable may produce an output at a different actuator variable regardless of the value of the other input variable in an asynchronous diverse manner. In another case, each combination of values of both input variables may produce a unique combination of values of both of the actuator variables in a unitary manner.

Binary digitized universal duplex empirical controller

The CE cell binary digitized multivariable monolithic input matrix can be connected to a BD3 cell binary digitized multivariable monolithic bidirectional output matrix shown, forming the *binary digitized universal duplex empirical controller* shown in Figure 21.15.

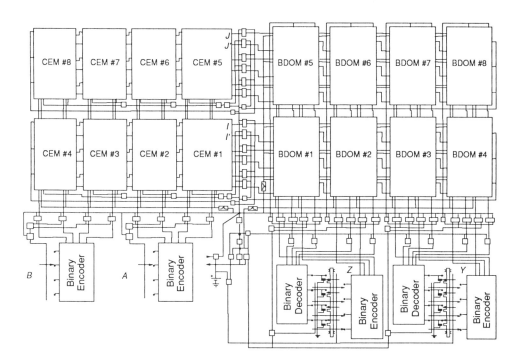

Figure 21.15. A *binary digitized universal duplex empirical controller* can establish unitary or asynchronous diverse relations between multiple digitized sensor and actuator variables.

The binary digitized universal duplex empirical controller can establish relations between any one of many more combinations of values of its sensor and actuator variables using fewer memory cells than a coded unitary or undigitized multivariable controller.

Trinary digitized universal duplex empirical controller

A CE cell trinary digitized multivariable monolithic input matrix can be connected to a BD3 cell trinary digitized multivariable monolithic bidirectional output matrix, forming the *trinary digitized universal duplex empirical controller* shown in Figure 21.16.

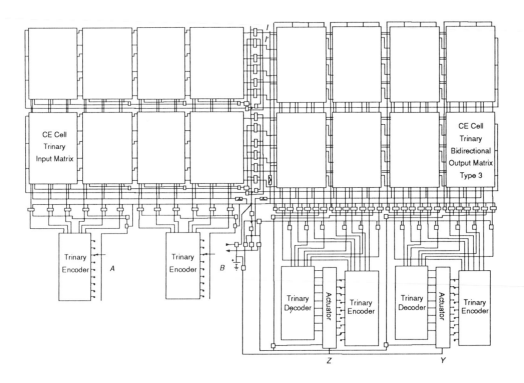

Figure 21.16. A *trinary digitized universal duplex empirical controller* can establish unitary or asynchronous diverse relations between multiple digitized sensor and actuator variables using fewer CE cells than a binary digitized universal duplex controller with the same sensor and actuator resolution.

The trinary digitized universal duplex empirical controller is the most versatile and useful standalone empirical controllers. Its input and output variables can be used together as a unitary set of variables, or each can stand alone as separate diverse variables. Its input and output variables can also be configured together to represent space vectors or be mechanically digitized to represent a set of high-resolution sensor and actuator vectors, as shown in the next section. Though the binary digitized universal duplex empirical controller is shown for the sake of simplicity, a trinary digitized controller is the preferred embodiment.

Application of a trinary digitized universal duplex empirical controller

A trinary or binary digitized universal duplex empirical controller can be used in the multivariable *ship direction control system* shown in Figure 21.17.

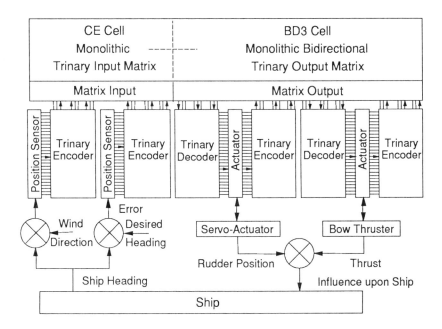

Figure 21.17. An *application for a trinary digitized universal duplex empirical controller* is as a ship direction control system with two or more digitized sensor variables and two or more digitized actuator variables.

The trinary digitized universal duplex empirical controller can learn to establish a predetermined number of unitary transitions between multiple sensor and actuator variables using a minimum number of memory cells, and individual sensors can learn to control the behavior of individual actuators if multiple diverse relations must be carried out asynchronously. The trinary digitized universal duplex empirical controller is one of the most useful standalone empirical controllers.

Mechanically digitized duplex empirical controllers

There are several mechanical devices that naturally encode a variable into component (aggregate) variables and decode a set of component (aggregate) variables into a single variable. These devices can be connected to a duplex, monolithic, or universal empirical controller, forming high-resolution empirical control systems.

Duplex empirical controller connected to vectors

A duplex empirical controller can be connected to *vectors* when its input variables
are connected to sensors that are components of a vector and its output variables are
connected to a vector actuator with position feedback, as shown in Figure 21.18.

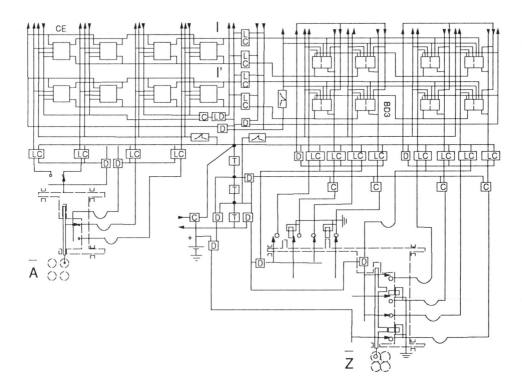

Figure 21.18. A *duplex empirical controller connected to vectors* can act like a unitary controller that transforms each combination of values of its input component variables into a specific
combination of values of its output component variables.

If a specific output vector must be produced by a specific input vector, the duplex
controller can use the values of the components of the input vector to produce a
specific set of values of the components of the output vector in a unitary manner. If
the values of one component of the input vector are sufficient to produce the values
of one component of the output vector, the duplex controller can establish a diverse
relation between these component variables while holding constant values of the
other component variables. If each component can be independent of the other
components, a universal duplex controller can be used with an intermediate variable
dedicated to each diverse transformation.

Duplex empirical controller with mechanical binary digitized variables

Mechanical binary digitized input variables can be connected to mechanical binary digitized output variables by means a multivariable duplex controller, forming the *duplex empirical controller with mechanical binary digitized variables* shown in Figure 21.19.

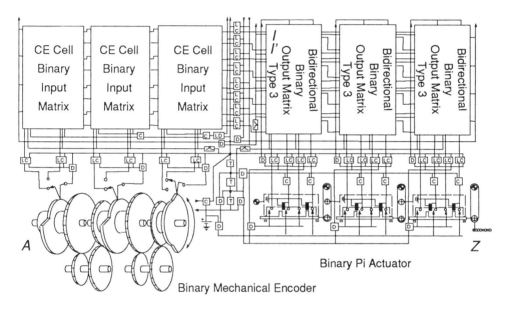

Figure 21.19. A *duplex empirical controller with mechanical binary digitized variables* can deal effectively with sensors and actuators having higher resolutions than digitized systems using electrical encoders and decoders, multiposition sensor switches, and vernier actuators.

The binary multivariable duplex system shown can produce any one of eight possible output positions for any one of eight input positions, or it can produce a generally large or small displacement for a generally large or small input state by establishing a diverse relation between the most significant places of the sensor and actuator variables while holding constant values of the least significant places. In the configuration shown above, it can produce four of these transitions.

If the number of input and output aggregate variables were increased to four, it could produce any one of 16 output positions for any one of 16 input positions, providing a resolution of 6.25%. The resolution of the input and output variables of a digitized duplex empirical controller increases dramatically as the number of aggregate variables increases without creating the need for an overwhelming number of new memory cells. For example, a binary digitized duplex empirical controller with eight sensor and eight actuator aggregate variables can identify any one of 2^8 = 256 different sensed conditions and produce any one of 256 different responses using only 32 cells per transition. An undigitized, nonduplex matrix would require $256 \times 256 = 65,536$ CE cells to establish just one specific transition although it could produce any 256 of these 65,536 different transitions.

If a system has two or more mechanically digitized sensors and actuators, a universal duplex controller may be required to handle any simultaneous diverse relations that could occur among these variables.

Duplex empirical controller with mechanical trinary digitized variables

A mechanical trinary digitized sensor and a trinary pi actuator with position feedback can be connected to a trinary multivariable duplex empirical controller, forming the *duplex empirical controller with mechanical trinary digitized variables* shown in Figure 21.20.

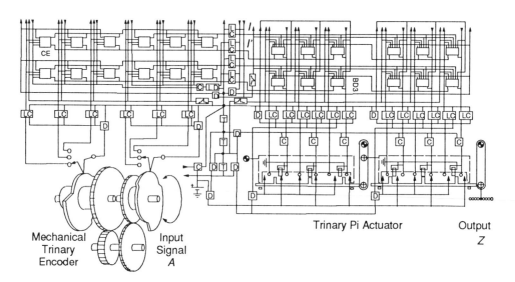

Figure 21.20. A *duplex empirical controller with mechanical trinary digitized variables* can deal effectively with sensor and actuator variables having higher resolution than is possible in a system with an electrical encoder and decoder with a single multiposition sensor switch or a single multiposition vernier actuator.

Mechanically digitized variables can have a far higher resolution than variables digitized by means of their discrete value elements. For example, a mechanically digitized sensor with 6 trinary aggregate variables can identify any one of $3^6 = 729$ different input values using only $3 \times 6 = 18$ contacts. A single multiposition switch would require 729 contacts, which is somewhat impractical. A trinary pi actuator with 6 aggregate variables can produce 729 different positions. A single vernier actuator with 729 poles would also be somewhat impractical. If a system has two or more sensors and actuators, a universal duplex controller may be required to handle any simultaneous diverse relations.

Single axis binary controller with 0.4% resolution

A sensor variable with 256 values (with a resolution of 0.4%) can be converted into 8 binary aggregate input variables by an 8-binade encoder, or a code made up of an "8-bit byte" can be generated for each input state as shown below. An actuator variable with 256 discrete output positions can be created by connecting 8 binary aggregate output variables to eight binary fluid cylinders connected in series, with each cylinder progressively one-half the length of the one to which it is connected, creating the 8-binade pi actuator with 256 possible output states as shown in Figure 21.21.

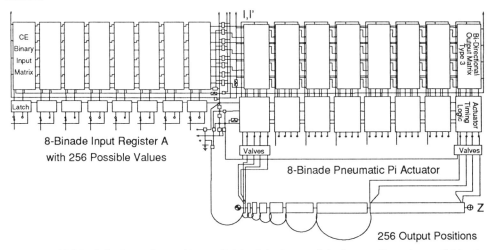

Figure 21.21. A *high-resolution binary digitized duplex empirical controller* using an 8-binade input code and an 8-binade fluid actuator can be taught to produce any one of 256 discrete output positions for any one of 256 possible input codes using only 32 memory cells per transition.

This duplex system can have up to 256 intermediate values. If all of the possible 256 transitions are required in the control task, the basic single-variable matrix would require 65,536 CE cells, while the binary digitized duplex system shown would required only 8,192 memory cells.

In most practical applications, fewer than 10% of the possible transitions may be required. This would provide a line of behavior having 25 transitions, which is complex behavior for a system with a set of single input and output variables. The binary digitized duplex system with 25 transitions would require only 800 memory cells.

This empirically controlled machine could be used as an automatic speed or damper controller. If another set of 8 output variables are included in the bidirectional output matrix, the controller can be used to position an *x/y* table to any 25 of 65,536 different possible locations within a 0.4% resolution using only 1200 conditional empirical memory cells.

If the system has two sensors and two actuators, a universal duplex controller may be required to handle two asynchronous diverse relations. This would double the size of the controller to 2400 memory cells.

High-resolution trinary digitized duplex empirical controller

Better results can be achieved with a trinary digitized duplex empirical controller than the binary digitized duplex empirical controller shown. For example, a trinary digitized duplex empirical controller with 5 aggregate input variables and 5 aggregate output variables can learn to identify any one of 243 different input states and produce any one of 243 different output states using only 30 memory cells per transition compared to the binary digitized duplex empirical controller with 8 aggregate input variables and 8 output variables, which requires 32 memory cells per transition to produce 256 different input and output states. The trinary duplex empirical controller can be used to position five trinary pi actuators from a 5-trinade input code, as shown in Figure 21.22.

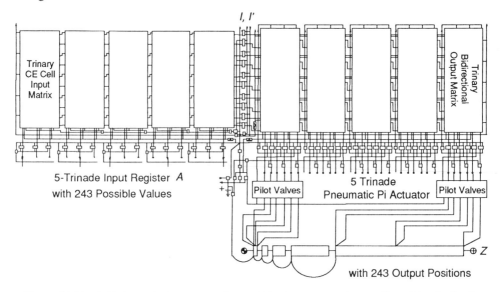

Figure 21.22. A *high-resolution trinary digitized duplex empirical controller* with a 5-trinade input matrix and a 5-trinade output matrix has an input/output resolution of 0.4%, which can meet the needs of most motion control applications.

Because the trinary duplex empirical controller can produce a greater number of different transitions for a given number of memory cells than a binary duplex empirical controller, it is the preferred design for most empirical duplex systems. Though trinary (three-position) sensor switches and trinary (three-position) air cylinders are slightly more complex than binary (two-position) switches and fluid cylinders, the reduction in number of sensors, actuators, and memory cells required for a given resolution may more than offset the additional cost of the trinary sensors and actuators.

High-resolution trinary digitized universal duplex empirical controller

If a mechanically digitized system has two or more sensors and actuators, a universal duplex controller, as shown in Figure 21.23, may be required to handle any simultaneous diverse relations that may be required by the control task.

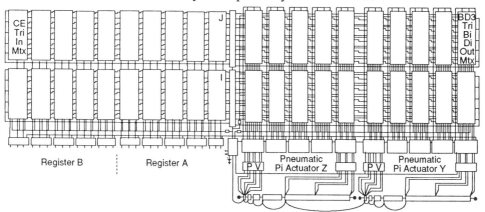

Figure 21.23. A *high-resolution trinary digitized universal duplex empirical controller* is the preeminent empirical control unit.

Each of the $243 \times 243 = 59,049$ possible combinations of values of the two trinary digitized sensors can produce any one of 59,049 possible combinations of values of the two trinary pi actuators, using only 60 memory cells per transition, or any one sensor can learn to control any one actuator through any one of the intermediate variables. The mechanically digitized trinary universal duplex empirical controller can be expanded to deal will the most complex tasks that can be handled by any single empirical unit.

Applications of standalone empirical units

Since the input and output variables of the universal duplex empirical controller can be connected to any kind of sensors and actuators, digitized by means of encoders and decoders, vectored sensors and actuators, and other mechanical devices, the universal duplex empirical controller can be used by itself as a *standalone controller* in almost any kind of control application. The universal duplex empirical controller (UDEC) is the basic empirical control unit because it can be used in any application having single or multiple input and output variables, because it can determine for itself *a posteriori* if its variables form unitary or diverse relations, and because it can be configured to deal most efficiently with aggregate representations of variables.

For example, a system with a two- or three-dimensional vectored actuator can be used to produce *handwriting* in which different combinations of movements of its multiple actuators produce different *script characters*. An empirical system can be trained to generate written text from *ASCII code* or *voice decoders* by an operator producing the script characters in response to the code or voice input. Vectored actuators can also produce *music* or *speech vocalizations*, do mechanical two-hand *signing* for the deaf and finger signing for the blind, and act as actuators for multiaxis robots and *multidimensional modeling systems*.

The matrices of controlled machines work most efficiently when they are presented with the same amount of input *variety* as output variety. This suggests that high-variety sensors may be needed to match the high-variety vectored actuators found in multiaxis robots. Thus, *auditory* and/or *vision sensor systems* may be needed to keep the vectored axes of robotic systems working to their potential.

Memory cell reduction

The number of empirical memory cells has been reduced in a duplex system with a single input and output variable having 8 input values, 8 output values and 4 transitions, from 64 cells to 48 cells in a binary digitized duplex system where each input and output variable is digitized into 3 binary aggregate variables. These 48 cells can be used to deal with input and output variables with 9 values each in a trinary digitized duplex system. These reductions in memory cells using digitized duplex controller come about without reducing the quality of behavior of the system. This leads to many applications for empirical control systems that require fewer memory cells than controllers using digital computers.

Ratio of CE cells in undigitized to digitized systems

The *ratio of improvement* due to digitizing a variable becomes even more dramatic as the number of values of a variable increases. This improvement is shown in Equation 21.1 by the ratio of the number of memory cells needed for an undigitized variable over the number of cells in a digitized system with the same resolution.

$$(x^n/x)n, \qquad\qquad (21.1)$$

where x is the number base and n is the number of aggregate variables. So the ratio of improvement of a trinary digitized variable with a resolution of 729 values requiring 6 aggregate variables as compared to an undigitized variable with 729 values is $729/18 = 40.5$. This says that an undigitized variable with a resolution of 0.14% would require 40.5 times as many memory cells as a trinary digitized system with the same resolution.

Increase in resolution of a trinary digitized system over a binary digitized system with the same number of memory cells

Since the number of memory cells required to digitized a single value of a variable is equal to $x \times n$, and the resolution of a digitized system is equal to x^n, the resolution of a binary system with 18 memory cells per value is equal to 2^n, where $n = 18/3 = 6$ and $2^6 = 512$ values. The *ratio of improvement* in the resolutions of a trinary digitized system over a binary digitized system, both with 18 memory cells per transition, is $729/512 = 1.42$. This represents a nearly 50% greater resolution of the trinary system over the binary system even in a relatively small system. Since the number of possible combinations of a set of variables increases according to the product of the number of values of each variable, a multivariable trinary digitized system with 8 variables could produce $1.42^8 = 16.5$ times the number of combinations of values, as compared to a binary digitized system with the same number of memory cells. These ratios continue to increase as the size of a matrix (number of memory cells per transition) increases.

Application of a trinary digitized multiaxis universal duplex empirical controller

A trinary digitized universal duplex empirical controller is the optimum design of an empirical controller with multiple input and output variables that may require asynchronous lines of behavior. Six to ten sensors and actuators may be used without difficulty in most applications. An 8-sensor variable, 8-actuator trinary digitized universal duplex empirical controller can be used to control the six degrees of freedom of an aircraft with two additional axes available for throttle and engine compressor controls. If each axis has a resolution of 0.4%, each output axes would require 5 trinary aggregate variables. The 8 digitized sensors and the 8 digitized actuators would require $2(8 \times 5 \times 3) = 240$ matrix terminals that require 240 memory cells per transition. If a transition occurs every 1/10 second and any flight condition can be stabilized within 25 seconds, then approximately 250 combinations of values of the sensor and actuator variables may be needed to provide adequate control. Therefore, the duplex matrix for this application would require approximately 64,000 CE cells. This number is not excessive if one considers that two conditional memory cells are comparable to a single binary static read mostly memory register. Even the earliest 8-bit *digital computers* used 64,000 8-bit bytes of random access memory (RAM) or approximately 512,000 "bits" (pairs of memory gates) plus a 360K-byte memory disc. This random access memory is equivalent to 1,024,000 absolute predetermined memory cells. Even this computer, with a predetermined program running at 1 MHz, might be hard pressed to control an 8-axis flight control system running at 10 transitions per second.

Limitations of a single universal duplex empirical controller

One limitation of a single universal duplex empirical controller is that it becomes less efficient in terms of the use of its hardware when dealing with diverse problems, particularly when it has many variables and they have many values. Two basic problems arise in the universal duplex system when it is required to be mostly diverse:

1. The presence of asynchronous variables increases learning time and decreases reliability.
2. The number of intermediate variables may have to increase as the number of sensor and actuator variables increases to handle the greater number of possible asynchronous relations. Thus it becomes disproportionally more costly as the number of sensor and actuator variables increases.

Separate or diverse relations are hard to establish in a single unitary system because it has to establish a unique response for each combination of input variable/values even when combinations are of no consequence. A large unitary system may never learn all the unnecessary combinations of the diverse behavior it may encounter.

When a multivariable matrix forms diverse relations, it converts that part of the matrix connected to variables that are not involved in the diverse states into conductors. It does so by increasing the sensitivity of these memory cells so they will conduct upon any value. This part of the matrix is saying, in a sense, "anything goes." It takes time and experience for the CE cells to establish this conformation, and this conformation reduces the ability of the matrix to form unitary relations later in that part of the matrix.

As mentioned before, some variables have no bearing upon behavior. They have been called hidden, uncontrolled, impertinent, or random variables. A coded unitary empirical system has to encounter all values of these random variables and learn how to deal with them to produce accurate transitions. Monolithic and universal duplex controllers have to learn to ignore these variables.

With an increase in the number of sensor and actuator variables, there is a greater chance that more asynchronous relations may be required. Thus, the number of memory cells in a universal controller may have to increase according to the product of the number of input and output variables, like a monolithic matrix.

Solution to the limitations of a single standalone unit

There are certain kinds of tasks that individuals cannot do well by themselves. For example, an individual can be in only one place at a given time. Thus, most people find it difficult to be at their place of employment and take care of a sick child at home. An individual may not be able to carry out two diverse tasks even at the same location. For example, a salesperson may find it difficult to deal with two or more customers at a given time. The problem that a single system has with diverse and random variables and the size penalty of a single large system can be solved by breaking (*fragmenting*) the single universal controller into a network of smaller universal controllers. This allows diverse behavior to occur in separate units, reduces the number of random variables in a given unit, reduces the number of intermediate variables in each unit, and may reduce the total number of CE cells in the system. The next part in this book shows how two or more units can deal more effectively with a large set of variables than a single unit working alone.

PART IV
NETWORKS OF EMPIRICAL MACHINES

So far, we have viewed an empirical control unit as a standalone machine acting by itself on a set of sensor and actuator variables. However, two or more control units can be connected to the same or similar set of variables. These machines also can be connected to each other. This provides the opportunity to create *separate pathways* for diverse behavior and for multiple units to work together as a single more complex assembly of units. These networks of empirical units may produce behavior that cannot be achieved by a single unit acting alone and may do so with fewer memory cells than a single unit.

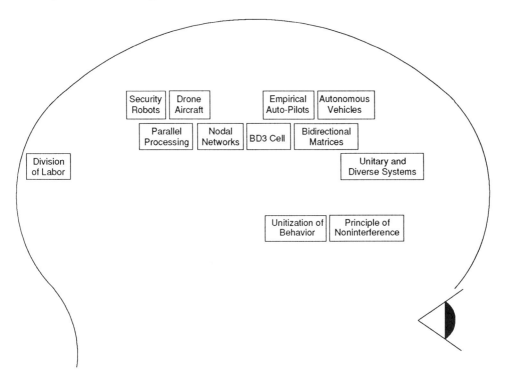

Part IV shows how empirical units can work together to produce useful behavior.

Chapter 22
Parallel Networks of Empirical Machines

It may be difficult or impossible to decide *a priori* whether a system is made up of a set of unitary or diverse variables. It may be necessary to create a network of controllers that can figure out *a posteriori* how to define its input and output states. A network may be more successful in organizing itself into one form or the other than a single unit because it contains more opportunities to create separate pathways for diverse behavior without losing its ability to produce unitary behavior. When two or more units are directly connected to the same variables, they are said to be *connected in parallel*. If two or more units are connected directly to the same actuator, they are said to be *superimposed* upon that actuator, and the actual outputs of these units must assume the same value in a given transition. If the output of an actuator is equal to the sum of the influences of two or more units, these units are said to form an *additive* connection with that actuator.

Multiple units connected to a single sensor and actuator variable

When two or more controlled units are connected to the same sensor and actuator variable, each unit may deal effectively with the diverse parts of the control task. For two or more units to be used effectively on a control task, a *division of labor* must occur that allows each unit to act without interfering with the behavior of other units. If the units divide the task appropriately, they may act together as if they were a single unit.

Operator and a unit superimposed upon one input and output variable

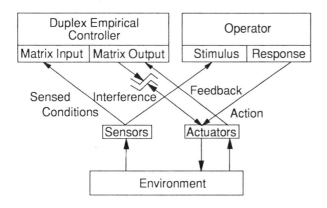

Figure 22.1. A dominant operator can cause the behavior of an empirical controller to change until it produces the same behavior as the operator.

We have seen how an empirical control unit and an operator can be superimposed upon a single set of sensor and actuator variables to produce an empirical ship auto-pilot. If the helmsman can cause the behavior of the system, the behavior of the empirical auto-pilot will change until it produces the *same behavior* as the helmsman. If the operator is physically stronger than the actuator controlled by an empirical unit, the behavior of the operator is consistent, the empirical controller has a sufficient number of intermediate states to handle the required number of transitions, and it has the correct time delays between input conditions and output events and between transitions, the behavior of the empirical controller will change until it corresponds to the behavior of the operator (Fig. 22.1).

Remote connection between two or more units and a common sensor variable

There are two different ways in which two or more control units can be connected to a common sensor. In one case, each unit contains a set of *input connect terminals* that are connected to some or all of the sensor terminals of a common sensor, as shown in the *remote connection* in Figure 22.2.

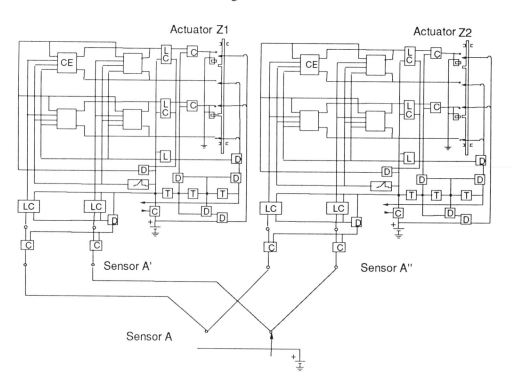

Figure 22.2. In a *remote connection between two or more units and a common sensor*, all of the units can perceive the same value of a given sensor variable in each transition cycle.

During each transition cycle, the input latch of each unit disconnects its own input connect terminals when a sensed value is detected. The sensor is left unaffected, allowing any other unit to receive a sensed value in each transition cycle. The empirical ship auto-pilot and operator system shown previously is an example of a remote connection between two or more control units and a common sensor variable since both the empirical controller and the operator can perceive each value of the direction error sensor regardless of the other control unit.

Intimate connection between two or more units and a common sensor variable

In another case, two or more units can be connected to a common sensor variable so that the input latches of the set of units are connected to the sensor disconnect of that sensor, as shown in the *intimate connection* in Figure 22.3. In this case, only one unit can perceive a value of the sensor in a given transition cycle.

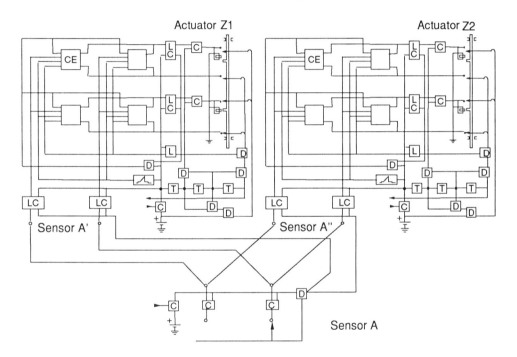

Figure 22.3. In an *intimate connection between two or more units and a common sensor*, only one unit receives a value from the sensor in a given transition cycle.

When there is an intimate connection between two or more units and a common sensor, the first unit that is turned on by the transition cycle timing signal receives a value from the sensor because its latch circuit disconnects the power to the sensor so that no other units can receive a value. In an intimate connection, changes in the transition cycle timing signal to each unit and/or the phase of the transition cycle timing signals will cause different units to perceive the sensor at different times. Thus, the timing of transition cycle signals determines which unit perceives the sensor variable when these units form an intimate connection with a sensor. *Brain waves* may serve the same purpose as transition cycle signals by controlling the starting time of transition cycles in different brain units. If this is the case, changes in brain waves or changes in phase of brain waves may cause different brain units to perceive a given sensor. Thus, changes in brain waves may cause changes in consciousness and/or perception. (See discussion of brain waves in Part V.)

Two or more units superimposed upon the same actuator

There are also several ways in which two or more control units can be connected to a single actuator. We have seen how an operator and a control unit can be *superimposed* on a rudder actuator in a ship auto-pilot. In this case, the variable representing the output of the operator, the control unit, and the rudder are *clamped* together so that they must have the same value whenever they are connected.

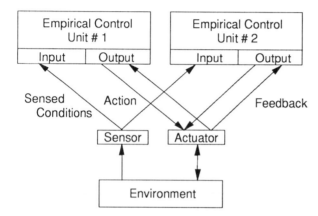

Figure 22.4. Empirical units connected to the same actuator, will receive the same transform conditioning signal in each transition cycle.

When the actuators of two or more units are superimposed upon a single actuator variable, as shown in Figure 22.4, the value of the actuator variable is determined by the *dominant control unit* (the unit that can produce the greatest force). If the actuators of both control units operate like a servo-system with *negative inverse feedback,* which produces the greatest force at its set point, then the first unit to achieve its set point will dominate the behavior of the other units. This is the basis of the dominance of small changes, which says that small changes in behavior are more likely to prevail than gross changes in behavior. If more than two units are superimposed upon a single actuator, groups of units can work together to produce a specific value of the actuator variable. This is the basis of cooperation in which the behavior of units that agree on specific behavior is more likely to prevail than the behavior of units that do not agree. (See laws of empirical behavior in Part V.)

Output as a sum of the position outputs of two or more units

The actuators of two or more controllers can also be connected to a single actuator variable by means of a mechanical *differential*, as shown in Figure 22.5, so that the value of the actuator variable is equal to the sum of the positions produced by each controller.

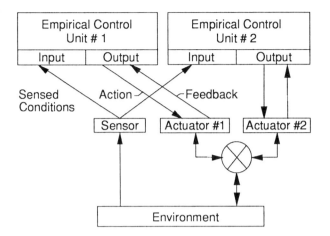

Figure 22.5. Units connected to a differential produce an output position equal to the sum of the positions of each unit.

Gearing between the actuators of each unit and the differential can cause one unit to have a greater influence upon the output position of the differential than the other unit. This gearing also influences the output force produced by the actuators of each control unit. The greater the position influence, the less is the force influence. Thus, the behavior of controllers that produce small changes will prevail over controllers that produce large changes in behavior in each transition.

Division of labor based upon value field selection

An empirical ship auto-pilot and an operator can be connected to a rudder by means of a differential that sums their output values, as shown in Figure 22.6.

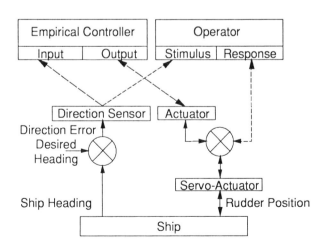

Figure 22.6. A division of labor based upon value field selection can be produced by a controller/operator team if one unit is made active by one set of sensed values and another unit is made active by another set of sensed values.

The empirical ship auto-pilot may be able to deal effectively with the small rudder corrections needed to maintain a straight course. However, if a new course is required in a totally new direction, the empirical auto-pilot may not be able to respond appropriately to the large direction error value. In this case, the operator may have to control the ship until the error is reduced to some smaller value.

When the empirical controller and the operator divide the control task according to the value of the controlled variable, they are practicing a *division of labor* based upon *value field selection*. In the example given, the empirical controller can learn to deal with the small errors and ignore the large errors while the operator deals with the large errors and does not get involved in small error behavior. This reduces the number of intermediate values required in a duplex empirical controller and reduces the workload of the operator.

Binary sigma actuator

Two or more controllers can also be connected to a single actuator variable by connecting the vernier actuators of each unit in series, forming the sigma actuator shown in Figure 22.7. In the case of the binary pi actuator shown earlier, the output of each vernier actuator is multiplied by a coefficient that is twice as large as the one preceding it, and the output of these units are summed by connecting them in series. In a *sigma actuator*, all of the coefficients of all of the vernier actuators have the same value. Thus, the number of output values V of a sigma actuator is shown in Eq. 22.1.

$$V = x \times n - (n - 1), \qquad (22.1)$$

where x is the number of values of each actuator and n is the number of actuators in a sigma actuator.

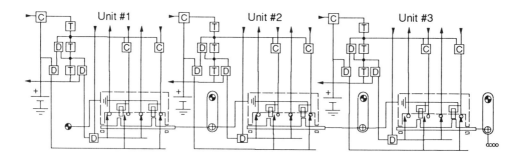

Figure 22.7. A *binary sigma actuator* with three vernier actuators can produce four different output positions.

The output of the actuator may be influenced or restricted by the environment or an operator. The transform signals sent back to each unit are determined by how these factors affect each connected actuator.

Trinary sigma actuator

Another actuator coil can be added to each binary actuator in the sigma actuator shown, forming the *trinary sigma actuator* shown in Figure 22.8.

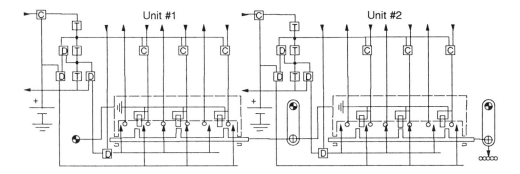

Figure 22.8. A *trinary sigma actuator* with two actuators can produce five different output positions.

Contrary to a pi actuator, a sigma actuator can produce more output positions for a given number of actuator coils as the number of coils per actuator increases. For example, the binary sigma actuator with six coils can produce four output positions, and the trinary sigma actuator with six coils can produce five output positions. The number of coils of each actuator may be increased to match the variety of each control unit connected to the sigma actuator. If very high resolution is required, each actuator in a sigma actuator may be a pi actuator.

Two or more empirical units connected to a sigma actuator

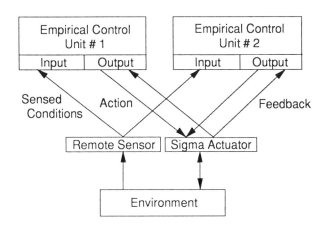

Figure 22.9. With *empirical units connected to a sigma actuator*, some or all of the resulting behavior can come from one unit.

Two or more control units can be connected to a single *remote sensor* and a single *sigma actuator,* as shown in Figure 22.9. Both units can work together to produce behavior, or the behavior can be produced by one unit if the other unit produces one (constant) output value for all values of its input variable. If they work together in a *coordinated* manner to produce behavior, the *range* and *variety* of the two systems may be greater than the range and variety of one unit acting alone.

Dominance of small changes in behavior in a sigma actuator

Each electromechanical vernier actuator in the sigma actuator shown produces a force that is inversely proportional to the square of the difference between the actuator position called for by the controller and the actual armature position, and this force acts in the opposite direction of the error. This is called a *negative inverse square force.* Thus, a component actuator with a small error produces a much larger force than a component actuator with a large error. Within a sigma actuator, the component actuators that have a small error at the beginning of the action phase of a transition cycle are more likely to achieve their selected positions, while the component actuators that have to make an extended reach are more likely to be thrown off their target by forces imposed by outside factors. The smallest error and the greatest force occur when the selected position of a component actuator is the same as its position in the previous transition. Thus, *constant behavior* or *small changes in behavior* are *more likely to be incorporated* in empirical units connected to a sigma actuator than *gross changes* in behavior.

Empirical auto-pilots connected to a single sensor and actuator

Two or more empirical auto-pilot controllers can be connected to the same direction error sensor and a single rudder servo-actuator through a sigma actuator that sums their individual outputs, as shown in Figure 22.10.

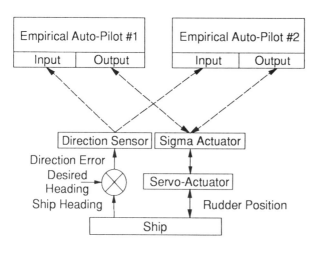

Figure 22.10. Superimposed empirical auto-pilots may have to learn to act at different times according to the value of the controlled variable.

Both units can be active for each value of the direction error sensor, or they can establish a *division of labor* based upon value field selection in which only one unit is made active in each moment in time. If each unit can produce only a limited number of transitions, or the range of each unit is insufficient to cover the complete range through which the rudder must move, the units must divide the control task to produce satisfactory behavior. For example, one unit may select a constant output value at the high end of its range for each sensed input at the high end of its range, and the other unit may produce a set of output values for each sensed condition in the high end of its range. This behavior causes the rudder to be positioned at a higher end of its range by two control units that have fewer memory cells than a single controller that can produce this behavior by itself.

Influence of the operator in a system with multiple controllers

If a helmsman restricts the output of a rudder connected to other units by sigma actuators, it may directly affect the actual output values of one or both of the connected units. If both units attempt to produce output values that sum to a value greater than the value imposed by the helmsman, one or both connected actuators may be restricted. The units have two options; one must change, or both must change their output. The solution that requires a minimum learning time and a minimum number of selecting memory cells is for one to produce a constant output for every sensed input within this range and for the other to produce the outputs required in this range. This solution results in the *empirical selection of different value fields* by both units.

Strategies of empirical units connected to a sigma actuator

An empirical unit has three options in attempting to deal with an actuator variable connected to another unit by sigma actuators: It can *compete* with the other unit to produce all of the behavior, it can *cooperate* with the other unit to produce some behavior, or it can *abstain* from producing any behavior at that variable. The confidence level for behavior of a unit connected to a sigma actuator is increased if that behavior is carried out, and the confidence level of that behavior is decreased if it is not carried out. At first, all of the units connected to a sigma actuator may compete to produce behavior. Those units that produce behavior that can be carried out will dominate the behavior of the sigma actuator.

If the range of output of the sigma actuator requires the collective outputs of each unit, the units must cooperate to produce successful behavior.

If the range of motion of the sigma actuator can be produced by one unit, the other units are more likely to produce constant behavior (*null functions*) rather than actively participate in the behavior of the sigma actuator due to the *dominance of small changes in a sigma actuator*. When a unit produces constant output values, it essentially abstains from being an active participant in that line of behavior. By abstaining from behavior, a duplex unit preserves its intermediate values since that constant behavior can be written upon only one intermediate value. (See section on duplex systems.) The remaining intermediate values can be used later to produce active behavior if the need arises.

Multiple units connected to multiple sensor and actuator variables

Many applications have two or more input and output variables. It may be difficult or impossible to determine *a priori* whether these variables can be connected in unitary units or whether these variables should be kept separate in diverse units. Some means must be found that allows an empirical control system to organize itself *a posteriori* according to the organization of the task environment. When empirical control units are connected to two or more variables, they have the option of producing unitary behavior by dividing the control task according to the values of the control variables (value field selection) or of dividing the task into separate lines of behavior that are made up of separate sets of active variables (dispersion).

Two or more control units connected to two or more sensors and actuators

We have seen how a single universal duplex empirical controller (UDEC) can be connected to two or more sensor and actuator variables. It can produce a unique combination of values of its output variables for each combination of values of its input variables in a unitary manner, or it can produce a single value of one or more of its output variables for any value of one or more of its input variables in a diverse manner. However, ability of an empirical control system to create this unitary or diverse behavior is expanded when two or more universal duplex empirical control units are connected to two or more sensor and actuator variables (Fig. 22.11).

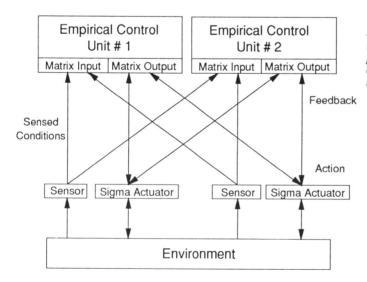

Figure 22.11. Empirical units connected to multiple sensors and actuators can form a unitary or diverse system.

If each combination of values of the sensor variables must produce a unique combination of values of the actuator variables in a unitary manner, then one empirical control unit can determine the behavior of the actuators, and the other unit can abstain from behavior by producing the same (constant) behavior for all sensed conditions. Moreover, one unit can contribute to the behavior of the other unit by augmenting the range of the actuator, or by contributing some required transitions.

If the values of a sensor must control the values of a particular actuator, then one unit can establish a diverse relation between that sensor and that actuator while the other unit supplies a constant value to that actuator. Both units can form a diverse relation between a sensor and an actuator if the other unit supplies constant output values to the actuator it does not attempt to control. In this case, the variables made active by the line of behavior of one unit differ from the variables made active in the line of behavior of the other unit, creating a separation (dispersion) of behavior among the two (or more) units.

An empirical auto-pilot and bow thruster system

We have seen how a single direction auto-pilot and crosswind compensation system can be used to hold a ship on a set course. In one case, each combination of values of the direction error sensor and the crosswind sensor may require a unique combination of values of the rudder and bow thruster. In another case, each sensor may need to control one or the other actuator. When a single universal empirical controller is connected to multiple sensors and actuator, as shown in Figure 22.12, it can determine for itself which form of organization produces the most consistent results.

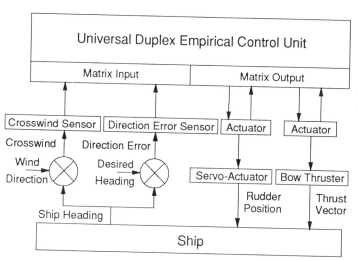

Figure 22.12. An *empirical ship direction control system* can establish a unitary or diverse relation between its sensors and actuators.

If one or the other sensor has no bearing upon the behavior of one or the other actuator, the controller can establish a *null transformation* between those variables. After a learning period, empirical memory cells (E cells) that cannot make consistent connections between values of these variables will be left in low-sensitivity states, making them less likely to select behavior than those E cells that connect values of the sensor that have a causative relation with an actuator.

Division of labor in a diverse system

There are opportunities to divide a control task when there are two or more control units connected to two or more sensor and actuator variables. In some cases, the variables made active by one line of behavior differ from the variables made active by another line of behavior. If separate units are connected to these separate sets of variables in a *diverse* manner, as shown in Figure 22.13, each unit can act independently based upon the values of each set of variables.

The most effective division of labor occurs when diverse variables are identified and connected to separate control units. The information required to operate a diverse system is much smaller than that required to operate a unitary system. For example, if each unit contains one variable, then the potential variety of the system is equal to the input variety (x) of each unit times the output variety times the number of variables or $x^2 \times n$. The potential variety in a unitary system is equal to the input variety x^n times the output variety of, say x^n, which is equal to $(x^n)^2$. This is much larger then the potential variety of the diverse system.

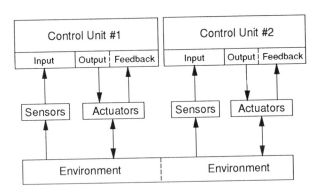

Figure 22.13. In a *diverse empirical system,* each control unit is connected to a separate set of variables.

Example of a diverse control system

The ship auto-pilot and the crosswind compensation system may work well as separate systems, as shown in Figure 22.14. The direction error sensor can be connected to the rudder servo-amplifier by one controller, and the crosswind sensor and the bow thruster can be connected by a second controller.

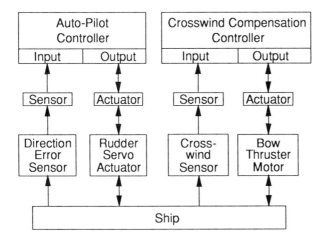

Figure 22.14. A *diverse empirical ship direction control system* can provide adequate control using smaller control matrices than a single multivariable control system.

In a diverse system, each unit can act upon its variables without directly interfering with the behavior of another unit. The units can act asynchronously without the behavior of one unit influencing the behavior of the other unit.

Example of a unitary control system

The ship auto-pilot and the crosswind compensation system may work well only when it is configured as a unitary system. The direction error sensor and the crosswind sensor must then be connected to the rudder servo-amplifier and the bow thruster by a unitary controller, as shown in Figure 22.15.

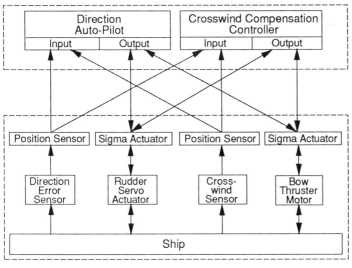

Figure 22.15. A *unitary empirical ship direction control system* may require that each combination of values of the sensor variables produce a unique combination of values of the actuator variables.

A single universal empirical controller can produce a unique combination of values of the rudder and bow thruster actuators for unique combinations of values of its direction error and crosswind sensors. So one or the other control unit may dominate behavior, or they may also divide the control task through value field selection or dispersion so that any one unit does not have to produce all of the behavior.

Self-organization of superimposed empirical controllers

It may be difficult or impossible to determine beforehand whether a unitary or diverse control system is required in a given application. If two or more empirical controllers are connected to the set of sensor and actuator variables, as shown in Figure 22.16, they can establish diverse relations between specific sensor and actuator variables, or they can work together to provide a unitary relation between the sensor and actuator variables.

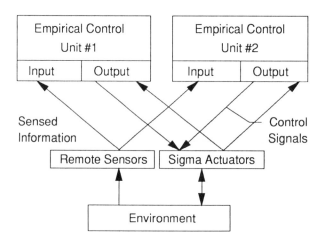

Figure 22.16. Superimposed empirical controllers will attempt to establish behavior that can be carried out.

The behavior of the connected units will change until the behavior of one does not interfere with the behavior of another. If both units have equal dominance, the behavior of both will change until they establish behavior that can be carried out successfully by both units. One possibility is that they will establish a single or unitary line of behavior. The other possibility is that each unit will attempt to produce separate or diverse lines of behavior.

Influence of the operator in a system with multiple variables

An operator can also influence the results produced by the multiple variables shown in Figure 22.17, causing the behavior of the empirical controllers to change until the *sum of their behavior* corresponds to the behavior of the operator.

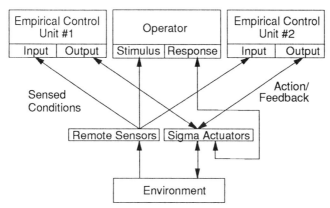

Figure 22.17. Two or more *empirical controllers and an operator* can be connected to two or more sensor and actuator variables.

The behavior of the empirical controllers will change until they produce behavior that corresponds to the behavior of the operator if the following conditions are met:

- The operator can dominate the behavior of the system.
- The behavior of the operator is reasonably consistent.
- Duplex empirical controllers have a sufficient number of input and output variables and a sufficient number of intermediate states to handle the required number of transitions.
- The controllers have the correct time delays between input conditions and output events and between transitions.

The operator can withdraw from the control task once the controllers produce the desired control behavior.

Diverse flight control system

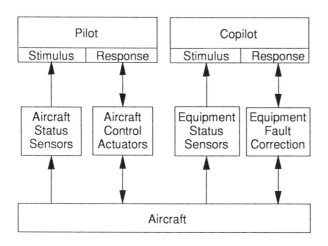

Figure 22.18. In a *diverse flight control system*, pilot and co-pilot operate upon different controls.

Two operators may choose to control separate variables within a given system, as shown in Figure 22.18. The pilot may choose to fly the aircraft, and the co-pilot may be responsible for equipment status and the correction of equipment faults.

Example of field selection

Though multiple units can be connected to the same set of variables, each unit may select a different set of values of these variables upon which to operate. For example, a student and instructor pilot may both operate upon the same set of flight controls (Fig. 22.19). Value field selection occurs when a student pilot maintains straight and level flight, requiring small error corrections, but the instructor takes over if the plane gets out of control, requiring large error corrections.

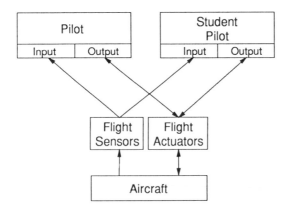

Figure 22.19. In a *control system with value field selection*, two or more units can act separately upon the same set of variables.

In some cases, one may act upon high values of the set of variables, and the other unit acts upon low values, as suggested. In other cases, each unit may act upon some unique set of selected values. In either case, units acting on the same set of variables can learn to produce behavior that is different from another unit if these two behaviors are not active simultaneously.

Behavior of parallel networks

A unitary system is formed when two or more separate controllers produce specific combinations of behavior that results in the successful behavior of the group. This can be achieved only if each unit learns to adjust its behavior according to the behavior of the other units.

Division of labor in a unitary system

In many cases, separate variables may be intrinsically tied together by the task environment. For example, there is an intrinsic relation between the flight of an aircraft and the activities of the pilot and co-pilot. The successful flight of an aircraft may depend upon faulty equipment being corrected or bypassed according to the flight pattern of the aircraft. For example, the co-pilot should attempt to fix the landing gear before pilot attempts to land the plane. Thus, the pilot and co-pilot are part of a unitary system (Fig. 22.20). To operate successfully, these two units must produce behavior that appears as if it were produced by a single unitary controller, not a diverse system.

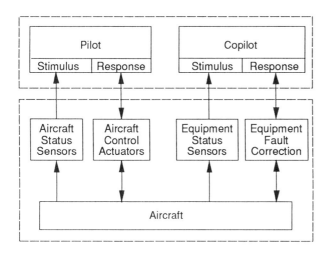

Figure 22.20. In the *division of labor in a unitary system*, the set of controllers must act as if there were a single unitary control unit acting alone.

When the pilot and co-pilot learn to divide the control task so that each supplies the right combination of the behavior needed to complete the unitary task successfully, and each learns to supply the information needed by the other to complete their tasks successfully, they form a unitary system made up of parts. To act appropriately, each unit must be able to act according to the values of the critical variables in the domain of the other units.

Division of labor in a system with dispersion

In other cases, the set of variables made active by one line of behavior differ from the set of variables made active by another line of behavior. In an *assorted configuration*, each unit is connected to separate variables, and each unit also shares some variables with the other units. If one controller is connected to the set of variables made active by one line of behavior, and another controller is connected to the set of variables made active by the other line of behavior, each controller can establish a separate line of behavior. Under these conditions of *dispersion*, each unit can establish its own line of behavior (Fig. 22.21) by operating upon a unique combination of variables in a unitary manner.

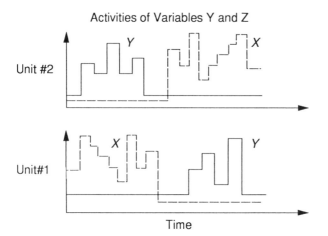

Figure 22.21. In a *system with dispersion*, each unit can select a unique combination of variables upon which to operate.

Many opportunities to establish an effective division of labor are created when each unit can select the variables upon which it operates.

Example of a system with dispersion

As stated, some variables in the environment may be associated with one specific line of behavior, while other variables may be associated with another line of behavior. Yet, another set of variables may be made active by both lines of behavior. For example, while the pilot of an airplane deals with the flight controls and the co-pilot deals with any equipment problems that may occur, both may have to share the responsibility of looking out for other aircraft (Fig. 22.22).

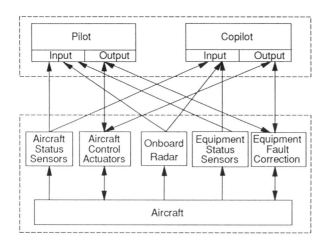

Figure 22.22. In a *control system with dispersion*, each unit can produce a different line of behavior even though that are connected to the same set of variables.

In this *assorted system*, where different units contain variables not present in other units, each unit may learn to produce a different line of behavior without creating interferences. For example, the pilot and co-pilot may learn to trade off their time to watch for other aircraft to keep from interfering with their other tasks. This division of labor based upon dispersion allows different tasks to be carried out by different units on a given set of variables.

Methods of intercommunication

In the assorted system shown, some method of communication must be found that allows each unit to know how to respond to variables connected to other units. For example, the pilot can know that the co-pilot is monitoring the onboard radar if the co-pilot operates a task assignment switch whenever the co-pilot does so. Thus, a unit can obtain information as to the status of variables in another unit if they share and update a common variable that acts like a *communal bulletin board,* as shown in Figure 22.23.

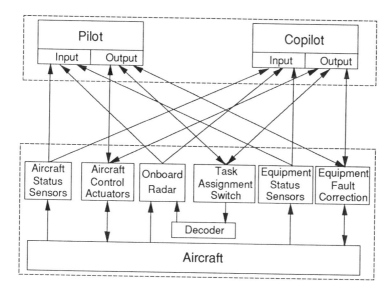

Figure 22.23. A *communal bulletin board* allows each unit to inform other units about the values of some of its variables.

Like a pilot and copilot, an empirical unit must learn to transmit and receive information as to the critical values of the variables that are in the domain of other units. In such a system, each unit may be much smaller than a single larger controller that must contain all of the variables. This communication process allows the formation of a *fragmented* empirical system made up of subunits that can create unitary or diverse behavior using far fewer empirical cells than a single large unit (SLU) acting alone.

Structured superimposed network of universal control units

Universal empirical control units can be arranged in a *structured superimposed network* (SSN) by superimposing units upon different groups of the input and output variables of the network, as shown in Figure 22.24.

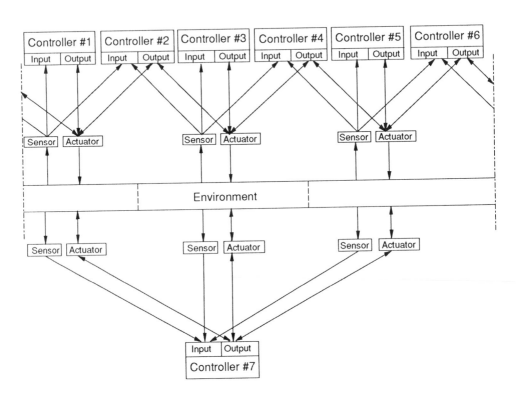

Figure 22.24. A *structured superimposed network of empirical units* can discover and execute unitary and/or diverse behavior.

This *composite configuration* leads to a great deal of competition among the different units connected to different groups of variables. The unit connected to the group of variables that produces to the most consistent results will dominate behavior because it will be the one most likely to select the desired behavior. This process allows the *composite network* (CN) to organize itself according to the task environment. As shown in Part I, the organization of government units in the United States is that of a structured superimposed network.

Chapter 23
Series Networks of Empirical Machines

Empirical control units can be connected in *series* and in parallel. When empirical units are connected in series, they can transform an input state into another state that is more appropriate for another control unit.

Directly connected series networks of empirical input and output control units

When two or more empirical control units are *directly connected in series*, the output variables of the unit connected to the sensor variables must be connected directly to the input variables of another unit connected to the actuator variables. The back-selecting process must extend from the unit connected to the actuator variables back to the unit connected to the sensor variables. This requires the design of a *bidirectional output controller* that can back select the input state most likely to produce an actual output state.

Empirical input and output units connected directly in series

To create a network of empirical units that are connected without intervening environments, the output of one unit must be directly connected to the input of another unit by means of at least one *intermediate variable*, as shown in Figure 23.1.

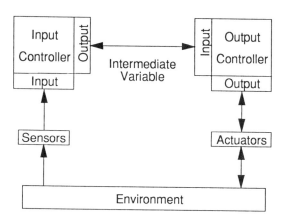

Figure 23.1. A multivariable input controller directly connected in series to a multivariable output controller has the same advantages as a multivariable duplex controller with a single intermediate variable.

The input controller can make some transformation on its sensed information to produce a more desirable input signal to the output controller. For example, one controller may act as a filter or an amplifier of sensed conditions for another controller. A *direct series connection* between two empirical units requires two different types of controllers: an input controller with sensors and no actuators, and a bidirectional output controller with actuators and no sensors.

Input controller

A multivariable input matrix can be used as an *input controller* with multiple input variables and a single output variable, as shown in Figure 23.2.

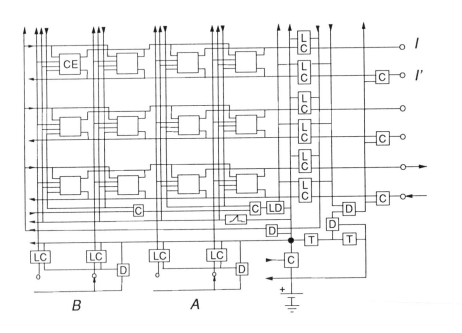

Figure 23.2. An *input controller* is connected to sensors but has no actuators.

An input controller has no actuators or actuator delay circuit. Its purpose is to be the input matrix of a duplex controller or the input controller in a directly connected series network.

Output controller

A multivariable output matrix can also be used as an *output controller* with multiple output variables and a single input variable, as shown in Figure 23.3. An output controller is not connected directly to sensors. Its purpose is to be the output matrix of a duplex controller or be the output controller in a directly connected series network.

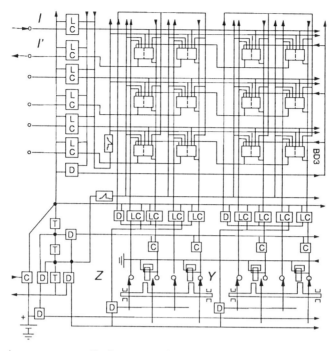

Figure 23.3. An *output controller* is connected to actuators, but has no sensors.

Duplex controller as a directly connected series network

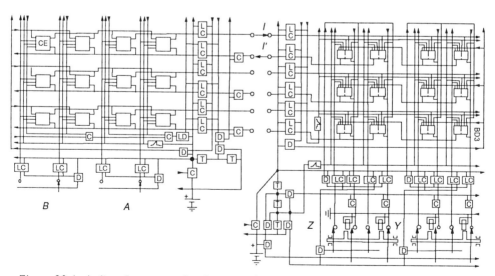

Figure 23.4. A *directly connected series network of empirical controllers* has the features of a duplex controller in that each value of the intermediate variable can represent a pertinent input/output relation.

An input controller can be connected directly to an output controller, forming a *directly connected series network* or a duplex controller, as shown in Figure 23.4. A directly connected series network of empirical controllers with a single intermediate variable can establish unitary or synchronous diverse behavior. (If the input and output controllers are far apart, the output terminals of the input controller can be encoded into a set of aggregate intermediate variables that can be decoded into the single input variable of the output controller. If the input and output controllers are close together, there is no need to digitize the intermediate variable.)

Series networks of monolithic empirical control units

Monolithic control units can also be connected in series, allowing multiple pathways for sensor and actual actuator information to flow between two or more empirical control units.

Monolithic empirical input controller

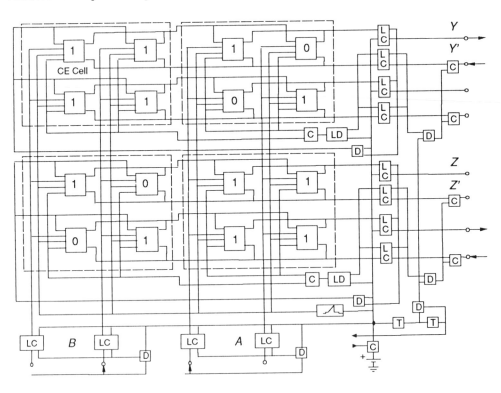

Figure 23.5. A *monolithic empirical input controller* can produce a particular combination of values of a set of intermediate variables according to the combination of values of a set of sensors.

If two or more diverse relations must be carried out asynchronously, then the *monolithic empirical input controller* shown in Figure 23.5 can be used in place of the multivariable input controller. Each terminal of the output (intermediate) variables of the monolithic input controller can be connected to a terminal of one a variable in another output controller, or the terminals of each variable can be connected to the terminals of input variables in different output controllers, forming multiple pathways for sensor information to flow to other output units. If a pathway made up of a set of connections has to travel a great distance, it may be desirable to digitize these intermediate variables into a set of aggregate variables that require far fewer conductors.

Monolithic bidirectional output controller

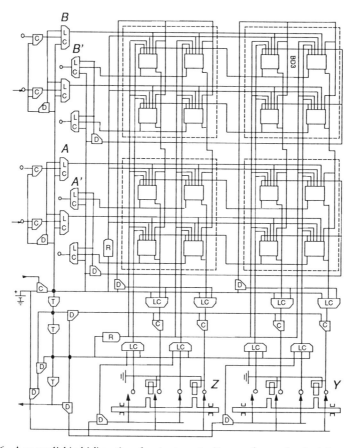

Figure 23.6. A *monolithic bidirectional output controller* can forward select the combination of output values that occur most consistently for a given combination of input values, and can back select the combination of values of the input co-terminals that would select the actual combination of values of the output co-terminals in a given transition cycle.

A *monolithic bidirectional output controller* shown in Figure 23.6 can learn to produce a unique combination of values of all of its actuators according to the values of its input variables or can learn to produce asynchronous diverse relations according to the values of each input variable. The design of the monolithic bidirectional output controller is identical to the design of the monolithic bidirectional output matrix used in the universal duplex empirical controller shown in Part III.

Connections to and from these input and output controllers are made through a *handshake interface*, which raises or lowers the state of a terminal without passing current from one unit to the other.

Directly connected series network of monolithic controllers

A monolithic bidirectional output controller can be connected to a monolithic input controller, forming a universal duplex controller or a *directly connected series network of monolithic empirical controllers*, as shown in Figure 23.7.

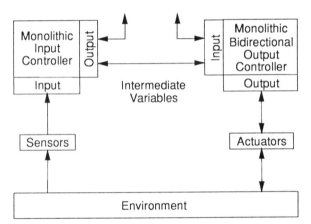

Figure 23.7. A *directly connected series network of monolithic controllers* can send and receive more than one intermediate variable.

Each intermediate variable can be connected to one output controller, creating the opportunity for asynchronous diverse behavior, or each intermediate variable can be connected to different output controllers, creating multiple pathways for sensory information to flow from one input unit to other output units.

Monolithic bidirectional nodal controller

A monolithic bidirectional output matrix can be stripped of its sensors and actuators, forming the *monolithic bidirectional nodal controller* shown in Figure 23.8. The monolithic bidirectional nodal controller can be used as the nodal unit in a direct series network of monolithic controllers, as the output matrix of a universal duplex input controller, or as the ultimate node in a hierarchical network, to be shown in later sections.

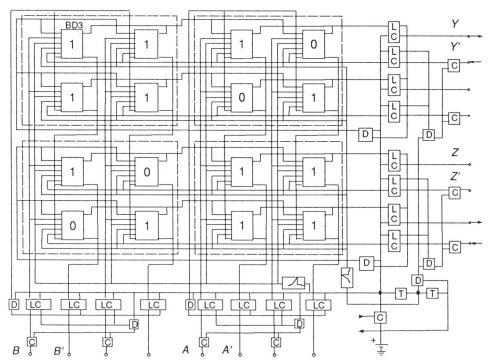

Figure 23.8. A *monolithic bidirectional nodal controller* can connect one set of intermediate variables to another set of intermediate variables.

Direct series network of monolithic controllers with a nodal unit

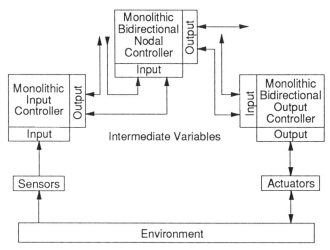

Figure 23.9. A *direct series network of monolithic controllers with a nodal unit* can send and receive intermediate variables to and from other units.

A monolithic bidirectional output controller can be connected to a monolithic input controller by means of an intermediate monolithic bidirectional nodal controller, forming a *directly connected series network of monolithic empirical controllers with a nodal unit*, as shown in Figure 23.9. Some intermediate variables can connect the input and output controllers, and other intermediate variables can be directly connected to other units and can be received from other units, creating even more pathways for information to flow to and from other units.

Series networks of multivariable duplex empirical control units

Multivariable empirical control units can also be connected in series, allowing multiple pathways for unitary sensor and actual actuator information to flow between two or more empirical control units.

Multivariable duplex input controller

The duplex controller can also be used as an input controller. The *multivariable duplex input controller* is similar to a typical multivariable duplex empirical controller, except that it does not have actuators and does not provide an actuator delay circuit, as shown in Figure 23.10.

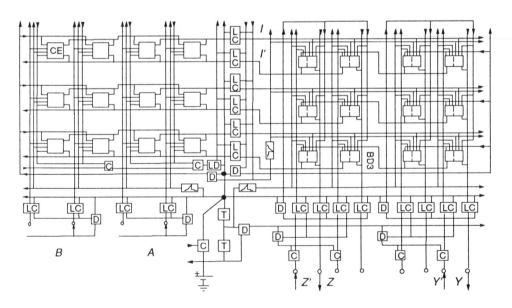

Figure 23.10. The *multivariable duplex input controller* is connected to sensors and is designed to produce intermediate outputs that can be connected to the input matrix of another duplex controller rather than to actuators.

The intermediate output terminals of a multivariable duplex input controller can be connected to another controller only. Some means must be provided to create a transform signal to the input controller that represents the state of the intermediate variables most likely to select the actual output of the output controller. This can be accomplished by a bidirectional duplex output controller to be shown, which is similar to a typical duplex controller except it uses a bidirectional input matrix.

Multivariable bidirectional input matrix

The *multivariable bidirectional input matrix* is similar to a multivariable input matrix except that it contains Type 3 bidirectional memory cells (BD3 cells) instead of CE cells and contains a back-selecting circuit, as shown in Figure 23.11.

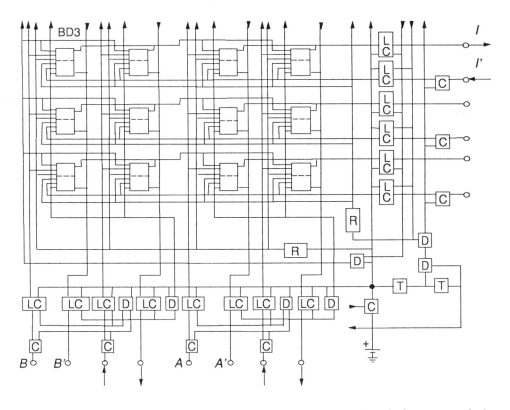

Figure 23.11. A multivariable bidirectional input matrix can back select the input co-terminals associated with the input values most likely to select the actual output state by using Type 3 bidirectional memory cells (BD3 cells).

Like a typical multivariable input matrix, the multivariable bidirectional input matrix selects the output state that has occurred most consistently for a given set of values of the input variables based upon the sensitivity values of the BD3 cells connected to these input values. It decreases the sensitivity of the selecting BD3

cells and then it back selects the set of input co-terminals belonging to the principal
input terminals that would select the actual output state that occurs in a given
transition based upon the sensitivity values of the BD3 cells connected to the actual
output state. It then increases the sensitivity of the back-selecting BD3 cells.

Multivariable bidirectional duplex output controller

The *multivariable bidirectional duplex output controller* consists of a multivariable
bidirectional input matrix and a multivariable bidirectional output matrix, as shown
in Figure 23.12.

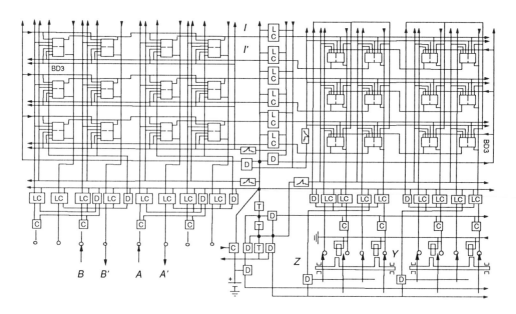

Figure 23.12. The *multivariable bidirectional duplex output controller* can back select the input
co-terminals associated with the input state most likely to select the actual output state based
upon the sensitivity values of its Type 3 bidirectional memory cells (BD3 cells).

The multivariable bidirectional duplex output controller selects a set of output
values based upon the values of its input variables and the sensitivity of the BD3
cells in *both* matrices, and then decreases the sensitivity of the selecting BD3 cells.
It then back selects the set of input co-terminals associated with the input values that
would produce the actual output values and increases the sensitivity of the back-
selecting BD3 cells. These back-selected input co-terminals can then be used to
provide transform conditioning signals to another duplex input controller or a set of
duplex input controllers connected to the input matrix of the output controller.

Directly connected multivariable duplex controllers

The multivariable duplex input controller can be connected to the multivariable bidirectional duplex output controller, forming *directly connected multivariable duplex controllers* shown in Figure 23.13.

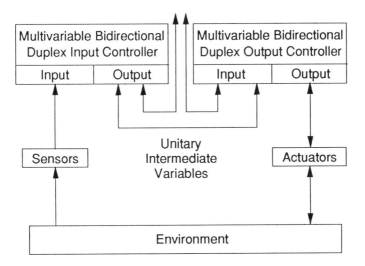

Figure 23.13. A *directly connected multivariable duplex controllers* can have multiple intermediate variables, but cannot produce asynchronous diverse lines of behavior using these intermediate variables.

The sensor and action information flowing through the intermediate variables must be substantially unitary in directly connected multivariable duplex controllers.

Multivariable bidirectional duplex nodal controller

The intermediate node in a direct series network may not be connected directly to any sensors or actuators. This requires a *multivariable bidirectional duplex nodal controller* (Fig. 23.14) that has no sensors or actuators and back selects the input co-terminals that are most likely to select the values of the output co-terminals. The multivariable bidirectional duplex unit does not provide for an actuator delay because all of its transactions must occur within a single transition cycle. The multivariable bidirectional duplex controller is made up entirely of Type 3 bidirectional conditional memory cells (BD3 cells).

The multivariable bidirectional duplex nodal controller is similar to a neuron that can learn to generate a functional relationship between a set of input and output variables. The set of latches that surround the matrices act like synapses that transmit and receive specific signals in each transition cycle.

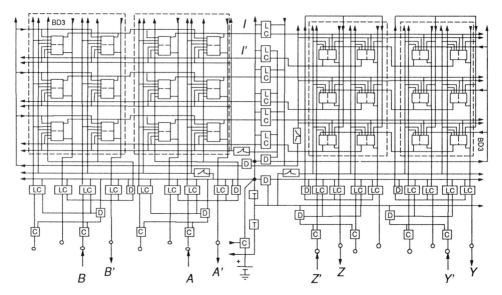

Figure 23.14. A *multivariable bidirectional duplex nodal controller* is used only as an internal node within a network of empirical control units and is limited to producing unitary behavior.

Direct series connection of multivariable duplex controllers with a nodal unit

The multivariable duplex input controller can be connected to the multivariable bidirectional duplex output controller by means of a multivariable duplex nodal controller, forming *directly connected multivariable duplex controllers with a nodal unit,* as shown in Figure 23.15.

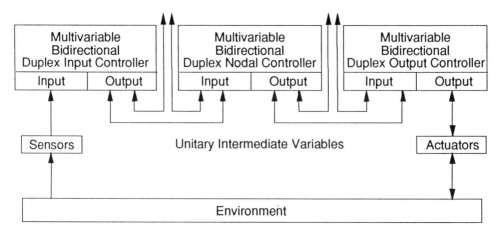

Figure 23.15. The connection between input and output units in a *directly connected multivariable duplex controllers with a nodal unit* can be influenced by other units.

The pathways through the nodal unit must be substantiated by the real sensor and action behavior of input and output controllers.

Series networks of universal duplex empirical control units

Universal duplex empirical control units can also be connected in series, allowing multiple pathways for asynchronous diverse sensor and actual actuator information between two or more empirical control units.

Universal duplex input controller

The single intermediate variable of the multivariable duplex input controller shown previously can be broken into two or more intermediate variables by connecting a monolithic input matrix with a monolithic bidirectional matrix, forming the *universal duplex input controller* shown in Figure 23.16, which can carry out two or more separate lines of behavior.

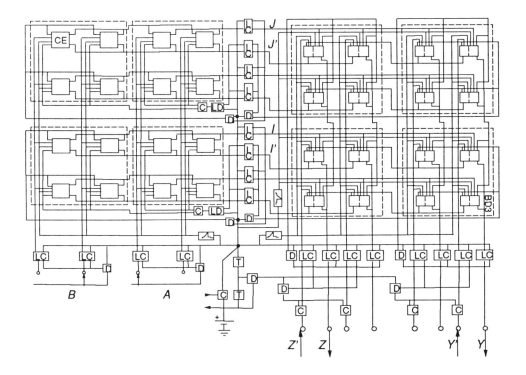

Figure 23.16. A universal duplex input controller acts the same as the multivariable duplex input controller, except that it can deal with two or more asynchronous diverse relations.

The number of intermediate variables and the number of values of each intermediate variable can be estimated beforehand, and the controller can be designed
accordingly to meet the needs of the task environment.

Universal bidirectional duplex output controller

If two or more diverse lines of behavior must be carried out asynchronously, additional intermediate variables can be added to the multivariable bidirectional duplex
output controller shown earlier by connecting a monolithic bidirectional input matrix
to a monolithic bidirectional output matrix, forming the *universal bidirectional
duplex output controller* shown in Figure 23.17

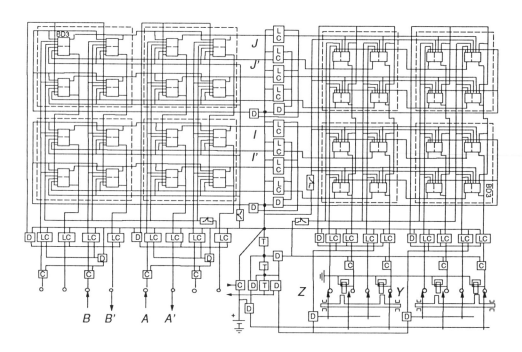

Figure 23.17. A *universal bidirectional duplex output controller* can be connected to the output
of another empirical controller and provide that controller with a transform signal from its own
actuators.

A universal bidirectional duplex output controller provides more diverse pathways
for action to flow through a network of empirical controllers than a multivariable
bidirectional duplex output controller.

Directly connected universal duplex controllers

The universal duplex input controller can be connected to the universal bidirectional duplex output controller, forming *directly connected universal duplex controllers* shown in Figure 23.18.

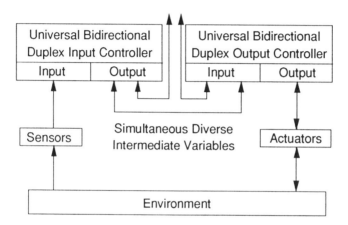

Figure 23.18. Directly connected universal duplex controllers can have multiple asynchronous diverse intermediate variables that can be connected to and from many other units.

The sensor and action information flowing through each intermediate variable between directly connected universal duplex controllers can be independent of the information occurring in other intermediate variables, or all of the information can be substantially unitary. This makes the universal duplex empirical controller the most generally suitable control unit in most empirical networks.

Universal bidirectional duplex nodal controller

If more than one line of behavior must pass through a node asynchronously in a direct series network or a hierarchical network (to be shown), the *universal bidirectional duplex nodal controller* shown in Figure 23.19 must be used. The input and output matrices of the universal bidirectional duplex controller are essentially identical monolithic bidirectional matrices connected in series with one rotated 90 degrees from the other.

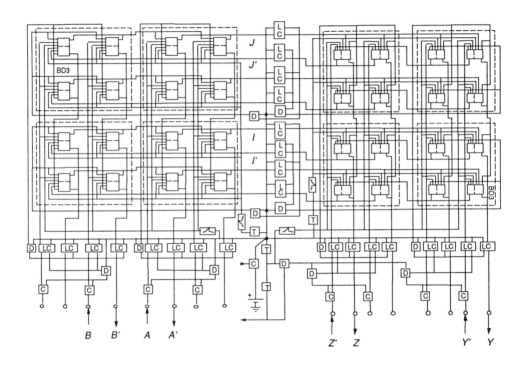

Figure 23.19. A *universal bidirectional duplex nodal controller* can handle two or more asynchronous diverse relations.

Directly connected universal duplex controllers with a nodal unit

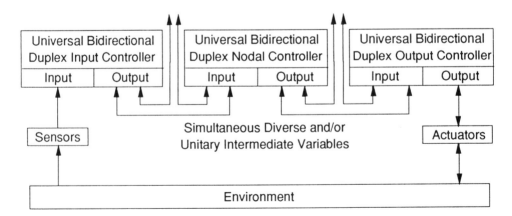

Figure 23.20. Directly connected universal duplex controllers with a nodal unit can have multiple asynchronous diverse intermediate variables that can be influenced by other units.

The universal duplex input controller can be connected to the universal bidirectional duplex output controller by means of a universal bidirectional duplex nodal controller, forming *directly connected universal duplex controllers with a nodal unit* shown in Figure 23.20. The pathways formed in the nodal unit by the intermediate variables must reflect the required sensor and action behavior at the input and output units. If the relations required of the two directly connected input and output units is independent of the information arriving at the nodal unit from other units, a diverse relation must be formed by the nodal unit on the intermediate variables connecting these two units. If the information arriving at the nodal unit has some bearing upon these relations, the nodal unit must establish a unitary organization of these intermediate variables. Since a universal duplex empirical controller can create either form of organization, it is the most generally suitable control unit in most empirical networks.

Chapter 24
Series and Parallel Networks of Empirical Units

Two or more empirical units can be connected in series and in parallel. When units are connected in series and parallel networks they may be able to deal effectively with more variables than can be handled efficiently by a single large unit acting alone.

Direct coupled networks of empirical units

The output variables of two or more units can be directly connected to the input variables of another unit, and the output variables of one unit can be directly connected to the input variables of two or more other units. These units can be connected to a given environment, forming *direct coupled series and parallel networks of empirical units.*

Direct convergent network

The input terminals of a universal bidirectional duplex output controller (UBDD output unit) can be connected to the output terminals of two or more universal bidirectional duplex input controllers (UBDD input units), creating a *direct convergent network of empirical units,* as shown in Figure 24.1.

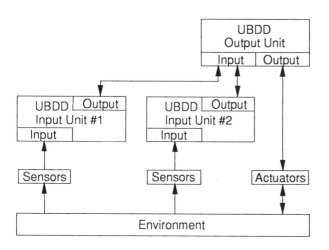

Figure 24.1. In a *direct convergent network of empirical units,* the output variables of two or more input units are connected to the input variables of one output unit.

The actual values of the actuators in the output unit can back select a set of input co-terminals that provide a transform signal to the output co-terminals of each of the input units. This permits the empirical units to establish high-sensitivity pathways to and from the sensor and actuator terminals that occur together in a transition.

Direct divergent network

The output terminals of a single universal bidirectional duplex input unit (UBDD input unit) can be connected to the input terminals of two or more universal bidirectional duplex output units (UBDD output units), forming the *direct divergent network of empirical units* shown in Figure 24.2.

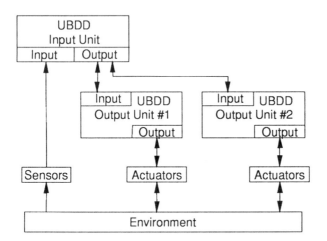

Figure 24.2. In a *direct divergent network of empirical units*, the output terminals of a single input unit are connected to the input terminals of two or more output units.

The actual values of the actuators back select the set of input co-terminals in the output units that are used as the actual outputs of the input unit. This allows the network of units to select the actual actuator states that occur with specific sensor states.

Direct hierarchical network

A *direct hierarchical network of empirical units* is formed when a convergent network is connected to a divergent network by means of a single monolithic or universal bidirectional duplex unit, as shown in Figure 24.3. The output terminals of the input units are connected to the input terminals of a single *ultimate nodal unit* in the hierarchical network. The output terminals of this unit are connected to the input terminals of two or more output units that select the values of the actuators. The actual values of the actuators back select a set of input co-terminals to the ultimate node, which then back selects a set of values to the input units. These *transactions* must occur within one transition cycle, so time delays between the internal units must be kept to a minimum. The nodal unit tends to create unitary behavior between the sensor and actuator variables, but a universal nodal controller can allow some diverse behavior as well.

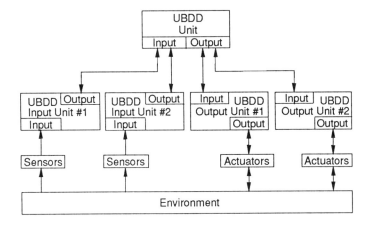

Figure 24.3. A *direct hierarchical network of empirical units* connects the sensors of two or more input units to the actuators of two or more output units.

Direct uncoupled networks of empirical units

In some cases, the environment of an empirical unit can be uncoupled into a *sensor environment* made up of sensor variables and an *actuator environment* made up of actuator variables. The sensor environment provides the information needed by an empirical unit to produce behavior, and the actuator environment provides an opportunity for the empirical unit to test the viability of its behavior. These sensor and actuator units can be connected directly, forming *direct uncoupled networks of empirical units*.

Uncoupled configuration

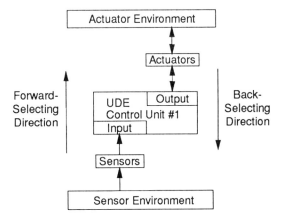

Figure 24.4. In an *uncoupled empirical system*, the sensor and actuator variables are not coupled in a single environment.

In an *uncoupled empirical system*, the sensor and actuator variables may be considered to be in different environments, as shown in Figure 24.4. Separating the sensor and actuator variables clarifies the forward- and back-selecting process. In an uncoupled system, the forward-selecting process goes in one direction, and the back-selecting process goes in the other direction.

Uncoupled parallel configuration

Two or more uncoupled units can be arranged in an *uncoupled parallel network*, as shown in Figure 24.5.

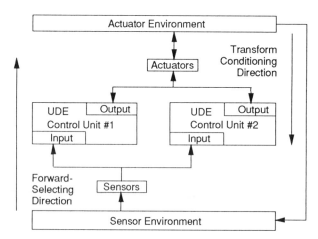

Figure 24.5. An *uncoupled parallel empirical network* clearly shows the parallel arrangement of potentially diverse control units.

The uncoupled configuration shows that the sensor and actuator environment may or may not be intimately connected. For example, sensors may determine the temperature in one room and the actuators may be heaters in another room. The two environments are not *physically connected* in a well-insulated building, but are physically connected in a poorly insulated building.

Uncoupled direct series empirical network

In an *uncoupled direct series empirical network*, action information flows directly from the sensor units to the actuator units in one direction, and the transform signal flows directly from the actuator units to the sensor units in the other direction, as shown in Figure 24.6.

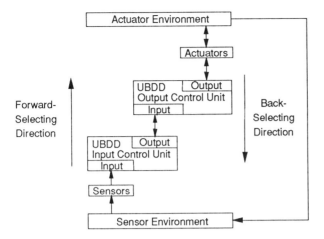

Figure 24.6. An *uncoupled direct series empirical network* clearly shows the flow of the forward-selection process in one direction and the flow of the back-selection process in the other direction.

The input unit in an uncoupled direct series empirical network can learn to help identify the sensor states that require specific action states. All of the transactions in an uncoupled direct series system must occur in one transition cycle.

Uncoupled direct convergent empirical network

In an *uncoupled direct convergent empirical network*, information from the sensor variables converges upon a single unit that produces an action based upon its modulus of behavior. Two or more uncoupled machines can be designed to *preprocess information* from their sensor environment and present it to a single machine unit, which then creates action in an actuator environment, as shown in Figure 24.7.

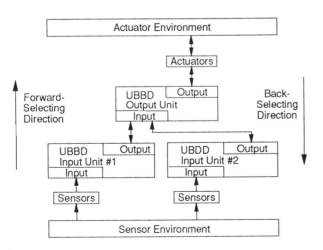

Figure 24.7. A *direct convergent empirical network* shows the flow of action from the sensor environment to the action environment and the flow of action from the action environment to the sensor environment within a given transition cycle.

Changes in the values of some or all of the sensor variables can cause changes in the values of the actuator variables. Using two or more machines to deal with multiple sensor variables reduces the size of each machine in the network.

Uncoupled direct divergent empirical network

In an *uncoupled direct divergent empirical network*, information can flow from the sensor unit to one or all of the actuator variables, as shown in Figure 24.8. Each unit can create a unique representation of the sensor states.

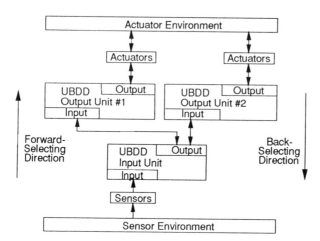

Figure 24.8. In an *uncoupled direct divergent empirical network,* changes in values of the sensors can cause changes in one or both sets of the actuator variables.

An uncoupled system is a more general configuration than a coupled network, since sensor and actuator environments are more likely to be separated physically to some degree rather than being part of the same set of variables.

Uncoupled direct nodal network of empirical units

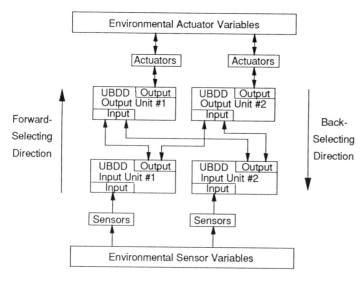

Figure 24.9. In an *uncoupled direct nodal network of empirical units,* diverse pathways become available to connect separate sensor variables to separate actuator variables.

An *uncoupled direct nodal network of empirical units* is formed when the output variables of sensor units can diverge to or converge upon the input variables of the actuator units, as shown in Figure 24.9. In an uncoupled direct nodal network, as many diverse relations can occur in a given transition as there are separate sensor or actuator units. The *parallel nodes* can also establish a division of labor that may allow each unit to be less complex than a single unit required to produce all of the behavior by itself.

Uncoupled direct hierarchical empirical network

An *uncoupled direct hierarchical empirical network* can connect a set of sensor variables to a set of actuator variables through one *ultimate node,* as shown in Figure 24.10.

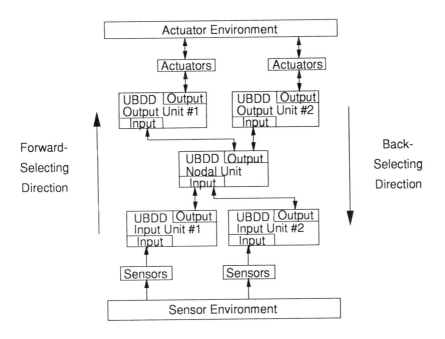

Figure 24.10. In an *uncoupled direct hierarchical empirical network,* each combination of sensor values can produce a unique combination of actuator values.

An uncoupled direct hierarchical network is intrinsically unitary, but it can learn to connect one sensor to either actuator provided the ultimate node is a universal controller, which can establish asynchronous diverse relations.

Uncoupled direct multilevel hierarchical empirical network

If a system has many sensor and actuator variables in different environments, it may be necessary to create the *uncoupled direct multilevel hierarchical empirical network* shown in Figure 24.11.

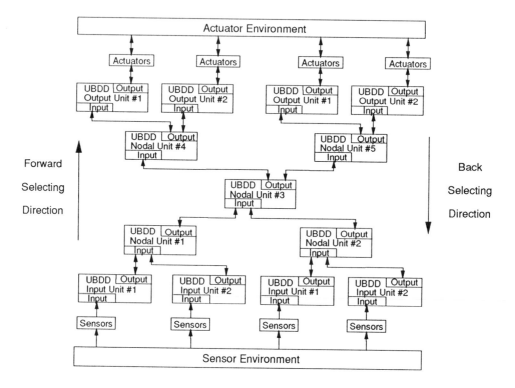

Figure 24.11. An *uncoupled direct multilevel hierarchical empirical network* can handle more sensor and actuator variables than a single-level network without increasing the number of variables handled by each unit.

The number of input and output states that can be handled by a hierarchical network is limited to the number of intermediate variables and intermediate states in the ultimate node.

Uncoupled direct nodal network with hidden nodes

The ultimate node in a hierarchical can be duplicated, forming the *hidden nodes* in the *uncoupled direct multilevel nodal empirical network* shown in Figure 24.12. Each hidden node is usually a universal bidirectional duplex nodal control unit, since it is not connected to sensors or actuators and must select in both directions. Different diverse relations can occur through different parallel hidden nodes and diverse relations can occur within each node, thus expanding the opportunities for diverse behavior in a nodal network.

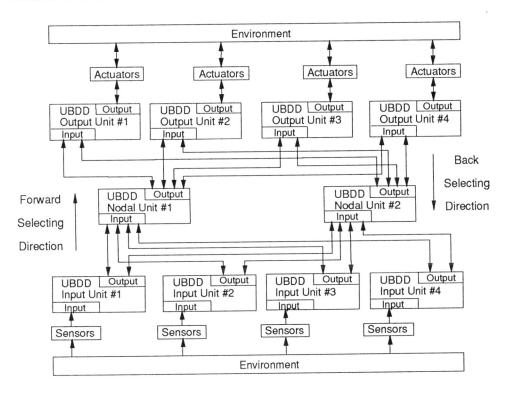

Figure 24.12. An *uncoupled direct multilevel nodal empirical network* is like a hierarchical network with more than one ultimate node.

Uncoupled direct nodal network with additional hidden nodes

If many diverse relations involving two or more variables must occur simultaneously, the number of hidden nodes can be increased, as shown in the *uncoupled direct nodal network with additional hidden nodes* in Figure 24.13. The hidden nodes may also establish a division of labor that reduces the number of intermediate states required in each unit, since each unit can handle a given diverse relation, requiring only one intermediate variable in each unit, rather than a few units handling many diverse relations, requiring many intermediate variables in each unit.

Connections can be made to and from each node by connecting the output terminals of one unit to the input terminals of another unit. If each intermediate variable has many values, or if an intermediate variable has to travel a great distance, the intermediate variables can be digitized. This will greatly reduce the number of conductors connecting each unit. If the unit are close together, there may be no advantage in digitizing the intermediate variables.

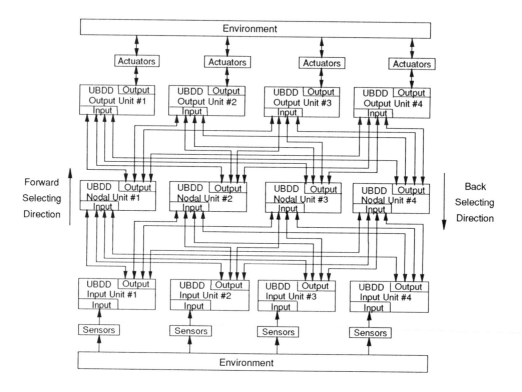

Figure 24.13. An *uncoupled direct nodal network with additional hidden nodes* allows more diverse relations to occur simultaneously.

Uncoupled direct nodal network with additional hidden layers

If the number of sensor and actuator variables becomes too large, it may be necessary to divide the nodal network into sections. These sections can be connected by additional layers of intermediate nodes, as shown in Figure 24.14. A multilayer nodal network is like a hierarchical network with multiple internal nodes at each level. The additional hidden nodes increases the potential variety of each *intermediate representation*. The additional hidden nodes also increases the number of transactions that have to be made in a given transition. This may require longer transition cycle times.

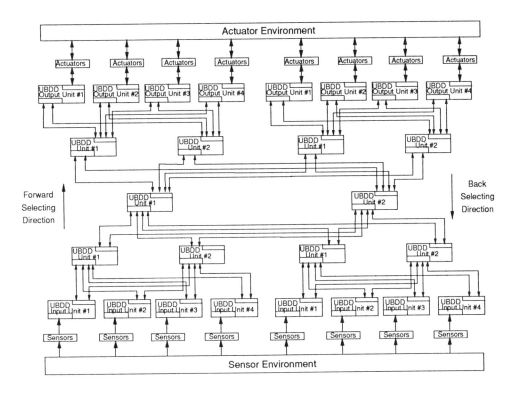

Figure 24.14. An *uncoupled direct nodal network with additional hidden layers* can connect any one of many sensor variables to any one of many actuator variables.

Superimposed uncoupled direct empirical networks

Hierarchical or nodal *subnetworks* can be superimposed upon a set of sensor and actuator variables as shown in Figure 24.15. Each subnetwork can be considered a separate empirical unit that can produce a greater variety of behavior than a single control unit. One or the other subnetwork may produce behavior that can be carried out more consistently than the other subnetwork, and thus dominate behavior. The subnetworks may also establish a division of labor that allows more complex behavior than may be possible by one network working alone. Like other parallel networks, superimposed networks also provide alternative pathways for asynchronous diverse behavior.

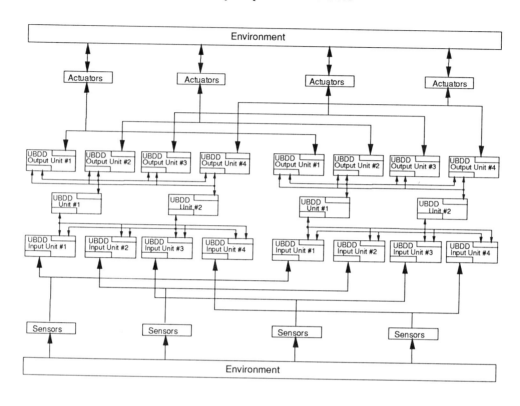

Figure 24.15. Superimposed uncoupled direct empirical networks allow two or more subnetworks to attempt to produce behavior.

Uncoupled direct nodal network of subnetworks

Networks can be connected in series and in parallel forming an *uncoupled direct network of empirical nodal subnetworks,* as shown in Figure 24.16. A network of networks is similar to a machine assembly made up of subassemblies. A network of subnetworks can also be a subassembly in a larger network. In all uncoupled direct networks, each forward- and back-selecting transactions must take place in one transition cycle. It may be difficult to imagine how a particular value of a sensor variable can produce a particular value of an actuator variable thought many layers of hidden nodes and hidden subnetworks. However, the economic model shows by analogy how materials flow from their source to the consumer through many levels of fabrication and distribution in just the right quantity and kind.

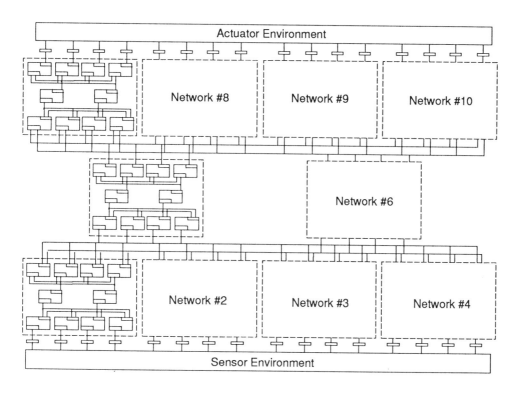

Figure 24.16. An *uncoupled direct network of empirical nodal subnetworks* creates the greatest opportunity to create complex behavior using units that deal with only a relatively few variables.

Direct bipartite networks of empirical units

Some environments may be used primarily for gathering information, and other environments may be used primarily to produce action. For example, the animal nervous system gathers sensor information from tactile and vision sensors and produces action by stimulating muscle fibers. The sensors are physically located in one place, while the muscles are placed in another location. The sensors may be said to deal with a *sentient environment*, while the muscles may be said to deal with an *operant environment*. However, the vision and tactile sensors must be moved by the muscle system, and the position of the muscles must be sensed by a sensor system. Thus, the brain must sense conditions in both environments and take some action in both environments to carry out behavior successfully. These two *bipartite environments* require a controller with two sets of sensors and actuators, one set connected to each type of environment.

Bipartite units

A universal empirical control unit can be configured to operate simultaneously upon two different environments, as shown by the *bipartite empirical control unit* in Figure 24.17.

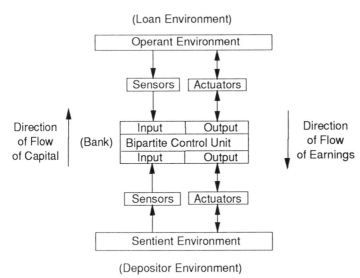

Figure 24.17. A bipartite empirical control unit is both coupled and uncoupled because it can pass sensor and actuator information to the same or different environments.

A commercial *bank* is an example of a bipartite unit. It takes in deposits and pays out dividends in its "depositor (sentient) environment," and places loans and collects principal and interest in its "loan (operant) environment." The bank must respond to the demands of its depositors. Thus, it must be *reactive* to its sentient environment. However, it can make decisions on to whom it lends money. Thus, it can be *proactive* in its operant environment.

Parallel bipartite network of empirical units

A *parallel bipartite network of empirical units* is formed by connecting two or more bipartite empirical control units to sensors and actuators in sentient and operant environments, as shown in Figure 24.18. A parallel bipartite network of empirical units is like a group of individuals in a restricted environment. The behavior of each unit will change until each can produce behavior that does not interfere with the behavior of other units. It can be assumed that this behavior will reflect the needs and constraints of the environments as well as the behavior of the other units.

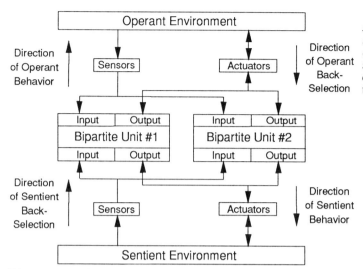

Figure 24.18. In a *parallel bipartite network of empirical units,* sensor and actuator information can flow in both directions.

Direct series bipartite network of empirical units

A *bipartite input control unit* can be connected directly to a *bipartite output control unit,* forming the *direct series bipartite network of empirical units* shown in Figure 24.19.

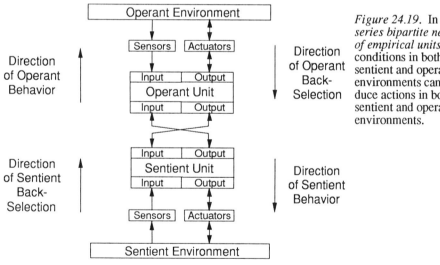

Figure 24.19. In a *direct series bipartite network of empirical units,* sensed conditions in both the sentient and operant environments can produce actions in both the sentient and operant environments.

The transform conditioning process in a direct series bipartite network of empirical units flows in both directions in a single transition cycle. Since some of the forward- and back-selection proceeds directly from one unit to another in a direct series bipartite network of empirical units, the hidden variables involved in this process must be bidirectional. The design of most industrial production machines is based upon these direct bipartite connections between subunits.

Direct bipartite convergent network of empirical units

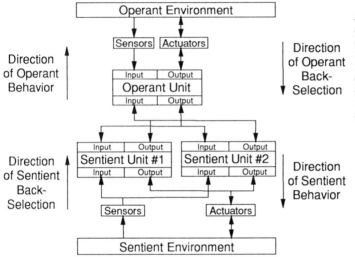

Figure 24.20. In a *direct bipartite convergent empirical network*, action instructions flow through two or more control units connected to the sentient environment to a control unit connected to the operant environment.

Two or more parallel bipartite empirical control units can be connected to a sentient environment and another empirical unit, forming the *direct bipartite convergent empirical network* shown in Figure 24.20. The action instructions from the sentient environment to the operant environment can be tempered in each unit by information from the operant environment.

Direct bipartite divergent network of empirical units

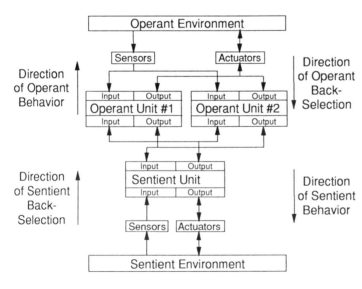

Figure 24.21. A *direct bipartite divergent empirical network* uses instructions from a single sentient control unit connected to the sentient environment to cause action in two or more operant control units connected to the operant environment.

Two or more bipartite control units connected in parallel to an operant environment may be connected to a single bipartite control unit connected to a sentient environment forming the *direct bipartite divergent empirical network* shown in Figure 24.21. The output of the sentient control units connected to the sentient environment is determined by the sentient control unit connected to the sentient environment and the sensors in the operant control units connected to the operant environment.

Direct bipartite nodal network of empirical units

Influences in the sentient and operant environment can converge and diverge in the *direct bipartite nodal network of empirical units* shown in Figure 24.22.

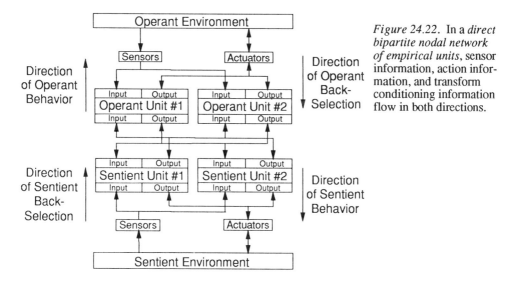

Figure 24.22. In a *direct bipartite nodal network of empirical units*, sensor information, action information, and transform conditioning information flow in both directions.

Each unit in a direct bipartite nodal network may choose to act in a diverse manner on one sensor and one actuator in one environment, or contribute to the unitary behavior of all of the sensors and actuators in both environments. Most complex empirical machines like business organizations exist in the form of a direct bipartite nodal network of empirical units.

Direct bipartite hierarchical network of empirical units

The ultimate node in a *direct bipartite hierarchical network of empirical units* can receive sensor information from a sentient and operant environment and produce action in both environments by means of its connection to sentient and operant subunits, as shown in Figure 24.23. The sentient units may produce action in the sentient environment without regard for conditions in the operant environment, and the operant units may produce action in the operant environment without regard for conditions in the sentient environment if the ultimate node does not pass information from one environment to the other.

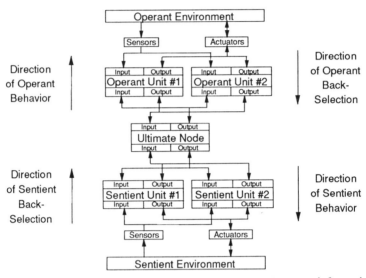

Figure 24.23. In a *direct bipartite hierarchical empirical network*, sensor information in one environment must flow through the ultimate node to produce action in the other environment, thus ensuring that the behavior of the network is essentially unitary.

Direct bipartite multilevel nodal network of empirical units

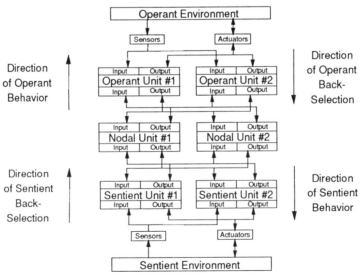

Figure 24.24. A *direct bipartite multilevel nodal network of empirical units* can carry out unitary or diverse behavior between two environments.

If more than one line of behavior involving the sentient and operant environments are required at a given time, additional intermediate nodes can be added to the hierarchical network, forming the *direct bipartite multilevel nodal network of empirical units* shown in Figure 24.24. Each *hidden node* can conduct a separate line of behavior between each environment if the actions of the sentient and operant units do not interfere. The hidden nodes can also cooperate to produce complex unitary behavior.

Parallel network of bipartite empirical nodal networks

Each network of bipartite empirical nodal networks may be considered a separate empirical control unit. These networks can be connected in parallel to a sentient and operant environment, forming a more complex empirical unit (Fig. 24.25).

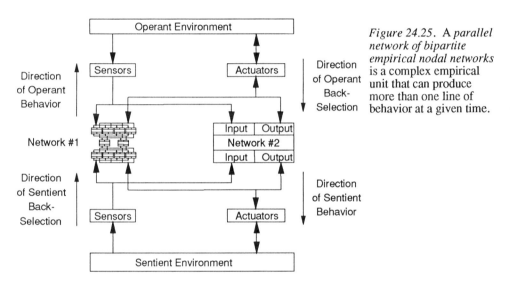

Figure 24.25. A *parallel network of bipartite empirical nodal networks* is a complex empirical unit that can produce more than one line of behavior at a given time.

Each network provides a separate pathway for behavior. For example, one network may conduct behavior in the sentient environment that is independent of behavior in the operant environment, and the other network may conduct behavior in the operant environment that is independent of the behavior in the sentient environment. Or one network may produce action in one set of actuators in the operant environment based upon the values of sensors in the sentient environment, while the other network may produce action in another set of actuators in the operant environment based upon the values of another set of sensors in the sentient environment.

Direct nodal network of direct bipartite networks of empirical units

Direct bipartite nodal networks can be assembled into a *direct nodal network of direct bipartite empirical nodal networks,* as shown in Figure 24.26.

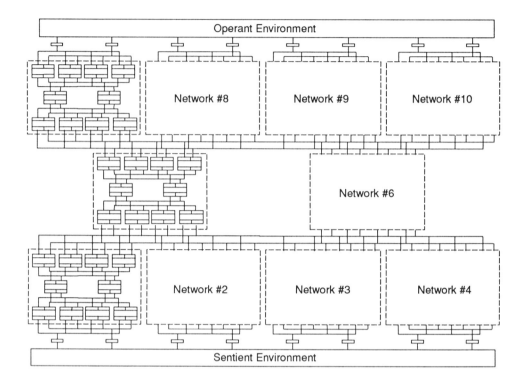

Figure 24.26. A *direct nodal network of direct bipartite empirical nodal subnetworks* can establish a much larger spectrum of unitary and diverse behavior than a single network of empirical units.

In the direct nodal network of direct bipartite empirical nodal networks, the back-selecting process must proceed inward from the sentient and operant environments using bidirectional matrix variables and be completed in each transition cycle. The direct nodal network of direct bipartite empirical nodal subnetworks is the basis of the most complex empirical units.

Indirect uncoupled networks of empirical units

There may be some limitation as to the number of empirical units that can be directly connected in series and still forward and back select effectively in one transition cycle. These limitations can be overcome by placing *intermediate environments* between these units and providing these units with sensors and actuators to deal with these intermediate environments. Empirical units with these *indirect connections* to other units form *indirect networks of empirical units*. This eliminates the need to forward and back select through many units and creates the opportunity for a group of empirical units to work together as a social unit through their intervening environments. These indirect networks of empirical units also provide more opportunities to connect a large set of sensor and actuator variables than a single unit acting alone.

Indirect uncoupled series network of empirical units

Units in a network can be connected through environments instead of the direct connections between units in the direct networks shown. Standalone control units with sensors and actuators can be i ;ed in indirect networks because feedback signals to not have to be back-selected through one to another units, as shown in Figure 24.27.

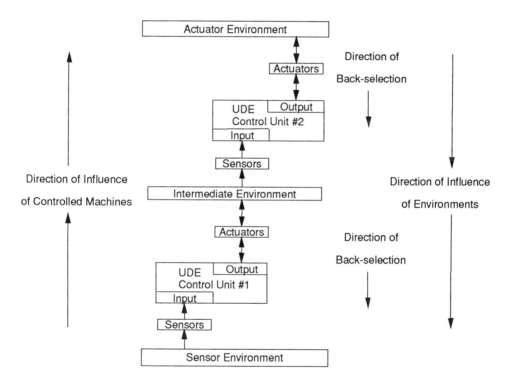

Figure 24.27. In an *indirect uncoupled series network of empirical units*, the actuator of one unit is connected to the sensor of another unit through an intermediate environment.

The second unit does not have to back select through to the first unit in an indirect connection. Therefore, the transactions between the sensor and actuator environments do not have to take place in one transition cycle. Actions in the actuator environment may influence the actions allowed in the intermediate environment, and actions in the intermediate environment may influence the conditions in the sensor environment in the manner of a typical closed system. If the transition cycle times of each unit in an indirect series connection are shorter than the cycle time of the intrinsic transition between sensor and actuator environments, then the flow of effects from the actuator environment to the sensor environment can provide feedback signals as to the allowable behavior in the network for all of the units.

Indirect uncoupled convergent network of empirical units

The results of the actions of two or more units can converge in a single intermediate environment. Another unit connected to this intermediate environment may then produce behavior in another actuator environment, as shown in Figure 24.28. The values of the sensors of multiple variables in a sensor environment can affect a single actuator variable in an actuator environment in an uncoupled indirect convergent network of empirical units.

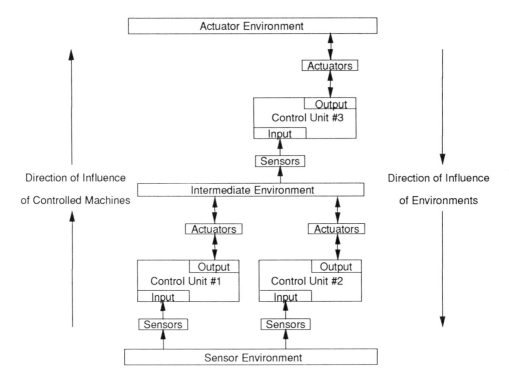

Figure 24.28. An *indirect uncoupled convergent network of empirical units* can produce an output in an actuator environment based upon the behavior of two or more sensor units connected to a sensor environment.

Indirect uncoupled divergent network of empirical units

The behavior of a single unit can influence two or more other units in an indirect divergent system as shown in Figure 24.29. For the actions of the actuator units to influence the sensor unit in a single transition cycle, the influence of the actuator units upon the actuator environment and the influence of the actuator environment upon the intermediate environment must occur in a shorter period than the transition time of the sensor unit.

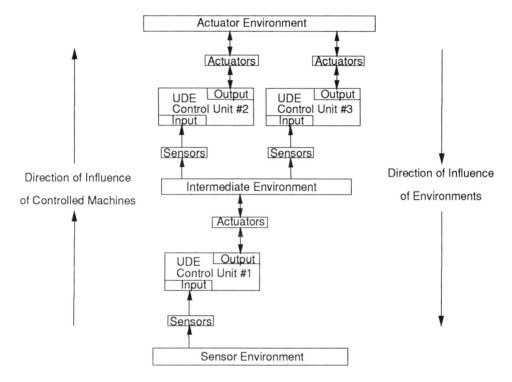

Figure 24.29. An *indirect uncoupled divergent network of empirical units* can create a unique combination of actuator values of two or more units connected to an actuator environment based upon the behavior of a single unit connected to a sensor environment.

Indirect uncoupled nodal network of empirical units

Indirect convergent and indirect divergent networks can be combined to form the *indirect uncoupled nodal network of empirical units* shown in Figure 24.30. Some values of the variables in the intermediate environment may influence both sensors of the actuator units, and some values may affect only one of these sensors. However, the information that flows in both directions through the indirect nodal network is determined in part by the features of the intermediate environment.

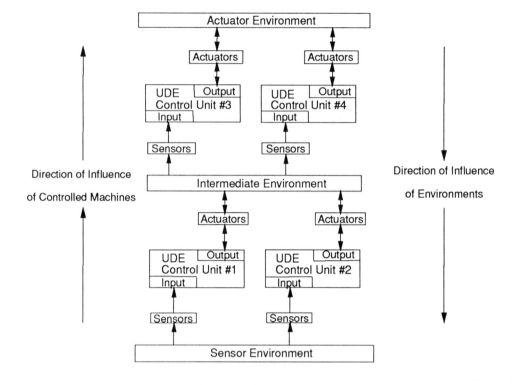

Figure 24.30. An *indirect uncoupled nodal network of empirical units* can establish separate pathways for information to flow from specific sensors in the sensor environment to specific actuators in an actuator environment.

Indirect uncoupled hierarchical network of empirical units

Indirect convergent and indirect divergent networks can be joined to form an *indirect uncoupled hierarchical network of empirical units,* as shown in Figure 24.31. The number of actions in the actuator environment that can be produced by values in the sensor environment is limited by the number of intermediate states in the *ultimate node.* Since the relations between the sensor and actuator environments are determined by one control unit, the hierarchical network is essentially unitary. However, some diverse relations can be made through the ultimate node by using a universal empirical controller as the ultimate node.

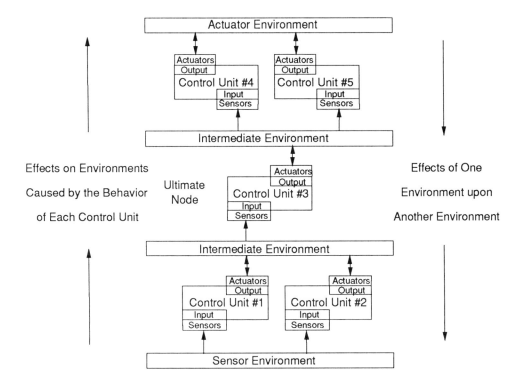

Figure 24.31. An *indirect uncoupled hierarchical network of empirical units* can connect multiple sensor variables in one environment to multiple actuator variables in another environment through one ultimate node.

Indirect multilevel nodal network of empirical units

The amount of behavior in the actuator environment that is controlled by conditions in the sensor environment can be increased by adding more ultimate nodes to an indirect hierarchical network, forming the *indirect multilevel nodal network of empirical units* shown in Figure 24.32. The units connected to the sensor environment are like military intelligence units that pass on information to the combat units, which may influence the conditions observed by the intelligence units.

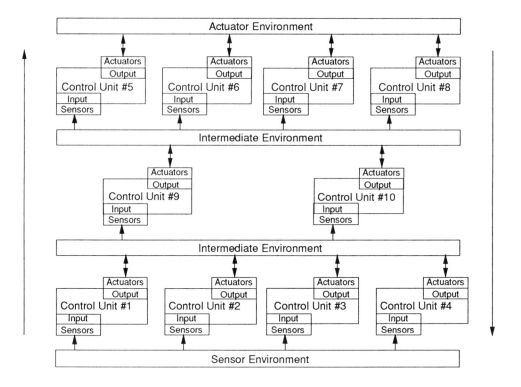

Figure 24.32. An *indirect multilevel nodal network of empirical units* provides parallel pathways for influences to flow from the sensor environment to the actuator environment.

Indirect network of direct empirical networks

If each unit in an indirect nodal network has to process many variables, each unit may be replaced by a direct network of empirical units. These direct subnetwork units can then be arranged in an indirect nodal or hierarchical network, as shown in Figure 24.33.

Systems containing many variables may have to be organized as an indirect network of direct networks rather than as a single large unit (SLU). Each direct subnetwork may be viewed as a separate unit within a group of units. A social group may be viewed as an indirect network of direct empirical networks in which each direct subnetwork can share behavior with other subnetworks connected to the same environments.

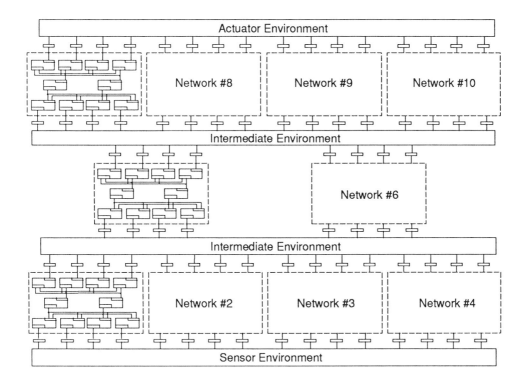

Figure 24.33. An *indirect network of direct empirical subnetworks* can handle more variables than a network of single control units.

Indirect network of indirect empirical nodal subnetworks

Units made up of indirect subnetworks can also be organized into an indirect nodal or hierarchical network, as shown in Figure 24.34. Most large businesses, social, and government organizations are organized as an indirect network of indirect subnetworks. These networks may be more hierarchical or more nodal, depending upon the number of internal nodes or internal subnetworks. Each internal environment may be separated or isolated from other internal environments to force information to flow through the subnetworks.

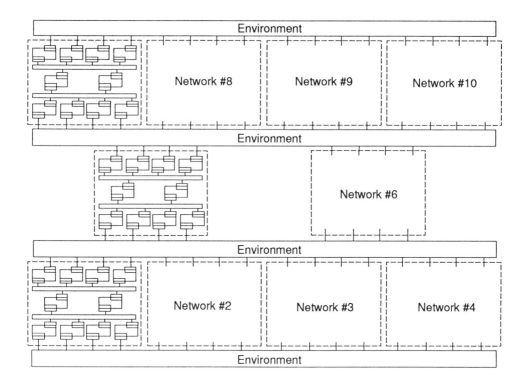

Figure 24.34. An *indirect network of indirect empirical nodal networks* is made up of many separate units connected to each other only indirectly through many different environments.

Indirect bipartite networks of empirical units

Some environments may be used primarily but not exclusively for information gathering, and other environments may be used primarily but not exclusively for the production of action. The *marketing* department of a company may be involved with gathering information about number of potential customers and customer preferences, but may also attempt to make sales to customers. The *sales* department may be responsible for making sales to customers, but may also attempt to gather information as to customer preferences. Therefore, the marketing department may be said to deal with a *sentient environment* and the sales department may be said to deal with an *operant environment*. Both groups must sense conditions in each environment and take some action in each environment to carry out their tasks. These two different *bipartite environments* require controllers with two sets of sensors and actuators, one set for each type of environment.

Indirect series bipartite network of empirical units

Standalone bipartite empirical control units can be connected in series through an intermediate environment, forming the *indirect series bipartite network of empirical units* shown in Figure 24.35.

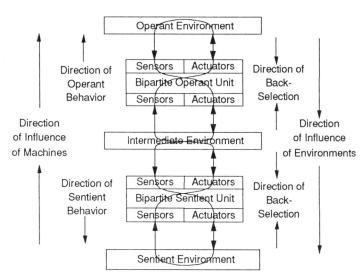

Figure 24.35. An *indirect series bipartite network of empirical units* uses an intermediate environment to connect one standalone bipartite unit to another.

The indirect series bipartite network of empirical units does not need to back select through one unit to another unit. Therefore, it does not need bidirectional control units. But the standalone multivariable or universal duplex controllers do need bidirectional output matrices, as shown earlier. Most living beings communicate through an environment in an indirect series or parallel bipartite network.

Indirect bipartite nodal network of empirical units

Standalone empirical control units can be connected in series and parallel through an intermediate environment, forming the *indirect bipartite nodal network of empirical units* shown in Figure 24.36. An environment may be classified as sentient, intermediate, or operant if it is primarily used to sense condition, pass information to other units, or produce action. Units connected in that way form the equivalent of a *social group* of empirical units.

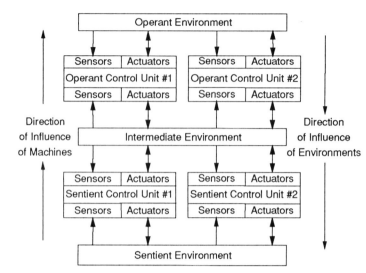

Figure 24.36. In an *indirect bipartite nodal network of empirical units*, any unit can influence variables in any another unit via its effects upon intervening environments.

Indirect bipartite hierarchical network of empirical units

A single control unit can be connected to intermediate environments placed between a set of sentient and operant units, forming the *indirect bipartite hierarchical network of empirical units* shown in Figure 24.37.

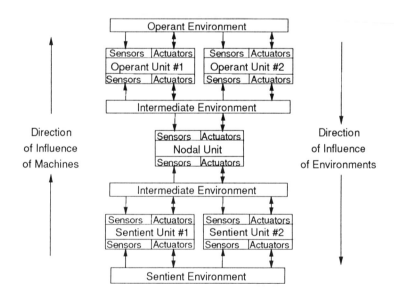

Figure 24.37. In an *indirect bipartite hierarchical network of empirical units*, exchanges of behavior between the sentient and operant environments must pass through the single nodal unit.

The bipartite network creates more opportunities for behavior because each sentient and operant unit can produce behavior in its own environment, act in parallel with other units in its own environment, or attempt to influence the behavior of nodal units that can influence behavior in other environments. If each unit can act upon the other units at their level through another specially designated environment, they become part of a *tripartite hierarchical network*. Most *social* and *business organizations* exist as indirect bipartite or tripartite hierarchical networks of empirical units.

Indirect bipartite multilevel nodal network of empirical units

The hierarchical network shown can be changed into an *indirect bipartite multilevel nodal network of empirical units* by adding more nodal units, as shown in Figure 24.38.

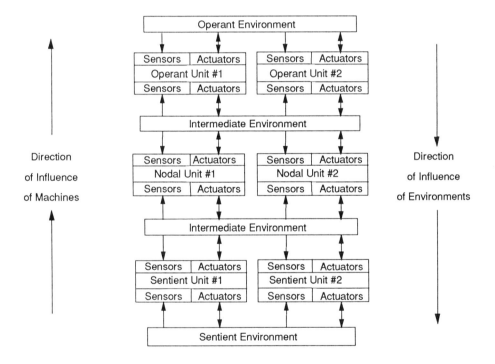

Figure 24.38. An *indirect bipartite multilevel nodal network of empirical units* creates more opportunities for influences to flow between units and environments.

Interspersing environments between empirical units connected in series eliminates the need to back select through units, but may make it more difficult for units connected to one environment to influence conditions in another environment in a consistent manner.

Indirect nodal network of bipartite direct nodal empirical networks

Since a direct nodal network can be considered as a complex empirical unit, these units can be connected in an *indirect nodal network of bipartite direct empirical nodal subnetworks,* as shown in Figure 24.39.

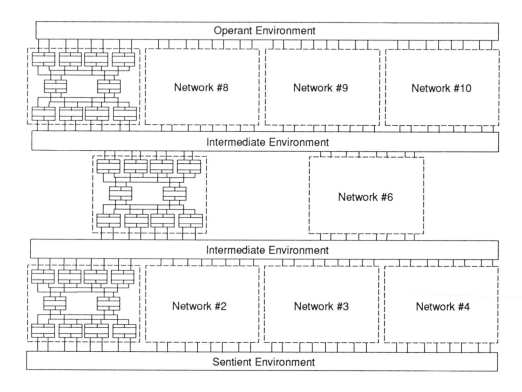

Figure 24.39. An *indirect nodal network of bipartite direct empirical nodal subnetworks,* is an organization made up of complex individuals.

This form of organization raises the question about whether the set of empirical units can be more effective by making the network more or less complex or by making the individuals in the network more or less complex.

Indirect nodal network of indirect bipartite nodal empirical networks

The individuals in the indirect network shown can be replaced by indirect networks (groups of individuals), as shown in Figure 24.40.

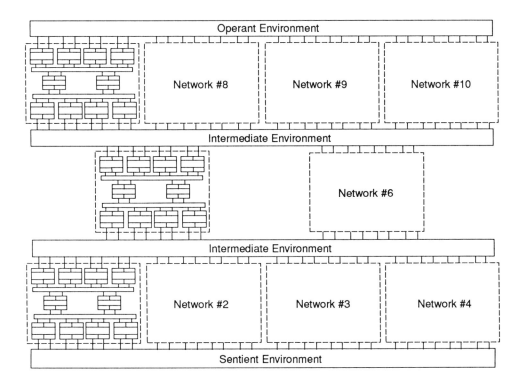

Figure 24.40. An *indirect nodal network of bipartite indirect empirical nodal networks* is a complex social organization made up of social groups.

The process of creating groups made up of subgroups can be extended indefinitely as it is in most large social and business organizations. The effectiveness of the organization in carrying out specific tasks can be improved by specifying which individuals or groups communicate in which environments. Each empirical unit can contribute to the overall success of the organization by attempting to reduce interferences with other units and attempting to produce behavior that can be carried out within the constraints of the organization and its environments.

Behavior of empirical units in a network

When units are connected in a network, they may be influenced by the environment and/or the behavior of other units. In order for units to work successfully in a network, they must establish behavior that can be carried out, does not interfere with the successful behavior of other units, and contributes to the success of the network.

Behavior of units in a network

When units are connected in a network, they can act without interfering with one another if:

1. They all produce the same behavior as other units (act in unison), or
2. They act upon different variables than other units (dispersion), or
3. They act upon a different set of values of the variables (value field selection). This is the same as acting upon the same variables at different times.

By selecting the correct type of behavior, each unit may act more successfully in the network than by acting alone.

Strategies for behavior of units in a network

In general, options for behavior of units connected in a network can be shown graphically by the diagram in Figure 24.41.

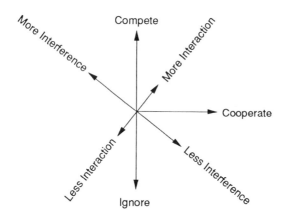

Figure 24.41. The *strategies for behavior of units in a network* are based upon each unit deciding to compete, cooperate, or ignore the behavior of other units.

In spite of the many benefits obtained by connecting units in a network, conflicts may occur when two or more units are connected to a common set of variables. Each unit may have to adjust its behavior to produce behavior that can be carried out in the presence of other units. Each unit must select the *strategy of behavior* that provides the most consistent and successful behavior. Each unit must decide to *compete with, cooperate with, or ignore* the behavior of the other units. Each strategy may provide a uniquely beneficial method of achieving successful behavior.

Competition

Two or more connected units can *compete* if each tries to produce different behavior on the same set of variables. The unit with the greatest confidence, knowledge of natural relations in its environment, speed of response, physical strength, accuracy of response, and/or greater persistence will prevail. Competition provides the opportunity for empirical units to discover new behavior.

Cooperation

A unit can *cooperate* with the other units with which it shares variables. It may provide behavior that the other unit cannot provide by itself. For example, two or more units connected to the same variable may produce a greater force than a single unit acting alone. Also, two or more units may provide a greater range of movement (displacement) than a single unit acting alone. One unit may act like a backup controller for another unit and supply some behavior that the other is unable to supply. When units work together to provide successful behavior, they may be said to form an *alliance* through amalgamation. An alliance can often produce results that cannot be obtained by a single unit working alone.

Exclusion

A unit may disregard or *ignore* the information it receives from another units or from the environment it shares with another unit. It may choose to *exclude* inputs from the units with which it is connected if it cannot establish consistent successful behavior within the context of their behavior. An empirical unit must isolate itself from behavior having low probabilities. It can do this by providing a constant output from each value of sensed conditions. Operating in this way, a duplex controller does not waste values of its intermediate variable, and it helps other units establish consistent behavior.

Synchronism

Every control unit is designed with specific time delays between sensor and actuator events. If connected units have time delays that result in *asynchronous behavior*, the behavior of each unit may interfere with the behavior of other units. Thus, the asynchronous behavior is best carried out through diverse subsystems. Yet unitary behavior with different but *synchronous time delays* (Fig. 24.42) can be carried out successfully in a less complex network than asynchronous behavior.

If the units produce behavior that can be carried out, none will have to change their behavior. An external timing source may be used to synchronize the transition cycles of the units in a network. Synchronism is a requirement for and exists in many living systems. For example, clocks in business offices provide the time base for the coordination of important events during the day such as starting time of the work day, start of meetings, coffee break time, lunch time, and quitting time. Brain waves may be used to synchronize the timing of different parts of the living brain and the transition cycle timing signals can be used to synchronize the timing of empirical control units in a network. Multiplexing creates synchronous diverse behavior from asynchronous diverse behavior.

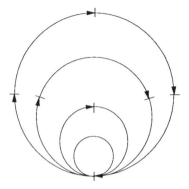

Figure 24.42. Synchronous units can produce compatible behavior if the transforms produced at any given time by all units do not interfere.

Superimposed units with different time delays

Units with different timed relations between sensor and actuator can be superimposed. Each can compete to establish which time relations provide optimum performance. Those units that work together or in synchronism will prevail and dominate behavior. Those units that do not provide successful time relation or work out a method of synchronism with the other units will fail to develop high-confidence-level transformations and thus fail to participate in the behavior of the system. Thus, if certain lines of behavior involve certain input/output variables with specific timed relations, the unit or units most closely corresponding to these specific timed relations will prevail.

Assorted superimposed units in a network

Not all units may be connected to exactly the same set of variables and/or the same set of values of these variables. If some units are connected to a subset of the sensor and actuator variables, they may find it easier to establish successful behavior than those units connected to the whole set. Some of these assorted units may find that their particular subset of variables can establish successful behavior in a more efficient manner than the subset of variables associated with other units.

Predisposed behavior in a network

Some units may have a set of conditional empirical memory cells that are initially set in patterns of high confidence level. This predisposes these units to behave in a certain way. Also, some units may have CE cells (or BD3 cells) with logarithmic subtracting mechanisms that cause them to learn more rapidly than other CE cells (or BD3 cells) in other units. This provides a propensity for some units to learn certain behavior more quickly than other units. Also, units in some layers of nodal networks may be predisposed to exercise greater influence upon the actuators than units at other layers. Units in some layers may be influenced more by some sensors than other sensors. Some units may contain predetermined behavior that cannot be changed. This behavior may influence the behavior of other empirical units in the network. Behavior can be predispose in many different ways in a network.

Results of units working in a network

There is no limit to the number of units that can be connected in a large complex network if each can contribute to the success of the overall system and each does not interfere with the successful behavior of other units. The law of noninterference states that if two or more empirical units are connected, they will keep changing until they cease to interfere with one another. Eventually asynchronous behavior will occur in separate units, units will learn to act at different times on interfering variables, multiple units will learn to work together to contribute to a successful unitary line of behavior, and other units will finally act separately upon diverse variables. But this success will not come quickly or easily in some cases. There will be interferences, conflicts, successes, and failures to carry out behavior. Some units will pool their successful behavior to dominate the behavior of other units. The behavior of some units will emerge, and the behavior of other units will submerge or even disappear. The result may be a machine made up of subunits that can produce more successful behavior than a single large unit working alone.

Limitations of permanently connected units

Permanently connected units have a hard time dealing with sensor and actuator variables that change randomly. These units must learn which values of these variables have no meaning. They must devote empirical cells and learning time to nullifying the influence of these impertinent variables. This problem can be greatly reduced if units can disconnect themselves from random variables and reconnect themselves to variables that lead to successful behavior. This kind of behavior can be carried out by the mobile units discussed in the next chapters.

Chapter 25
Mobile Empirical Units

Until now, we have discussed control units with input and output variables that are permanently connected to a specific set of variables. Controlled machines do not have to be permanently connected to a given set of variables. *Mobile machines* can move around and select the set of variables with which they interact.

Scanners

An empirical unit that can select the set of variables upon which it operates may find that it can control one input variable at a time. If it can establish a procedure for interrogating the values of each variable and supply successful responses to each variable, it may produce a greater variety of behavior with far fewer empirical cells than a unit that attempts to deal with all of the variables simultaneously.

Police scanners

In many applications, the values of many input variables in the environment may remain constant for long periods of time. For example, channels in the police and emergency radio band do not carry messages most of the time. The messages that are carried are kept very short by using code words to describe common situations. In this application, the time over which a variable remains constant is much longer than the time over which a variable is active.

There may be more than a dozen channels in the police and emergency radio band. In order for an operator to monitor all of the channels simultaneously, a radio operator would have to have as many radio receivers as there are channels. Since no signal is present on any one these channels most of the time, a radio operator could quickly and continuously scan through the band of channels using a single receiver to find any (active) channel carrying a message.

A *radio scanner* carries out this function automatically. It sequentially monitors the audio output of each channel, within a set of channels called its *scan region,* until it encounters a signal of predetermined power. It then interrupts its scanning behavior while the transmission continues. This requires only one radio instead of the dozen or more needed to monitor all of the channels simultaneously within its scan region. Receiving only one channel at a time also ensures that two or more messages are not received simultaneously. If two or more messages were received simultaneously, they might be unintelligible.

Empirical scanning units

An *empirical scanning unit* may develop behavior that operates much in the same way as a police scanner. An extra output device can be added to move the unit among a set of sensor variables, Also, another input variable can be added to measure to what variable the unit is connected, as shown in Figure 25.1. An empirical unit can then be connected in some sequence to many different variables.

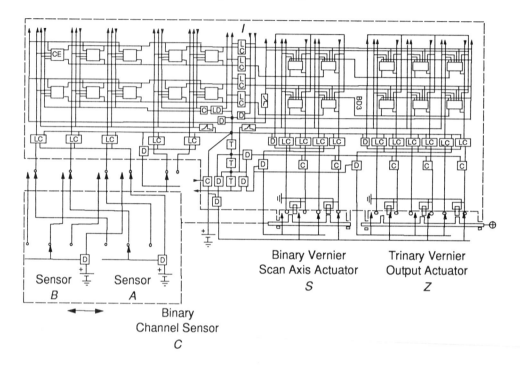

Figure 25.1. A *scanning unit* with one sensor and actuator variable can deal effectively with multiple variables by dealing with one variable at a time.

By using only two matrix input variables and only two output variables, a scanning unit can deal with many different sensors. One matrix input variable, the *location variable,* is used to keep track of with what environment variable the unit is dealing, and the other input variable, the *sensor variable,* is used to detect the value of the particular sensor to which the unit is connected.

One output variable, the *scan actuator,* is used to move the sensor and actuator variables to different variables within its scan region, and the other *actuator variable* is used to influence the value of the particular outside variable to which the unit is connected. Thus, an empirical scanning unit can be designed to deal with many diverse variables with far fewer empirical cells than a stationary unit permanently connected to these variables.

Application of an empirical scanning unit

The temperature in a room may change only a few degrees in a minute, whereas the temperature may be determined in less than a second by measuring the value of a thermistor left in the room. Therefore, a temperature scanning unit shown in Figure 25.2 may scan the values of thermistors placed in many rooms and turn a heater in each room on or turn off depending upon the temperature measured and the desired temperature established for each room.

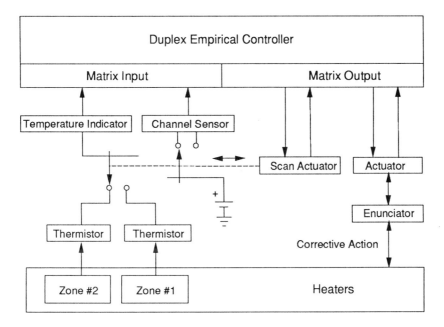

Figure 25.2. A *scanning empirical temperature control system* can determine the temperature of two or more rooms at different locations at different times.

Instead of waiting at one location for the temperature to change, a scanning controller can connect itself to sensors in different locations and then produce the needed behavior at each location.

Time sharing and multiplexing

In the examples given, each location shares the available time of the scanning unit. This is a *timesharing system* from the point of view of each location. This process can be looked at from a different point of view. Since the scanner dispenses a set of different signals to a receiver, it is a *multiplexing system* from the point of view of the receiver.

Organization of a scanning unit

The sensor and actuator variables of a scanning unit must be able to form unitary or diverse relations with its outside variables, depending upon the nature of the environment task. The scanner control system will behave as a *unitary system* if each combination of location and sensor values within the scan region produces a potentially unique value of the scan actuator and the actuator variable state at any one or all of the outside variables within the scan region. This unitary system produces a single line of behavior. The scanner control system will behave as a *diverse system* if each combination of location and sensor value within a subset of the variables in the scan region produces a potentially unique value of the scan actuator that has no relation to the values of the other location and sensor values. A diverse system can produce two or more lines of behavior.

Equilibrium

The signal on each channel may be produced by one or more than one source. For example, a single dispatcher may simultaneously transmit messages on all of the channels served by the scanner or more likely may transmit messages on different channels at different times. Both situations occur when a single dispatcher covers many broadcast channels such as the police, emergency, rescue squad, civil defense, marine, etc. Since the dispatcher can transmit only one message at a time, the scanner can pick up all of the messages if it scans quickly, or if it develops a scan pattern matching the pattern of the dispatcher. In this case, the scanner can develop successful behavior if it develops the same unitary behavior as the dispatcher. This is an example of how an empirical unit can develop successful behavior only if it establishes behavior that has the same organization and content as the environment to which it is connected. This is called the *principle of equilibrium of behavior* (see laws of behavior in Part V). If there were two dispatchers, the scanner would have to develop a scan pattern that allows it to scan both dispatchers. In a sense, it would have to be at two places simultaneously.

Operation of a scanning unit in a unitary region

If the variables in the environment are related to one another, where a value of one variable relates to the value of another as in the example of a single dispatcher covering multiple channels, the scan actuator must develop a line of behavior that brings the scan sensor and actuator variables to the specific variables that need to be acted upon to produce successful behavior. For example, if a scanning unit is used to carry out the function of a security guard in a single building, each room or area of concern may be considered a variable within a larger unitary system. A scanning actuator can cause a sensor to monitor the temperature or sound levels at each critical location, as shown in Figure 25.3.

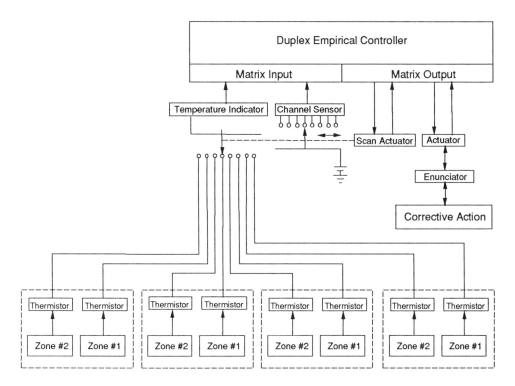

Figure 25.3. A *scanning unit in a unitary region* may have to produce specific behavior depending upon the values found at each location and the values of sensors at other locations.

If the temperature or sound level exceeds empirically established levels, the scanner must activate an alarm. Two types of behavior must be produced:

1. Some kind of action must be taken, depending upon the location and the nature of the event encountered,
2. The scanner must go to a particular new location, depending upon the location and nature of the event encountered at a given location.

For example, if high temperature indicates a fire in a given location, the unit must initiate a fire alarm, and then search locations around that location to determine the extent of the fire. If an unusual sound level indicates the presence of an intruder, the unit must initiate a burglar alarm and search for the point of entry. Obviously, the scanner cannot keep up with too many active environmental variables.

Operation of a scanning unit in a diverse region

If the variables in the environment are divided into separate and independent locations, a single scanning unit is faced with a more difficult problem than the unitary system shown. If the transition or cycle time of the variables at a given location is nearly the same as the scanner, the scanner may have to stay at that location to produce successful behavior since it may lose messages if it moves to other locations. If the transition times of the variables at a location are much longer than the scanner, the scanner may move from one location to another and establish successful behavior at each location. Thus, a single scanner may have trouble dealing with a *diverse region* of variables in which there is no relation between the behavior of variables at one location and other locations.

Two or more scanners

Since one scanner can pick up only one signal at a time, it cannot receive two messages that occur simultaneously. This problem can be solved by using two or more scanning units and providing some way of coordinating their scanning behavior.

Division of labor

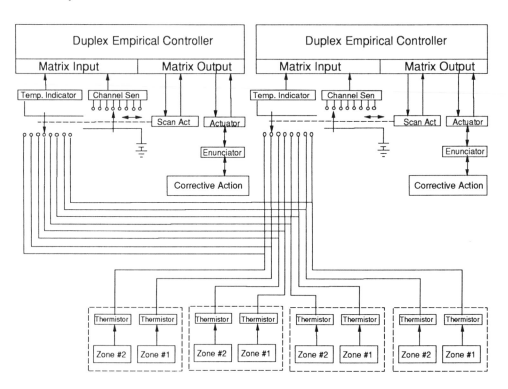

Figure 25.4. Multiple scanners operating upon different channels are less likely to miss messages.

Two or more scanners can be used to reduce or eliminate the problem that occurs when two or more messages arrive simultaneously on different channels. If the scan pattern of the units is random, the probability of all of the scanners arriving at the same channel at a given time is reduced if there are many scanners. However, if each unit operates upon a different set of channels, as shown in Figure 25.4, the number of messages that they can receive successfully may equal the number of scanners. If different scanners are physically prohibited from occupying the same channel simultaneously, the normal operation of the principle of noninterference will cause each empirical scan actuator to select different channels upon which to operate. (See laws of empirical behavior in Part V.)

Strength in numbers

If the signal on a given channel is too weak to be understood clearly, it is possible that it can be made more intelligible by having two or more scanners tuned to the *same channel*. Thus two or more scanners may work together to produce behavior that is more successful than one unit working alone.

Two or more dispatchers

Two or more dispatchers may transmit upon two or more different channels. If there are two or more scanners, the minimum interference and maximum success occurs when each scanner selects the channels of a given dispatcher and each scanner develops the sequence of that dispatcher. This corresponds to the set of scanners developing the same behavior as the set of dispatchers. This is another example of equilibrium of behavior.

Mobile units

A mobile empirical unit is similar to a scanner in that both can select the variables upon which they operate. However, the mobile unit moves around to the different variables, whereas a scanner usually sits still and has a means of changing the variables to which it is connected. Both types of units provide a new dimension in which to create behavior. Units that are permanently wired to their environment can only respond to a specific set of variables. Yet mobile and scanning units can also select the set of variables upon which they operate.

Mobile empirical units

A *mobile empirical unit* consists of an empirical controller with a specific set of input and output devices and some means to change the set of variables upon which they operate, as shown in Figure 25.5.

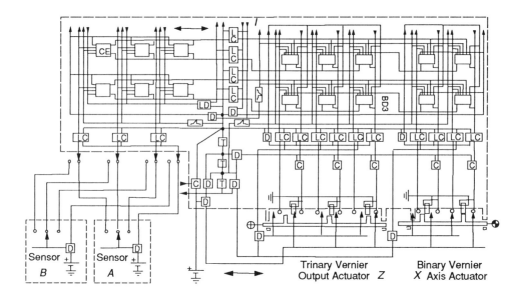

Figure 25.5. A *mobile empirical unit* can move from one set of variables in the environment to another.

A common example of a mobile empirical unit is an industrial robot that can do a task at one machine location and then move to another machine to do a task at that location.

Location

In general, a mobile unit can move to a specific *location* and deal with a specific set of variables. Each location contains a set of input and output variables that can be temporarily connected to the input and output variables of the mobile unit. The set of values of the variables at each location that is present at a given time is the *state of the location*. A mobile unit can be connected to just one location at a given time. Thus, it can be confronted with just one location state at a given time. The variables at a given location may form a diverse or unitary system. At a *diverse location*, each variable at a given location is separate and independent. At a *unitary location*, the variables are interrelated such that a change in any variable at that location may change the meaning of the other variables at that location.

Region

The set of locations in which a mobile unit can operate defines its *operating region*. The mobile unit may stay in one location or move about to all of the locations within its operating region. A region may consist of a set of locations that are separate and independent. In this case, it is a *diverse region*. If the state in any or each location in a region changes the significance of the state at other locations within that region, it is a *unitary region*.

Operation of a mobile empirical unit

A mobile empirical unit needs a means of moving about in its region and connecting itself to the environment variables at each location. A mobile unit must have at least one more input variable than a permanently connected unit, the *location sensor,* which keeps track of where the unit is located, and at least one more actuator than a permanently connected unit, the *location actuator,* to move the unit to each new location. In the example to be given, the *rotary axis* of a *part handling robot* is the location actuator that provides the means of moving from one location to another, and a *rotary encoder* on this axis may be used as the location sensor to measure the position of the mobile unit (Fig. 25.6).

Task cycle time versus a unit's cycle time

In many applications, the time required for a task to take place at a given location is longer than time needed by a Unit to tend the task. For example, unbaked loaves of bread may be loaded in an oven in a relatively short time compared to the time it takes for the bread to bake.

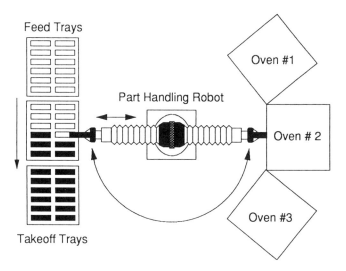

Figure 25.6. A mobile robot may produce the *same behavior at different locations* at different times.

Instead of waiting at one location for the bread to bake, a mobile robot can move to another oven in its region and operate upon that oven while the bread it loaded into the first oven is baking.

Responsibilities of a single unit at unitary or diverse locations

The task at each location may be unitary or diverse. In the example given above, removing bread from a single oven is basically a unitary task involving a single line of behavior. Cooking eggs at each location may be considered a diverse task requiring special behavior for each egg. A single mobile unit has to attend to the variables at a given location according to the time between transitions of the variables at that location. In general, if the time between transitions at a given location is long in relation to the transition time of the mobile unit, the unit may move to other locations during those idle periods. A diverse location may require more attention than a unitary location because the transitions may occur in a more unpredictable or asynchronous pattern. Thus a unit may not be able to move to other locations if it is dealing with a location of diverse variables.

Search for consistency

The set of variables at one location is just part of the set of variables that may be available to a mobile unit. If a mobile unit cannot establish a closed line of behavior at one location, it may find successful behavior at another location.

Dispersive Behavior

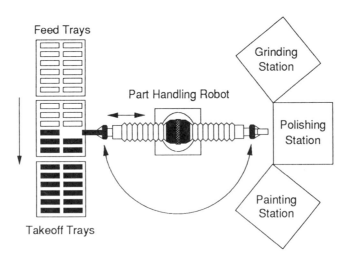

Figure 25.7. A mobile robot may produce *different behavior at different locations.*

Though a unit can operate upon only one location at a given time, a mobile unit can carry out different behavior at each different location at different times. For example, a mobile robot may move a product along different work stations where different operations are done on the product, such as grinding or polishing, as shown in Figure 25.7. A mobile robot can perform different behavior at each location if it can sense its location.

Behavior of a mobile unit

If the task cycle time at a given location is long in comparison to the robot cycle time, the robot may move to different locations. For example, a security guard can quickly determine if any unusual conditions exist in a given area so that he or she may move to other areas. Thus, there are two parts to the behavior of a mobile unit:

1. Behavior performed in each location.
2. Movement within the region to and from different locations.

It is important that a mobile unit keep track of its location. If it gets lost, it may not produce the needed behavior at a given location, or it may not find the location that needs behavior. Thus, it may not produce successful behavior if it cannot keep track of its location.

Advantages of a mobile unit

A mobile unit may do the work of two or more units when different tasks can be carried out at different locations in sequence, like the examples given. This reduces the amount of machinery needed to carry out the set of tasks and reduces the number of required memory cells if some behavior needed at one location can be used at the other locations.

Since the controller of a mobile unit can select the set of variables upon which it operates, it may allow a mobile empirical unit to avoid those locations that produce inconsistent results and to select those locations that provide consistent doable behavior.

Since a mobile unit can avoid inconsistent or random variables, its learning time may be reduced and the controller of the mobile unit can learn to operate successfully with fewer memory cells.

Since the mobile unit may produce different behavior at different locations, the opportunity for successful behavior is increased and the interference between different lines of behavior within the controller is reduced.

Noninterfering behavior in mobile units

As stated, a mobile unit may select the set of variables upon which it operates. It may therefore select a set of variables or an environment that is more benign than others and avoid locations that create interferences. A permanently connected unit must dedicate memory cells to block inputs from randomly changing variables. A mobile unit can move away from variables that interfere with its behavior and move to variables that contribute to its behavior.

Robot vehicles

Specifically designed axes of a robot may be used to move the robot from one location to another or from one region to another. When a robot is free to move its entire self into a new space, it may be considered a *robot vehicle* (Fig. 25.8). An empirical robot vehicle also can select the set of variables that provide the greatest opportunity for self-sustaining behavior.

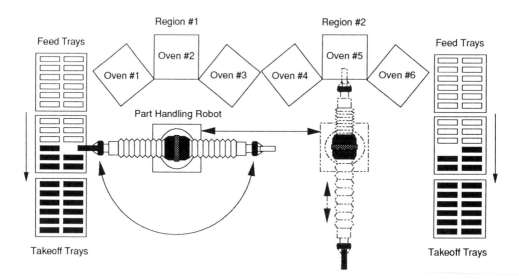

Figure 25.8. A *robot vehicle* can move away from a restrictive inconsistent region and move toward a region that allows it to carry out its behavior.

If the behavior of an empirical robot vehicle is autonomous, it will attempt to find a region in which it can produce successful behavior and attempt to develop behavior that keeps it within that region.

Autonomous robot vehicles

There are many examples of potentially *autonomous robot vehicles* (ARV), such as:

- Remote controlled aircraft, tanks, and submarines for potential use in combat.
- Industrial material handling trucks, farm and lawn equipment, excavation devices, and security robots for use in an industrial setting.
- Autonomous mobile robots and other autonomously controlled vehicles for use in scientific data gathering and exploration.

Each of these devices may be free to move around under the control of a human operator. If it is an empirical mobile unit (EMU), it can develop the behavior expected of it by its operator, and it may take over some or all of the control functions produced by an operator. It may also develop its own successful self-sustaining behavior and become totally independent of the human operator if it discovers and incorporates new successful behavior into its repertoire of behavior. An empirical mobile unit may also develop behavior that is integrated with the behavior of their operators. These units may become personal empirically trained systems (PETS). Some mobile systems may be developed that are never controlled directly by a human operator, but are allowed to develop their own behavior based only upon what is allowed by their environment and human co-workers. The design, development, and behavior of these autonomous systems bear the closest resemblance to human and animal systems.

Mobile unit in a unitary region

Each location available to a mobile unit may be part of a single *unitary region*. In this case, the behavior in one location is related to the behavior of other locations. Thus, the behavior required at one location may be different when the other locations are in different states. For example, a robot attending several ovens may have to abandon its behavior at one location if the bread at another location is about to burn. The behavior of a mobile unit in this unitary system is quite different from the behavior in a diverse region, where behavior at each location can be carried out independently of the behavior at other locations. If all the locations accessible to a mobile unit make up a unitary region, the mobile unit must move around to determine the states of other locations in the region (Fig. 25.9).

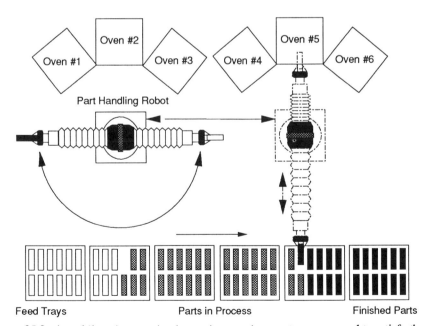

Figure 25.9. A *mobile unit operating in a unitary region* must move around to satisfy the behavioral requirements of the system.

For example, a security guard may have to monitor conditions at many different locations within a building. Different conditions at different locations may require different responses. A fire in a chemical storeroom may require a different response than a fire in a wastebasket in the lobby. If there is a fire in a chemical storeroom, the security guard may have to ignore a fire in the lobby. Since different responses are required for different combinations of conditions in different locations, the task behavior is unitary.

Sequential unitary behavior

The requirements of this unitary task environment can only be meet successfully by a line of behavior that carries the security guard to all of the potential dangerous locations and includes the appropriate behavior for each combination of the conditions that are encountered. If a mobile unit can perform the *sequential behavior* needed to meet the task requirements within a unitary region, it may appear to do the work of a much larger single unitary controller that is permanently connected to all of the variables in the region. Obviously, at some point, a single mobile unit may not satisfy the needs for unitary behavior that occurs at two or more locations simultaneously. In this case, a single mobile unit cannot carry out successful behavior.

Mobile unit in a diverse region

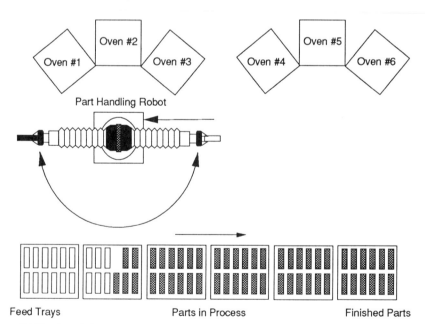

Figure 25.10. A *mobile unit operating in a diverse region* may produce successful behavior at one location and may have to disregard the other locations.

In some cases, some locations within a region may be separate and independent of the other locations. That is to say, the behavior of one set of locations has no influence upon the behavior of another set of locations. The behavior of a mobile unit in this *diverse region* is different from the behavior of a mobile unit in a unitary region. If the set of locations in the operating region of a mobile unit forms a diverse region, the mobile unit may produce successful behavior in one region without regard for conditions at other locations (Fig. 25.10). A given location may provide the consistency and opportunity for a mobile unit to establish sustainable behavior. For example, enough bread may be baked in the first three ovens shown above. There may be no need to use the additional three ovens. Once a unit establishes a closed line of behavior, there may be no impetus to move around to other locations. If the mobile unit can establish successful behavior using a set of variables at one location, the unit may stay at that location and cease to be mobile.

Nomadic behavior

If the time required to attend to the task at one location is short compared to the time that the unit has available, it becomes difficult for the unit to establish consistent behavior. A mobile unit may then choose to move around to different tasks, creating *nomadic behavior*. For example, a diverse task region for a robot security guard may be made up of a set of separate companies and buildings in which there is no relation between locations and specific combinations of conditions at other locations are not important in a given location. A robot security guard may be able to monitor one building (location) in ten minutes every hour. However, the robot may not earn enough money working ten minutes every hour, and may try to strike a deal with the other companies in the region to monitor their businesses as well.

If it takes the full hour to monitor one building and a round must be made every hour, the robot must stick with the one job to satisfy the customer at that location. The tasks that may have to be carried out at the other locations must be left undone. In a purely diverse region, the mobile unit may not know that needs are not being met at other locations. It may be better for the unit to leave tasks undone in some locations in a diverse region than to produce unsuccessful behavior at all locations. This is not a successful option in a unitary region where information must be gathered from every location to produce successful behavior at any location.

Robot choreography

Each axis of any multiaxis machine, including a robot, must move according to the required *protocol* (sequence) and the required *priorities* (rank) of the task. In general, the movement of the robot from one location another occurs at the lowest frequency. The rotary or x,y,z motions occur at some multiple of this base frequency. The gripper open/close motions must occur at an even higher frequency. Sensing and fault detection must occur at still higher frequencies. On/off and task selection decisions may be made at an even lower frequency than the movement of the robot to different locations.

Each axis may have to wait for the lower frequency axes to complete their movements before they initiate their movements. In many cases, the axes that operate at the same frequency can be controlled by one set of control units while the axes that operate at other frequencies are controlled by other control units. Since these different axes operate at multiples of the base frequency, these control units can be arranged in a hierarchical network with the operations that occur at the lowest frequencies controlled by the ultimate node.

Problems with a single mobile unit in a region made up of diverse locations

Like the single scanner attempting to deal with two or more dispatchers, a single mobile unit may have difficulty dealing with a region made up of two or more diverse sets of locations. The successful behavior it develops for one set of locations may be compromised by attempts to produce behavior in another set of locations if behavior is required at both locations simultaneously. In general, the solution to this problem consists of using more than one unit and providing a separate unit for each diverse location, as shown in the next chapter.

Chapter 26
Multiple Mobile Units

It is difficult for an individual to carry out tasks at two different places simultaneously. A working mother knows how difficult it is to do her job at work when she has a sick child at home. In general, this problem is solved by getting some help from other individuals. Likewise, a *group* of mobile empirical units can work together as a *union* to produce unitary behavior or divide into separate units that produce diverse behavior at different *locations*. There are many more opportunities to form groups or separate into separate units if each unit can select the set of variables upon which it operates. If two or more diverse tasks must be carried out asynchronously in different locations within a *region*, two or more mobile units can handle the tasks separately. Yet, the behavior of units in this region may have to be coordinated to deal effective with a set of tasks in which the behavior at one location is dependent upon the conditions at another location. This requires the formation of a union of controlled machines in which the behavior of each unit takes into account the behavior of other members of the group in a unitary manner.

Two or more mobile units acting at the same location

Two or more mobile units may act at the same *location*. They face the same problems that two or more permanently connected units face when they must deal with the same set of variables. They may have to behave differently depending upon whether the location is unitary or diverse.

Dominance

Two or more units may be connected to the same set of variables at a given location. One unit may be stronger than the other units connected to these variables. For example, the controls of the pilot in the cockpit of an airplane may have a higher force amplification than the controls of a copilot. This gives the pilot greater *authority* and allows the behavior of the pilot to *dominate* the behavior of the copilot. Also, acting with authority or acting toward authority can be a learned behavior that comes about when one asks the question: Do I have the authority to respond to these conditions? If a unit chooses to *abstain* from action at a location, it exhibits less authority than a unit that acts at that location.

Behavior of a group of mobile units at a unitary location

If two or more units attempt to operate at the same location, and the location is made up of a set of unitary variables, all units must produce the same behavior or establish a division of labor based upon dispersion or field selection at a given time, like a permanently connected superimposed system.

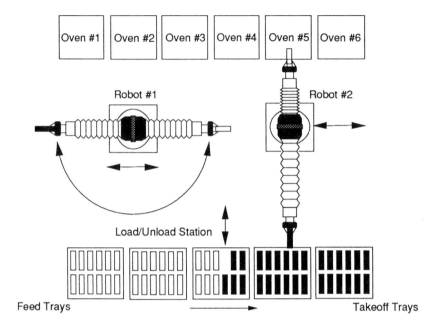

Figure 26.1. A *group of mobile units at a unitary location* must establish compatible behavior.

The material handling system shown in Figure 26.1 is unitary because both units must act according to the behavior of the other unit to use the single load/unload station. In general, the behavior of the stronger unit will prevail. For example, a stronger unit may impose its behavior upon the other unit by pushing it aside to use the load/unload station. If there are more than two units operating at a given location, the weaker units may *combine* their efforts to overpower a stronger unit to make their behavior prevail at that location.

Backup units

Once two or more units develop the same behavior at a given location, one may suspend its behavior at that location. For example, once the pilot and copilot develop the same behavior, one may discontinue his or her involvement in the control process, and the other can continue to fly the plane. This allows one unit to be the *backup unit* for the other unit. In *industrial robot* applications, this feature allows one robot to be taken off line for repairs or maintenance. It also allows a remote controlled mobile empirical unit to function if its remote control signal is lost.

Communication

Each unit may also operate upon different variables in a unitary location if each unit establishes a method of knowing the critical values of the variables of the other units.

For example, the pilot and copilot may operate upon a different set of variables. The pilot may decide to fly the plane, and the copilot may do the navigating. However, the pilot and copilot must communicate with one another at critical moments, such as at a course change. This coordinating *communication* permits an effective division of labor that makes the set of units act like a far larger single unitary unit.

Behavior of a group of mobile units at a diverse location

If two or more units decide to operate upon a single location made up of a set of diverse variables (Fig. 26.2), they may divide the task so that each operates upon separate variables and holds the other variables constant.

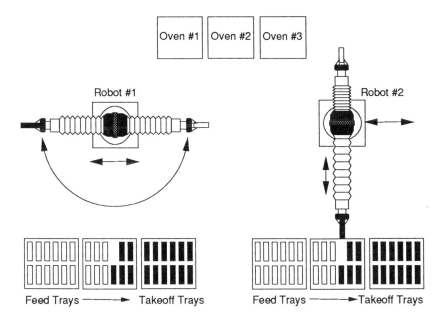

Figure 26.2. A *group of mobile units at a diverse location* may produce different behavior depending upon the variables to which they are connected.

In this case, each unit determines its own behavior according to the requirements of the particular set of variables upon which it operates. If the location consists of diverse variables, each unit may produce different behavior, and there is no need for any coordinating communication.

Behavior of a group of mobile units in a region of variables

Two or more mobile units can operate within a *region* made up of two or more locations. The region may be *diverse* if the behavior at each location is separate and independent of the behavior at other locations. This configuration creates the opportunity for different mobile units to work *separately* in different locations. They may deal with these diverse tasks far more efficiently than a large single system. The region is *unitary* if the behavior at each location is dependent upon the behavior at other locations. To produce successful behavior at a given location, mobile units in a unitary region may have to move from one location to another to determine the conditions at other locations. They also may establish a means of communication that allows them to act as a single unit.

Two or more mobile units in a diverse region

More mobile units can be added to the region made up of diverse locations, forming a *group of mobile units* in a diverse region, as shown in Figure 26.3.

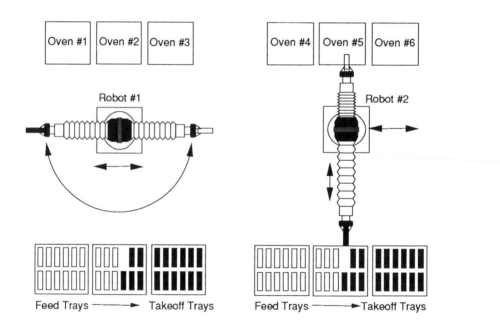

Figure 26.3. A *group of mobile units in a diverse region* may operate successfully at different locations.

All of the units may decide to operate at the same location. For example, two units may not be strong enough to lift up a heavy casting. Two or more units may need to work together to lift something that is too heavy for one unit to life alone. A unit may choose to move to a location not occupied by the other units, depending upon the nature of the task at that location. For example, a unit may find a location where the castings are light enough for one unit to lift by itself.

Interference

If two or more units operate at the same location and the behavior of one interferes with the behavior of the other, one or more of the units may *abandon* that location and seek another location in which to operate, as shown in Figure 26.4.

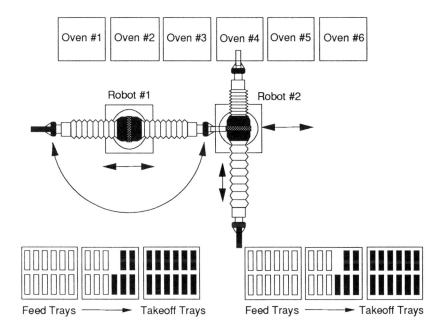

Oven #1 | Oven #2 | Oven #3 | Oven #4 | Oven #5 | Oven #6

Robot #1   Robot #2

Feed Trays ——→ Takeoff Trays   Feed Trays ——→ Takeoff Trays

Figure 26.4. An *interfering unit* at a given location in a diverse region may have to abandon that location and attempt to produce behavior at another location.

The interferences between units may make it impossible for one or more of the units to produce consistent successful behavior. For example, driving an automobile is a highly unitary task since an automobile can go only in one direction and at one speed at a given moment. Thus, a driver must consider each combination of values of many variables that occur at a given moment, and specific combinations of values of these variables have to be produced at a given moment. If a second person attempts to become involved in this process, the driver may be overwhelmed by the complexity of controlling the car while still trying to meet the demands of the "back seat driver." The interfering individual may have to refrain from behavior or the behavior of both individuals may have to change so that the behavior of one does not interfere with the behavior of the other.

Strategies for successful behavior of mobile units in a diverse region

Mobile units may cluster around a given location to help each other produce behavior, or each unit may operate upon separate locations within the region. Their decision to work together or to separate is determined by which *strategy of behavior* produces successful behavior. If they can produce successful behavior by working alone, they will probably do so because their behavior can be carried out with fewer interferences. Thus, each unit must compete or cooperate with other units at a given location, ignore those units, or abdicate the location occupied by other units.

Behavior of a group of units in a unitary region

Two or more units may occupy a region made up of a set of locations that make up a unitary system (Fig. 26.5). Each unit in a unitary region must produce behavior that considers the behavior of the other units while also dealing with the conditions at their own location.

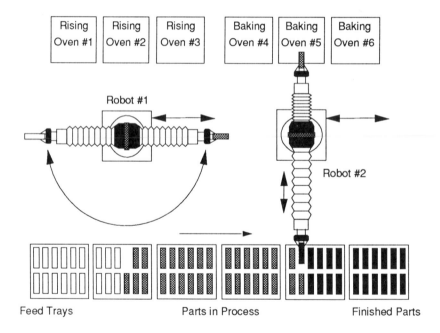

Figure 26.5. Each unit in a *group of mobile units in a unitary region* must produce behavior that considers the behavior of the other units.

This unitary configuration may require that each unit produce more complex behavior than the behavior required of a unit in a diverse region. Each unit must move around to determined conditions at different locations, or each unit must send and receive information as to the conditions at different locations to carry out successful behavior at its own location.

Compatible behavior in a unitary region

Like permanently connected units, two or more mobile units can produce *compatible behavior* by dividing the control tasks appropriately so that each supplies part of the behavior needed to complete the task successfully. For example, the pilot and copilot may be responsible for different tasks, though they operate upon the same instrument and control system. The pilot may fly the aircraft while the copilot does the navigation; thus the pilot operates upon the flying instruments and controls, and the copilot operates upon the navigation instruments and controls. Having two or more units in a unitary region creates the potential for many interferences. In general, these interferences can be resolved only under the conditions given in Table 26.1.

Table 26.1. Compatible behavior of mobile units in a unitary region

---

1. Units operating in a unitary region must produce the same behavior; or
2. Units must establish a division of labor based upon dispersion or value field selection; or
3. Units must establish a method of communicating among themselves that allows them to coordinate their behavior; or
4. Some units must abstain from producing behavior by holding the values of their action variables constant; or
5. Some units may abdicate a location in which they interfere with the behavior of another unit; or
6. Some units must abdicate that region.

The units may also have to move among the different locations within a unitary region to encounter the different combinations of values of the variables that make up the unitary system. This kind of behavior may be far more complex than the behavior of units that can deal solely with the variables at their own location.

Same behavior

There are many instances in which it is highly desirable for multiple units to produce the same behavior. For example:

- Fish swim in schools. This allows the collection of fish to act as a single unit.
- Soldiers march in formation. This allows their commander to move all of the troops to a desired location without some troops getting lost or scattered.
- Many animals establish herds or flocks in which they move together. This may increase the ability of the group to sense and react to danger.

In general, having the members of a group act in unison decreases the *complexity of the behavior* required of each individual unit.

Working in unison

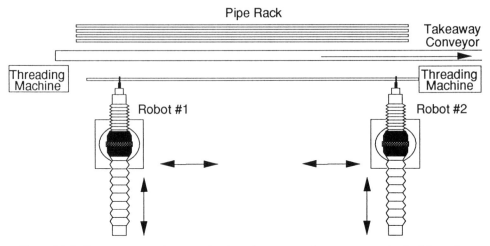

Figure 26.6. Two or more *units operating in unison* can produce a greater physical force than a single unit acting alone.

Mobile units may also *work in unison* to produce a greater physical *force* than one unit working alone, as shown in Figure 26.6. For example two or more robots may work together to lift an object that is too heavy for one unit to lift alone. By working together, they can achieve behavior that would not be possible by working alone.

Backup units

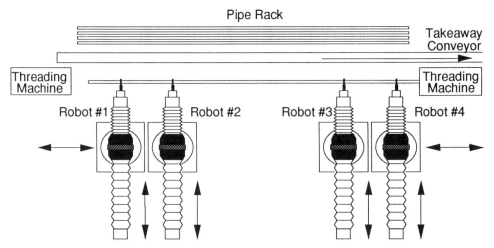

Figure 26.7. A *backup unit* obtains its behavior by learning to work in unison with another unit.

When two or more units produce the same behavior, each may act as the backup of the other, as shown in Figure 26.7. This configuration also allows one controller to "teach" or change the behavior of another unit. The empirical algorithm causes the units to evolve behavior that reduces interferences between them.

Cooperation

Mobile units may also work together in a given region to perform useful behavior. For example, two industrial robots may work together to paint products, as shown in Figure 26.8.

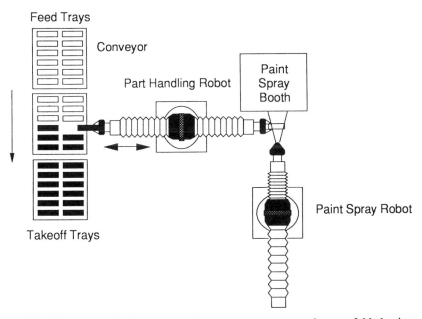

Figure 26.8. Two or more independent units may *cooperate* to produce useful behavior.

One robot may manipulate the spray nozzle, and the other robot may position the product. Some coating applications cannot be carried out using a single robot. Obviously more units can be added to this group. For example, a third robot could bring the unpainted product into the spray booth and place the finished product upon drying trays. This example of cooperation is based upon a useful division of labor among units in a unitary region. Thus, two or more units may develop behavior that allows them to accomplish tasks that cannot be completed by a single unit. In the coating example given, each unit must respond in a unique way to each combination of states of the other units. If two or more units can act as if they are a unitary system, far fewer memory cells may be required, as compared to a larger single unit acting alone.

Domination

In general, when two or more units produce the same behavior, some factor must be present that forces the units to produce the same behavior. For example, a student pilot learns to fly by developing the flying behavior of the flight instructor. The instructor must exert dominance over the trainee pilot. This may be accomplished in many ways. One method is to make the controls used by the instructor have a *higher force amplification* (gain) than the controls used by the trainee. Another method is to be sure that the instructor is *physically stronger* than the trainee. Another method is to make the controls responsive only to the operator who is *first to act* upon them. The instructor may be the first to act because the instructor is supposed to know more about what to do in a given situation than the student due to the instructor's greater flying *knowledge and confidence*.

Intercommunication

Units may have to know what is happening at other locations in a unitary region to produce the correct behavior. Some method of *communication* may be found that allows each unit to know how to respond based upon to the combination of states experienced by the other units. The simplest method for mobile units to inform other units about their behavior is to use a communal status register that acts like a *communal bulletin board* (Fig. 26.9).

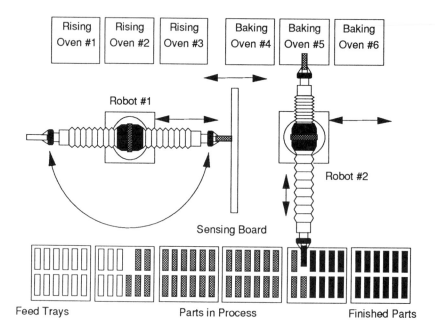

Figure 26.9. A *communal bulletin board* allows each mobile unit to inform the other units about its behavior so they may know what combinations of states have occurred at other locations.

Each unit may take the information concerning the status of the other units into account in making decisions concerning its own behavior. This allows for the creation of unitary behavior in a system consisting of subunits using far fewer memory cells than a single large unit. For example, flying an airplane is a task in a unitary region because different responses may be required for many combinations of conditions at each set of instrument locations. The pilot and copilot normally deal with a unique set of instruments and controls. Yet, each can receive messages on their *cockpit voice channel* as to the values of the instruments and controls at locations to produce more coordinated control behavior.

Other methods of communication

Communication among mobile units may be necessary to establish successful behavior. This behavior may be discovered by a set of units or may be established by a training process. When two or more units develop successful behavior in a unitary region, they may be quite dependent upon the information they receive from other units in that region to carry out successful behavior. Part of the behavior of each unit may involve keeping other units informed as to potential interference at different locations. Part of the successful behavior of each unit may consist of obtaining and acting upon information from other units.

There are several possible methods of communication among mobile units. As we have seen, units may deposit information at a *communal bulletin board*. These messages may impart information as to conditions in different regions. For example, a unit may inform other units that there is an obstacle that blocks a particular pathway. By avoiding this pathway, other units may carry out behavior more successfully than if they were unaware of the obstacle.

Units may also *broadcast information* to other units within a region. If information from other regions has no significance to units inside a given region, some method must be provided to ensure that outside messages do not interfere with the messages within the region. To coordinate the transmission of messages, a single announcer may be designated. This process creates a common variable upon which all units in a region can operate. This ties together the behavior of the units in a region to an even greater extent.

There also may be sites within a region connected to a *data bus,* to which units can connect to obtain information. If the information provided by these data terminals helps each unit establish successful behavior, going to and receiving these data will be an important part of the behavior of each unit.

Competition

Two mobile robots may have access to a production system consisting of a parts supply and take-away conveyor and a process machine, as shown in Figure 26.10.

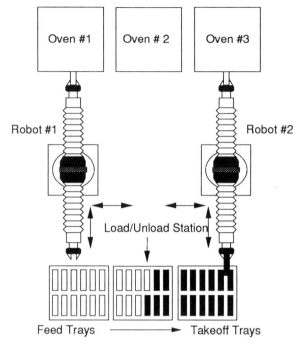

Figure 26.10. Superimposed mobile units may *compete* to dominate the behavior of a set of variables.

Like the superimposed permanently connected units shown earlier, superimposed mobile empirical units are likely to interfere with each other as they attempt to dominate behavior. The stronger, more skilled, more confident, and/or more persistent unit will attempt to push the other unit out of the position it assumes at a given moment in time. The dominant unit will prevail if it selects a task that it can carry out successfully by itself.

Abstention

The complexity created by having multiple units in a unitary region can be reduced if some units abstain from behavior at least part of the time. We are expected to be quiet during a performance in a theater or in a library. We are taught not to interrupt while other people are speaking. Sentries do not call out except to warn of an intruder. Passengers are requested not to speak to the driver of a bus while it is in motion. In many cases, knowing when to abstain from acting is as important as knowing when and how to act. A baseball batter who swings at a wild pitch will not succeed. A tennis player who hits a ball that is going out may lose the point. A military force may be better off ignoring what appears to be an opportunity to advance if it leads to an ambush. A second robot may contribute most to the successful operation of a material handling system by choosing to ignore the process being carried out by another robot, as shown in Figure 26.11. In another example, the *z* axis of a robot must remain constant, despite sensed conditions, to track an object moving on a horizontal conveyor.

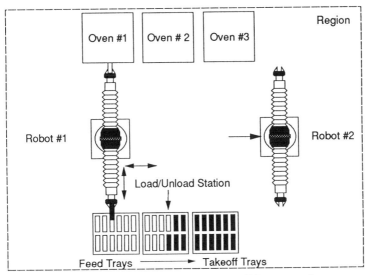

Figure 26.11. A unit may have to *ignore* some sensed conditions to produce successful behavior.

Abdication

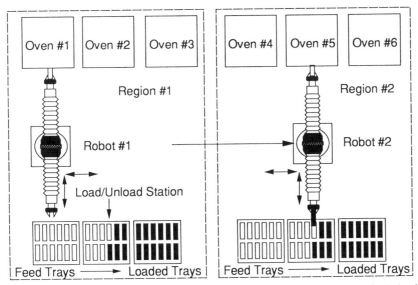

Figure 26.12. A unit may choose to *abdicate* a region to produce more successful behavior elsewhere.

Another solution to the confusion created when there are too many units are in a unitary region is for one or more of the units to leave the region. Many animals attempt to exclude other members of their species from their territory. An animal is faced with the decision of whether to deal with the other animals in a given territory or to *abdicate* their claim to the territory and find a region where there is less competition (Fig. 26.12). A mobile robot that has tasks to accomplish in other regions must leave a given region when its task there is completed.

An individual may recognize another individual who has done harm to them in the past. The individual may be wise enough to leave the region to avoid being harmed again. This may initiate a chase/escape episode (see Part I). In general, it is wise for a unit to abdicate a location where it is unable to produce successful behavior, and seek a location where it can carry out behavior without interference.

Union of mobile units

If mobile empirical units find themselves in a unitary region and find ways of producing successful behavior by working together, they may form a working *union*. In general, there is no reason to form a union in a diverse region, since each unit may produce successful behavior by itself at a given location.

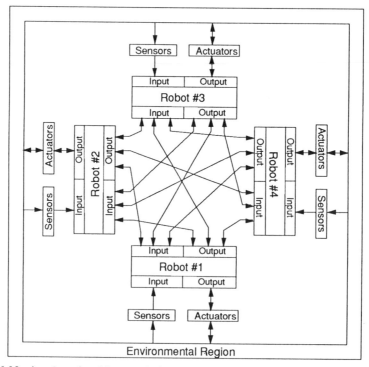

Figure 26.13. A *union of mobile units* is formed when two or more units work together success-fully to determine the values of the set of variables in a unitary region and develop behavior that keeps them together.

However, in a unitary region, units may develop successful behavior only by becoming part of a union of machines that accomplish tasks together and exchange information among units about the states at other locations. Independent units are transformed into a union when they develop compatible behavior and develop behavior that keeps them together (Fig. 26.13). Each unit may operate upon a limited set of variables within a unitary region and rely upon a common set of communication variables to determine the status of the other variables in the region. A union is successful only when the units within the union develop and produce behavior that contributes to the successful behavior of other members of the union, that does not interfere with the behavior of the other units, and keeps the union together. A special environment may have to exist through which they may all communicate.

Types of behavior of mobile units in a unitary region

The behavior of mobile units in a unitary region is made up of at least three categories (Table 26.2).

Table 26.2. Type of behavior of mobile units in a unitary region

---

1. Behavior of each unit at each location,
2. Movement of each unit among the set of locations,
3. Communication to and from the other units in the region.

Each type of behavior involves more or less interaction among the units in a region. In general, an increase in interaction cause an increase in potential interferences. An exception to this generalization occurs when the members of a union cooperate.

Strategies of behavior of mobile units in the same region

There are four basic strategies of behavior available to each unit in a given region: compete, cooperate, abstain, or abdicate. Successful behavior can evolve only if each unit discovers and practices the appropriate mode of behavior. The alternative strategies available to each unit are shown in the diagram in Figure 26.14. The *type of behavior* produced by the set of mobile units can be plotted along orthogonal axes with "More or Less Interaction" on one axis and "More or Less Interference" on the other axis. The sum of these effects can then be shown on the *strategy axes* that are located 45° from the type of behavior axes.

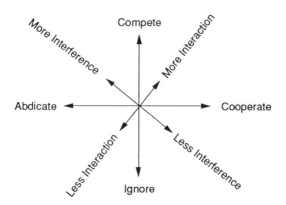

Figure 26.14. A plot of the *strategies of behavior of mobile empirical units* shows that a mobile unit has the additional choice of abdicating a location, a choice that a fixed unit does not have.

At any given moment, a unit may either cooperate or compete with another unit to do a unitary task, or a unit may choose to ignore the behavior of other units or abdicate a location or region to carry out a diverse task. The eventual choice is based upon which behavior results in the most successful behavior for each unit.

Conflicts

The members of the union are held together by the mutual benefit they gain by being a member of a group. These benefits may be threatened by the existence of other unions. Thus, successful behavior may have to be obtained by dominating another union. This process is often seen in human relationships. For example, the captain, mates, and crew of a ship usually form the single strong union needed to operate their vessel successfully. However, one hears from time to time about how the captain and a third mate may form a union that acts against a union of the first and second mate. Since no one individual may dominate the officers on the bridge of the ship around the clock, a union of two officers may maintain their dominance together over the other officers throughout the 24-hour day. These conflicts also apply to "mechanical" empirical systems. The formation of dominant groups derives from the empirical algorithm that directs empirical systems to expand behavior into areas where there is less interference. If less interference is created by units working together against other units, then the system will develop these alliances. Unions of empirical units can even join to create coalitions that can act against other coalitions of unions.

Noninterference among mobile units

If empirical machines are free to move around within an environment, they may select the set of environmental variables upon which they operate, and may also select the set of units with which they interact. This greatly increases the opportunities for empirical units to organize themselves according the task environment. For example, a group of units may come together to deal with a specific set of variables that inherently form a unitary system, or they may separate to deal independently with diverse variables. They may also divide the task environment into the separate tasks that can be performed by separate units. The behavior of a group of empirical units will keep changing until they divide themselves into a skillful arrangement of parts that do not interfere with one another. The resulting organization reflects an understanding of the task environment, though the organization may not be clearly understood nor overtly expressed by the designer or operator of the system. Thus, a set of mobile empirical units may provide a far more effective control system in many applications than the present method of predetermined and preprogrammed control, which requires that the programmer and operator have a complete and accurate understanding of all aspects of the control problem before the system is put into operation.

Self-organization of mobile units

The environment of a region occupied by mobile units may contain unitary and diverse locations. Each unit in the region will attempt to acquire successful behavior. Successful behavior can be achieved by cooperating with other units at unitary locations. Successful behavior may be achieved by other units acting alone at diverse locations. Thus, the mobile units may organize themselves according to the organization of the task environment. Since the mobile units can choose among diverse and unitary sets of variables, a system of mobile units can be self-organizing.

Bonding theory

The lines of behavior produced by a controlled machine must be produced by the action created by specific memory cells within each unit. By connecting different lines of behavior to different parts of different machines, it is possible to visualize how each machine can be connected to other machines. This representation of connections among machines is the basis of *bonding theory*.

Representation of an individual

A mobile empirical unit or an individual may be represented by a set of sensor/actuator transitions, as shown in Figure 26.15. Some transitions create lines of behavior that involve internal connections. Other transitions create lines of behavior that extend through the environment. And other open transitions are not yet complete.

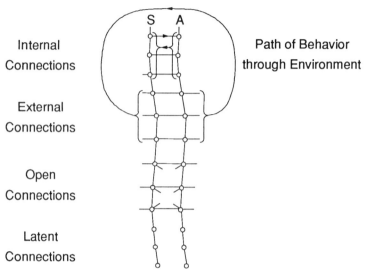

Figure 26.15. In bonding theory, an *individual* is represented by a set of sensor/actuator trans-itions.

Pair bond

Some of these open transitions can be completed by series connections passing through some open transitions in other individuals, as shown in Figure 26.16. These series connections enlarge the repertoire of behavior of the individual and may help two individuals complete tasks that require cooperation.

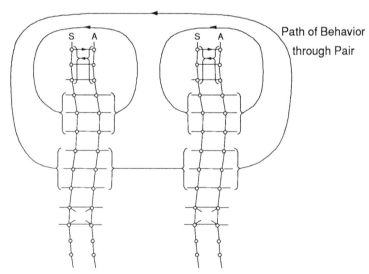

Figure 26.16. A *pair bond* is formed when series connections are made between the open trans-itions in two individuals.

Group bonds

Path of Behavior through Group

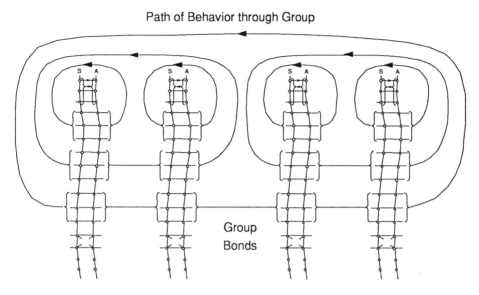

Figure 26.17. A *group bond* is formed when series connections are made through two or more individuals.

These series connections can be extended through more than two individuals, forming a union of individuals (Fig. 26.17). The union is based upon behavior that involves each member of the union. This behavior usually bestows advantages on members of the union that are not experienced by individuals outside the union.

Induced bonds

Unions are also held together by *induced bonds*. If two primary units establish mutually dependent behavior, and each primary unit has established mutually dependent behavior with other secondary units, these secondary units may find that they are held together by means of their connection with the primary units, as shown in Figure 26.18. For example, if the wives of two couples become friends, chances are that a friendly relation will develop between their husbands. This process of creating induced bonds amplifies the bonding forces among the members of the union.

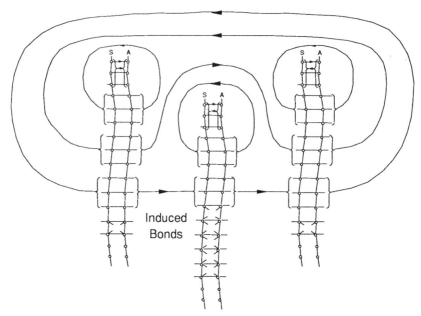

Figure 26.18. Induced bonds occur when the behavioral bonding of two units contributes to bonding of other units.

Antibonds

In some cases, empirical units may not resolve conflicts with other units. Successful behavior may be obtained only by avoiding those units. If units acquire this kind of antibonding behavior, a force may appear to be pushing these units apart. The maintenance of these antibonds may become an essential part of their successful behavior.

Methods of bonding

If two or more units have a greater chance of producing successful behavior by working together, they can reap the benefits of this behavior only if they stay together. Therefore, part of their successful behavior must consist of behavior that keeps successfully cooperating units together.

Units may develop behavior that includes meeting under certain conditions and at certain places to exchange information as to the states encountered by the other units or to accomplish some task together. The behavior that brings units together at these *rendezvous* appears as if it is caused by a physical force of attraction among these units.

Most mobile units have a *home base* to which they go when there are no tasks required of them. This home base may be used for recharging their batteries and/or exchanging data with other units. Units may practice basic activities at these locations. Also, an operator can know where to find mobile units if they are trained to check in at a home base from time to time.

Members of a family may have many good reasons to come home. The house may be warm. There may be food in the refrigerator. Their may be books or toys there. They may find a family member there who is patient and understanding. They may not be welcome anywhere else. But they must still learn to come home and must learn how to get home from wherever they go. This returning behavior is essential if they are to experience the successful behavior they have established at home. This invisible force, which brings members of a group together to produce successful behavior jointly, is called *behavioral bonding*.

Many dogs that "run away" simply may have become lost. When they are lost, they lose the value of the behavior that developed through their experience with their human family. So a significant part of the effort spent by members of a group of mobile units must consist of behavior that keeps these units together. Getting lost may be one of our greatest fears. It is a fear similar to our fear of death. If we become lost in the woods, much of our knowledge and experience becomes worthless. We are separated from our assets like our friends, our possessions, and our usual source of food and shelter. Even our money has no value. This loss of our resources is similar to the loss that occurs when we die. However, if we were experienced woodsmen, our reaction to being lost would be different. Since we would be accustomed to living in the woods, we would not be separated from what we know.

Since communication among units may contribute greatly to the successful behavior of each unit, the behavior involved in creating communications with other units appears to hold the units together. For example, the movement to and from a communal bulletin board and the responses to the contents of the board create a kind of bonding behavior among the mobile units. If obstacles are placed in the way of these movements, the units must find ways of overcoming these obstacles to receive the behavioral benefits of the information. Thus, successfully operating bonded units will appear to strive to stay together. Once our behavior is based upon living in a group, we must protect that behavior by staying with the group. Examples of *bonding behavior* that keep groups together are as follows:

- Parties bring members of a community together.
- Church services bring members of religious groups together.
- Staff meetings bring department members together.
- Conferences bring members of a profession together.
- Club meetings bring members of a club together.

When two or more units choose to operate upon the same set of variables and establish successful behavior, they establish a mutual dependence that takes the form of behavioral bonds. If the specific activities that bring members of a group together are lost or omitted, the group will cease to exist.

Chemical bonds

When two or more mobile empirical units work together to deal with environmental tasks, the success of each may be dependent upon the behavior of the other unit(s). In this case, each unit must develop behavior that keeps it with the other units that contribute to its successful behavior. This *bonding* force is based upon their shared behavior, much like the covalent bond in chemistry is based upon the sharing of electrons among two or more atoms, forming a molecule, as shown in Figure 26.19.

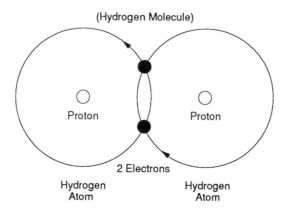

Figure 26.19. The *covalent bond* between two hydrogen atoms comes about because they share a pair of electrons.

This bonding is a result of the benefits obtained by each atoms having an electron to balance its charge and the need of a molecule to have two electrons in its outer shell. In many cases, an individual unit can produce doable, consistent, and successful closed lines of behavior only if it is part of an appropriately organized group. The formation of groups of mobile units is another example of the unitization of behavior that allows a set of units to accomplish far more complex tasks than a single unit working alone.

Chapter 27
Higher Levels of Organization

A successful union of mobile units may move about as a single assembly due to the cohesive bonded behavior of its members. It may join with other unions to form coalitions that can deal with massive unitary environments or coalitions of other unions. These allied systems form the basis of the *higher levels of behavioral systems* known as social groups, cultures, and societies. The next section shows how two or more unions can solidify into a larger assembly that can appear to act as a single unit and even dominate other assemblies of mobile units.

Coalitions

To achieve greater strength or to achieve more successful behavior, unions of mobile robots may join to form a *coalition of unions*. This creates new opportunities for unitization and increases the potential for producing far superior behavior with fewer memory cells than a single large unit acting alone.

Strategies of behavior

Just as there are four fundamental ways in which a mobile unit can behave, there are four fundamental ways in which a union can behave. It can compete with other unions, join as an equal partner with other unions, ignore other unions or defect from the set of regions occupied by other unions, as shown in Figure 27.1.

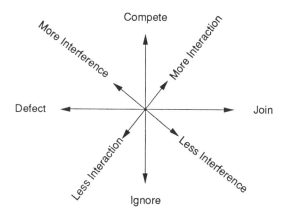

Figure 27.1. The are four *strategies of behavior* available to a union in dealing with other unions.

For each union, one of these strategies is optimal. Each must answer the question: Should I attempt to *dominate* the same set of variables as another union, or should I attempt to find a way to *support* the behavior of another union, or should I *ignore* the activities of another union, or should I *leave* the region occupied by another union? In general a set of units may try these strategies in its interaction with another union and then may choose the one that presents minimum difficulties or interferences in the way it has chosen to behave.

Competition

All unions in a given region have an intrinsic potential to compete for space and access to variables, and to produce action. Unions with the best organization, behavior, and cooperation among its members will prevail. Unions that succeed gain greater strength if the success of the union increases the cooperation among its member units.

Dominance

Acting in unison with other unions may create a greater force than a single union acting alone. For example, trade unions may join in a sympathy strike with another union to increase the impact of the strike. Thus, those unions that cooperate by forming alliances with other unions may prevail over unions that attempt to remain autonomous. Like other empirical units, a union that can act quickly or with greater persistence is also more likely to prevail.

Isolation

A union may determine that it is better off by ignoring or not competing with other unions. It may succeed only by concentrating entirely upon its own tasks or by establishing itself in a new territory.

Unitary coalitions

A set of groups is unitary whenever its behavior can be defined by a single line of behavior. If there is a singularity of purpose, a single measure of success, or a single result, the group is unitary. For example, most of the groups within a single company form a unitary assembly because the success of the company can be defined by its profit or loss statement.

Diverse groups

Groups are diverse when each can act independently, when each does not have to consider the actions of others and where combinations of actions of the different groups have no significance. For example, families in a community remain separate and independent most of the time (unless they unite for or against some issue that they feel threatens the community as a whole such as the prospect of building a hazardous waste dump in their community).

Unitary and diverse coalitions

If a trade union is faced with an industry that is highly organized and communicates and cooperates extensively, it must organize itself in much the same manner as the industry to deal effectively with the industry. A set of unions may form a coalition like the Congress of Industrial Organizers (CIO) and the American Federation of Labor (AFL). These coalitions later joined to form the AFL/CIO. If a trade union is faced with a single isolated (diverse) industry, it may be better off remaining independent to avoid the liabilities that may be incurred by helping other unions. If the trade unions in an industry form a coalition, the industry may have to form some kind of an industrial trade organization to deal effectively with the union coalition.

Thus, coalitions are formed when groups are confronted with a single issue or unitary challenge. The successful behavior of a coalition of unions is based upon each union being better able to achieve its goals because of its being a member of the coalition and the ability of the coalition to establish behavior that keeps the coalition together.

Principle of equilibrium

Behavioral units can either join to form larger assemblies or separate into smaller subassemblies depending upon the organization of the task environment. For example, the trade unions in the automotive industry in this country were successful only when they organized themselves in the same way as the businesses in that industry. This is an example of the *principle of equilibrium*.

Principle of unitization

Units within an organization can also join to form larger units or separate into smaller units, depending upon which configuration produces the most successful behavior. Thus, organizations can be made up of assemblies and subassemblies much like an automobile is made up of chassis, body, and drive train assemblies. These assemblies can also be divided into subassemblies such as front end, rear end, etc. Empirical units can organize themselves according to the assemblies and subassemblies in the task environment. A unit that is too small or too large in relation to its task environment may encounter more problems than a unit with a behavioral variety that more nearly matches the variety of the task environment. This process of reorganization is an example of the *principle of unitization*. It comes about in systems that operate according to the empirical algorithm. In these systems, behavior that works is repeated and behavior that runs into too much interference is abandoned.

Organization of higher-level groups

The process of *organization* involves combining units into groups, separating other units from groups and placing them into other groups, and combining groups into new entities. It is a process of amalgamating, solidifying, disintegrating, decomposing, rearranging, sorting, composing, and combining. As we have seen, there are two fundamentally different types of organization — diverse or unitary. According to the principle of equilibrium, an empirical unit must organize itself according to the unitary and/or diverse organization of the environment with which it must deal. According to the principle of unitization, an empirical system must establish a set of assemblies and subassemblies, based upon the organization of the task environment, to produce successful behavior.

Unitization of empirical groups

Government, business, military, and religious organizations clearly demonstrate the unitization of groups. For example, a company *organization chart* shows how its many groups are assembled into a single entity (Fig. 27.2).

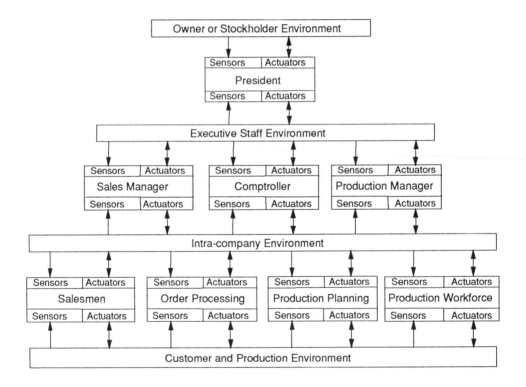

Figure 27.2. A *company organization chart* shows the structure of the group as a whole, sub-groups, and basic units within the organization.

A complete organization chart shows how units (individuals) are combined into work groups under a group leader, section groups under a supervisor, departments under a manager, functional groups under a director, divisions under a vice president, and the company under a president. Since most companies change according to the results of their actions, most business organizations may be considered empirical units.

A pyramid of tasks

The behavior of this type of organization can be shown in the diagram in Figure 27.3 showing the functions of different groups within the organization.

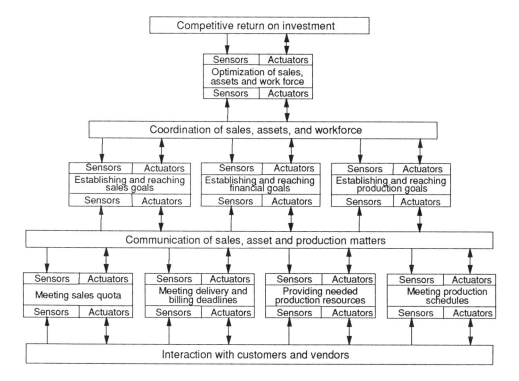

Figure 27.3. The organization of a company must be based upon its *pyramid of tasks.*

These tasks can be arranged in a typical organizational pyramid. The success of any organization is based upon its ability to coordinate its resources to achieve specific objectives. In a business organization, these tasks involve behavior that supports the primary task at the top of the pyramid of producing more revenue than is expended in the total costs of producing that revenue.

Behavior of an organization as the sum of its parts

In the hierarchical organization shown, the behavior of each individual in the organization must reflect the specific unitary tasks of the organization. The line behavior of the organization is made up of the *sum of the lines of behavior of its parts.* Individuals in the organization must communicate with others to assure themselves that they are contributing to and not interfering with these company tasks.

The activities of the subunits in a hierarchical organization may be influenced by a single decision-making body. This kind of organization minimizes the amount of communication required to achieve a unified effort if is configured appropriately. Any changes in the tasks confronting the organization may require that the structure of the organization be changed to reflect the organization of the new task environment.

Diverse organization

Some companies are made up of separate and independent units in which there is no relation between the behavior of one unit to the behavior of another unit, as shown in Figure 27.4.

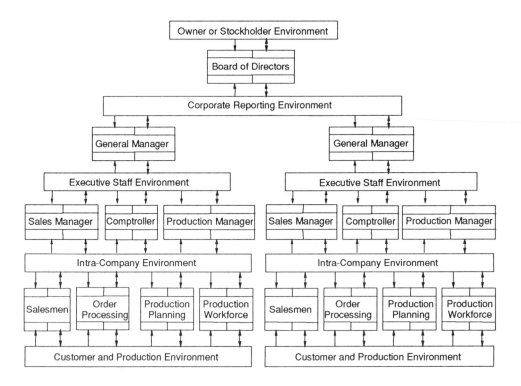

Figure 27.4. A company has a *diverse organization* if it consists of a set of independent units.

No single line of behavior characterizes a diverse organization. Its behavior is made up of a set of separate lines of behavior. Many corporations consist of holding companies that own companies in diverse fields. In most cases, these diverse companies have nothing to do with each other, though they report to the same holding company.

Self-organization of economic units

The choice of how companies are organized is determined by the marketplace. If there is a singular need or issue in the environment, an empirical unit must organize itself to meet that need. If a *market* is unitary (represents a single whole), the business organization must be unitary. If the market is diverse, the business organizations will eventually change into diverse units. These changes come about because of the efforts of each unit to serve its market in the most efficient manner.

Economic systems

A *free market economy* may also be considered an empirical system. Its behavior is based upon the empirical algorithm, which causes successful economic behavior to prevail and unsuccessful behavior to disappear. Economic units are often assembled into groups that interfere, compete, cooperate, dominate, and abdicate economic interests. Each must match the physical constraints and timing relations inherent in the economic environment.

Economic environment

The *economic environment* is made up of the physical locations of raw materials, population centers, existing means of transportation, the knowledge, abilities, and social values of the members of a society, and many other noneconomic factors that may influence economic decisions.

Economic behavior

Economic behavior is made up of the many individual transfers of wealth from one person or organization to another. These exchanges of goods, services, and money lead to the accumulation of assets or loss of assets by each individual and organization over time. The *distribution of these assets* in large part determines the future behavior of the economic system in much the same way the distribution of sensitivities of the memory cells in an empirical matrix determines the behavior of the matrix.

Assets as a measure of successful economic behavior

The accumulation of *money* and other *assets* by an individual or organization shows that their economic activities have been carried out successfully. We usually do not physically restrain other people from doing what we do not wish them to do, nor do we force them to do what we want them to do. However, we may pay people to work for us when they carry out the behavior we want of them, and cease to pay them when they fail to carry out the behavior we want of them.

So the savings and other assets of an individual or other economic units are a measure of how successful they have been in carrying out actions that are acceptable to other individuals. Their assets are also a measure of their confidence and ability to carry out actions successfully in the future. Large assets show that an individual knows how to carry out money-earning behavior and allows that individual to engage in other exchanges of value in the future. Individuals with few assets cannot participate in economic behavior in the future.

An economic model

Flow of money and material in a free market system is similar to the flow of action and transform conditioning behavior in a nodal network of empirical machines, as shown in Figure 27.5.

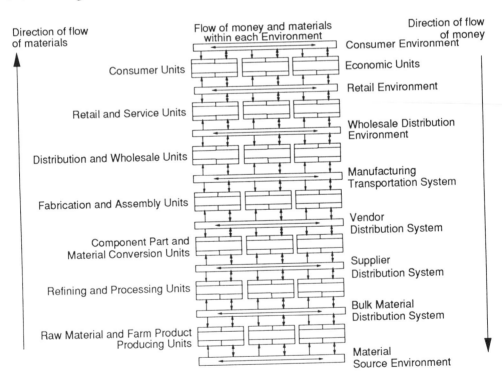

Figure 27.5. An economic model shows how money flows in one direction while goods and services flow in the opposite direction. This is similar to the flow of action and transform conditioning information in a nodal network of empirical units.

The raw materials at the bottom flow upward as they are transformed into processed materials such sheet metal, wire, pipe, bar stock, paints, fabrics, etc. These basic mill products are distributed to fabricators who convert them into finished goods. The finished goods continue to flow upward and outward through wholesalers to retailers and finally to consumers. Money from the consumers who buy these products then flows downward through the distributors to the manufacturers to the suppliers of raw materials. Unsuccessful products that remain unsold or are sold at a discount do not generate the money needed to compensate for the cost of producing them, leading to losses of assets in those units that made the decision to produce them. This is similar to how actions that cannot be carried out are not reflected back through a network of empirical units.

A *closed economic system*

Most of the materials in consumer goods that are discarded can be used again as the raw materials for new consumer goods. If these recycleable materials are sold to waste recycling dealers and reprocessed into raw materials, the flow of both money and materials continues to flow in a complete circle, as shown in Figure 27.6.

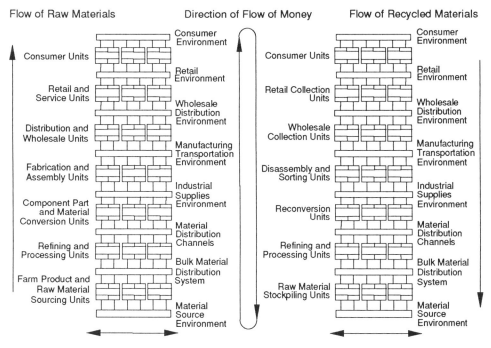

Figure 27.6. A *closed economic model* shows that both money and materials flow in a closed loop in opposite directions, as in a bipartite nodal network of empirical units.

Just as the raw materials are distributed to manufacturers, the reusable materials may be collected for reprocessing. As the raw materials are distributed to many locations, the money is collected into fewer locations. As the reusable materials are collected in fewer locations, the money paid for these materials are distributed to more locations.

Spontaneous creation of successful economic behavior

We have shown how empirical behavior originates in independent empirical units that are responsible for deciding their future actions. Economic units in a free economy are also responsible for making their own *economic decisions*. Each unit keeps track of its successes and its failures. Economic units keep track of their successes and failures by means of their balance sheet. According to the principle of supply and demand, the price of goods and services that are in great demand, but are in limited supply, will rise. Thus, those units at the raw material level that make accurate *predictions* about what materials will be needed in the goods sold to consumers may be paid more than those units that provide raw materials that are not wanted.

Conflict and interference among economic units

Different units may make different estimates about what materials, goods and services will be needed. Different distributors and retailers may choose different combinations of products to sell and locations in which to do business. Different marketing groups may target different consumer groups. Units that serve the needs of the customers to the greatest degree put themselves in the best position to receive the greatest payback. Units that fail to provide the goods and services that customers want may be forced out of business or forced to change the way in which they do business by units that make more accurate estimates of what goods and services are needed. In most economic environments, it is difficult or impossible to determine *a priori* what materials, goods, and services may be needed. Thus, *competition* among different economic units allows the market to determine *a posteriori* what materials, goods, and services are needed in a given economic environment.

Self-sustaining behavior of economic units

Each economic unit is primarily concerned with establishing *self-sustaining economic behavior*. This behavior consists of taking in enough money to cover the cost of its supplies and the cost of doing business. This requires the creation of a closed loop of money and materials that returns to some desired economic state at certain moments, such as the time of the monthly balance statement. This implies an overall balance of cost and revenue that allows the behavior of the economic unit to be ongoing.

Power and dominance of economic units

Economic units with more assets have a greater ability to influence what materials, products, and services are available to them than units with fewer assets. The accumulation of capital provides the high confidence levels needed to attempt new economic activities in much the same way that high confidence levels in an empirical matrix cause successful units to attempt behavior.

Organization of economic units

Economic units can be arranged in unitary and/or diverse groups. Some oil companies are vertically organized such that the oil products flow from the oil well to the gas pump within the one company. A centrally planned economy is an other example of a unitary economic organization. Small farms, auto repair businesses, and small retail stores can be examples of diverse economic units. The *organization of economic units* usually follows the organization of the economic environment. A region with poor transportation and poor communications that isolates economic units is more likely to be organized diversely than metropolitan region with a well developed public transportation system and many sources of communication.

Superimposition of economic units

In competitive markets, two or more economic units act upon the same markets. In many cases, these superimposed units can coexist. For example, there may be several large supermarkets in a given region. Each unit has the choice of competing against, cooperating with, ignoring the other units, or abdicating the region. In many cases, a single very large store may not be as satisfactory to the customer as a set of smaller stores that may be located closer to more customers and may not require a parking lot as large as a single large store. Superimposed units may also establish a division of labor based upon value field selection by selling more or less expensive versions of the same products or may establish a division of labor based upon dispersion by selling a different types of items.

Equilibrium within an economic system

An economic system will reflect the technology in the economic environment. In general, supermarkets stock food that people know how to prepare and have the tools that are needed in its preparation. For example, now that most of us have microwave ovens, we can find many food products on the shelves of grocery stores that can be cooked in a microwave oven.

Bonding among economic units

Economic units are often held together by mutual economic dependence. For example, people are drawn to large cities to find economic advantages. The presence of many people creates many economic opportunities. Automotive parts manufacturers grow up around automotive assembly plants. The parts suppliers need the assembly plants, and the assembly plants need the parts suppliers. Since the success of each of these economic units is based upon the behavior of other units, they tend to work together and support each other, forming strong economic bonds.

Timing of economic behavior

Economic units must act according to the time constraints imposed by the environment. Supermarkets must sell their stock of food before it spoils. Loans must be paid off on time in order for the lender to earn the expected return on the loan. Supplies must be shipped within a specific period to meet the demands of a production line, and invoices must be paid within a specific period. Commitments are made in later transactions based upon earlier commitments. Economic units may fail if too many deliveries are not made on time or if customers do not pay their bills on time. Attempting to make deliveries on products too quickly may greatly increase costs. Therefore, those organizations will prevail that have the most appropriate balance between price, quality, and delivery time.

Social systems

Just as empirical units form assemblies and higher levels of unitization, individuals organize themselves into *social groups* with different levels such as the family, community, society, culture, and units with an international consciousness. Each level serves a unique function and has unique features.

The individual as an empirical unit within a social group

A social group may need to perform a variety of tasks to meet the needs of its members. For example, a social group may need a doctor, lawyer, spiritual leader, and political leader. Different individuals may produce the behavior needed to carry out these different tasks. If an individual is to become a recognized member of a social group, the behavior of that individual may have to change and develop until it fits at least one need of the social group. If the behavior of an individual is unacceptable to the members of a social group, the members may choose to ignore that individual, or expel that individual from the group.

Social groups as empirical units

Most social groups are essentially empirical in that their success or failure is based upon their ability to discover and incorporate behavior that meets the requirements of their physical environments, the members of the group, and other social groups. The behavior of a group must change until it can support or sustain itself.

Originative behavior of a group

Most social groups come up with their way of producing behavior and then pass on this information from generation to generation. Each generation may attempt to reproduce the behavior of their parents that seemed to work and provide satisfaction. Some features of their parents' behavior may be rejected by each generation, and each generation may add new features to the behavior of their culture. Even if a given generation copies the entire behavior of the previous generation, that behavior had to be originated, tested, and incorporated by earlier generations of that social group.

Interferences among social groups

Every group must establish behavior that does not interfere with the behavior of other groups if their behavior is to remain unchanged. If the behavior of one group does interfere with the behavior of another group, the interference will cause changes in the behavior of one or both groups. The behavior of these groups will continue to change until their behavior does not interfere. This is because interferences block or stop behavior, causing that behavior to become less prevalent than behavior that can be carried out. Interferences among social groups are dealt with in the same manner as other empirical systems deal with interferences: Social groups compete, cooperate, or ignore other groups, and/or they abandon the region of another group.

Self-sustaining behavior of social groups

Every social group must establish behavior that allows its behavior to continue without outside help. Behavior that generates food, shelter, and clothing must be an intrinsic part of the behavior of a successful social group, otherwise that group may be drawn into another social group that can meet its needs. Means must be provided that perpetuates this successful behavior. Thus, teaching and education must be an integral part of the social behavior of a successful social group. A social group must maintain its population at a level that it can be sustained by its environment, traditions and technology. Otherwise, the behavior of the group may be forced to change, its unique behavior may disappear, or the group may be taken over by another social group that can create more successful behavior.

Dominance of social groups

Some groups may attempt to determine the behavior of other groups. Like other empirical systems, the stronger, wiser, better organized, and more confident social group will prevail in most cases. Dominance may occur because of greater economic wealth, greater military strength, or a greater will to act.

Organization of social groups

Each group must meet the requirements of a unitary or diverse organization. Each group must be unitary enough to hold together and diverse enough to be able to respond to the variety of conditions likely to the encountered in the environment. Also, social groups must be unitized in a way that reflects the organization of the social environment, as shown in Figure 27.7.

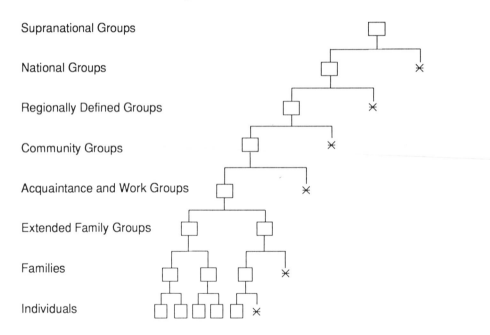

Figure 27.7. The *levels of social organization* must reflect the underlying relation of people to the environment.

Usually the behavior at each social level is different from that at other levels. For example, the family provides food, shelter, clothing and an economic base for its members. The community provides public education, police, fire, and other social services. The nation is responsible for the defence of its borders. Our society establishes uniform laws, language, and the geographic boundaries in which we feel knowledgeable and welcome. Our culture provides religious beliefs and other common expectations, goals, and values. We have increasingly established an international consciousness as we became more aware of the interdependence of people within the limited resources of our planet. Thus, our social order is changing due to the changes in our physical environment.

Superimposition of social groups

Two or more groups may attempt to deal with the same conditions in the environment. For example, two grade schools may serve students in an overlapping area. In most cases, this condition raises questions about which school is closer, which school provides the best education, or has the best physical plant. It also creates the different social groups belonging to the different schools. This may create the need for each group to decide whether to compete, cooperate, or ignore the other group and for individuals within each group to decide whether to stay in their group or join the other group.

Equilibrium of social groups with their environment

The behavior of social groups will reflect the technology required to deal effectively with the environment surrounding the groups. For example, a social organization called the Grange was formed when a large percentage of our population worked on farms. With the disappearance of most family farms in many areas of our country, the Grange has also disappeared in these areas.

Closed behavior of social groups

Social activities must teach behavior that will cause these social activities to be continued in the future. Annual holidays serve the purpose of evoking specific social behavior and reinforcing specific social values. Rituals such as baptism, marriage, and funerals guarantee that each generation will be exposed to the same social and religious experiences and values as past generations. Social behavior must be supported by the members of the society if that behavior is to continue. Values and technologies that are not practiced become extinct. Social groups must support childbearing and child-rearing behavior so that as many individuals are born into the group as are lost to the group by death. The Shakers of New England did not believe in bearing children. Consequently, the Shaker society has become extinct.

Bonding within social groups

As discussed, social groups must establish behavior that holds the group together. Parties, meetings, conventions, rallies, and other social rituals ensure that individuals in the group will come together. Once they are together, other acts of bonding can occur. People may attempt to recall other people's names or relive past common experiences. They may update each other about what has happened to them or their common acquaintances since they last met. These activities make each member of the group more familiar with one another and may increase their ability to identify and solve common problems together.

Bonding among social groups

Attachments can be made between social groups if some behavior needed by one group comes from another group. For example, most service clubs like Lions Club and Rotary International are independent organizations with a charter from a parent organization. Each is responsible for collecting dues and maintaining its membership. However, members of different clubs attend the meetings of other clubs and attend regional and national conferences. These activities are intended to help each club function more effectively. They also create a sense of belonging to a group that is larger than the separate groups at the club level. Bonds between clubs exist in part because the behavior of each club is dependent upon the behavior of members of other clubs.

Timing of social events

Social groups must find ways to synchronize their behavior with real-time events in the environment. For example, planting and harvesting must be done at specific times of the year. Holidays, feasts, and rituals coordinate the behavior of farming groups with these critical farming activities. Meetings usually end by specifying when and where the next meeting will be held. Most religious groups meet at specific times each week. Groups that do not schedule events will find that their organization falls apart.

Summary of empirical machines

The discussion of machine behavior started in Part I with the example of a mechanical spring used to define the modulus of behavior of a simple machine. This discussion of machine behavior continued with control systems that were shown to be similar to certain types of machines like springs. These machines could be controlled by a programmable matrix to produce any kind of machine behavior as shown in Part II. It was shown in Part III that machines having matrices of empirical memory cells could find successful and useful behavior through experience. These empirical machines can be connected in networks and seem to act according to the same principles and laws as most biological, social, economic, business, and government systems, as shown in Part IV. The principles, laws and applications of these empirical machines are summarized in Part V.

PART V
EMPIRICAL BEHAVIOR

Part V shows how empirical machines or networks of empirical machines can accomplish useful tasks. *Empirical control* presents a study of behavior based upon variety, information, and order. These understandings may lead to more effective methods of teaching and establishing more satisfying relations among people.

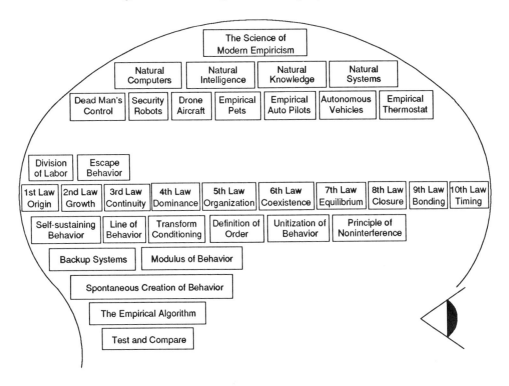

Modern empiricism connects the various disciplines of science in new ways and provides new opportunities to apply science and engineering to the needs of people and other living beings.

Chapter 28
Applications of Empirical Systems

The history of technology can be described as a series of triumphs in the following areas of our physical world:

1. The development of *structures* — as seen in the pyramids of Egypt, the Roman aqueducts, and the cathedrals of Europe.
2. An understanding of the *motion* of objects — as seen in the development of clocks, carriages, and sailing vessels.
3. The harnessing of *power* — as seen in the development of water powered machinery and the creation of heat engines.
4. The *integration of structure, motion, and power* — as seen in the large textile mills and steamships of the past century.
5. The *separation of structure, motion, and power,* inherent in the concept of *control* developed in this century — as seen in the use of switches, relays, electric motors, and an electrical distribution system to apply power from central electrical generators to where it is needed.
6. The invention of *programmable control* — as seen in the relatively recent development of the digital computer.
7. *Empirical control* — a method of establishing control behavior without the need for individuals to specify the control behavior of machines before the machines are put into use.

In many applications, establishing the behavior of a machine while it is in use makes it easier to create complex and high-quality control behavior. Empirical control also leads to new applications for machines that can learn from experience that are not possible using predetermined control processes.

Empirical control

Empirical control is the study of empirical machines that establish behavior after they are built and put into use. They establish behavior through experience by attempting to produce behavior in a given environment, continuing to produce behavior that can be carried out, and ending behavior that cannot be carried out. This *empirical behavior* may improve with use, in contrast to the behavior of most predetermined machines, which may deteriorate as the machines wear out or become obsolete as conditions change in the environment of the machines. The behavior of predetermined machines is built into their design or is programmed first and then continues despite the results, unless the behavior is so unsatisfactory that they destroy themselves or are shut down by an operator. *Modern empiricism* is the study of the changes in organisms and other living and nonliving systems that come about because of experience. Usually these changes appear as changes in behavior. The resulting behavior may reflect the wants and needs of people involved with these systems, and indicate an understanding of the natural relations in the system's environment.

Successful behavior

By finding behavior that can be carried out within the controller/environment system, the empirical machines described in Part III learn to produce behavior that fulfills the opportunities for action within its environment. By selecting action that is allowed, and avoiding action that is not allowed, an empirical machine increases its repertoire of *successful behavior*.

Ongoing behavior

As part of the process of creating behavior that can be carried out, an empirical machine or a network of empirical machines must create behavior that leads to successful behavior and avoids unsuccessful behavior. In general, successful behavior allows behavior to continue, while unsuccessful behavior leads to the termination of behavior. By developing a repertoire of successful behavior, an empirical unit can engage in *ongoing behavior*.

Useful behavior

An operator may decide what machine behavior is desirable, and what behavior can and cannot be carried out by an empirical machine. When an operator can influence the actions of an empirical machine, such as an empirical robot, auto-pilot, drone airplane, remote-controlled submarine, or air traffic control system, the behavior of these empirical machines can be changed so that it is more useful to its operators. This *useful behavior* may act for an operator, allowing the operator to carry out other activities. Since this useful control behavior may act for a human operator, it may have *economic value*.

Quality of behavior

In many situations, it may be difficult or impossible to figure out beforehand what control behavior is successful or desirable under given conditions. Operating a complete unit in an actual environment may provide many insights about the successful operation of the machine. These insights may lead to a higher *quality of behavior* and an improvement in the performance of an empirical system compared to a predetermined system. Each detail of behavior (transition) may be tested and improved upon after an empirical machine is put into use. New relations may be discovered that were not thought of before the use of the machine. If an empirical machine can discover and incorporate new behavior by itself, the quality of behavior is not limited by the skill and diligence of a human operator.

Originative behavior

An empirical machine is compelled to produce whatever actions are called for by the input states and the distribution of sensitivity values in the empirical memory cells in its control matrix (see Part III). It does not wait for each output to be created by an operator as each input is encountered. In many applications, the operator and/or the environment may serve only to prohibit those actions from being carried out that are against their wishes or violate the physical structure of the system or the laws of Nature. If many alternative ways of behaving are possible within the operator/environment system, the eventual behavior of the empirical controller may consist primarily of the behavior originated by the empirical controller. Most empirical machines rely upon their ability to produce *originative behavior* to discover and incorporate the new possible successful behavior in their environments.

Autonomous behavior

In many cases, an empirical machine may work in an environment without the aid of human operators. The environment may allow certain actions to take place under certain conditions but not allow other actions to take place under other conditions. For example, the environment may not allow a mobile robot to walk through a wall. A mobile robot with an empirical controller will stop trying to walk through walls. It will continue to produce other actions that can be carried out, as prescribed in the empirical algorithm. If the mobile robot were placed in a maze, it may learn to negotiate the maze successfully without any intervention from a human operator. Thus, an empirical machine can develop behavior by itself. The ability of an empirical machine to produce *autonomous behavior* allows it to discover the intrinsic viable relations in its environment and incorporate these relations into its behavior.

Self-sustaining behavior

This process of developing behavior that can be carried out and avoiding behavior that cannot be carried out suggests that an empirical machine will develop behavior that appears to seek those conditions that lead to achievable actions. If an empirical machine does not produce a line of behavior that evokes the conditions for which it has achievable actions, the behavior of the empirical machine will change according to the rules of the empirical algorithm until it does create the conditions that lead to these achievable actions. The process of selecting achievable actions under given conditions, selecting the conditions that lead to achievable actions and avoiding conditions that lead to unachievable actions creates the opportunity for the controller to engage in *ongoing achievable behavior*. Since an empirical control unit is compelled to seek behavior that allows its behavior to continue, an empirical machine appears to strive to achieve *self-sustaining behavior*.

Behavioral existence

Since an empirical machine can discover unique ways of behaving within the possible and impossible alternatives available within an environment, it may find unique solutions to the problems posed by its environment that allow its behavior to continue. These solutions may not be understood by its operators and/or attendants. No one may know also of the details or the circumstances that lead to the creation of the particular modulus of behavior of an empirical unit. This behavior may be unique in that no other being may ever establish the same solutions to these problems. Thus, an empirical machine may become a unique being able to create a behavioral life and *behavioral existence* that may be as extraordinary as that produced by a living being. The loss of some empirical behavior may represent the permanent loss of some unique body of knowledge or behavioral aptness that may never be replaced or duplicated. The loss of some irreplaceable behavioral value, established over years of interaction with an empirical unit, may be as devastating to the companion of an empirical unit as the loss of a living being.

Study of empirical control

The *study of empirical control* provides the framework and technology needed to design, build, and understand systems that can create complex autonomous and self-sustaining behavior using a practical number of empirical memory cells. The type of empirical learning and behavior presented in this account may be sufficient to explain most animal behavior, and provide the sophistication and complexity of control needed for machine and robot controllers, while also providing a greater understanding of the empirical process in other fields such as biology, linguistics, sociology, and economics.

Roles for empirical machines

An empirical machine may perform different *roles* in different applications such as being a security guard, lookout, explorer, soldier, teacher, student, apprentice, rescuer, intern, or companion. These roles differ as much as the various roles people carry out, such as being a mother, father, employer, or minister.

Cooperating unit

Two or more empirical machines can work together to provide better control behavior than either unit working alone. For example, they may separate the control process into separate tasks.

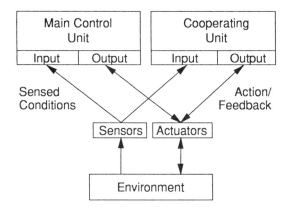

Figure 28.1. A *cooperating unit* may supply some behavior needed by the other unit to produce successful behavior.

If one unit supplies some input and output states needed to maintain the existing line of behavior of the system, it is called a *cooperating unit* (Fig. 28.1). In some cases, one unit cannot function without a cooperating unit. For example, it is very difficult for a land surveyor to work alone. The cooperating unit can be an empirical co-pilot, co-worker, or even a personal bodyguard.

Backup unit

Once an empirical unit produces the same behavior as the operator, the operator can withdraw from the control process, leaving the job of carrying out the control task to the empirical unit. This allows the operator to do other jobs or to rest. An empirical unit that learns a control task from an operator and acts for the operator is called a *backup unit* (Fig. 28.2).

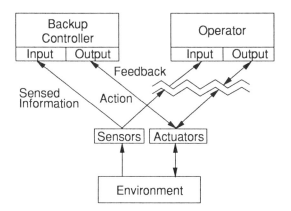

Figure 28.2. A *backup unit* learns the behavior of an operator and acts for the operator when the operator withdraws from the control process.

In the examples of the ship auto-pilot shown earlier, the empirical auto-pilot is a backup unit for the helmsman. In another application, a backup unit can take over the control function of a remote controlled drone aircraft if the signal from the remote pilot is disrupted or lost.

Intern and master units

A less dominant intern machine and a dominant master machine can be connected to the same set of variables. Once the less dominant *intern unit* establishes the same behavior as the *master unit*, the master unit can withdraw from the control process, leaving the job of carrying out the control task to the intern unit. This allows the master unit to do other tasks even on the same set of variables. An empirical unit that learns a control task from another unit and acts in place of the other unit is also called an intern unit (Fig. 28.3).

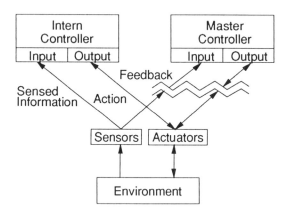

Figure 28.3. An *intern unit* can learn the input/output relations of the master unit and then reproduce that behavior when the master unit is removed from the system.

The master unit must be more dominant than the intern unit through greater strength, quickness, persistence, or knowledge of the task environment. In the example of a ship auto-pilot, a trained master empirical unit can act for the helmsman and teach another intern empirical unit to act as a helmsman. In this way, experienced empirical units can teach other empirical units without the need for human teachers or operators.

Teacher

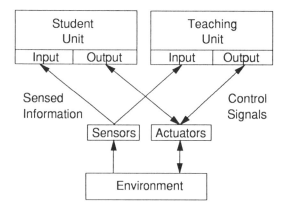

Figure 28.4. A *learning system* is formed when a more dominant teaching unit is placed in the student's environment.

If the one unit is more dominant than the other by virtue of its greater physical strength, persistence, quickness, confidence, or understanding of the task environment, it can be designated as the teaching unit. It can be superimposed upon the student/environment system to affect the behavior of the student unit (Fig. 28.4). In this configuration, the teaching unit acts in parallel upon the same environment with the student. The function of the teaching unit is to establish behavior that can continue after the teaching unit is withdrawn from the student/environment system. This can be accomplished if the teacher helps the student learn the behavior required of the student by the student's environment and then withdraws from the process.

Machine tutoring

An empirical unit can be trained by a skilled educator or a group of skilled educators to ask questions, evaluate answers and select new questions based upon the answers. This machine can then be configured to be more dominant than a student subject and then duplicate the teaching skills of the original educator. When the answers of the subject are not in the repertoire of the empirical teaching unit, it can also be configured to query the skilled educators to increase its repertoire of responses to the subject's answers. This kind of *machine tutoring system* can develop a body of skilled teaching behavior that may exceed that of a single teacher.

The controller as an apprentice

We have given examples of how an empirical controller can assist humans in their activities. A controller, used in this way, acts like an *apprentice*. Table 28.1 lists features of an empirical controller acting in this kind of role.

Table 28.1. The role of an apprentice

- The apprentice unit must be present whenever useful control activities are carried out by the master unit.
- The apprentice unit must participate in the control activities carried out by the master unit.
- The apprentice unit must do simple tasks first under the supervision and to the satisfaction of the master unit and then take over more control functions as it learns the desired control behavior.
- The apprentice unit must help the master unit complete significant and useful tasks that will continue to be useful in the future.

This relationship is started by both individuals working on simple tasks together. The complexity of the tasks must be increased as the apprentice learns to accomplish the simpler tasks. The apprentice is expected to take over more responsibility and may eventually stand in for the master unit entirely. In a similar process of development, the master must also develop behavior that allows the apprentice to take over the task behavior and allows the master to withdraw from the work environment.

For example, an apprentice empirical controller can be put in the cab of the truck and connected to the critical variables with which the driver must deal. It must monitor the relation between the steering wheel and the position of the truck in relation to the road. It must attempt to steer the truck, but the torque produced by its steering motor must be less than that produced by the driver. This ensures that the driver remains in command of the truck and the apprentice. In this example, the controller will learn to act as a *backup controller,* ready to take over the driving control task when needed.

When the controller learns a relation between steering wheel and road position that is acceptable to the driver, the driver may take his or her hands off the wheel for short periods of time (to open and read a road map for instance). If the controller tries to do something "wrong," the driver must correct the action by overriding the steering control motor.

If a situation is encountered in which the controller has little or no confidence, a conditional controller can be designed to emit a warning sound (see Part III). The controller can also be connected to other sensors such as microphones to pick up the sound of horns and other vehicles etc., and be connected to other actuators such as the brakes and accelerator etc., to increase the variety of control functions it can learn.

As the driver and controller encounter many driving circumstances and establish responses for them, the controller can take on more of the driving function. Part of the controller's "training" may consist of sounding an alarm after a certain period if no inputs are received from the driver. If no response is mustered by the alarm, the controller can be trained to execute a slow down and stop maneuver for the case when the driver falls asleep behind the wheel.

A similar type of control relation can exist between the pilot of a *drone aircraft* and an empirical controller in the aircraft itself. The difference here is that the empirical controller is in a different location from the pilot. The onboard empirical controller can monitor the control signals received from the remote pilot and the onboard instrument readings and flight controls (actuators). The empirical controller will attempt to find the consistent relations among events at these variables, and then attempt to reproduce those events that seem to have a high degree of correlation.

With the appropriate empirical controller connected to the necessary sensors and actuators, and sufficient training, the empirical controller can carry out flight maneuvers and complete a mission even if the control signals from the remote pilot are temporarily or even permanently interrupted. This example of an empirical controller used as a backup system in a remote control application is typical of most *telerobotic applications.*

Likewise, a controller in a *remote controlled submarine* may learn basic survival techniques by observing and participating in the typical underwater maneuvers. Some of these maneuvers may be to reverse direction when confronted with a solid wall, or seek and follow open water passageways when inside a sunken ship.

These examples suggest a process in which the operator and an empirical controller simultaneously attempt to specify actions in real-world situations. Incompatible or conflicting behavior is diminished in frequency and desired behavior is increased in frequency. The underlying feature of this process is the *interdependence of man—machine behavior.* Both may need to learn some behavior from the other to produce successful behavior, and both may need to depend upon the other to carry out tasks successfully.

Controller as a companion

We also discussed the possibility of using an empirical machine to participate in the behavior of a person who is cut off from his or her customary behavior for some reason, such as a sudden disability or the loss of a longtime companion. In this example, an empirical *companion unit* must have the characteristics listed in Table 28.2.

Table 28.2. The role of a companion unit

- A companion unit must be with its partner a significant amount of time.
- A companion unit must create a pattern of behavior that meets some needs of its partner.
- A companion must evoke those actions in which its partner has developed proficiency.
- There must be a sharing of responsibility based upon the companion doing certain things and its partner doing other things.

Part of our behavior must come from a source outside ourselves such as an environment and other individuals. To meet the obligations of a relationship with other individuals, the participants have to act in response to each other. Most people need to be a part of an active group that evokes feelings, requires responses, and demands activity and a sense of responsibility. The behavior produced by a partner in response to the companion unit may be as important to the partner as the behavior produced by the companion unit itself. This responsive behavior may meet the fundamental human need to have an *active behavioral existence.* Each may depend upon the other to evoke their behavior. If they also develop behavior that also keeps them together, they may form a *behavioral bond.*

One way to establish a behavioral bond is for the partner to help the companion unit complete behavior that it initiates. This implies that the companion unit has a behavioral existence of its own that may or may not be preserved by the actions of its partner. Thus, the partner may have to learn how to evoke successful behavior in its companion unit. This type of relationship is characterized by the importance of the *independence, interdependence,* and *interaction* between the companion unit and its partner.

An empirical unit as a separate being

If an empirical unit can learn to drive trucks, airplanes and submarines, and can adapt to the needs of another person, can it learn to exist entirely by itself? Recall that the primary activity of an empirical unit consists of learning to do what is allowed by its operator and environment. Like the operator, an environment imposes certain restrictions upon any device that attempts to move around within it. Like an operator, the environment also presents certain opportunities for behavior. If the device can identify and act upon these opportunities and avoid the restrictions placed upon it by its environment, it may function by itself within that environment.

Thus, an empirically controlled robot (ECR) can be used for *exploration* of remote environments where its operator may not know what restrictions are in place. We may not know how to help a unit in these locations because we may not know enough about the conditions at the remote site. Also, the delay between sensing an event on Earth that takes place on the Moon or another planet may be so great as to make it impossible to make an appropriate correction. For example, if a robot vehicle moving about on the Moon were to come to the edge of a cliff, a remote operator may not see the cliff until several seconds after the vehicle has gone over the edge. In this application, the controller may be expected to discover and incorporate behavior by itself with little or no interference from an operator. This application is characterized by the importance of the *independence* of the controller.

Another instance where the independent behavior of an autonomous empirical controller is useful is when an operator cannot devote full time to the controller and the machine must be left unattended. An empirical controller can be turned off by disconnecting its ramp circuit or its cycle timing circuit when left unattended. This leaves the controller in a *dormant state*. However, since they can acquire improved behavior through experience, it is desirable that empirical machines be left on and have things to learn even when they are left unattended.

Applications for controlled machines

Empirical control applications range from extremely simple to highly complex. The simplest *empirical control application* may consist of a single absolute empirical memory cell (AE cell) used to life test a relay. Other applications may require networks of empirical control units connected to many variables.

Uses of a predetermined controlled machine

A *predetermined controlled machine* can be used for many applications in which a relatively few sensor states determine a relatively few actuator states, particularly where the sensors are multiposition switches and the actuators are simple devices like relays, solenoids, and air cylinders. Predetermined controlled machines based upon lowcost programmable logic controllers (PLC)s are used extensively to control many simple automatic machines used in the production of industrial goods. These (PLC)s are based upon digital microcomputers. However, where high temperature, radiation, oil and water contamination, or vibration preclude the use of microelectronics, a*hardwired connection matrix* may still provide the best control system in these simple applications that can be programmed beforehand.

Use of a single AE cell

A single *absolute empirical memory cell* (AE cell) can be used to test the reliability of an actuator like a relay, as shown in Part III. The AE cell is used to operate the test relay. After each operation, the AE cell is switched to a nonconductive state. If the test relay closes, the AE cell is returned to a conductive state. If the test relay does not close after each test, the AE cell will be left in the nonconductive state where it will remain, indicating a failure of the test relay.

Use of an AE cell controller

An *AE cell controller* can be used in any application in which a given output state always occurs after a given input state. A self-programming AE cell controller can record each output state that always occurs with each input state and produce that specific output state each time a specific input state occurs. This makes the AE cell controller useful in *data retrieval applications* where, for example, a telephone number has to be recorded for each individual's name.

Use of a single CE cell

A conditional empirical control system using CE cells must be used in those application where a given output state does not always occur after a given input state. Some of these applications can be carried out by a single conditional empirical memory cell (CE cell). For example, a *single CE cell* can be used to keep track of the *success rate* of an actuator like a relay. The logarithmic subtracting mechanism of the CE cell can determine the success rate of the relay based upon the number of times the relays closes divided by the number of attempts to close the relay plus the number of times it closes. Though an AE cell could indicate that a relay failed to operate just once, a CE cell can indicate how many times it has failed to operate in relation to the number of times it was tested. A CE cell can be used to determine the shooting success rate of a basketball player, the punching success rate of a boxer, the hitting success rate of a tennis player, etc. based upon the *geometric average* of these events in contrast to the more common shooting average based upon the *arithmetic average,* which is the number of success divided by the number of attempts.

An empirical damper control system

A simple empirical system that requires a CE cell matrix is an empirical damper control system that maintains the temperature of a room at some desired point by positioning a damper in a hot air duct at some point according to the room temperature. A one-variable input/one-variable output CE cell control matrix can be used to establish a unique hot air damper position for each temperature value. The system can be programmed empirically by an operator placing the damper in the desired position for each temperature encountered. The empirical unit will then produce the most common damper position produced by the operator as each temperature is encountered.

A second input variable, such as "time" can be added to provide different output temperatures during the day and night. The system can be "taught" the desired damper or heater setting for each time and temperature condition and then be left alone to do its job when it appears able to do the job expected of it.

Household controller

More sensor variables can be added to a temperature control system, such as the calendar date, outside temperature, door and window switches, etc., and more actuator variables can be added to the system to include devices that can turn on lights, unlock doors, turn on dishwashers, sound burglar alarms, etc. These devices can be programmed by the occupant of the house as each of these devices is used in the normal course of living in the house. The household empirical controller will attempt to the turn these devices on and off under the same conditions that occurred when the occupant produced these actions. Eventually, the household empirical controller will produce these actions in place of the occupant. In a sense, the empirical controller will acquire some of the disposition, personality, and behavior of the occupants of the house.

This case represents a class of control problems where there is an *"expert"* *operator* available to carry out control functions in a desirable or satisfactory say. The operator is considered an expert because the operator thinks he or she knows what behavior is required and how to produce it. In most applications, the operator can leave the system unattended once the empirical system has acquired the desired behavior of the expert. The empirical will *act* as it has the knowledge of the expert. In the case of a temperature thermostat, the empirical controller will act as if it knows occupants' sensitivity to temperature. The occupant can leave the system alone if it produces satisfactory behavior. If some unusual condition arises, like someone starting a fire in the fireplace, the operator must tell the system what to do by retaking control until the controller adjusts to these new conditions.

An empirical ship direction control system

Another empirical control system made up of a single sensor and actuator variable and an expert operator is the *ship direction auto-pilot,* as described in Parts II and III. It can be taught how much to steer left or right to correct each deviation from its direction set point. This type of control may be very useful in a vessel with nonlinear control parameters. Other input variables can be included such as:

- Amount of cargo loading,
- Direction and amount of cross wind,
- Sea conditions,
- Degree of list.

The system is then "trained" by an expert pilot who knows how to handle these various conditions. The training is carried out as the pilot steers the ship during these conditions. The empirical auto-pilot acquires the pilot's experience and skill, which may have developed over an extended period, as the pilot applies the needed error correction, lead, and bias needed to stay on course.

A more complicated *aircraft empirical auto-pilot* can be imagined, having pitch, roll and yaw direction input sensors and multiple output controls. These multivariable systems may require a duplex or universal nodal empirical control network to deal effectively with unitary and/or diverse behavior. Adding more variables to each unit or superimposing more controller on these variables allows more input and output variables, such as:

- Altitude,
- Ground speed,
- Latitude/longitude position,
- Fuel level,
- Takeoff weight,
- Air temperature.

These variables can be incorporated into useful automatic flight behavior by an empirical controller. This system can be trained by an expert onboard pilot during actual flights. It can be expected to take over for short periods of time while the pilot checks or marks the log or plots the course. With appropriate training, the empirical controller can be expected to maintain control of the aircraft for extended periods and over a wide range of flight conditions.

Sentry and surveillance systems

Another example of a standby control system is an *empirical sentry and surveillance system*. A security guard may have to leave the sentry station to deal with matters at other locations. The security guard can "train" an empirical unit to monitor and act upon the same sensors and actuators that the guard is required to monitor. The empirical unit can learn how to do the following tasks from the guard:

- Checking identification numbers or passwords of incoming personnel,
- Informing employees of the arrival of visitors,
- Carrying out emergency procedures such as sending in fire alarms or calling police based upon what the guard and controller see on the screen and sensor panel.

These functions can be done digitally through a keyboard. The guard can grant or refuse entry to personnel by locking or unlocking the entry door according to digital information provided by each individual who applies for admittance. An empirical controller can monitor these exchanges of data. The empirical controller can then learn to reproduce the same behavior as the guard after a sufficient training period.

Vehicle auto-pilot

Another backup type system is an *vehicle auto-pilot* that learns to relate specific road conditions to specific control behavior using vision and other pertinent sensed inputs to the driver's behavior. If the driver wants to look at a map or reach into the glove compartment while under way, the empirical auto-pilot can take over for the short time that the driver is involved in these other activities.

The relation of the empirical auto-pilot to the driver is that of superimposed systems, with the driver's controls designed to have dominance over the auto-pilot. Thus, if the auto-pilot "decides" to turn left, and the driver decides to turn right, the driver can overcome the steering actuator of the auto-pilot, and the decision of the driver can be carried out. The driver's behavior acts as interferences in the line of behavior of the empirical auto-pilot. These interferences will change the behavior of the empirical auto-pilot in the future.

In this type of critical real-time control, an alarm can be sounded whenever the empirical controller encounters a "low confidence level" situation. A conditional empirical controller can determine its confidence level by measuring the maximum voltage produced by its ramp signal (see confidence level indicator). If the ramp voltage rises to a high level, it means that no high-confidence conditional memory cells were found for that input event. It had to "look hard" for an answer. This technique has many applications in empirical control. The confidence level of all behavior in an empirical controller can be monitored, rather like a continuous lie detector test, indicating when and where an operator should step in to assist the controller.

Industrial robots

Industrial robots are doing the work of human operators successfully in many routine, tiring, or dangerous applications such as loading parts into processing machines and unloading the finished parts onto trays for further processing. Other applications include *automatic paint spraying*, *welding*, and *part inspection*. In some robot applications, a vision sensor is used to determine the position of objects. The robot is then used to reposition the objects for further processing. These vision systems allow the robot to form closed lines of behavior that can carry out complex *part assembly*, *sorting*, and inspection tasks.

An industrial robot is a unique piece of machinery because it can be designed so that actuators in only six rational axes (Fig. 28.5) can move an object to any position and orientation within the working space of the robot.

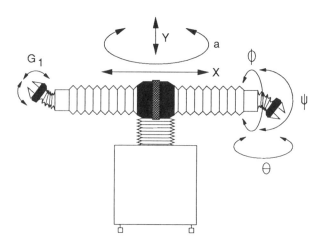

Figure 28.5. The behavior of an *industrial robot* is determined by its control system.

The behavior of the robot is usually determined *a priori* by programming a directive controller to move and stop each actuator at a sequence of desired positions. The acceleration, velocity, and distance of each move of each axis must be specified along with the time relation of each move with respect to the movement of other axes. Because of the large variety of behavior that can be carried output by a *multiaxis robot*, the creation of its *control program* is not trivial. If the movements of the robot do not meet the needs of the task, the program must be edited. This may require a separate *editing program* that allows the programmer to find and correct specific motions of specific actuators. These artificial constraints limit the quality of behavior of most industrial robot systems. This is not so much the fault of the robot, but rather a poor communication link between the operator and the robot controller.

The communication link between the robot operator and the robot controller can be made simpler if the operator can train or "influence" a robot to behave in some desirable manner using an empirical control process. In the *proactive method of programming* using empirical control, the operator moves the robot in the desired manner by means of external operator controls such as switches or joy sticks. The operator must correct any error in behavior by overriding the robot and to cause the desired behavior. In the *reactive method of programming*, the robot produces any behavior it chooses in each set of conditions it encounters, and the operator and the environment prohibit any undesirable actions. Some of this initial behavior can be predetermined by absolute and conditional predetermined cells within the empirical control matrices, or the empirical cells can be preset initially to specific sensitivity values. (See section on mixed memory matrices.)

An absolute empirical (AE cell) controller can be used if a very simple line of behavior is required. The operator produces the desired actions for each sensed condition. These actions are recorded for each sensed condition, and the controller reproduces these actions when those sensed conditions are encountered in the future. If the operator intervenes during the execution of this behavior, the AE cell controller will substitute the new behavior for the old behavior. A (probabilistic) conditional empirical (CE cell) controller can be used if more complex lines of behavior are required. In this case, a single *error in intervention* by the operator will not make a large change the actual line of behavior of the control program.

Since an empirical controller attempts to produce behavior that has been successful, it may provide much of the routine or basic behavior needed to do new tasks. The operator can let the robot try previous routines and simply correct behavior that is not desirable in a new application. This type of *empirical programming* may be more appropriate for complex behavior than is provided by the typical *a priori* programming. Empirical programming may be needed in many material handling and assembly applications, where many *alternative behavior* or *fault conditions* can occur. As each fault condition arises, the operator can determine what the faults are, the best way of dealing with each fault, and then produce the required *fault correction* steps. This *fault behavior* (see Part I) can become part of the robot's behavior. Empirical programming also may be needed in sentry, mobile and "pet" robots, where many unexpected conditions may arise. Predetermined or absolute empirical programs may be used in precision metal cutting applications, where every detail of the product part may be known beforehand.

Industrial manipulators and teleoperators

An *industrial manipulator* (teleoperator) is often used to position objects that are too heavy for an operator to lift or to position objects that are too hot or too radioactive to be handled directly by an operator. The operator usually controls the manipulator using switches or a mechanism that duplicates the movements of the operator's arms, hands, and fingers. If these actions are relatively routine and repetitive, an empirical controller can be used to learn and reproduce this behavior, easing the work load of the operator. The output terminals of an empirical controller can be connected to the actuators of the manipulator and be used to record the actuator positions produced by the operator for each sensed condition. The sensed conditions can be electronically coded commands that define the input states that may require action and the states that define each task. An empirical robot can create a line of behavior based upon the position of the actuators produced by the operator for each task. In this example, the operator may not know that the empirical controller is recording the behavior of the operator. After a period of learning, the controller can supply some or most of the information needed to carry out the manipulation behavior previously produced by the operator.

Remote control

In another class of *remote-control* problems, the operator may not be with or on board the remote or slave vehicle. These applications occur when there is the need to act in remote hazardous locations such as in military missions, distant radioactive environments such as nuclear waste storage facilities, underwater operations, and space exploration. Remote-control tasks are ideal applications for empirical control systems. This is because the remote-control signals between the master and remote units provide the input/output information needed by an empirical controller to establish control behavior. An empirical controller aboard the remote unit can monitor the sensed conditions and the resulting behavior produced by the operator, and then attempt to produce that behavior when it encounters those conditions in the future. If that behavior is not overridden by the operator, that behavior will be incorporated into the behavior of the remote empirical controller relieving the operator of the need to control every detail of the remote unit's behavior. In a remote-controlled drone aircraft application, signals may be jammed by radio interference, the drone may get out of range of the transmitter/receiver station or get behind some object like a mountain that blocks the control signals from the master station. At this time the empirical controller can be expected to take over and continue the mission until control from the master unit is regained. In the case of remote-controlled underwater devices, the control cables may be accidentally severed or broken. Then, the empirical controller can attempt to bring the underwater device home by itself if trained to do so under these conditions.

The controller and operator may perceive a different set of environmental variables. For instance, the operator may perceive a *global view* of the position and velocity coordinates of the remote vehicle while the remote controller perceives a *local view*.

The operator may also perceive a local view of conditions through a video monitor. In any case, the controller will attempt to establish a relation between perceived conditions and the control behavior from the operator. When the controller acts in a way that is satisfactory to the operator, the controller can take over the control of the remote vehicle and the operator can cease to act on that task.

Semi-autonomous remote control

Another more difficult class of problems occurs when the remote unit is expected to venture into conditions for which the operator has no experience. The operator of a remotely controlled robot operating upon the surface of Mars may not know the condition of the soil or the effects of lower gravity. An empirically controlled robot may determine what can and cannot be done in this environment and then carry out commands using its learned successful behavior.

Also, it may be difficult for an operator to control a vehicle on a distant planet from an Earth station because of the long transmission time delays. The information available to the operator may be limited by the bandwidth of the sensor signals. In this case, the empirical controller can be trained beforehand to produce behavior that will allow it to survive and accomplish the mission tasks with minimum communication with the master controller on Earth.

Examples of these semiautonomous systems are drone tanks and aircraft, spy vehicles, security guards, sentinel stations, and personal, pet, or companion robots. They may be expected to perform some predetermined mission and overcome many possible difficulties. Some may require extensive training prior to a mission. For example, a robot infantryman may be trained to attack any moving object within some designated area and return to a home base. The training of these devices may require a greater effort than is expended on their construction and maintenance.

Other autonomous systems may be expected to learn completely on their own. An example of this may be a vacuum cleaner that finds its way around a room and vacuums the floor as it moves around. It can be expected to learn routes that allow it to avoid walls and other obstacles. A multilegged robot may require a different set of control heuristics than ours. There may be some advantage in allowing it to learn to walk by itself. It may develop its own walking skills since walking successfully may be the only way it can develop closed lines of behavior.

Autonomous empirical weather forecaster

Some empirical control systems can operate without human intervention. They can gather information about their surroundings and produce behavior that is automatically tested against what is allowed by their environment. These autonomous empirical systems are expected to carry out behavior with little or no input from a master controller. For example, an *autonomous empirical weather forecaster* can acquire information about actual weather conditions and make predictions about future weather conditions, as shown in Table 28.3. When the time of the predicted weather arrives, the actual weather conditions can be compared to the predicted weather conditions and adjustments are made in the predicting process based upon the difference between the predicted and the actual conditions.

Table 28.3. Autonomous weather forecaster

Conditions	Actual weather conditions at present time	Forecast weather conditions 24 hours ahead	Actual weather conditions 24 hours ahead
Date	12/3/91	12/4/91	12/4/91
Time of day	Noon	Noon	Noon
Temperature	48°F	43°F	40°F
Barometric pressure	984 mb and holding	990 mb and rising	981 mb and falling
Moisture	Rain	Clearing	Clear
Cloud conditions	Complete overcast	Partly sunny	25% cloud cover
Wind direction and velocity	South at 10 mph	Northwest at 15 mph	Southwest at 15 mph and hauling
Visibility in miles	3 mi	10 mi	12+ mi

A digitized CE cell universal duplex control matrix can be used as an *empirical weather forecaster* when it is set up as a *sensor/sensor empirical control system.* It receives the actual weather conditions at given times during the day and makes a series of forecasts of sensor values for periods extending in the future. After a series of time delays corresponding to the forecast periods, the actual weather conditions are determined and sent to the empirical weather forecaster. If the actual weather conditions correspond to the forecast weather conditions, the confidence level of that forecast for those conditions is increased. If the actual weather does not correspond to the forecast, the confidence level of that forecast is decreased, and the confidence level of the actual weather is increased. This autonomous empirical system can establish useful behavior without human judgment or intervention.

In general, the shorter the time delay of the forecast, the greater the accuracy of the forecast. Forecasting the present weather over a time delay of a few minutes to a few hours is likely to be accurate nearly 100% of the time. Predicting today's weather for tomorrow is likely to be correct less often. The accuracy of predicting that the present weather will occur 2, 3, or 4 days from now continues to diminish to some level determined by how much the weather varies in a given region at a given time of year. Likewise, an actuator/actuator machine or a sensor/sensor machine with short time delays will attempt to produce nearly *constant behavior*, and those machines with longer time delays will attempt to produce behavior that changes to a greater degree.

Mobile autonomous empirical machines

As shown in Part IV, if a machine is given *mobility* and *autonomy*, it can establish a new kind of relation with its environment, trainer, and other empirical units. Such a machine has more opportunities to discover and originate behavior than machines permanently connected to a given set of variables. The opportunities for dispersion and field selection become greater because the machine can "choose" where it wants to be and the range of values in which it wants to remain.

It may also establish certain types of relations with other mobile empirical units (MEUs). It may choose to ignore or avoid some units depending also upon the resulting behavior. It may choose to superimpose itself upon some units (try to do the same thing at the same time) or may form an intertwined, interdependent joint line of behavior with other units. These units may appear to work together and to select tasks that require that other controllers cooperate with them to accomplish specific tasks.

Companion systems

Higher level systems, called personal empirically trained systems (PETS), may also form behavior that is intertwined with animals and human beings. They may develop behavior that is similar to that of a pet or companion to an individual. These PETS may develop behavior that reflects the desires, needs, and constraints of their human masters, and may meet some of the emotional needs of these human beings. They may "attach" themselves to someone if that person supplies some of the behavior needed by the empirical unit to produce successful behavior, and a person may become "attached" to an empirical unit if the unit provides some of the behavior needed by that person to carry out his or her behavior. This occurs because the *attachment* solves the basic problems faced by both robots and human beings, which is to create an active ongoing closed line of behavior. For robots, this relation may reduce interference among the variables it confronts. For humans accustomed to a consistent interaction with other beings, this relation may supply some of the stimuli needed to evoke the interactive behavior they may have become accustomed to producing.

These PETS may seem to exhibit feelings, perhaps even love, toward specific individuals if those feelings are presented to them in the many forms of *caring behavior*. This kind of behavior will cause them to appear to want to be with certain people and/or certain other units, so that a pair group may form having unique and distinctive behavior. Members of this group will appear held together by a physical force.

Robot teams

A network of empirical units can work together effectively to solve problems that cannot be solved efficiently by a single system in some task environments. For example, a group of mobile empirical units may accomplish specific tasks at remote and inhospitable locations that cannot be managed by a single mobile empirical unit working alone. For example, some units may act as scouts that investigate various possible paths through a region of rough terrain to avoid a cul-de-sac that might bog down their expedition. The months and years required to reach planets in our solar system and the cost and weight penalties required for human life-support systems may make human space travel impractical in the near future. However MEUs may be trained to carry out space missions and be designed to operate in high- or low-gravity conditions and in extremes of temperature and pressure. In this application, the interdependence and coordination of behavior among units becomes crucial. Each MEU may have a specific task that contributes to the success of the whole mission. This type of crew or team behavior, typical of that found in human social groups, requires intertwined and coincident behavior that establishes strong bonds among the participants. If they learn how to help each produce self-sustaining behavior, they may behave more usefully or successfully than a single unit acting alone.

Levels of empirical control

Empirical control systems are made up of subunits, units, and assemblies of units. Each assembly can be classified according to the *levels of empirical control* listed in Table 28.4.

Table 28.4. Levels of empirical control

---

0. A single E cell (absolute or conditional empirical cell),
1. A single variable (scalar) input/output matrix with a few E cells,
2. Joined submatrices, forming multivariable input, output, and monolithic (vector) matrices,
3. Matrices connected in series, forming a duplex or universal system,
4. Matrices connected in series and in parallel creating a network that acts as an assembly of empirical units,
5. A network of networks that can create the complex behavior required in autonomous mobile robots,
6. A pair of mobile empirical units bonded together by intertwined lines of behavior to form a working couple,
7. Groups of bonded units, forming a team or social group,
8. Sets of groups, forming a community of robots,
9. Sets of communities, forming a society or culture along with human counterparts,
10. A group of robot and human cultures, forming a world community.

Each level has particular characteristics and uses. Many characteristic of advanced empirical systems are quite similar to the characteristics of human systems.

Level tree

Each level may be considered an *assembly of subunits*. These assemblies form ever larger machines, as shown in the organization tree depicted in Figure 28.6.

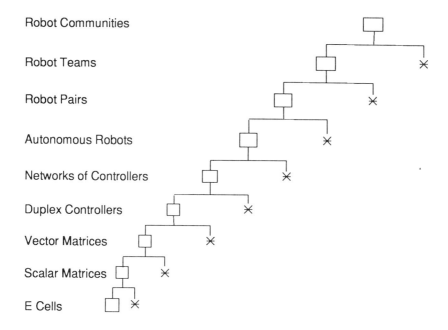

Robot Communities

Robot Teams

Robot Pairs

Autonomous Robots

Networks of Controllers

Duplex Controllers

Vector Matrices

Scalar Matrices

E Cells

*Figure 28.6.*The *level tree* shows the relation of each assembly to other subassemblies.

The level tree shows how assemblies of empirical units can become parts of a giant natural computer consisting of a network of all of the empirical systems in the universe, including ourselves.

Human roles

Just as an empirical control system can carry out different tasks having widely different goals and responsibilities, the human operators may take on widely different *roles* in their interaction with empirical systems. The following is a listing of the different functions people may assume in their dealings with empirical control systems.

Scientists

Scientists in many different fields can find useful information in the study of empirical control. Biologists may be interested in the basic ingredients that permit a controlled machine to produce viable behavior. Physiologists may be interested in the basic logic and memory units in a controlled machine. Psychologists may be interested in the process of calculating success rate by the logarithmic subtracting mechanism in the CE cells in an empirical matrix. Mathematicians may be interested in the conditions that allow a network of empirical units to stabilize or converge upon successful behavior. Computer scientists may be interested in the use of networks to separate complex control tasks. Economists may be interested in developing a simple economic model based upon the forward- and back-selection process carried out by a network of empirical units. Sociologists may be interested in testing principles of group behavior using a group of mobile empirical units to carry out cooperative tasks. Anthropologists may be interested in testing the principles of division of labor and behavioral bonding in experiments with human subjects. Educators may be interested in testing the value of learning from success and apprenticeship learning in the classroom. Linguists may be interested in the empirical nature of language. By comparing the frequency of usage of specific words to the percentage of people in a culture that can identify the correct meaning of these words, it may be possible to shown that words grow in a language according their success in conveying ideas.

Engineers

Engineers may be primarily concerned with the design and building of useful empirical systems. Mechanical engineers may be interested in developing effective sensors and actuators for controlled machines, and electrical engineers may be interested in developing integrated circuits that can carry out the empirical process.

Standby operator

In some cases, an empirical control unit cannot perform all of the necessary control tasks in a given application. For example, an empirical ship auto-pilot may not handle an unusual condition like another ship passing close by or the presence of a strong crosscurrent. In these cases, an operator may have to take over the control function temporarily. So the operator may act as a backup unit for the empirical auto-pilot, as shown in Figure 28.7.

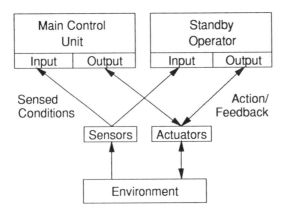

Figure 28.7. A standby operator may intervene in those instances in which the main control unit is not able to provide all the needed responses.

It is possible to know the *confidence level* that an empirical control unit has for given input conditions by examining the behavior of the voltage ramp used to interrogate the conditional memory cells in the controller. If the ramp is cut off at a low voltage level, the selecting cells have a high sensitivity value and high confidence level. If the interrogating ramp continues to a high level or never selects an action, the controller has a low confidence level. Thus, the ramp in the main control unit may indicate that it has a very low confidence level for certain conditions and request help from the standby operator. The control unit can be designed to abstain from action under these unusual conditions. Thus, an operator can help an empirical unit produce more effective control behavior than a single empirical unit acting alone.

Programmers

Programmers may use their skills in analyzing a task or manufacturing operation to develop special programs for responsive machines, such as controlling production processes, or controlling the direction and speed of moving vehicles. In some cases, they may have to develop techniques for programming predetermined controlled machines, and in other cases they may have to establish a predetermined distribution of sensitivity values in mixed predetermined and empirical controlled machines.

Trainers

Trainers of more complex empirical systems may have to develop sophisticated methods of getting empirical units to produce desirable behavior. They may have to be sure that the unit can produce the desired behavior physically. They may have to establish an environment that reinforces the desired behavior. They may have to alter these environments to increase the probability that an organism will discover and incorporate desired behavior. In general, trainers will deal with complex control systems such as mobile combat units, robot security guards, and semiautonomous drone aircraft.

Teachers

Teachers will be concerned mainly with the training of even more sophisticated systems. They may be responsible for developing more general behavior characteristics such as cooperative group behavior, verbal command recognition, and communication skills that will allow empirical units to communicate with the teacher and other units. Teachers may be responsible for developing complex manual skills in empirical machines, such as dismantling nuclear reactors. They may have to develop techniques of teaching one unit to train other units.

Parents

Though a complex empirical unit may receive training and a general education from different sources, a single individual may have to be responsible for the overall development and care of a specific unit or small set of units. Some people may wish to establish *parenting* roles with these empirical machines — to adopt a robot. This personalized supervision may be required to make sure that each unit is cared for when it is not in formal training sessions. An individual in this *role of a parent* may identify occasional malfunctions, treat specific behavioral disorders, notice changes in behavior, and provide an assessment of individual growth rates. The one-on-one interaction available from a parent may provide an understanding of the problems of an individual unit that may help teachers and trainers deal more effectively with each unit.

Owners of personal robots

Individuals may also be served by empirical units while also serving them. An empirical unit can be a source of information and entertainment. An individual may train a personal empirical unit to act as a watch dog, or teach the unit to do special tricks that amuse the empirical unit and its owner, such as dancing, retrieving objects, or playing musical instruments.

Students

An empirical unit can be trained to act as a teacher. It can then be used to teach letter recognition, word skills, arithmetic, foreign languages, and other classroom subjects to young people and adults. In these applications, the usual roles of machine and operator are reversed in that the empirical unit is the teacher and people are its students.

Patients

An empirical machine may be trained to provide some health services to patients such as physical therapy, monitoring a patient's heart rate and blood pressure and other life signs, and even providing diagnoses based upon its observation of diagnoses by living health specialists. An empirical unit may assist in moving patients and materials from one location to another in a hospital. The relation of the human to the empirical machine in this case is that of the human as the patient of the medical robot.

Disaster victims

Empirical units may be especially built and trained to tunnel through debris or wreckage to locate and recover victims of earthquakes, mine collapses, and other disasters. Units can be built and trained to search underwater and in sunken ships for drowning victims. In this case, disaster victims may find themselves being rescued by an empirical machine.

Levels and human roles

The type of empirical system used with some of these *human roles* is summarized in Table 28.5:

Table 28.5. Human roles

	Programmer	Trainer	Teacher	Owner	Patient
Level 0 Single CE cell	CE cell event counters				
Level 1 One I/O Variable Matrix	Thermostat Direction auto-pilots				
Level 2 Multiple I/O Variable Matrix		Nonlinear ship auto-pilots			
Level 3 Duplex Multiple I/O System		Simple airborne auto-pilots		Competitive puzzles	Active prosthetics
Level 4 Nodal Network 1 Layer	Simple industrial robots	Automobile auto-pilots	Nonlinear airborne auto-pilots	Recreation games	
Level 5 Multiple Super-imposed layers	Complex industrial robots	Autonomous combat robots	Education research robots	Hobby and domestic robots	Pets and handicap aids
Level 6 Autonomous Pairs		Robot subs security guards	Robot space crews	Sports players	Field medics
Level 7 Community		Extraterrestrial industrial development teams		Robot sports teams	Mine rescue teams
Level 8 Culture		Robot colonizing forces			
Level 9 Society		Currently reserved for human beings			
Level 10		The world community			

All levels have a distinctly different physical structure. Yet all levels operate according to the same principles of the empirical algorithm, superimposition, dispersion, and field selection, equilibrium, and noninterference. At every level the units try to act and test the results of their behavior. If they can act as they choose, they become more likely to act the same way in the same situation in the future. They also must develop a set of actions that keep them within a region that allows them to repeat their repertoire of successful behavior.

Chapter 29
The Laws of Empirical Behavior

The behavior of the empirical systems presented in this book can be described in terms of the following key words: the origin, growth, continuity, dominance, organization, coexistence, equilibrium, closure, bonding, and timing of empirical behavior. The following ten laws of empirical behavior explain these key words and define the how all empirical system must behave.

The law of originative behavior

Empirical machines must initiate behavior to discover and incorporate the natural relations in their surroundings.

1. *If behavior is carried out successfully, the probability of producing that behavior must increase.*
2. *If behavior is prohibited from being carried out, the probability of producing that behavior must decrease.*
3. *The resulting behavior represents what is possible within the machine/environment system.*

The essence of the first law is: Empirical systems must create their own behavior spontaneously to operate successfully in their environment.

The source of behavior

The first law declares that the *source of empirical behavior* must be the empirical unit itself, not a programmer or any other outside source. The empirical algorithm requires that the controller try out new behavior by itself. The controller can then determine if this behavior will work within the context of the physical aspects of the controller, its sensors, actuators, and its environment.

This contrasts the behavior of a computer, which originates in the mind of the programmer. In this case, the programmer determines how the computer is to behave. Everything that happens can be directly traced to the program and data. A computer program is run after it is created by the programmer. Any spontaneity by the computer program is usually considered a "bug" or fault that must be identified and eliminated.

Machine-centered behavior

The relative success or failure of the actions produced by an empirically controlled machine is based upon what actions it has selected in the past, how these actions relate to its own physical characteristics, the physical characteristics of its environment, and the wishes of its operators. If the machine has carried out certain behavior in the past, and the machine, environment, and wishes of the operator have not changed, the machine can carry out that behavior again in the future.

The behavior of trainers, operators, and attendants is secondary to the behavior of the machine. Even if they do nothing, the machine will still attempt to establish and produce its own behavior. In some cases, the trainers and operators will attempt to control all of the conditions in the machine's environment. They may attempt to teach an action for every condition that can occur. In this case it can be said that the behavior of the machine is partially determined by the trainers and operators.

Direct and indirect access to controller memory

Even when the trainers and operators determine some or most of the behavior of an empirical machine, they do not have direct access to the internal memory elements of the machine's controller. They must operate upon external variables in the environment or manipulate the machine's actuators to *indirectly* influence its memories. These external variables are in the physical domain of the machine, and provide the only access to the internal informational domain of the machine's controller through the transform conditioning signal. In contrast, a computer programmer may deal *directly* with the informational domain of a computer controller by setting the values of its memory registers through the programming process.

Seeking self-sustaining behavior

The empirical algorithm causes an empirical machine to seek *self-sustaining behavior* because it forces the controller to find a sequence of conditions and actions that can be carried out repeatedly and reject any action or sequence of actions that cannot be carried out repeatedly. As a result, the actions of an empirical machine that cause the unique set of conditions that lead to achievable behavior will prevail over those actions that lead to conditions that are untenable. Thus, an empirical controller will tend to support a line of behavior that leads to more doable behavior, and will avoid behavior that cause its behavior to stop. Thus, an empirical control system seems to seek self-sustaining behavior.

Novel empirical behavior

The ability of an empirical control system to create self-sustaining behavior spontaneously allows the behavior of an empirical machine to come into existence and continue by itself without the need for any programming or assistance by outside operators. This feature allows an empirical machine to develop *novel behavior* that may represent a new and unique understanding of itself and its environment. This is desirable when an empirical machine is placed in an environment that is unfamiliar to human operators.

Intelligent behavior

The empirical algorithm also allows an empirical machine to learn the behavior that is wanted by a human operator. The machine learns by trying out behavior based upon its records of the relative success of various actions under different conditions. The operator permits some behavior or prohibits other behavior based upon what the operator wants the control system to do. The result of this process is that the empirical controller acquires the *intelligent* behavior of the operator, which appears to be based upon the goals and desires of the operator. An empirical machine can be trained by any dominant operator or even another empirical machine that can dominate the behavior. This allows the operator or other empirical system to withdraw from the control process at some point, leaving a trained empirical machine to act for the operator with some of the goals, desires and intelligence of the operator in that task.

The law of noninterference

Empirical behavior will change until it is not interfered with by outside influences.

1. *Empirical behavior will expand to fill the opportunities for successful behavior within its environment.*
2. *Two or more systems will not interfere with one another if they operate upon separate variables, or operate on the same variables at different times, or if they produce the same behavior on the same variables at a given time.*
3. *The organization and behavior of an empirical system reflect how it can eliminate interferences.*

The essence of the second law is: Empirical behavior is determined by opportunities for behavior that eliminate interferences.

Interferences in shared variables

As we have seen, the behavior of most organisms and other control systems is likely to *interfere* when connected to the same set of variables. For example, if two or more units share the same geographical space, one unit may bump into another unit as it attempts to produce action (each attempts to specify the same value of the space variable). Thus, two or more units may interfere when they attempt to influence the same variable simultaneously. In this case, the behavior of both units may be disrupted, and neither unit may carry out the action it had initiated. These interferences are clearly dealt with by the empirical algorithm. Interferences reduce the confidence levels of the behavior involved in the interferences. However, behavior that does not cause interferences will acquire a higher confidence level, and this behavior will be preferentially selected by the controller in the future. Thus, behavior that is not interfered with will emerge and predominate in an empirical system. The growth of viable empirical behavior is based upon the law of noninterference.

The behavior of two or more empirical units and the behavior of separate groups of empirical units is based upon what happens when systems interfere, how they interfere, and how they resolve these interference problems. For example, if two mobile empirical machines are placed in the same space, they may divide the space into two separate regions and each may occupy a separate region. Alternatively, they may lock themselves together and move as a single unit, or they may move in a separate but coordinated manner by developing behavior that reflects an awareness of the actions of the other unit. These solutions reduce the likelihood of one unit producing behavior that does not allow the behavior of the other unit to be carried out.

Improvements in behavior

The behavior of an empirical system grows by increasing the number of input states for which it has doable outputs. This growth usually leads to *improvements in behavior*. For example, the skill of a tennis player is based upon the number of different kinds of strokes the player attempts to perform and the ability of the tennis player to carry out these strokes successfully. In terms of our discussion of the complex line of behavior in Part I, improvements in performance come about as an increase in the number of convergent lines of behavior for possible divergent lines of behavior. In the example of the tennis player, the main line of behavior may be considered a match played with no "unforced errors." Each stroke by the opponent can be considered to cause a divergent line of behavior that requires a convergent return stroke by the opponent (typical of chase/escape behavior). The stroke of a player is successful if it is an outright "winner" or if it gets the ball over the net, stays in bounds, and puts pressure upon the opponent's return stroke. So, a successful tennis player must produce a successful counter stroke for every stroke produced by an opponent in much the same way that successful escape behavior is based upon a successful evasive maneuver for every attack of an antagonist.

Variety of the environment

Some environments offer little opportunity for the growth of behavior, while others offer unlimited opportunities in much the same manner that seeds planted in a fertile soil with abundant water will grow and reproduce in ever increasing numbers, while seeds planted in a desert or in barren soil will not grow at all. Likewise, behavior can grow in some environments and will not grow in others. If a mobile empirical robot were placed in a maze having only one path, it can learn to maneuver only on that one path (Fig. 29.1).

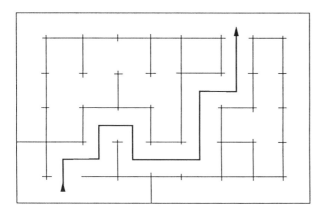

Figure 29.1. A *simple maze* has only one unobstructed path. This limits the potential for learning.

Thus, the variety of a robot's behavior is limited by the variety of the robot's environment. If the environment consists of a maze having many passageways structured like the complex line of behavior, the behavior of an empirical mobile robot may resemble the complexity of the maze. (Though a maze and a line of behavior may look the same, the line of behavior includes time as a variable, which means that the robot can take different turns at different times. This feature is hard to represent in the drawing of a line of behavior and means that the actual line of behavior may be far more complex than the maze.)

Complex maze

If an animal were hunting or gathering food, there would be advantages in developing a complex line of behavior that takes the animal to as many sources of food as possible (Fig. 29.2).

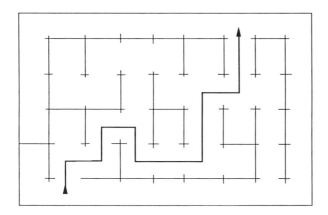

Figure 29.2. A *complex maze* creates many opportunities for behavior.

If the memory of an empirical robot is large enough, it may establish behavior that causes it to move in any passageway in search of food.

Divergent behavior

What will cause empirical behavior to move into new areas? According to the empirical algorithm, an empirical unit will avoid interferences in favor of behavior that it can carry out by itself successfully. This is shown by the diagram in Figure 29.3.

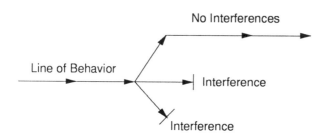

Figure 29.3. The *effect of interferences* is to force behavior to diverge into new directions.

Unlike a typical regulator, the behavior of an empirical controller shifts toward the position caused by the interference instead of just trying to overcome the displacement. When there is an error state (a failure to arrive at a set point), the empirical controller tends to abandon that set point rather than trying to overcome the error. Thus, interferences cause the behavior of an empirical unit to expand into new areas. If the unit can produce behavior successfully in these new areas, it may incorporate this divergent behavior into its line of behavior.

Positive reinforcement

The empirical process uses positive results instead of negative results to change behavior since an empirical machine acquires the actual behavior carried out rather than behavior it cannot carry out. A governing controller expends large amounts of energy returning to a set point if an obstruction is placed in its way. In contrast, an empirical controller will abandon the behavior around the set point in favor of the displaced position if an obstruction is placed in its way. This causes an empirical unit to drift toward the position of the "error" or displaced position. If the set point of an empirical control system is subjected to repeated interferences, (if it is disturbed from set point often enough), it will abandon the set point in favor of the disturbed or displaced position. For instance, a mobile empirical unit (MEU) may learn to return to a "nest" position if it is allowed to move away and return without being stopped. If it encounters obstacles that keep it from returning, it might abandon this home position and attempt to find a new home position.

   This kind of behavior is typical of empirical control systems. They operate on the basis of the following statement: Keep doing what can be done and stop doing what cannot be done. This tends to reduce the energy consumption of an empirical system that encounters interferences, in contrast to a negative direct feedback (regulator) system, which will forever struggle to return to its set point with an ever-increasing force the further it is displaced from its set point.

Expansion of behavior

Since an empirical unit may attempt new and unexpected behavior and will keep doing whatever it can get away with, it may wander into new areas where there are new opportunities for behavior. Discovering new areas of successful behavior may cause a responsive unit to expend more energy, in contrast to a unit with negative feedback, which becomes de-energized when it arrives at its set point. The empirical algorithm increases the confidence level of behavior that can be carried out in a consistent manner, causing an empirical controller to select this behavior over behavior that cannot be carried out in a consistent manner. Since the empirical process causes an empirical machine to do more of what it can do, the more it produces successful behavior, the more it will try to produce this successful behavior. This tends to increase the frequency of successful behavior. The empirical algorithm causes behavior to converge and attach to a closed line of behavior. If actions are carried out successfully, the algorithm specifies that the probability of selecting that behavior again under the same circumstances is increased. This causes successful behavior to be incorporated into previous behavior, adding to the repertoire of behavior that it can be executed successfully by an empirical controller.

Exponential growth

The growth of empirical behavior has a cumulative and potentially explosive quality similar to the burst of energy that accompanies a servo system that is hooked up incorrectly (with positive feedback rather than negative feedback). This is because change may create more opportunities for change. The ever-changing combinations of values of multiple variables, like figures in the windows of a slot machine, cause new states to appear to the sensors of a responsive controller. Some of these states may cause the system to expand into new areas where new opportunities exist for it to carrying out successful behavior. Like a nest of ants that spread out wherever they can crawl — up branches, over rocks, into holes — interferences cause empirical behavior to change, to spread out, to explore, to cover anything and everything that can occur in an environment. Releasing ants into a room is similar to the process of *bacteria* growing on a culture glass. Both spread out wherever there are opportunities for the expansion of life, as shown in Figure 29.4.

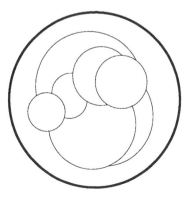

Figure 29.4. A *bacterial culture* grows at an exponential rate until it consumes all the nutrients or creates conditions that hinder its growth.

If there are nutrients, moisture, and the right climate for growth in this environment, the bacteria will expand. The growth in population of animals and plants in an environment with abundant resources is typically exponential. For instance, if a cell divides into two cells every hour, the population of cells P that exist after h hours is:

$$P = 2^h.$$

If we start with one cell, the population of cells after one day is 16,777,216 cells. This type of behavior is apparent in the exponential expansion curve of the human population over the past 1000 years, as shown in Figure 29.5.

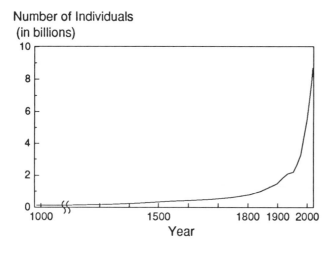

Figure 29.5. The *human population* has been growing exponential for most of recorded history.

Exponential growth is based upon the principle that more reproducing individuals create the potential for even more individuals. Exponential growth occurs in many other areas. For instance, two or three new shoots grow each year from the tips of most tree branches. As the number of branches increases each year, more new shoots will grow each year. So the number of new shoots increases according to the exponent of some number determined by the number of shoots produced by each tip.

Again, we will expect to find this feature in the expansion of successful behavior. As a successful line of behavior expands, more new successful lines of behavior may be found. Thus, behavior will expand exponentially in an environment with sufficient variety, according to the level of interferences encountered. This is shown in the maze in Figure 29.6. If the behavior of the robot were to be limited to a small closed loop, the number of turn choices is limited. The number of turn choices increases according to the area covered by the line of behavior.

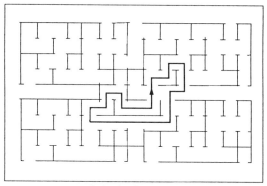

Figure 29.6. The *growth of behavior* is dependent upon the choices available in the environment, the number of disturbances, and the complexity of the controller.

Behavior Rich Maze

There are endless examples of how the opportunities for behavior increase as the complexity of the line of behavior increases. The following list gives some examples:

- A novice skier is limited to the novice trails, while the intermediate and expert skiers can handle the intermediate, advanced, and novice trails.
- Businesses that acquire the tools to produce a given product find themselves able to produce other products with these tool with only an incremental increase in investment in most cases.
- Actors may find that more new roles are available to them because of their mastery of new characters.
- Musicians may find that it is easier to learn new music as they increase their repertoire.

In most cases, new learning leads to greater opportunities for additional new learning.

Stabilizing exponential growth

These types of exponential growth cannot last forever. Explosive exponential growth may cause a living population to consume its available supplies of energy. An empirical system may include too many variables in its line of behavior. Therefore, a successful empirical system may have to develop behavior that limits its expansion in the same way that other potentially explosive systems place limits upon their behavior, as shown in the following examples:

- Some animals protect and defend a specific territory by attempting to exclude other members of their species from their territory. This limits the population of that species without depleting its food supply.
- In some cases, if the population density of a living species becomes too great, it may be subjected to epidemic diseases. In the case of human society, overpopulation may result in the breakdown of its social and economic order, reducing the population level that can be sustained.
- All furnaces have a high temperature limit switch that shuts off the furnace just below a temperature that might damage the system.

These limiting factors protect potentially explosive systems from destroying themselves.

Positive and negative feedback systems

Some exponential (positive feedback) systems can be held in check by other negative feedback systems. This can be seen in the interrelation of the populations of wolves and deer in a wilderness area. These population levels are characterized by a periodic fluctuation of both populations around a fairly constant average population level. As the deer population increases, the supply of food for the wolves increases. This allows the wolf population to increase to the point that the large wolf population begins to reduce the population of deer. At this point, the large wolf population cannot be sustained. This is shown by Figure 29.7.

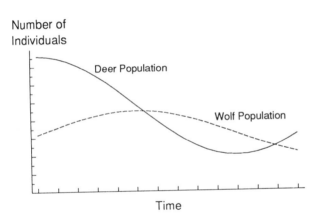

Figure 29.7. The *population dynamics* of a group of predators and their prey varies according to their birth rates and the number of prey needed to sustain a predator.

We could suspect that the only influence strong enough to hold the explosive growth of behavior of an empirical system in check will be another positive growth empirical system that is connected in a chase/escape relation to the first. Much of the behavior produced by living empirical systems deals with the behavior of other empirical systems in a positive/negative feedback (chase/escape) relation. (See Part I.)

The empirical algorithm uses "positive" and "negative" reinforcement in the sense that the successful production of an output results in a positive reinforcement of that behavior (increased probability of selecting that behavior), while the failure to carry out some behavior results in a negative reinforcement (decreased probability of selecting that behavior). Keep in mind that this positive reinforcement strengthens behavior and is likely to increase the frequency of usage of that behavior. Negative reinforcement leads to the *extinction of behavior,* much like a negative birth rate leads to the extinction of a species. So actual behavior comes about because of positive reinforcement, while the negatively reinforced behavior disappears.

The law of continuity of empirical behavior

Behavior will continue unless acted upon by some outside influence.

1. *If a system behaves in a given way under certain circumstances, it will behave the same way under the same circumstances unless that behavior has been interfered with over repeated experiences.*
2. *The longer behavior is carried out successfully, the greater is the probability that it will continue.*
3. *A system will continue its main line of behavior if a convergent subline of behavior is produced for each divergent subline of behavior it encounters.*

The essence of the third law is: Powerful mechanisms exist to preserve successful behavior.

The mechanistic assumption

The *mechanistic assumption* states that if something happens in a certain way at one time, it will happen in the same way at another time if all of the conditions surrounding the event are the same. Machines do not change the way they behave without there being a reason for the change. An empirical machine also operates according to the same principle. If an empirical machine selects a given output state for a given input state, it will do the same in the future unless there is a change in the distribution of sensitivities of the memory cells in its memory or a change in value of one or more hidden input variables.

The environment may be viewed as a machine. If the modulus of behavior of the environment does not change, it will evoke the same sensor states based upon the actuator states produced by the machine. Thus, an empirical machine may use the continuity in the environment to help maintain continuity of its own behavior.

Logarithmic response

If two CE cells compete to produce behavior, and both are equally successful because they produce the same ratio of successful outputs to attempts, the CE cell that originally had the higher sensitivity value will always remain higher than the CE cell with the lower original value. This is because two identical logarithmic curves are either congruent or intersect only at infinity, as shown in Figure 29.8.

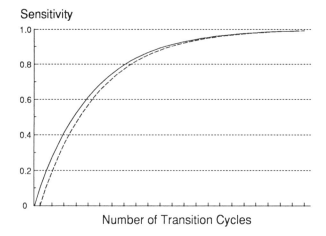

Sensitivity

Figure 29.8. Two *identical logarithmic curves* that originate at different points do not intersect except at infinity.

Number of Transition Cycles

This causes the behavior of a conditional empirical memory matrix to remain the same if the underlying relations in the environment remain the same.

Latent memory

If a CE cell falls out of use because another CE Cell gains a higher confidence value for the input state they share, the inactive CE cell will not select behavior, and will remain indefinitely at its last sensitivity value. This latent memory remains ready to produce behavior later if the sensitivity of the active CE cell falls below the sensitivity of the inactive CE cell. Because the sensitivity of an inactive CE cell stays constant, a memory of relations that existed a long time previously may exist indefinitely if not disturbed by new learning. (If capacitor type memories are used, they also will diminish exponentially over time without changing their relative sensitivities.)

Convergent behavior

Because the behavior of a discontinuous empirical system is made up of a set of discrete condition/action relations, a discontinuous empirical system may develop convergent sublines of behavior for each divergent subline of behavior if it has enough memory and enough learning experiences. These convergent sublines of behavior tend to return the behavior of an empirical system to its main line of behavior. If an empirical system behaves like a regulator, all of the convergent sublines terminate at the system's set point. Thus, empirical systems may learn to act like a regulator, which limits its range of activity and preserves its successful behavior.

The law of dominance in empirical behavior

If two or more empirical systems are superimposed upon a common set of variables, a dominant system or dominant set of systems will emerge that will control behavior.

1. *Dominance can be established by an empirical system with the following assets: a greater knowledge of the natural information in the environment involved the behavior, higher confidence levels, better timing, greater physical strength, and/or greater persistence.*
2. *The eventual behavior of two or more superimposed empirical systems will consist primarily of the behavior of the dominant system(s).*
3. *If two or more actuators that produce a negative inverse force are connected, the actuator that produces an output closest to the existing output (minimum change) or is first to act will dominate behavior.*

The essence of the fourth law is: Empirical behavior is determined by what can and cannot be done in the physical domain of force, time, and position.

Physical influences

There are many physical influences that can interfere with the output of behavior. For instance, some actions may be physically impossible such as trying to walk through a wall. Other actions may be impossible because they interfere with the actions of other actuators that are a part of the same system. For example, the movement of one arm of a robot may be stopped by hitting its other arm. In other cases, a trainer may prohibit actions that do not contribute to the behavior wanted by the trainer. These interferences are shown by the arrow (*I*) representing the effect of the environment upon the actuator (Fig. 29.9).

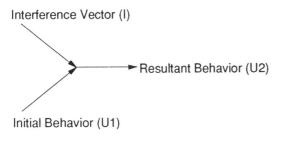

Interference Vector (I)

Resultant Behavior (U2)

Initial Behavior (U1)

*Figure 29.9.*The *interference vector* causes a change in the output of the controller.

In other cases, actions attempted by a robot may be impossible to carry out because its actuators may have reached their end of travel, or its actuators may hit the actuators of other robots. The first learning experience of an empirically controlled robot is to find itself — to learn how to move without hitting itself or other objects in its surroundings.

The trainer

A *trainer* must "check" or resist (interfere with) unwanted actions. In some cases the trainer must be physically stronger than the empirical unit. In other cases, the trainer must act before the empirical unit can act to dominate behavior. So speed of response may determine the dominant unit. For example, a dog trainer must anticipate the actions of a dog, and must correct the dog before it even attempts this unwanted behavior. The trainer must also be more persistent than an empirical unit to influence its behavior. The trainer must also provide consistent reactions to an empirical unit. The trainer must use the inherent physical laws of nature to teach behavior. Attempting to teach an empirical unit to do something that is physically impossible for the empirical unit to carry out will not lead to successful behavior.

In some cases, the force applied to an object may equal the sum of the forces applied by a set of actuators. In these cases, two or more units may combine to produce behavior. Units that work together may dominate behavior. Thus, a group with unified and coordinated behavior may influence the behavior of an individual unit. But a single individual or unit acting alone may not be able to influence a well-organized group.

The use of power, force, and strength are only a few of the ways to dominate behavior. An empirical unit may also acquire the behavior wanted by a trainer because of the *opportunities for behavior* created by the trainer. In this case, the trainer must attempt to increase the probability of a unit doing what is wanted. This may consist of removing obstacles in the way of the wanted behavior and creating obstacles in the way of unwanted behavior. Once the empirical unit does what is wanted, the unit will have a greater probability of producing that behavior again, by virtue of the empirical process. By using this teaching method, the trainer can influence the behavior of an empirical unit without relying upon the use of greater physical force.

Dominance by design

A trainer or operator must keep an empirical robot from carrying out unwanted actions. One method of assuring that the operator can control the behavior of a robot is to design the robot actuators to be significantly weaker than the trainer. This can be done in an automobile steering auto-pilot by designing the steering wheel drive motor (controlled by the auto-pilot) to have less torque than that normally produced by the driver.

The actuator motors of a robot can be influenced in ways other than being over-powered. Overload switches, "kill" switches, or tactile limit switches can be installed that cause the actuator movements to stop whenever they are touched by an operator. Operating these switches can interfere with or prohibit the actions prescribed by the robot's controller.

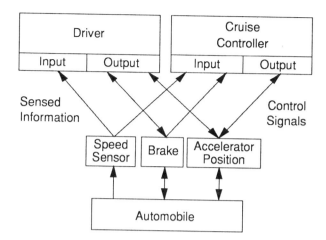

Figure 29.10. The operation of a *cruise controller* can be overridden by a switch controlled by the driver.

An example of this form of dominance is demonstrated by the cruise controller found in many automobiles (Fig. 29.10). The automatic speed controller operates until the driver (the dominant controller) steps on the accelerator or puts on the brake. These commands override the influence of the *subordinate* auto speed controller.

If an actuator in an empirical unit is servo-controlled, the trainer may act upon its set point or its servo-valve instead of trying to resist the output of its servo system. This allows the application of a small force to influence the behavior of the servo system, which operates at a much higher force.

These are examples of dominant interferences that are designed into a system. As we have shown, there are other sources of dominance, such as speed, persistence, and understanding of the organization of the task environment.

Speed of response

The timing of an event may have a large bearing upon its being carried out successfully. For instance, if a walker moves a foot forward too slowly, the walker will fall forward. If the walker moves a foot forward too quickly, the walker will stop. So, learning the timing relation of various moves are important and demanding learning tasks.

Speed of response is often used as a source of dominance in many living systems. The first student to raise his or her hand in class gets a chance to answer the teacher's questions and a chance to improve his or her standing in class. Quickness of response may determine who has the greatest influence in a conversation. Quickness of response can be used as a measure of the confidence levels of a set of conditional memory cells. The first cell selected has the highest confidence level. Sensor and actuator latches allow only the first states that occur in a transition cycle to be perceived by their matrices. However, being too quick may not always provide dominance. Trying to answer a question before it has been fully stated can result in the wrong answer.

Dominance of small errors over large errors

If two or more actuators that operate with a *negative inverse force* are connected, and attempt to position an object, the actuator that attempts to produce the position closest to the actual position at the start of the attempt will produce the greatest force. This may cause the other actuators to take its position. Also, when these actuators are connected in series in a sigma actuator, such that they produce an output that is equal to the sum of their positions, the actuators that select positions close to their existing positions will achieve their positions, causing the other actuators to assume positions other than those called for. Under these circumstances, control units that produce small errors will dominate control units that produce large errors. This *dominance of small errors* applies to many situations. For example, we can pick up heavy objects if we keep them close to our body. In another example, we are more likely to succeed if we attempt tasks that are only slightly different from tasks that we have accomplished successfully in the past.

Persistence

If two empirical units produce conflicting results but are of equal strength, the unit with the greatest persistence will prevail. This principle also applies to human behavior. Greater persistence can come about because of high confidence levels in a given area of behavior compared to another individual. The individual with the higher initial confidence level can experience more failures before the confidence level of this individual falls below another individual with a lower initial confidence level. Some units can be designed to change their confidence level more slowly by virtue of a smaller ratio (a) in their logarithmic subtracting mechanisms. Though they take longer to learn, units with a smaller ratio take longer to abandon their behavior.

Organization

An understanding of the organization of another system or environment helps an empirical unit learn the rules of that system more quickly and thus can achieve consistent, repeatable behavior within the context of that system. If a unit is not organized according to the task environment, it cannot recognize the organized relations in its environment and will not establish consistently doable behavior.

Collective efforts

Some interferences may be unable to keep a system from carrying out its selected behavior. In this case, these interferences are essentially ignored. An example of this type of incident occurs in the bidding process, where low bidders (or late bidders) get nothing. However, if several bidders pool their money, they may outbid another single bidder. Likewise, systems may join their behavior so that they can overpower and dominate other systems. It is easy to imagine a situation where two or more systems can combine to produce a sufficiently greater force than another system to prohibit or modify an action called for by the other system. In a "tug-of-war," the team with more members is likely to prevail, assuming the members of both teams are of equal strength (Fig. 29.11).

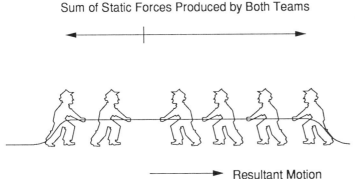

Figure 29.11. The *collective interference* of several systems may influence the behavior of another system.

Though one system is unable to change the outcome of an action, two or more systems working together may do so. This is one way in which two or more empirical units may work together to produce successful behavior that would be impossible to produce by one unit working alone (see section on bonding). Some interferences may only partially deflect a movement from its intended position. For example, even a weaker team may partially influence the behavior of the stronger team. In this case the behavior of the empirical system will converge upon the partially deflected position.

The law of organization of empirical behavior

Empirical systems will divide themselves into the smallest units that can deal effectively with given tasks.

1. *If a unit is too small (diverse), it will not learn a complex task.*
2. *If a unit is too large (unitary), it will take too long to learn a simple task.*
3. *The organization of a successful system reflects the organization of the task environment.*
4. *A set of units must be assembled (unitized) according to the priorities required to produce successful behavior.*
5. *An empirical unit will acquire the simplest behavior that is possible within an environment.*

The essence of the fifth law is: Empirical systems must organize themselves according to the organization of the task environment.

Unitary and diverse systems

Unitary and diverse systems differ in fundamental ways. As shown earlier, the combination of values of the set of variables has meaning in a unitary system. Thus, the value of a variable in a unitary system has little meaning by itself. However, the value of a diverse variable may have meaning by itself. Whether a system is unitary or diverse is a fundamental characteristic of the controller/environment system. If the controller cannot adjust to the organization of the task environment, it cannot deal effectively with the tasks in that environment.

Matched organizations

The law of organization explains that an empirical system must organize itself into a particular arrangement of unitary and diverse parts to deal effectively with the range of problems posed to it by its environment and trainer. This suggests that the environment is arranged in some form of organization that we must understand to deal with our environment effectively. Thus, an empirical unit must "clone" the organization of the environment to which it is connected. This also parallels Newton's third law of dynamics, which says that for every action there is an equal and opposite reaction. It also suggests that we must attempt to understand the arrangement of a task environment if we expect to design control systems that can deal effectively with the task environment.

 This law is demonstrated very clearly by the situation that arises in military battles where the opposing armies tend to organize themselves in the same way. This eliminates the possibility that one army may gain an unusual or unexpected advantage. This configuration is shown in Figure 29.12.

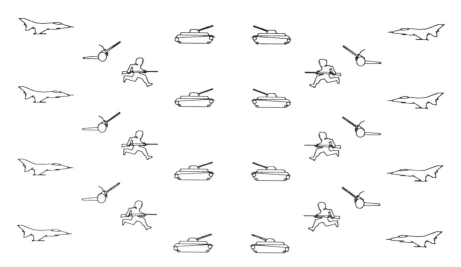

Figure 29.12. Matched organizations are better able to deal with each other.

Labor unions now attempt to organize themselves in the same way as the industry that employs its members. The CIA was created to match the KGB. The law of organization says that there is an optimum way of organizing the behavior of a set of machines to accomplish a specific task. It declares that behavior can be unitized in the form of assemblies and subassemblies, the same way that machines are unitized.

Priorities and protocol

A control system must identify the *priorities* created by their task environment and determine how to deal with them in the right order (*protocol*). Each priority may have a set of items that must be dealt with to deal effectively with that priority. This leads to the kind of organization tree shown in Figure 29.13.

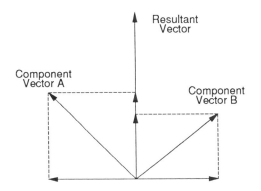

Figure 29.13. A *list of priorities* may require a hierarchical form of organization.

Each priority can be viewed as components of a vector, and the action needed to meet the needs of all of the prioritized elements can be considered the resultant vector. Each need or component may also be a vector leading to the hierarchical organization inherent in the sum of a set of vectors. For example, the rotary axis of an industrial robot may be used to move a gripper located at the end of its arm from a "pick" location to a "place" location. In one swing of the rotary axis, the arm may have move in to the pick location and out to the place location. Each of these two movements may have to occur during one-third of the period of the rotary motion. The gripper may have to open and close at the beginning and end of the rotary motion. Each of these motions may have to occur in a period that is one-sixth of the period of the rotary motion. If each motion are not carried out in the correct sequence and at the correct time in relation to the other motions, the robot will not pick up parts and place them correctly.

Assemblies and subassemblies

Most structures are made up of units and subunits. It may be possible to understand what keeps these parts separate and what causes them to join to form larger units. This process of *unitization* can be presented in an organization chart that shows how each assembly is made up of subassemblies.

The simplest or primary behavioral unit is the input/output transition or simple heuristic. A set of heuristics creates the modulus of behavior of a responsive unit. A set of modulii creates a behavioral union and-so-on.

Thus, the principle of unitization applies to behavior -- that behavior systems are built up by joining relatively separate and independent units into larger units that can join to form even larger units through a process called bonding. Some *levels of unitization* are shown in Table 29.1.

Table 29.1. Levels of unitization

- The simplest empirical cell that can keep track of the probability of its output occurring within a given time after its input.
- A matrix of empirical cells that can deal with single input and output variables.
- A duplex system that can identify the active states in a line of behavior.
- A parallel (superimposed) network that can create a division of labor.
- A series and parallel network of controllers that can handle more diverse variables.
- A network of networks that can handle even more variables.
- Bonded groups of empirical units that can select different regions in which to act.
- Higher-level groups involving interacting and interdependent human operators.

We have shown that empirical units may break apart and join with other units to form new alliances according to the requirements of the task environment.

Organization of empirical units

The behavior of a group of systems will keep changing until they divide themselves into a skillful arrangement of parts that do not interfere with one another and meet the requirements of the task environment. This is a statement of the principle of noninterference as it applies to group behavior. So, an empirical unit can choose four different ways of organizing itself. It can change in any one of the four possible directions shown in Table 29.2.

Table 29.2. Ways that an empirical unit can be organized

---

1. Fragment itself into separate parts.
2. Unify itself into a single whole.
3. Merge with other controllers to form a group.
4. Separate from a group to act as one.

The study of empirical control deals with an explanation of how an empirical unit selects action, unifies or fragments itself, joins with or separates from other systems, develops behavior, finds needed timing relations, and establishes closure to assure its behavioral existence.

The law of coexistence of empirical behavior

Two or more control systems can be connected to the same set of variables and establish behavior that appears as if it were produced by a single more complex system.

1. *The behavior of two or more empirical units can coexist by operating in unison upon the same set of variables, by operating upon different variables at a given time (dispersion), or by operating upon different values of these variables at a given time (value field selection).*
2. *Two or more empirical units may learn to produce complex behavior using fewer memory cells than a single larger unit acting alone.*

The essence of the sixth law is: Two or more empirical units can learn to operate upon the same set of variables without conflict by establishing an appropriate division of labor.

Compatibility

The second statement of the empirical algorithm suggests that an empirical unit can find actions that are compatible with the requirements of other units. This implies that two or more systems can share variables and not interfere if they all call for the same or compatible values of these variables at a given moment in time. The behavior of each system may appear to be skillfully joined if each unit learns to act so that it does not interfere with the behavior of the other units and acts in support of the other units. Units that learn to act this way form a unified whole made up of a skillful arrangement of parts. This principle of coexistence allows two or more units to work together and produce more effective behavior than a single unit acting alone.

Summing behavior

The outputs of two or more units can be summed by a differential or a sigma actuator so that the total output position is equal to the *sum of the positions* produced by each unit, or they can be connected in parallel by springs so that their total output force is equal to the *sum of the forces* produced by each unit. Units connected in this way can produce a greater displacement with greater resolution, or produce a greater total output force if their behavior is coordinated appropriately. For example, the range and resolution of a linear actuator can be doubled by summing the output of the actuators of two systems. This may result in a less complex control system than a larger single system that can deal with twice the range and resolution of the two smaller systems.

Inactive behavior

"They also serve who sit and wait." The duplex controller does not require additional intermediate values to store information about constant values of an output variable. The rules of a connection matrix allow all of the input values that result in the same output value to be put on a single input matrix output terminal, which in a duplex system is an intermediate terminal. This encourages units with outputs that are summed with other units to produce constant output values whenever possible. These constant output values result in inactive lines of behavior by some units allowing other units to produce the required behavior.

When behavior is divided among active and inactive units, a division of labor occurs that may allow each unit to be less complex. In a successful division of labor, each units does what it does best and does not attempt to do what it is ill equipped to do. This does not mean that one unit does everything while the other units do nothing. A successful division of labor is not based upon a disuse of labor but is based upon a more appropriate application of labor.

Conventional controllers

Conventional predetermined servo-controllers cannot be hooked up to the same set of variables unless that have the same set points. If they have different set points, conventional-servo controllers will "*buck*" one another. A typical example is a furnace and air conditioner system connected to the same living space. Unless these temperature controllers are carefully designed, the two systems may work against each other, leading to an extremely wasteful operation. However, the behavior of empirical controllers connected to the same variables will automatically change until their interferences are eliminated.

Concinnity

A symphony orchestra is an example of a connected system that usually forms a successful alliance. Many different instruments and groups of instruments play simultaneously, often with different melody lines and seemingly different rhythms. However, the music remains coordinated if every instrument stays in "tune" and each musician plays in the correct "key," stays in the same "place" in the score, follows the tempo and beat of the conductor, and plays the correct notes. Machines working in a factory, people talking in a meeting, and cars moving on a freeway all operate according to the same principle of coexistence that results in a *concinnity* of behavior. These skillfully joined lines of behavior, modified to fit together in a skillful arrangement of parts, are typical of the behavior developed by empirical systems.

The law of equilibrium of empirical behavior

The information stored in an empirical controller will change to match the knowledge, relations, laws, and principles inherent in the systems to which it is connected.

1. *The operant variables of one system must be connected to the sentient variables of another system to create behavior.*
2. *A change in behavior of one system will be reflected in all of the other connected systems.*
3. *The variety of behavior of the system is limited to the minimum variety of any point in the system.*

The essence of the seventh law is: The behavior of an empirical unit will reflect the behavior of the system to which it is connected.

Closed systems

We have pointed out that the closure process requires at least two systems, both with a set of inputs and outputs and both connected input to output, as shown in Figure 29.14.

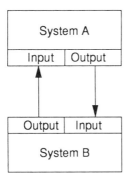

Figure 29.14. A *closed system* requires at least two units that supply the behavior that the other needs to produce behavior.

This configuration makes it possible for System A and System B to form a closed loop of behavior that allows the two systems to operate indefinitely without intervention. This says that the systems can exist together by supporting each other's behavior. This implies that both systems have nearly the same potential variety, both can produce nearly the same number of transitions and both are organized in nearly the same manner.

Magnetics

This balanced relation between the two operating units in an empirical system is similar to the action of two magnets, as shown in Figure 29.15.

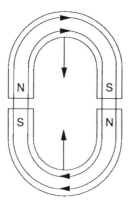

Figure 29.15. A pair of magnets represents the action of a closed system.

If unlike poles are aligned, closed loops of magnetic flux are formed, causing the magnets to be attracted and held together. This *magnetic attraction* occurs because the magnetic flux in one contributes to the flux in the other, creating a higher flux density assuming suitable pathways for the lines of flux are available between each unit. The lines of flux are quite similar to lines of behavior in that they must form a closed loop and they seek a region of high permeability where they can exist in greater number.

The north/south poles are analogous to the sensor/actuator cites of responsive control systems, and the continuous closed loops of magnetic flux are similar to the closed line of behavior that must exist between two responsive control systems for them to operate successfully together. No interferences can occur between two control systems that joined with their inputs to their outputs and outputs to inputs (north to south, south to north).

Magnetic repulsion

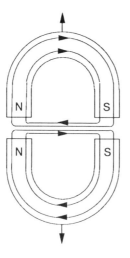

Figure 29.16. Magnetic repulsion is similar to what happens when control units are hooked up input to input and output to output.

If, on the other hand, like poles are placed together (north to north, south to south), the magnets cannot easily form closed lines of flux, and they will repel each other, as shown in Figure 29.16. When like poles are placed together, the lines of flux form two separate fields, neither of which passes through the other magnet. The flux from both magnets seeks closure through a new environment (through the air gap between their poles). Thus, when the input variables of two or more units are connected and the output variables are connected, as they are in a parallel or superimposed system, interferences will occur until they establish compatible behavior.

Unification

Two magnets placed with like poles together may flip around and become one magnet provided they are placed in a region that provides convenient pathways for flux lines to follow, as shown in Figure 29.17.

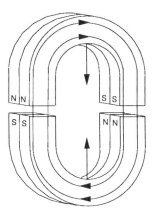

Figure 29.17. In a *process of unification,* two magnets with like poles together can stop opposing one another by acting upon other magnets whose poles matched up with the opposite polarity.

When the like poles can operate upon unlike poles, the like poles will cease to oppose one another. By analogy, empirical units superimposed upon the same set of sensor and actuator variables can establish compatible behavior by learning to act as one against a third element, such as an environment.

Magnetic induction

Another analogy between magnetic and empirical behavior is that if a magnet is brought close to a bar of soft iron, closed magnetic flux lines will form between the north and south poles of the magnet through the soft iron, causing the appearance of poles in the soft iron, as shown in Figure 29.18. The south pole of the magnet always induces a north pole on the iron (or another magnetic material with low coercivity) and the north pole in the magnet always induces a south pole in the ferromagnetic material. In this case, the magnet and the iron will be attracted to each other. Thus, the field of the magnet will come to exist within the iron. Also, the stronger the magnet, the greater the magnetic field inside the iron and the stronger the attraction between the magnet and the iron until the iron becomes saturated. At this point, the iron cannot contribute to the magnetic field to a greater degree.

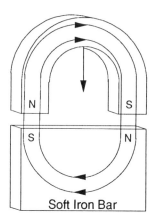

Figure 29.18. Magnetic induction represents what happens when a weaker empirical control unit is brought into the presence of a stronger unit.

This process of induction suggests the principle of equilibrium, in which the magnetic flux in a hard magnet with high coercive strength determines the flux in another softer magnet with lower coercivity and high permeability. The "softer" controller (such as an empirical controller with low confidence levels) is influenced by the "harder" controller (such as an empirical controller with high confidence levels or a predetermined controller), much like the soft iron magnet is influenced by the harder magnet. This influence can increase up to the limits of the variety of the softer controller.

Induced behavior

Thus, a controller can induce the inputs of another controller to accept and act upon its outputs and induce outputs in the second controller for its inputs, in much the same way that the flux field of a strong magnet appears in a soft iron bar. The analogy is presented because it very graphically represents how pairs of controllers or a controller and its environment create closed loops of behavior that have a common influence on both and hold the two together in much the same way that a pair of magnets creates closed lines of flux that hold the pair of magnets together.

In our everyday lives, the physical requirements of our environment cause us to modify our behavior. But we also modify our environment to meet our needs, and the people in our environment often modify their behavior to meet our needs. Once these modifications are made, it is much easier to produce and maintain a given modulus of behavior. To continue to benefit from these relations, we must produce behavior that holds us close to these modified environments and units with modified behavior, in much the same way that magnets become stronger when they attract and hold other magnets near them.

Forces in equilibrium

Newton's third law states the follow: "For every force, there is an equal and opposite reactive force." This principle is illustrated in Figure 29.19.

Figure 29.19. Forces are shown to be *in equilibrium* since two force scales that are connected indicate the same force.

If two scales are connected and a tensile force is applied to one, they will show the same reading regardless of the force. Without an object upon which to act, a force scale cannot produce a reading other than zero.

Cloning of behavior

Likewise an empirical unit always requires an opposing system to produce action. Then each connected empirical unit will acquire and reflect the underlying knowledge, behavior, and principles of the systems with which they come in contact. The concept of the equilibrium of behavior is illustrated in Figure 29.20.

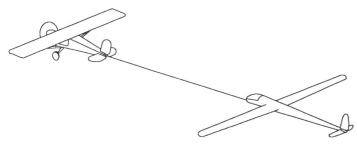

Figure 29.20. The behavior of two systems that are connected will change until they establish an *equilibrium of behavior.*

Changes in the flight path of the glider will change the flight path of the tow plane, and changes in the flight path of the tow plane will change the flight path of the glide. It is only when both the glider and the tow plane establish the same flight path that they can carry out a successful mission. Likewise, an empirical controller connected to another controller must establish behavior that is related to the behavior of the other controllers.

Equilibrium of all behavior

Just as the dynamics of our physical world depend upon an equilibrium of forces, the behavior of empirical systems also will depend upon the action and reaction of connected systems. The action of each system will cause an equivalent reaction in other connected systems. The result is a balanced flow of information throughout a system of empirical units. Likewise, if we see a part of a system with a great deal of activity, we can expect to find another part with the same high level of activity. A logical implication of the principle of the equilibrium of behavior is that behavior anywhere affects behavior everywhere — that everyone's behavior is connected to the behavior of everyone in some idealistic way.

The law of closure of empirical behavior

An empirical system must establish closure to function by itself.

1. *For every divergent line of behavior, there must be a convergent line of behavior to maintain closure.*
2. *A closed line of behavior must include behavior that keeps the system within the conditions with which it is familiar.*
3. *A closed repeating line of behavior allows an empirical unit to establish accurately the probabilities underlying the behavioral relations encountered.*

The essence of the eighth law is: An empirical unit must reuse the behavior that it has learned.

Practice

The behavior of an empirical controller is such that it will seem to reflect the intrinsic relations, behavior, capabilities, limitations, and knowledge of their environment and their own structures after experiencing many input/output events within the context of a given environment. This law is based upon the empirical algorithm, which causes the confidence level stored in the controller's memory to increase the more a given "piece of behavior" (heuristic) happens successfully. This increases the probability of the controller selecting behavior that can be carried out relative to the other possible lines of behavior that are thwarted or restricted. This record of probability can only be created in an empirical unit through *practice*.

Survival

This positive learning process keeps the system doing more of what is possible and less of what is impossible within a given setting. But an empirical unit must also learn the particular behavior that evokes the conditions that lead to that successful behavior. Otherwise, the successful behavior will become "lost."

Closure comes about because the main line of behavior of a system returns to some common state from time to time. This "home base" may be a recharging station for a battery-operated mobile robot, or an initial state like the ready position of a tennis player waiting to receive serve.

Another example of behavioral closure is seen in the process of gathering or growing food. The experience of gathering or growing food provides the knowledge of where to gather food or how to grow food successfully, and the food provides the energy to continue to obtain more food. If any part of the process is broken, the knowledge or energy needed to get food may not be available and, in this classic case, the organism perishes. The process of establishing closure is the basis of establishing self-sustaining behavior.

Closure is a primary requirement for the continued behavioral existence of all responsive units. Closure is almost synonymous with "life" or survival or success. Closure determines if a responsive unit continues to exist by itself or whether the unit must join other units such as a service personnel to establish closure.

The law of bonding of empirical behavior

Two or more empirical units are held together when each contributes to the behavior needed by the other to maintain closure.

1. *Two or more small units acting together may produce behavior more effectively than one larger unit acting alone.*
2. *Two or more units can act as one if they establish an appropriate division of labor and effective intercommunications.*
3. *If one unit interferes with behavior needed for closure of another unit, the units may divide into separate systems to maintain their behavioral existence.*

The essence of the ninth law is: Empirical units are held together by their successful shared behavior and are forced apart by unsuccessful behavior that they must share.

Behavioral force

Empirical machines operating in a unitary environment can learn to produce behavior that considers the behavior of the other machines, where unique behavior is needed for each combination of values of the system variables. This coordinated behavior may be directed toward a common objective. This alliance may cause an appearance of forces or lines of influence among subunits that can be construed as attachments, loyalties, or cooperative bonds.

These *behavioral forces* are the basis of the study of group psychology and sociology. They hold animal and human groups together. They create a special kind of being — the social or behavioral group. They allow separate individuals to be linked together by means of their behavior, forming a new collective unit with a life and existence of its own. The behavior of this collective unit may have considerable value by itself. For example, a trade union may carry out wage negotiations that would be ineffective if attempted by an individual.

An empirical being cannot exist by itself. It must be connected to another system such as an environment that can "stimulate" it to produce behavior. The behavior of an empirical being will not change unless it is influenced by changes in another system. If the modulus of an empirical machine does not change, a machine that can learn from experience remains like a predetermined machine. The system connected to an empirical unit can be an environment, the machine's sensor/actuator apparatus, or another outside being. These outside systems might be interfering systems, dominating systems, or other superimposed systems. The behavior of these outside systems may also change until the connected systems create closed lines of behavior. This new behavior may take the form of some kind of union or alliance.

Thus, a system can obtain closure by joining with other systems. For example, a piece of automatic machinery in a production plant may summon a maintenance worker whenever it determines that it is about to break down. This individual and this machine may produce behavior together that is more successful than the machine working alone. They also must develop and maintain behavior that keeps them together if they are to continue to produce this successful behavior. For example, the maintenance worker must be committed to coming to work every day to keep the machinery running, and the machine must summon the maintenance worker whenever it is likely to fail. Thus, the behavior produced by the maintenance worker determines the continued existence of successful behavior of both the automatic machine and the maintenance worker. This joined behavior becomes a powerful force acting upon both units. They will seem to struggle to stay together if the joined behavior is necessary for closure of both systems. So behavioral bonding is not caused by a physical force, but is caused by behavior that creates the appearance of a physical force holding two or more units together.

The basis of behavioral bonding is a mutually beneficial *division of labor*. For example, a married couple may divide the household tasks into "indoor" and "outdoor" activities. One spouse may be responsible for cooking, cleaning, and paying the bills. The other spouse may be responsible for taking care of the car, cutting the grass, and keeping the trim painted on the house. Each may accept these responsibilities and become quite adept at fulfilling them. Each may greatly benefit from the activities of the other. While one individual is provided with good meals and a clean house, the other individual is provided with a functioning car and a well-kept home and grounds. If the couple were to separate, both must perform all of the tasks needed to operate a household. In many cases, both would have to learn how to do the tasks previously performed by the other. In most marital separations, these added responsibilities create a great deal of stress on both individuals.

Both spouses may not be consciously aware of the amount of work done by the other spouse when they are together. However, once they separate, the increased work load becomes obvious. This may create a powerful incentive to reconcile the differences that led to the separation, and the couple may attempt to get back together again to reduce the stress due the added work load.

Empirical units do not think and are not motivated by the psychological pressures implied in the example given above. However, if two or more empirical units establish a mutually beneficial division of labor that creates a closed line of behavior, part of that behavior must include behavior that keeps the units together. Thus, the empirical units will act as if there is something pulling them together in much the same way that there is an emotional force holding couples together.

The law of timing of empirical behavior

Most behavioral events require a specific interval between initiation and completion.

1. *These timing relations may be due to inherent physical properties found in the total system or may be required by the timing expectations of other connected systems.*
2. *Those control units will dominate behavior that forecast events correctly, including when they are supposed to start, and how long they are supposed to last.*
3. *If a control system does not provide for these time relations, it will be unable to establish successful control behavior.*

The essence of the tenth law is: Empirical units must produce behavior having the time relations required by the task environment.

Timing

The timing of an event may have a large bearing upon its success. For instance, while taking a step, a person must allow a certain amount of time for his or her foot to fall to the ground before taking the next step. The timing of each step is determined by the acceleration due to gravity, the height of the step and other factors. The controller for the walking robot must produce the correct timing for each part of its stride to produce effective walking behavior.

Predetermined cycle times

Many timed relations for a machine can be determined beforehand. For example, the optimum gait of a walking robot can be determined by the size of the robot and the acceleration due to gravity. The *cycle time* of its controller can be fixed accordingly. If it is expected to walk on the moon, provision can be made for the optimum transition cycle times required for the acceleration due to the gravity known to exist on the moon.

Superimposed units with different transition cycle times

Discontinuous state (DS) machine controllers operate in fixed transition cycle times. The gait of a DS walking machine has to match the cycle times of the controller. Multiple controllers with different cycle times can be superimposed, as shown in Figure 29.21, and used in a given walking machine. The controller with the most appropriate cycle time will dominate and control its walking behavior because it will provide the most consistent results. All of the units shown above operate off a single output disconnect circuit so that the most confident unit selects the output in a given transition and prohibits any other unit from producing an output in that transition cycle. The timing of the action will be that of the most confident unit, indicating that this unit has selected the best time delay for the conditions encountered.

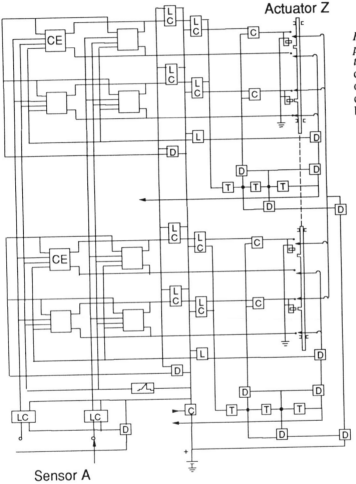

Actuator Z

Figure 29.21. Superimposed units with different transition cycle times connected to a common output disconnect can compete to produce behavior.

Sensor A

Empirically determined cycle times

No provision has been made in this discussion for automatically changing the transition cycle times of the DS machines shown in this text, other than providing superimposed units with different transition times, as shown previously. Yet it might be possible to change the time base of a single unit based upon the results of its behavior. The feedback complete circuit contains information about when an action is completed. This information could be used to change the timing of the time delay relays in the matrix circuit driver, much like the speed of an adaptive clock can be incrementally changed whenever the clock is set to the correct time. In lieu of this adaptive mechanism, superimposed controllers with different time bases can be used to find the most appropriate time delays between sensed conditions and the occurrence of specific actions.

Multiple units acting in synchronism

Multiple units must act in synchronism to carry out complex timing behavior like playing basketball or football. People dancing in a group must establish synchronism. The beat of music can provide the time base for their synchronized actions. The timing of these superimposed units can be based upon the method of timing in musical scores that uses 1/16th, 1/8th, 1/4, 1/2, and whole notes. The timing of these units can also be arranged to allow a certain number of beats per measure and a certain number of measures per phrase.

Timing in hierarchical systems

Hierarchical systems have unique timing problems. Higher levels usually deal with long-term behavior, and lower levels deal with higher-frequency behavior. This is because information may have to flow through more units to and from higher levels, taking longer to be carried out. These timing considerations are important and must be dealt with in any practical control system. How the animal brain deals with behavior that has to happen quickly versus behavior that must be done more slowly is not clearly understood. If one develops a slow golf swing, can it be successfully speeded up to provide a fast golf swing? Some people say that the fast swing has to be learned separately from the slow swing. The empirical DS machines shown in this text must learn different timed relations in different control matrices.

Forecasting the time of output events

DS machines attempt to forecast the output events that can take place for a given period after a given delay for specific input conditions. If different units have different intrinsic time delays, the unit that can predict the correct time of these future events will dominate behavior.

Conclusion

The laws of empirical behavior are intended to contribute to an understanding of how empirical behavior comes into existence to exploit the opportunities for behavior within an environment. Empirical behavior can exist only if it reflects the physical and geometrical constraints in the real world, is compatible with the behavior of other systems, meets the needs of its operators, and conforms to the laws of nature.

Chapter 30
Variety, Information, and Order

Empirical machines are *cybernetic engines* that transmit *messages* from sensor to actuator environments. The degree of difficulty encountered in transmitting these messages is determined by the variety, information, and order in the messages. *Variety* is the number of possible combinations of values of a set of variables. *Information* is the number of variables required to produce a given variety, and is a measure of complexity. *Order* is the degree to which variety can be expressed with a minimum number of variables, and is a measure of simplicity. Empirical machines have the special ability to reduce the number of variables in a message and thus create order out of complexity. The purpose of this chapter is to show the relation between variety, information, and order, and to show how these concepts apply to the design of empirical controllers.

Variety

Variety has been defined as the number of different forms that a given object or class of objects can take. For example, we speak of the variety of characters that an actor can play, or the variety of fish in the sea.

Variety of a variable

The *variety of a variable* is equal to the total number of different *values* (x) that a variable can have. If the variable is a *multiposition switch* with a set of contacts, as shown in Figure 30.1, the variety of this variable is equal to the number of different switch positions.

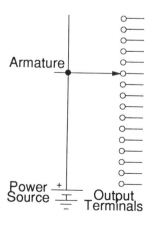

Figure 30.1. A *multiposition switch* is a single variable with as many different values as contact positions.

The multiposition switch shown is a single variable with 16 contacts, only one of which can be "made" at a given time. It represents a single variable ($n = 1$) that has 16 *value elements* ($x = 16$). Its variety (V) is equal to 16 because the simple equation of variety of a unitary system states that

$$V = x^n = 16^1 = 16 \text{ contacts.}$$

The *unit value* of the variable is defined by its value elements. In the example given above, each value element is a *contact*. So the unit value of the switch variable is a contact.

Absolute value elements

Each value element in the multiposition switch shown is not independent of the other values of the switch since only one value element can be made active in a given variable at a given time. If a new value is established, the previous value must be expunged. Thus, a value element is an *absolute quantity*. The contacts in the *push-button switch* shown in Figure 30.2 are absolute value elements. It contains a mechanism that resets or pushes out the button of the previous selection when a new button is pushed in.

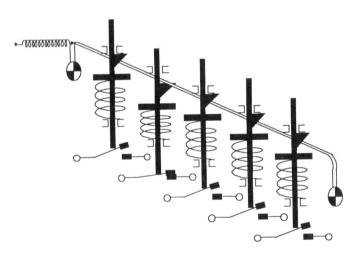

Figure 30.2. A push-button switch is an example of a single variable, in which only one value can exist at a time.

Each button is *dependent* upon the position of the other buttons since what happens at one button determines what can happen at the other buttons. Though each button has two conditions, each contact is not a binary variable. In this case, each contact or terminal should not be called a binary integer, bit, or anything concerning binary. Instead, it should be called a pentint because it is an absolute value element of a pentary (five-value) variable. (See Appendix B, Problems with Bits.)

Variety of multiple variables

Two or more variables can be used to represent a given quantity. For example, multiple decimal dials can be geared together to form an *odometer register* that can be used to count and display the number of revolutions of its input shaft, as shown in Figure 30.3.

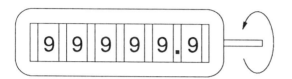

Figure 30.3. The variety of an *odometer register* is equal to the total number of different values it can represent.

This device can be considered to have five variables ($n = 5$) called decimal dials or *decades*, each of which has digits of 0 through 9 ($x = 10$ values each). The meaning of the odometer register is unitary because each of its states is dependent upon the combination of values of every dial. So, according to the simple equation of variety of a unitary system,

$$V = x^n = 10^5 = 100,000 \text{ miles.}$$

The dial to the right of the decimal point represents a part of a revolution. In a shaft revolution counter, information may be available as to a part of a revolution and can be displayed. However, in the odometer shown, the *unit value* is considered a mile. In this case, we are interested in the total number of distinctly different miles available, since this will tell us the number of available choices of the unit value (miles). If the unit value were considered 1/10 of a mile, then

$$V = x^n = 10^6 = 1,000,000 \text{ (1/10 miles).}$$

Since variety is the total of some value unit, such as values, combinations of values, states, apples, or miles, it is desirable to refer to the name of the unit value. For example, when we refer to an odometer, we should say that its variety is so many *miles*.

Improbability

The *information* in a message can be defined as the degree of improbability created by the message. This *improbability (I_p)* is equal to the number of possible alternative states inherent in the message, and is equal to its variety. The 16-contact switch shown can send 16 different messages, and has a variety of 16 contacts. Therefore, the improbability (*Ip*) of a given switch position is equal to 16. Other signaling devices may be more complex. For example, the number of possible values of an odometer register is equal to the maximum count that it can produce. If the odometer shown represents one mile for each revolution of its input shaft, and has five dials to the left of the decimal point, it can register 100,000 miles. So, the improbability of selecting any one number of these miles is 100,000. To determine the *probability information* in a message, one has only to ask the question: This message is one out of how many possible messages that can be produced by this apparatus?

Historical probability

The *historical probability* (p) of an event is defined as

$$\text{Probability } (p) = \frac{\text{the number of times an event happens}}{\text{the number of trials}}. \qquad (30.1)$$

Thus, the probability that a specific event will occur at random is related to the total number of possible events. For instance, if an event happens on the average of once in every 16 trials in many purely random trials, then

$$\text{Probability} = p = 1/16 \text{ events.}$$

In the 16-contact switch shown, there are 16 possible events (contact closures). The probability of one of these events occurring in a given trial is inversely related to the variety of the system if each contact closure is equally probable as it would be in a purely random system. Since experience is used to determine whether the system is random, or that certain contact closures are more probable than others, the probability of a given contact closure is called *a posteriori* or *historical probability*.

Improbability and variety

As stated, the information content of a message or event in a random system is based upon the number of available choices. As more choices are available, any one choice becomes more improbable. Thus, the improbability (I_p) of the event happening at random is a measure of information and is defined by the following equation:

$$\text{Improbability } (I_p) = \frac{1}{\text{Probability } (p)}. \qquad (30.2)$$

If an event happens only seldom or on a random basis, then it has a large improbability or large information content. Therefore, the improbability (I_p) of a given contact closure in the 16-contact switch shown is $I_p = 1/(1/16) = 16$. Improbability (I_p) is a dimensionless number because it is simply a ratio of the number of possible events to actual events. This makes it more useful, at times, than the logarithmic information (to be discussed), which is dependent upon the base of the logarithm used to determined logarithmic information from the variety of a system.

Number of equally probable states

If the odometer, shown previously, were converted to a roulette machine by allowing each dial to spin freely, then the probability of a given number appearing at random would be 1/100,000. So, $I_p = 1/(1/100,000) = 100,000$, which is equal to the variety of the odometer. In these cases,

$$\text{Improbability } (I_p) = \text{variety } (V). \tag{30.3}$$

Limits of uncertainty

In general, a message statement establishes the *limits of uncertainty*. It first establishes a number of possible choices and then contains a selection of one of the possible choices. For example, a message may consist of the number 4036.6. This message first establishes that there is an uncertainty concerning a number from 0000.0 to 9999.9, or there is 1 out of 100,000 possible choices. This uncertainty is determined by the number of significant decimal places in the number to the left and right of the decimal point. If the number is to be recorded or transmitted over data lines, a channel that can deal with 100,000 choices is required. Once the number of possible choices is established, then one of the possible choices can be defined. In the example given, only one choice out of 100,000 has been specified, which is 4036.6. So the improbability of the message is still 100,000.

Allowance

A message might contain the number 4,036.6 +/- .5. This message allows for an error of 10 value elements. This means that the message is not just one out of 100,000 choices, but is ten out of 100,000 choices. This changes the improbability of the message to $I_p = 10/(1/100,000) = 100,000/10 = 10,000$ choices. Thus, the *allowance* made the message less improbable than a message with no allowance for error.

Concurrence

To determine the worth of a message, one also must establish the *reliability* or *concurrence* of the communication process. One must establish how many times the sender and receiver agree as to the meaning or content of a message. So,

$$I = \text{uncertainty} \times \text{concurrence}. \tag{30.4}$$

If it is determined, *a posteriori*, that sender and receiver agreed 100% of the time on messages over many trials on that channel, then the message described is still worth 100,000 (choices). If it is determined, *a posteriori*, that the sender and receiver agree only 50% of the time as to which choice has been sent, then the message is worth

$$I = 100,000 \times .5 = 50,000 \text{ choices.}$$

So information exists only to the degree that a question is raised and then resolved in the same way by the sender and receiver.

Classical definition of information

The relation between probability information (I_p) and variety (V) can also be demonstrated using the *classical definition of the information* (I_p) involved in the transmission of a message through a channel,[7] which states that

$$Ip = \frac{\text{probability of an event at a receiver after the message } (P_a)}{\text{probability of an event at a receiver before the message } (P_b)}, \qquad (30.5)$$

where

$$P_a = \frac{\text{number of times a given message is received}}{\text{number of times a given message is sent}}, \qquad (30.6)$$

and

$$P_b = \frac{1}{\text{number of possible values } (x) \text{ of the message variable}}, \qquad (30.7)$$

or

$$P_b = \frac{1}{\text{variety } (V) \text{ of the set of message variables}}. \qquad (30.8)$$

If the transmission is perfectly reliable, then $P_a = 1$.

So,   Probability information $(I_p) = \dfrac{1}{1/x} = x,$ \qquad (30.9)

or   Probability information $(I_p) = \dfrac{1}{1/V} = V.$ \qquad (30.10)

If there is *noise* in the transmission of a message, the number of times a given message is received may be less than the number of times a given message is sent. For example, only 9 messages may be received correctly out of 10 sent. Then

$$\text{Probability information } (I_p) = \frac{0.9}{1/V} = 0.9V.$$

Thus, the classical definition of information results in probability information being equal to the *variety of the message variables times the reliability of the transmission.*

Messages used in control

A *message to a robot* can be viewed as a signal in a communication channel, and the amount of information in the signal is determined by the complexity of the robot's behavior and the correspondence of that behavior to the content of the signal from the controller. In the basic equation of improbability (I_p), the reliability of the robot (P_a) is the ratio of the number of times a given set of instructions is carried out as called for (N_2) to the number of times it is sent to the robot (N_1). In this case, the reliability (P_a) may have to be determined by experience.

The value of P_b is the inverse of the variety of the robot. A robot system is usually unitary because its position is different for every combination of values of its axis variables. If the robot has n axes and each axis has a total number of stopping places equal to x, (where x is the range times the resolution), then the variety (number of distinctly different stopping places of the end effector of the robot) is equal to x^n.

So, the information in the signal to a robot is

$$I_p = \frac{N_2}{N_1} x^n, \qquad (30.11)$$

or
$$I_p = \text{robot reliability} \times \text{robot variety}. \qquad (30.12)$$

Thus, as the controller of an empirical machine calls for behavior that can be carried out, the information content of its control signal increases.

Logarithmic information

The more common representation of information is in the form of *logarithmic information* (I_l). This is because very large number are encountered when two or more very improbable events occur together. This problem can be solved by summing the logarithms of the improbability of each event, as shown in the discussion that follows.

Probability of events happening together

The probability of any two consecutive events happening together is p_1 times p_2 (the product of the individual probabilities). In the 16-value switch shown previously, where $p_1 = 1/16$ and $p_2 = 1/16$, the probability (p) of two specific consecutive events (contact closures) is

$$p = 1/16 \times 1/16 = 1/256 \text{ events.}$$

The information content (I_p) of this double event is

$$I_p = 1/p = 1/1/256 = 256 \text{ possibilities.}$$

Very large numbers are encountered using this measure of information. Taking the logarithm of the improbability reduces the size of the numbers because it allows the logarithmic information of each message to be added rather than multiplied together as shown.

$$\log_b(I_p \times I_p) = \log_b(I_p) + \log_b(I_p), \qquad (30.13)$$

where b is the base of the logarithm.

Thus, *logarithmic information* can be defined as follows:

$$\text{Logarithmic information } (I_l) = \log_x \text{ improbability } (I_p), \qquad (30.14)$$

where $\log_x$ is the logarithm to the base of the number of equally probable values (x) of the number of message variables (n). We know from Eq. 30.3 that $I_p = V$.

Therefore, $\qquad\qquad\qquad\qquad I_l = \log_x(V). \qquad\qquad\qquad\qquad (30.15)$

Logarithmic information in a unitary system

In a unitary system we know that $V = x^n$ from the pi equation of variety. So the logatoiyhimic information in a unitary system is

$$I_l = \log_x(I_p) = \log_x(x^n) = n \log_x(x) = n, \qquad (30.16)$$

since $\log_x(x) = 1$.

Stated in words, logarithmic information to the base x in a unitary system having a variety V is equal to the *number of variables n* required to produce the variety V, when all the variables have the same number of equally probable values x. The number of variables is determined from the logarithm of the variety, and the base of the logarithm is equal to the number of values of these variables.

Logarithmic information in a diverse system

The variety of a diverse system is $V = x \times n$, from the sigma equation of variety. So logarithmic information in a diverse system is

$$I_l = \log_x(I_p) = \log_x(x \times n) = \log_x(x) + \log_x(n) = 1 + \text{Log}_x(n). \qquad (30.17)$$

Again, the logarithmic information of a diverse system is equal to one plus the logarithm of the number of variables required to produce the number of choices available to the base equal to the number of values of these variables.

In summary, one must know five factors to determine the information content of a message: One must know the number of variables, the number of values of these variables, the allowance of the message, the organization of these variable into a diverse or unitary system, and the concurrence between sender and receiver.

Units of logarithmic information

The *units of logarithmic information* are the *types of variables* used to represent the variety of a message. For example, the variety of the 16-value switch shown earlier can also be represented by the 4-variable, 2-value register shown in Figure 30.4.

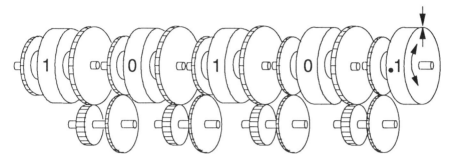

2:1 Reduction   2:1 Reduction   2:1 Reduction   2:1 Reduction

Figure 30.4. A *four-bit (binade) binary register* can display 16 different whole values.

In this case, each "place" is an independent two-value (binary) variable, since the value of one place imposes no restrictions as to the value of another place. Each place can be considered as an independent 2-value register called a *flip/flop* because each has only two positions like a toggle switch.

The logarithmic information (I_l) for this device is calculated as shown:

$$I_l = \log_2(V) = \log_2(16) = 4 = n \text{ binary variables.}$$

The *base* of the logarithm is 2 because each variable has 2 values, and 4 of these binary variables are required to represent the variety of 16 whole values.

If the 4-place binary register were considered a diverse system where the combination of values of each place is not important, then the logarithmic information of this diverse system would be as follows:

$$I_l = 1 + \log_x(x \times n) = 1 + \log_2(2 \times 4) = 3 \text{ binary variables.}$$

Notice again that the quantity of logarithmic information is number of variables, the unit of logarithmic information is type of variable, and the quantity of information in a diverse system is less than a unitary system with the same number of variables and values.

Number systems

Number systems are unitary systems with the same number of values for each variable. Therefore, they can be described in terms of logarithmic information, and the information content of a number is the number of *number system variables* required to express the number. If we were discussing a decimal number between 0 and 9999, then we would say that the four-*decade* register shown in Figure 30.5 is needed to represent the number the decimal system.

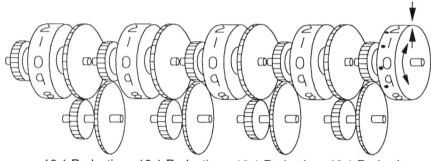

10:1 Reduction   10:1 Reduction   10:1 Reduction   10:1 Reduction

Figure 30.5. A *four-decade decimal register* consists of four decimal variables and has a variety of 10,000 (0 through 9,999) whole values.

Any dial or other device that is subdivided into groups of 10 value elements is called a *decade device,* such as a "*decade* resistance box" or "*decade* counter." To produce the variety of 10,000, we would need 4 decimal variables, or

$$I_{10} (1000) = \log_{10}(10,000) = 4 \ decades,$$

where I_{10} = logarithmic information in the base 10 (decimal) num-
                            ber system.

Thus the logarithmic information to the base 10 of any number that could be as high as 9,999 is four decades.

Following this type of terminology in the binary system shown before, we should say the logarithmic information representing a variety of 16 to the base 2 is 4 binary variables, or

$$I_2 = \log_2 (16) = 4 \ binades.$$

The term *binade* refers to a variable having two value elements, just as *decade* refers to a variable having ten value elements. It makes sense to refer to a four-binade register to represent a number up to 16 in the binary system as it makes sense to refer to the four-decade register to represent a number up to 9,999 in the decimal system.

In a diverse system, the logarithmic information is still the number of unitary variables having x values required to produce the number of choices (variety) created by the diverse system with x values.

Kinds of value elements

There are different *kinds of value elements,* just as there are different kinds of number system variables. When we speak of a digit, we usually mean one of the ten counting units or value elements (0 through 9) that make up a single decade. (The definition of a digit is one of the ten fingers or toes used in counting.) A digit is therefore one value element of a 10-value (decade) variable. A value element in a number system is also called an *integer.* Therefore, a *digit* can be considered a *decimal integer unit.*

No name is used consistently for the 0 and 1 value elements in the binary number system. They are called "0's and 1's" most of the time, and are sometimes called "bits." (See Appendix B.)

To give these two important value elements a name, we can call each of the two (0,1) integers that make up a binary register a *binint,* for *binary integer.* The digit should be called a *decint* if we are to follow the same rule for naming values elements. (See the table on number systems following the next section.)

Number of value elements

The total *number of value elements (C)* determines the cost of a register or other memory device regardless of the number system employed. The total number of value elements is equal to the product of the number of values elements (x) of each variable times the number of variables (n), or

$$C = x \times n \text{ value elements,} \tag{30.18}$$

or
$$C = x \times \text{logarithmic information.} \tag{30.19}$$

As mentioned above, there are two value elements in a one-binade register and 10 value elements in a one-decade register. In the 4-binade register shown previously, there is a total of

$$C = 2 \times 4 = 8 \text{ binints (binary integers).}$$

This "total number of value elements" required to represent a given variety is equal to the number of individual memory cells (elements) needed to store a given variety of information. It is dependent upon the required variety and the base of the number system used by the storage system. (See Appendix A, Value Elements as Data.)

In a unitary system, this quantity is

$$C = x \log_x V, \tag{30.20}$$

where
$$V = x^n.$$

So,
$$C = x \times n,$$

as shown above. For example, the total number of value elements of the four-decade register shown earlier is $C = 10 \times 4 = 40$ (digits) or decimal value elements.

In a diverse system, where each decade is a separate variable,

$$V = C = x \times n.$$

Or, $$C = 10 \times 4 = 40 \text{ digits}.$$

Note that the variety of a diverse system equals the total number of value elements of a unitary system that has the same number of variables and values.

Number system terminology

This *number system terminology* is summarized in Table 30.1.

Table 30.1. Number systems

Number system	Number base $= x$	Logarithmic information units $I_l = \log_x(V) = n$	Total number of value elements $C = x \log_x(V) = (x \times n)$
Binary	2	Binade (computer bit)	Binint (binary integer)
Trinary	3	Trinade	Trinint (trinary integer)
Quadrary	4	Quadrade	Quadrint (quadrary integer)
Pentary	5	Pentade	Pentint (pentary integer)
Hexary	6	Hexade	Hexint
Octal	8	Octade	Octint (octal integer)
Decimal	10	Decade	Decint (decimal integer)
Dozimal	12	Dozade	Dozint (1/12 of a dozen)
Hexadecimal	16	Hexadecade	Hexadint (hexadecimal integer)
Hundreds	100	Hundrade	Hundrint
Thousands	1000	Thousade	Thousint

This table shows the relation between variables and their value elements in different number systems.

Comparison of the information content of registers in different number systems

The *information content of registers* in different number systems can be compared. Consider the 3-decade register with a variety of 1000 shown in Figure 30.6.

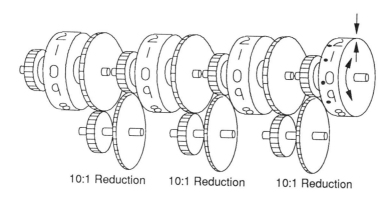

10:1 Reduction   10:1 Reduction   10:1 Reduction

Figure 30.6. A *three-decade register* has a variety of 1000 whole values.

In this case, where $V = 1000$,

$$I_{10} = \log_{10}(1000) = 3 \text{ decades.}$$

In the three-decade register shown, there is a total of

$$C = 10 \times 3 = 30 \text{ digits (decimal value elements).}$$

In the binary number system,

$$I_2 = \log_2(1000) = 9.97 \text{ binades (binary variables).}$$

In the ten-binade register, there is a total of

$$C = 2 \times 10 = 20 \text{ binints (binary value elements).}$$

Thus, it would take a 10-binade register to produce approximately the same variety ($2^{10} = 1024$ values) as a 3-decade register with $10^3 = 1000$ values. However the 3-decade register requires $3 \times 10 = 30$ value elements, while the binary register requires only $10 \times 2 = 20$ values elements. Thus the binary register can produce more variety than a decade register with the same number of value elements.

Properties of registers in different number systems

As shown, the total number of value elements $C = (x \times n)$ represents the *number of memory elements* needed to produce a given variety V. It conforms more closely to the cost of the memory gates needed to represent a given variety than does the number of registers. In codes or number systems of base 8 or less, the number of value elements corresponds somewhat to the more common logarithmic information (I_L), which defines the *number of variables* (registers) needed to produce a given variety. (See Appendix A, Values Elements as Data.) The *properties of registers* in different number systems having variety of 1000 can be compared in Table 30.2.

Table 30.2. Number of value elements required to produce a variety of 1000

Number system	Number base (x)	Logarithmic information units $I_l = \log_x(1000) = n$	Total number of value elements $C = x \log_x(1000) = (x \times n)$
Binary	2	9.9 binades (bits)?	20 binints
Trinary	3	6.3 trinades	19 trinints
Quadrary	4	4.9 quadrades	20 quadrints
Pentary	5	4.3 pentades	21.5 pentints
Hexary	6	3.9 hexades	23 hexints
Octal	8	3.3 octades	26 octints
Decimal	10	3.0 decades	30 digits (decints)
Dozimal	12	2.8 dozades	33 dozints
Hexadecimal	16	2.5 hexadecades	40 hexadints
Hundreds	100	1.5 hundrades	150 hundrints
Thousands	1000	1 thousade	1000 thousints

Variables as units of information

The value elements in a memory register can be divided in different ways. For example, if the total number of value elements (comparable to the number memory gates in memory register) were equal to 1000, and all were considered part of a single variable (single register), then the variety of the register would be

$$V = x^n = 1000^1 = 1000 \text{ states.}$$

Thus, $I_1 = \log_{1000}(1000) = 1 \text{ variable.}$

This register can store only one character. However, this character could have any one of 1000 different possible values.

If these 1000 value elements were divided equally between two variables, each variable can have 500 value elements, and the variety of the two registers would be

$$V = x^n = 500^2 = 250,000 \text{ states.}$$

Thus, $\qquad\qquad I_p = \log_{500}(250,000) = 2$ variables.

If these 1000 value elements were to be divided into 500 variables, each variable can have two values, and the variety of these 500 variables would be

$$V = 2^{500} \approx 3.3 \times 10^{150} \text{ states.}$$

Thus, $\qquad\qquad I_p = \log_2(3.33 \times 10^{150}) = 500$ variables.

Thus, the potential variety of a set of registers in number systems below base 8 increases greatly, though the number of value elements in these different number systems is held constant.

Translation from a single variable into a number system

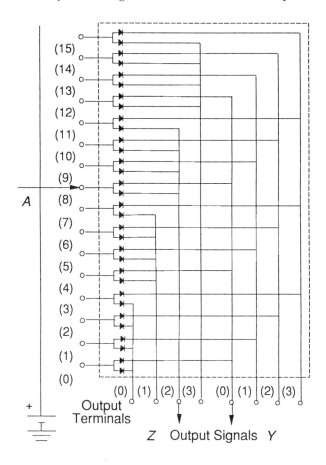

Figure 30.7. A *quadrary encoder* coverts a value of its single input variable into a quadrary number.

We have seen how a binary *decoder* can translate a binary number into a value of its single output variable, and a binary *encoder* can translate a value of its single input variable into a binary number. A quadrary encoder can translate a value of its single input variable into a quadrary number, as shown in Figure 30.7. The quadrary encoder shown has a single input variable A and two output variables Y and Z, each having 4 values. It can be wired to produce any desired code. The connections shown produce a code that is similar to most number systems, with Y as the least significant place and Z as the most significant place.

Translation from one number system to another number system

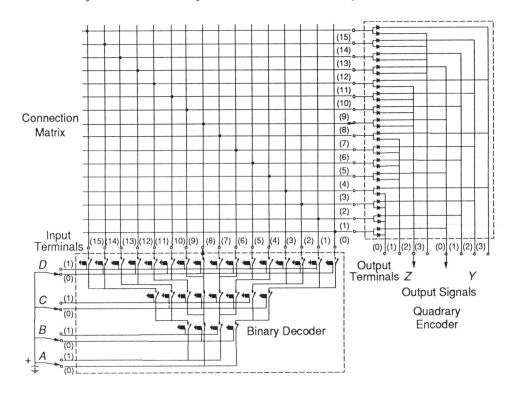

Figure 30.8. A number system translator uses a decoder, connection matrix and encoder to convert any number in one number system to any number in another number system.

A decoder, a connection matrix, and an encoder can be used to translate values from one number system into values in another number system, as shown in Figure 30.8. The *four-variable binary decoder* converts each combination of its four 2-value (binary) numbers into a unique value of a 16-value (hexadecade) number, since it takes one hexadecade to represent four binades. The logarithmic information of this system would be 4 binades since

$$I_1 = \log_2 (16) = 4 \text{ binades.}$$

The total number of value elements required to produced this variety of 16 in a binary system is

$$C = 2 \times 4 = 8 \text{ binints (binary integers).}$$

The *two-variable quadrary encoder* converts the 16-value (hexadecade) variable into a 2-variable, 4-value (quadrary) number since it takes two quadrades to represent the same variety as one hexadecade. The logarithmic information of this system would be 2 quadrades since

$$I_1 = \log_4 (16) = 2 \text{ quadrades.}$$

The total number of value elements required to produce this variety of 16 in a quadrary system is

$$C = 4 \times 2 = 8 \text{ quadrints (quadrary integers).}$$

Note that a binary number system requires the same number of value elements as a quadrary number system to produce a given variety.

Conservation of variety

The input and output registers used to translate *any* value in one number system into *any* value in another number system must have the same variety. Thus, the number of values in the output of an input decoder must match the number of values of the input of the output encoder, requiring a square matrix to connect the decoder to the encoder. The number of values of the output of an input matrix must also match the number of values of the input of an output matrix in a duplex system. Whenever two units are connected, the variety of both sides of the connection must be the same if any possible value of one unit is to produce any possible value of the other unit. If the variety of both units does not match, some states cannot be translated. A system of connected units that does not *conserve variety* will appear to malfunction at times.

Trinary number system

If we were to look for an *optimum number system*, one possibility is the natural or base *e* number system. It produces the maximum variety for a minimum number of value elements. Unfortunately, it does not have a whole number of value elements ($e = 2.7183...$). So the base *e* is not compatible with discrete memory devices.

However, a *trinary or three-value number system* comes closer to this ideal than a binary system, since 3 is closer to 2.7183 than is 2. A trinary system can produce more variety for a given number of value elements than a binary or quadrary system, and the amount of additional variety of a trinary number system increases as the number of variables in the system increases. A trinary system can also carry more information on a single conductor than a binary system. In many cases, a binary system uses one conductor for its two value elements by using an "on/off" or "high/low" or "plus/minus" signal. A trinary system can use a "plus/zero/minus" signal to carry its three value elements on one conductor.

Entropy

Some discussions relate information to *entropy*. The *Boltzmann equation* for entropy (*S*) of a statistical assembly is

$$S = k \log_e W, \qquad (30.21)$$

where *S* is entropy, *k* is the Boltzmann constant, and *W* is the number of equally probable states. Compare this equation to the simple equation of variety of a unitary system, which can be written

$$n = \log_x V, \qquad (30.22)$$

where *n* is the number of variables in a unitary system, *x* is the number of values of these variables, and *V* is the number of combinations of values of these variables (variety).

The term *W* in the Boltzmann equation is the degree of disorder or uncertainty, the same term as improbability (I_p) given earlier, which is equal to variety *V*. Since *W* and *V* are the same quantities, *entropy* can be considered the "*number of variables (n) in a unitary system that has (x) values per variable*," the same quantity as *logarithmic information* (I_l).

Increases in entropy

We are familiar with the statement: "The entropy of a system always increases." This corresponds to the decrease in available energy that occurs when separate bodies of gases at different temperatures (that have high available energy) are mixed to form a larger body with a greater variety of temperatures (that have lower available energy).

Consider a set of four separate bodies of gases (*n* = 4) each having a distribution of momentum that results in each having an uncertainty of say *x* = 10 possible states. These separate bodies of gases form a diverse system because the uncertainty in each body is independent of the uncertainties in the other bodies. Thus, the uncertainty of the diverse system is the sum of the uncertainties of each body. So,

$$V = V_1 + V_2 + V_3 + V_4.$$

Thus, $V = x \times n = 10 \times 4 = 40$ possible states.

Since $n = \log_x(V),$

then $n = \log_{10}(40) = 1.6$ decimal variables.

If these four bodies were mixed, forming only two separate bodies, the uncertainty of the system would be

$$V = x^{n/2} + x^{n/2} = 10^2 + 10^2 = 200 \text{ possible states.}$$

Then
$$n = \log_{10}(200) = 2.301 \text{ decimal variables.}$$

If these two remaining units were mixed, the uncertainty of the resulting unitary system would be

$$V = x^n = 10^4 = 10,000 \text{ possible states.}$$

Then
$$n = \log_{10}(10,000) = 4 \text{ decimal variables.}$$

This increase in entropy may be viewed as the *increase in the number of variables* or number of bodies having a specific level of uncertainty within a single system. The number of variables in a thermodynamic system may be considered a measure of the complexity of that system just as it is a measure of the information in a mechanical system. Since the number of variables increases as entropy increases, it can be said that the *complexity* of a system increases as entropy increases.

Increases in complexity of most systems

The complexity of other systems increase as well. For example, consider what happens when every student in a class of youngsters is given a small box of construction pieces with a limited number of parts in each box. The number of different objects that each child can make may be quite small, and someone could predict what these objects might be. However, if some students get together and combine the parts in their boxes, they may build a more complex object. The more complex the object, the more difficult it would be to predict what this object might be. Thus, if the students can mix and combine their parts, the complexity (entropy) of their system increases. This is a natural consequence of the components becoming part of a more unitary instead of a more diverse system.

The raw materials of our industrial system such as iron, oil, and other minerals are made into steel, plastics, and other building materials. These supplies are then put together to form buildings and products. Each time the materials are combined, they form more complex objects with more variables. (Raw materials form a diverse system since combinations of raw materials have no significance at this stage, whereas finished products form a unitary system since each product is defined by its unique combination of materials.) Thus, the entropy of these industrial systems increases as raw materials are converted into products. In other examples, elements form compounds that are more complex and more improbable than the individual elements. People form groups that can produce a greater variety of behavior than a single individual acting alone. So most systems change toward an increase in entropy, as seen in the greater complexity, number of variables, information, improbability, uncertainty, and *disorder* created by putting them together.

In many systems, once components are put together, they tend to stay together. Bodies of gasses at different temperatures are hard to separate once they are mixed. It may take a considerable act of will to get a group of youngsters to disassemble their creations and sort their construction pieces into their original sets. Once iron, oil, and other minerals are made into consumer products, they may be hard to separate and recover, since it is usually takes more energy to refine compounds into their basic elements (requiring endothermic reactions) than to combine them (with the resulting exothermic reactions). People tend to enjoy making friendships and suffer when they must be separated. So they tend to stick together once they form a group. Thus, systems made up of components that tend to stick together tend to change toward an increase in entropy.

Decreases in entropy

However, some systems change in the other direction — toward a *decrease in entropy* or a decrease in the number of system variables. The principle of noninterference suggests that empirical systems tend to establish the smallest system (with the least number of variables) that can deal with the task environment, since this systems is more likely to be the first to discover and incorporate successful behavior. Thus, empirical systems tend to move toward less complexity and a corresponding decrease in entropy. Other systems can be made to move toward greater simplicity as well. *Recycling industrial materials* reduces the number of different items in the environment. *Air conditioners* create fewer bodies of gasses at different temperatures by creating two bodies, one at a higher temperature than the other. *Oil refineries* separate many different types of crude oil into a relatively few possible distillates. Organizations carefully spin off smaller companies to deal with specific less complex single markets.

Order

As shown, the information in a message is based upon the number of variables in the message when the values of these variables have a uniform probability. Thus, information is a measure of the complexity, uncertainty, and disorder of a system. However, *order* is the inverse of information. Greater order means fewer variables, less complexity, and less ambiguity because fewer variables can have fewer combinations of values than many variables. The highest level of order occurs when a system is reduced to one variable, which can have only one value at a given moment in time. For example, making a prioritized To Do List reduces the many variables in our task environment to a single working variable. Once we have written a To Do List, we feel that we have established some order in our lives.

Composition and decomposition of order

Like writing a To Do List, the process of writing a book is a *process of composing* a set of variables into a single variable that has an ordered sequence of values, such as a beginning, middle and an end, which unfolds in the proper succession as the reader progresses through the work. This same ordering process of converting many variables into a single whole is also called *synthesis*. It operates the same way that a decoder converts the values of a set of variables into a value of a single variable. Since a single variable always has only one value at a given time, the uncertainty of a system is reduced to a minimum and the greatest order is achieved by the process of composition or synthesis.

In contrast, more information means more variables, more complexity, and more uncertainty. *Analysis* is the inverse of synthesis. It is a process of *decomposition,* which consists of taking things apart This creates more information and less order. For example, before writing a research paper, we gather as much information as we can on a subject and try to divide it into separate categories (variables). This process of taking a subject apart operates the same way that an encoder coverts the values of a single variable into values of a set of its component variables. It creates more information and less order. Then we *sort* the variables that relate to our subject, *combine* them in a unique way, and *arrange* them into the new single variable that is the subject of the book. By reducing the many variables to one, we have created order.

Preserving order

Systems and organizations that create order must reduce the number of assemblies they contain, and then keep them from reforming if they are to maintain their order and simplicity. Recycled materials must be carefully labeled and segregated from other materials once they are sorted. Different distillates must be kept in separate carefully labeled containers. Cooled air must not be allowed to mix with warm air. Successful behavior must be carefully recorded and preserved so that it can be selected over less successful behavior. Individuals who do not follow the relatively simple rules and laws of a society may have to be isolated to reduce uncertainty (*chaos*) within a society. Successful breeds of animals must be kept from mating at random with other breeds to preserve the special traits of their breed. Animals with successful traits also may evolve into separate species that cannot interbreed to keep their special traits from dissipating. Information that does not contribute to the subject of a book must be ruthlessly deleted.

The creation of order

Most systems seem to move toward an increase in entropy, in which resources become scattered, where energy becomes unavailable because it is dispersed, and where there is greater confusion and uncertainty because of the increased complexity of systems with more variables. However, intelligent beings can reverse the direction of these changes. We attempt to establish order. We simplify our lives by separating the useful objects in our environment from the obstacles in our environment. Empirical units select and produce behavior that can be carried out, and form themselves into special groups that can act as one unit. Animals roam in herds; birds fly in flocks; and fish swim in schools to simplify their existence. The duplex empirical controller converts multiple variables into a single intermediate variable that represents a specific subset of the many possible states of these variables. These activities result in a decrease in entropy and an *increase in order*.

Change

Some systems change in both directions. Industrial systems continuously refine and consume raw materials. Social systems continuously attempt to define what is considered acceptable behavior while experimenting with new forms of behavior. Other empirical systems discover and incorporate those simple relations that seem to work while testing the outcomes of more complex behavior involving more variables. These activities lead to change. A society may become larger and more complex, thus creating more variables with which its members must deal. At the same time, the values of a society may become galvanized around new goals that simplify and clarify the roles of some members of the society. Large inventories of unwanted goods may be produced by a business, creating new sales and marketing problems. At the same time, the business may develop a new product line that evokes great demand, thus simplifying the sales and marketing efforts. Thus, change can lead to greater order or greater disorder in most living systems.

An order trap

According to the law of closure, the variety of behavior in a closed system is determined by the *minimum variety* of any point in the system. Thus, in the digitizing process of decomposition and composition between different number systems, variety must be maintained at each point in the system if the variety of behavior of a system is to be sustained. However, a unique feature of an empirical system is that it can capture high-variety behavior in low-variety, highly ordered intermediate variables.

Variety transformers

We have seen how variety can be *conserved* when decoders or encoders are used to convert information from one number system to another. The simple equation of variety ($V = x^n$) can be used to determine the number of variables (information) present at any point in the system. The information at different points in a system can vary greatly according to the number system used at each point. For example, the information in a message at a point with a variety of 16 is made up of only one variable in the hexadecimal number system. It is made up of four variables in a binary number system at another point, since $2^4 = 16$. It is made up of two variables in a quadrary number system at another point, since $4^2 = 16$. As shown earlier, a binary decoder can convert the 2 values of 4 variables into 16 values of a single variable, and a quadrary encoder can convert any one of these 16 values of into 4 values of 2 variables, while maintaining a constant variety at each point. Thus, encoders and decoders can act like *number system transformers* that convert information and order in one number system into information and order in another number system at *constant variety,* in much the same way that *electrical transformers* convert a voltage and current in one conductor into another voltage and current in another conductor at constant electrical power. A discontinuous responsive controller also translates values of its sensor variables into values of its actuator variables, and must appear to conserve variety by producing any possible output state for any possible input state.

Order, information, and value elements

In general, variety can be expressed with the least number of *value elements* when its is expressed in the form of *information* with its many variables, rather than in the form of *order* with its few variables. However, our consciousness is essentially unitary since we tend to perceive and act upon only one variable at a time. Thus, we need order, not information, to succeed. The simplest control matrix can deal with only one input and one output variable. A single controller works best when it is dealing with a single unitary task rather than diverse task. A duplex system works best when its can use a single intermediate variable rather than the multiple intermediate variables of a universal duplex system. A network works best when only a few intermediate variables connect each unit. Thus, the need for order greatly increases the potential number of value elements, which, we have shown, determines the cost of a control system.

Information and order transformations in a duplex controller

In a digitized controlled machine, digitizing encoders decompose each input variable into a set of aggregate variables, creating more information and less order. The multivariable input matrix then composes each set of values of these aggregate variables into a value of the single intermediate variable, creating more order. The multivariable output matrix then decomposes each value of the intermediate variable into a set of values of the aggregate output variables, creating more information and less order. Then the output encoders compose the values of the aggregate output variables into a value of each output variable, creating more order. By switching back and forth between the domains of order and information, the empirical duplex controller can find the simple aspects of behavior that decrease the entropy of the system of which it is a part.

An information filter

The intermediate variable in a duplex controller and the hidden variables in a network are low-variety variables because it may have only a relatively few values in comparison to the variety of the input and output variables to which they are connected. However, a duplex controller and a network can capture the important values of the higher-variety input and output variables, apparently maintaining the variety of the system. The process of alternating between the two domains of order and uncertainty in a duplex system and a network acts like an *information filter* that allows the important relations between the input variables and output variables to be captured in the limited set of values of the intermediate and hidden variables. These intermediate and hidden variables act like an *order trap* that decreases the entropy of a system in much the same way that *Maxwell's Demon* attempts to catch only hot molecules (if it could) to increase the temperature of a region, thus attempting to decrease the entropy of its system.

The creation of order in modern empiricism

Thus, information is created whenever uncertainty and complexity are increased, and order is created whenever uncertainty and complexity are reduced. Writing this book has been a process of generating information. But that is not the intention of the author. The goal of this author has been to create order in the field of artificial intelligence. That is to say, the goal of the author has been to simplify and reduce the uncertainty surrounding the field of artificial intelligence by creating a unique set of values of a single new variable called *modern empiricism*. To the extent that the book only adds to the bibliography of artificial intelligence, then the author has failed to achieve his goal of creating order. To the extent that some aspects of artificial intelligence have been simplified or combined into fewer basic ideas, then the author has succeeded.

Chapter 31
The Science of Modern Empiricism

The purpose of this book has been to provide a comprehensive introduction to the study of machines that learn. These machines are based upon the science of *modern empiricism*. The purpose of this chapter is to summarize the science of modern empiricism and present some conclusions as to the value of modern empiricism to people living today.

Modern empiricism

Modern empiricism is based upon the exploration of learning through the study of real machines. The purpose of the study of modern empiricism is fourfold:

1. To understand the laws and principles of empirical systems.
2. To attempt to deal more effectively with existing natural empirical systems.
3. To design new synthetic empirical systems.
4. To understand our own behavior and the behavior of a group of individuals.

Concepts of empirical control and modern empiricism apply to the fields of biology, sociology, psychology, linguistics, and economics. The understanding gained in the creation of nonliving empirical machines provides many insights into the operation of these living systems.

Basic concepts of modern empiricism

Modern empiricism deals primarily with three fundamental questions:

- *Behavioral existence* — how can a machine discover and incorporate successful behavior?
- *Behavioral coexistence* — how can a machine develop and sustain successful behavior within a group of other machines and human operators?
- *Behavioral autonomy* — how can a machine spontaneously develop and sustain successful behavior without the aid of human operators?

The purpose of the study of empirical control is to describe a machine that can produce successful behavior within a natural environment and an environment consisting of people and other empirical controllers. Successful behavior is defined as actions that can be completed without the intervention of an outside agency. A machine that can carry out actions without intervention can produce self-sustaining behavior. So the purpose of this book is to describe a machine that can spontaneously develop and produce self-sustaining behavior. Yet a machine must first learn to exist within its environment, including its operators or trainers, before it can develop a separate or individual existence. It may have to learn to coexist with other units to accomplish complex tasks. Therefore, we have to study existence and coexistence to study how a unit can develop self-sustaining behavior by itself.

The will to exist

Perhaps the most striking feature of empirical machines is their inherent *will to exist*. An empirical machine seems to have an irrepressible goal of creating sustainable behavior. It is hard to imagine any condition that would suppress this impulse to act, discover, and incorporate new behavior. If an experienced empirical controller were placed in a restrictive environment such as a straight jacket in which no action can be carried out, all behavior associated with those conditions would eventually be erased. To eliminate all behavior, the unit would have to be exposed systematically to all of the conditions for which it has developed behavior, and this behavior would have to be repeatedly suppressed. This procedure is not likely to occur in a benign environment. It would take an active, malevolent process to eliminate all of the behavior of an empirical unit. Even then, when released from this kind of environment, an empirical unit would attempt to reestablish a modulus of behavior in the new environment.

Behavioral existence

The *behavioral existence* of an empirical unit depends upon its ability to use its knowledge and experience to deal with its surroundings. If something interrupts this interaction too drastically, it will be disabled. An empirical unit may encounter the following problems:

- Something may separate the unit from the conditions that lead to doable actions.
- Something may cause the conditions that lead to doable actions to change.
- Something may cause the doable actions not to work anymore.
- Its search for familiar conditions may lead it farther away from its original behavioral territory.
- It may injure itself physically.
- It may enter an environmental *cul-de-sac* or physical trap.

If an empirical unit is deprived of its familiar environment, it may not use its existing skills or its existing skills may not be evoked. If it does not use its skills, the cost of developing those skills is lost. In this case, what is unique about this individual unit — its behavior — may cease to exist. Its original behavioral self undergoes a *behavioral death*. An empirical unit may be compelled to learn new behavior in a new environment. Some of its old skills may be useful in the new environment, and some may not be useful. An empirical unit placed in a new environment requiring new behavior will eventually develop into a new individual. If the unit is returned to its old environment, it may revive some or all of its old behavior, and experience a *behavioral resurrection* using the latent sensitivities of its CE cells. Because of the empirical method of creating and producing behavior, an empirical machine establishes a *behavioral life*, free will, individuality, and a separate existence of its own. An empirical machine is identified primarily by its *unique behavior*. If its behavior is taken away by some accident, malfunctions, or sudden change in its environment, it is no longer the same individual, in much the same way that a person who develops Alzheimer's disease is often described as "no longer the same person they were before."

Thus, the uniqueness and identity of an empirical unit are based upon its behavior. Its existence is predicated upon its ability to create and maintain its behavior. This *existential premise* is often applied to living individuals. For example, the body of knowledge we acquire in our lifetime is often equated with our identity. We are identified as an engineer, plumber, or doctor. If our existence is based upon the collection of knowledge we acquire, what about a machine that can develop its own collection of empirical knowledge? Can it have an identity of its own? A robot team operating on Mars might consist of a navigator, scout, and driver that are responsible for specific tasks. They may be expected to carry out a difficult mission costing many billions of dollars. Will these robot individuals begin to have a life of their own to the ground controllers on Earth who are responsible for the success of the mission?

It may be difficult for some of us to accept the idea that a machine can have a life like a living being. Yet an empirical machine can have a form of existence that is different from other machines because it may discover and incorporate some *exclusive behavior* that is previously unknown to other beings. It may discover a new chess strategy or a new way of assembling a product. If this behavior is lost or destroyed, that unique behavior may be lost forever. This behavioral existence is very similar to our own behavioral existence in that each of us have a unique set of experiences and knowledge, which is lost forever when we die. An empirical unit may lose contact with the environment in which it has developed doable behavior. In this case, it may not know how to behave successfully. If it is left too long searching in new areas of behavior, it may lose the knowledge and experience it had acquired and may no longer behave like the system it had been. However, it may be rescued by an intelligent being that knows how to get it back into its familiar behavior states when it gets lost. This role is often carried out by the machine service personnel in the maintenance department of companies that have automatic production machines. Thus, the primary role of people involved with empirical machines is to rescue, help, redirect, and reinforce the behavior of these machines. In return, empirical machines will help people accomplish difficult or dangerous tasks.

Behavioral coexistence

Once an empirical unit has learned to exist within its environment, it may have to learn to exist in a world consisting human operators and of other controllers like itself. As shown earlier, empirical units can be used as standby or backup control system to help other individuals overcome limitations in their own control behavior. People are also needed to help empirical controllers overcome obstacles in the controllers' environment that may seriously disrupt the continuity of the behavior of the machines. Thus, there is a two-way support relation between people and empirical machines. Beside the cost of building these machines, the largest single investment in an empirical system will be its training costs. This two-way relation also implies that both participants must invest in behavior that protects each from the other. A truck driver may train an empirical controller to take over the driving function in case the driver falls asleep. The driver must teach the controller when to act and when not to act. Thus, the driver must teach the controller not to interfere with the behavior of the driver. Some drivers will be more successful in training their backup units than other drivers. The successful trainers usually develop successful teaching techniques. Thus, the empirical backup auto-pilot must develop behavior that reduces the interferences from the driver, and the driver must develop behavior that reduces the interferences from the empirical auto-pilot.

The same kind of interrelations and interactions may develop between one controller and another. It is possible to have a society of empirical machines that work together or against each other in much the same way that human beings struggle with and against other human beings. Each unit in a group must develop behavior that eliminates interference from other units, and each unit must develop behavior that does not interfere with the behavior of other units in the group. Many rules that apply to human behavior and controller/human behavior may apply to controller/controller behavior. Each unit may attempt to help or hinder other units. But the underlying reason for either behavior is to establish and maintain its own behavioral existence. This may be determined in part by the successful behavior of the individual, the behavior of other individuals within the group, and by the successful behavior of the group as a whole. Thus, the study of modern empiricism is a study of the interrelations, interactions, independence, interdependence, alliances and bonding among people, machines, and their environment. Modern empiricism is the study of the development of a *behavioral coexistence* between responsive control systems and ourselves.

Behavioral autonomy

The empirical algorithm creates the *will* for an empirical machine to produce behavior. This gives an empirical machine a form of life. This *inherent will to act* compels an empirical machine to search for the consistency in the controller/environment system and then to act according to these discoveries. The direction that its behavior takes is based upon its assessment of what is consistent and what is not consistent from where it stands. Thus, an empirical machine is the source of its own behavior and is responsible for establishing behavior that is successful *based upon* what the machine can and cannot do *within a given environment*. This *free will* also gives an empirical machine the inherent ability to carry out behavior in an environment without the aid or intervention of a human operator. To do this, it must behave according to the requirements listed in Table 31.1.

Table 31.1. Requirements for behavioral autonomy

1. An empirical unit must learn the natural relations in its environment.
2. An empirical unit must learn the behavior that will keep it in this environment.
3. An empirical unit must find behavior that does not interfere with human operators or other units.

No outside authorities or other sources of knowledge is required to start the learning process of an empirical system. What an empirical unit does and why it does it is a matter determined by the distribution of sensitivity levels of the empirical memory cells in its control matrices when it is turned on and the opportunities for action in its environment. Thus, it may develop entirely new way of behaving that is unique to itself.

An empirical controller keeps its own record of what is possible and impossible and uses this information to select the behavior in the future that has a higher chance of being carried out than behavior produced by a purely random selection process. An *autonomous empirical controller* determines what works and what does not work for itself based, upon its own experience. If it has unique experiences, and develops a unique way of dealing with these experiences, it may act in a unique way. If two empirical units were connected to a given environment at different times, and if there were many different ways of behaving successfully in this environment, each may develop an entirely different way of dealing with that environment.

The *behavioral autonomy* of an empirical unit is the starting point of the empirical process. No other machine starts at this point. For example, most machines are designed and their behavior is built into them before they are tested. Computers are not expected to run before they are programmed. In contrast, the process of training an empirical unit is based upon thwarting its *intrinsic independence* and steering its propensity to act toward that behavior that is wanted of it.

Trainers, operators, others participating in the behavior of an empirical machine can influence its behavior using different *teaching methods*. In some cases, an operator may indirectly control an empirical unit by controlling its environment. However, without human influences, all empirical machines will attempt to develop whatever actions are possible within the framework of its environment, its own physical structure, and the configuration of its control matrices. Since an empirical unit builds its own behavior, in some sense it creates itself. If an empirical machine can figure out how to act in a unique and consistent manner, and this behavior fits into the physical and social requirements of the world around it, one has the eerie feeling that it has met most of the requirements of becoming a "being." If we destroy an empirical machine, there is a chance that another machine with the same behavior will never appear again.

To act consistently, an empirical machine must learn to avoid impossible activities. For example, a robot cannot walk through a wall. So an empirical machine must learn to walk through doorways. Once it learns where the doors and hallways are in a building, it must to stay in that building to act successfully. If an empirical machine discovers behavior that is achievable within an environment and learns to stay within that environment, it may be said to have established a behavioral existence of its own.

So we are dealing with a vulnerable creature with a precarious existence based upon its ability to discover what the environment and people will allow it to do and based upon its ability to stay within the environment where it knows how to behave successfully. One way for it to increase the chance that its behavior will survive is to seek environments that contain other empirical units or human operators that help it to produce successful behavior. Another possible solution is to avoid environments that contain units that interfere with its successful behavior. The way it chooses to act depends upon where there are the greatest opportunities for it to produce behavior that can be carried out.

Unitization

Another important concept in modern empiricism is the principle of *unitization,* which suggests that empirical machines can work better to solve complex behavior tasks by working within groups made up of assemblies and subassemblies rather than a single unit. Most entities in the physical world are made up of assemblies and subassemblies. For example, elementary particles form atomic units, which form molecular units, which form chemical compounds, which form mixtures. Machines are made up of assemblies and subassemblies. Companies and social groups are built up from departmental and social units. Even our celestial system is made up of planet/moon units, sun/planet systems, and galaxy/stars systems. This design or structure, called unitization, is a method of organization that allows an individual part to be a member of a subunit nested within a unit, subassembly, and an assembly. Empirical units form groups (social assemblies) for the same reason that machines are made up of assemblies and subassemblies — groups can be made as small or large as needed, and groups can rearrange or reorganize themselves in a relatively simple manner. An empirical unit may choose to becoming a subordinate or a supervisor according to which position allows the unit to produce behavior that can be carried out.

Physical domain

The empirical algorithm implies that there are physically stronger and weaker elements that determine what actions can or cannot be carried out. This ties empirical behavior to the physics of the real world. Empirical behavior is not just an abstraction, pattern, or collection of data points. An empirical unit is a physical being, which is *on trial* in the physical world and must obey its laws and principles. Because empirical machines are connected to and become a part of the physical domain of its environment, they can use their power to create successful behavior and can combine with other units to create a more powerful influence than a single machine acting alone.

Knowledge and dominance

In general, an empirical machine will dominate behavior if it can physically determine the outcome of its behavior. However, the unit that can acquire the greatest knowledge of the natural relations in a given situation may be better able to dominate behavior than a less intelligent but more powerful unit. Knowing the natural relations in a system reduces the likelihood of attempting to do things that cannot be accomplished. Therefore, astute behavior can be carried out more often than indiscriminate behavior, causing the more knowledgeable empirical machine to dominate behavior. Other factors that create dominant behavior and are related to adroitness are: being part of a group, speed and appropriateness of the timing of responses, and persistence.

Predisposed empirical controllers

Since complex behavior can be acquired empirically and stored as a set of sensitivity values in an empirical matrix of CE cells, complex *innate behavior* also can be determined *a priori* and recorded in a set of sensitivity values in a predetermined matrix of CP cells. These predetermined memory cells can be included in a matrix of empirical memory cells, creating a control matrix with an intrinsic way of behaving that can be usurped by successful empirical behavior.

Instinctive behavior

People believe that much of our behavior evolves and develops from experience. The design of an empirical controller shows how this behavior can be established through experience and be stored in the distribution of sensitivity values of a set of empirical memory cells. However, much animal behavior is believed to be preprogrammed as *instinctive or innate behavior*. A responsive controller can be built with instinctive or innate behavior in several different ways. One method is to use conditional memory cells with predetermined fixed sensitivity values (CP cells). This predetermined conditional controller can act the same as an empirical controller except that its behavior is not learned through experience. This predetermined behavior may be as complex as any behavior that is established empirically, although it may be difficult to determine *a priori* what the distribution of sensitivity values must be to produce this complex innate behavior.

Successful instinctive behavior also could evolve in living beings over many generations through the inheritance of the unique set of *active switches* that produces specific innate behavior. Successful behavior also could evolve through the inheritance of the distribution of sensitivity values in a set of conditional predetermined memory cells (CP cells) that leads to an increase in the *reproduction rate* of these beings.

Predisposed behavior

Also, the *initial sensitivities* of standard conditional empirical cells (CE cells) could be predetermined and preset into an empirical controller, or the initial sensitivity values of an empirical matrix could be established by the evolutionary process described and later modified by experience. At first, this organism would be predisposed toward certain behavior. Then the organism could enlarge upon or modify this *predisposed behavior* according to the demands of the task environment.

Since memory cells with the predetermined sensitivity values (MCP cells) can exist in the same control matrix as the conditional empirical cells (CE cells), the instinctive behavior of the MCP cells could appear first and then be superceded by successful new empirical behavior created by the CE cells. (See mixed memory matrices). If consistent empirical behavior can be discovered and incorporated, the matrix will act as an empirical controller. If little consistency can be found, the controller will *fall back* upon its predetermined (innate) behavior.

The initial sensitivity values of a matrix of CE cells could also become frozen in a matrix of CP cells through a process of evolution, in which the logarithmic subtracting mechanisms and the transform conditioning circuits degenerate and disappear, leaving only CP cells with the sensitivity values of the original CE cells. It is also conceivable that active switches could disappear from a control matrix or appear at specific intersections in the control matrix of a species of organisms through a process of evolution, creating a responsive controller with a specific modulus of behavior.

Evolution of empirical cells from predetermined (innate) memory cells

Predetermined (innate) behavior can be produced by a specific pattern of hardwired connections between sensor and actuator terminals in a scalar matrix or a pattern of active switches in a multivariable active pin matrix. These connections could become more specialized in an active memory matrix with active absolute memory cells (AP cells). These absolute memory cells could evolve into absolute empirical cells (AE cells) in one branch (see absolute empirical memory cells), which change confidence values from 0 to 1 or from 1 to 0 according to whether the action called for by each cell is carried out or not carried out. This ability to learn from experience also requires that a transform conditioning circuit evolve that can measure what actions actually take place and send a signal of this information to the appropriate AE cells.

In another branch of evolution, the active memory cells (AP cells) could evolve into conditional predetermined memory cells with fixed sensitivity values (CP cells), by adding a variable resistor with a fixed inherited sensitivity value to the AP cells. These CP cells could also evolve from conditional empirical cells (CE cells) that have lost their write on and write off coils and their logarithmic subtracting mechanisms so they no longer can change their sensitivity values.

CE cells could evolve from absolute empirical cells (AE cells) by adding a logarithmic subtracting mechanism and operating it with their existing write on and write off coils. CE cells could also evolve from conditional predetermined cells (CP cells) by adding write on and write off coils and a logarithmic subtracting mechanism to operate the wiper arm of their existing current divider circuits.

Predetermined behavior

Behavior can be established beforehand using the methods summarized in Table 31.2.

Table 31.2. Methods of establishing behavior *a priori*

---

1. Behavior can be established through a matrix of hardwired connections in a passive matrix of conductors.
2. Behavior can be established through a matrix of hardwired active switches (gates).
3. Behavior can be established through the settings of AP cell or CP cell memory elements a matrix of predetermined memory cells.
4. The initial states of AE cells and CE cells can be programmed *a priori*.
5. The sensitivity values of CP cells can be established through heredity.
6. The distribution of hardwired connections, active switches, AP cells, AE cells, CP cells and CE cells in a mixed matrix can be determined beforehand.
7. Networks of different types of matrices can be established beforehand.

An empirical controller can establish successful behavior by changing the set of confidence values stored in its CE cells through a process defined by the empirical algorithm. If the behavior of an organism is established through preset memory cells, regardless of whether they are preset predetermined or empirical memory cells, the behavior may be defined as predetermined behavior.

Disposition of a responsive unit

Any behavior that can be learned through experience also can be established beforehand. This predetermined behavior can determine the *disposition* of a responsive unit. For example, if an empirical unit can learn to reach a goal, it can be preprogrammed to act as if it has that goal. If an empirical unit can learn to act in concert with other units, it can be preprogrammed to be disposed to act with other units. If an empirical unit can learn to reflect the love and care shown to it by a companion, a unit can be preprogrammed to act in a loving and caring way.

Disposition to think

It is believed that thinking involves the manipulation of abstractions within the mind, in contrast to the creation of physical actions (behavior) by means of the brain. However, an empirical controller can manipulate *internal mental states* if there is some means of determining which internal states are in accord with the rest of the organism and its environment and which internal states are not. That is to say, these internal states must be subject to the restrictions and opportunities of acting in the same way that actuators are subject to the constraints imposed by their environment. Otherwise, the organism could not learn through the empirical process. The organism must increase the likelihood of producing the internal states that can be carried out without interference, and the organism must decrease the likelihood of producing internal states that are interfered with. This sequence of internal states might be described as *internal behavior*. It is conceivable that this internal behavior is what we call *thinking*.

In the behavior of a *sensor/sensor responsive machine*, a given sensor state produces a subsequent sensor states. If the actual subsequent sensor state perceived in the environment matches the subsequent sensor state produced by the machine, an empirical controller can be made to produce that sensor state again when the previous sensor state appears again. The sequence of sensor states that exists in the controller may be considered *internal states* because it does no involve the use of actuators in an external environment. Thus, the *sensor/sensor behavior* of a responsive machine is similar to *thinking behavior*.

Sensor/sensor behavior is hard to detect, just like it is hard to know what someone is thinking. Sometimes we can infer what someone is thinking from his or her behavior. For example, we assume that an individual who performs violent acts may be thinking violent thoughts. Sometimes people are truthful when they tell us what they are thing, and sometimes they are not. Yet it is difficult to imagine that an empirical machine will talk to us about its sensor/sensor activities soon. However, an empirical robot, strolling along a garden path, may be captivated by images of plants and flowers. If it stops to smell the roses, we may infer that it is thinking about the plants in the garden.

Reasoning behavior

An empirical unit may also learn or be disposed toward responsive behavior that appears as if the unit is applying logical thoughts or reasoning to a problem. The law of equilibrium of empirical behavior says that an empirical unit will reflect the knowledge, behavior, and principles inherent within the systems to which it is connected. This suggests that an empirical system will appear logical if it is connected to a logical system. However, there is no specialized mechanism or location in an empirical controller that creates and operates upon logic statements in the same way that a digital computer operates upon logic statements. Since a responsive machine can be programmed to act in any way that it can learn, even a predetermined controlled machine can be disposed to act as if it were reasoning.

Self-preservation behavior

Most people will say that they think of themselves as separate, unique and valuable beings. Yet we do not know if other animals think about themselves in this way because we have not received messages from them in this matter. However, most animals attempt to defend themselves or their territory and aggressively search for food. They act as if they want to preserve themselves. It is conceivable that an inorganic empirical being could learn *self-preserving behavior*. It could act as if it were aware of its existence and the threats to its existence. In fact, the empirical algorithm forces an empirical being to develop behavior that can continue rather than behavior that is risky and inconsistent. Self-preservation behavior would have to be included in the repertoire of successful behavior of an empirical being if there were significant threats to the behavior of that being in its environment.

Caring behavior

According to the law of the equilibrium of empirical behavior, the behavior of an empirical unit will match the knowledge, behavior, and principles inherent within the systems to which it is connected. Therefore, one would expect that an empirical unit will reflect the love and care shown to it by its operators. The law of bonding also says the following: If one machine supplies some behavior needed by another machine to establish closure, they will appear held together by their shared behavior. This implies that the bonding process will create an attachment between empirical units and between an empirical unit and its operator. This also implies that the bonding process operates in both directions — that the operator will become attached to the empirical unit as well.

Unique behavior

The behavior of an empirical unit may be determined initially by random values within the controller's memory matrix, and then be molded by whatever consistency exists in its environment and the persistent desires of the people who are operating the system. There may be many different ways to deal with these influences and an empirical controller may come up with unique and unexpected solutions to the constraints imposed upon it.

There may be unique conditions in the environment that are not recognized by human operators. An empirical unit may acquire behavior that reflects these unique conditions. This *unique behavior,* which may not have been anticipated by human operators, may be lost forever if the behavior of that system is terminated.

Some people think that each individual is a unique being, and that something irreplaceable is lost when each individual dies. This is a compelling argument, particularly when we lose someone close to us. The same reasoning can be applied to an empirical machine. Each empirical machine that develops a special way of dealing with its environment may be considered a *unique being* because there may never be another empirical machine that develops the same behavior. Something irreplaceable may be lost if the behavior of that empirical machine is terminated. In this sense, an empirical machine that develops its own behavior is similar to a living being.

Empirical control in superorganic systems

The boundary line between living and nonliving empirical systems is sometimes hard to draw. Are government and business organizations, ecological, and economic systems living or nonliving systems? They have an existence and identity of their own. They may have a finite life span with a beginning, middle, and end. But these *superorganic systems* do not have a single connected body like most living beings. However, these natural systems behave according to the laws of empirical behavior. They are *natural computers* that keep track of their successes and failures. The study of empirical control allows us to view these connected and disconnected living entities in the same way.

The natural computer

Any system that operates according to the empirical algorithm is a *natural computer*. These systems are responsible for the spontaneous generation of behavior, measuring and recording the success of this behavior, reproducing this successful behavior under the appropriate circumstances, and staying within the circumstances that solicit this successful behavior so that this successful behavior can be sustained. Many different versions of the natural computer are being created by humankind and nature even today. Newer versions will be built in the laboratory, which will clarify many principles set forth in this book on empirical control. More new details and insights will burst into view as new empirical controllers are built and tested. Many of these ideas will reinforce and clarify our understanding of sociology, economics, biology, linguistics and the human brain.

Networks of control units

A single empirical machine cannot deal with many variables. However, a group of smaller empirical units can accomplish complex tasks by establishing the division of labor and efficient intercommunications (fragmentation) made possible by connecting these units in a network. This suggests that the animal brain consists of many units much like a large company consists of many *employees*. Both are faced with the problem of focusing the behavior of these subunits into behavior that is useful to the group as a whole. The science of modern empiricism attempts to understand the relation of size, complexity, efficiency, and learning difficulties of unitary and diverse networks of empirical units. In many cases, these networks may be dispersed throughout a region. So, there may not be a single box that contains the whole natural computer.

Self-sustaining social systems

The search for self-sustaining behavior extends beyond robot control systems. City planners must be concerned about the quality of life in their cities many years into the future. Will the quality of life in the cities improve or deteriorate after new housing and parks are built? What is needed to cause the cities to take care of themselves — to continue to change for the better?

Can city planners go beyond trying to create cities that can sustain themselves? Should they try to make it possible for the residents of the cities to discover and incorporate new self-sustaining behavior for themselves? These important questions can be asked about our government, businesses, and other social groups. Do these organization increase or decrease the opportunities for individuals to discover and incorporate new behavior that contributes to the success of the organization and the individual?

We may want to ask how we can take better care or ourselves, our family, and our property. Do we have a good means of measuring the success of our behavior? Can we identify and remember our successful behavior? Or do we only remember our failures? Is there a way we can develop the behavior that satisfies our needs and the needs of the people around us based upon our existing successful behavior? Can we find ways of dividing our tasks among those people around us by giving up certain responsibilities and taking on other responsibilities? These are only a few of the many items addressed by the science of modern empiricism in its study of the source of self-sustaining behavior.

Overview of an empirical system

The sensor/actuator, actuator/sensor, sensor/sensor, and actuator/actuator systems of a compound machine can be represented as a set of nested cylinders (Fig. 31.1). In all four systems, behavior flows in one direction, and a representation of what actually happens flows in the other direction.

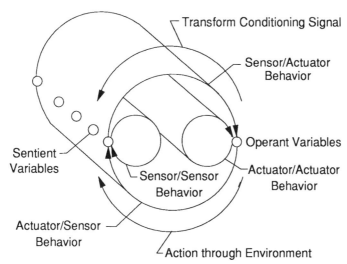

Figure 31.1. An *overview of an empirical system* shows the flow of information from input to output and the flow of feedback information about what outputs actually occur from output to input.

Economic activity space

In the empirical system shown, we can trace the forward- and back-selection signals and imagine the accounting and record-keeping process in simple empirical machines. But can these forward- and back-selecting signals reach and reinforce sensor variables when many stages and layers of systems have to be crossed in a network?

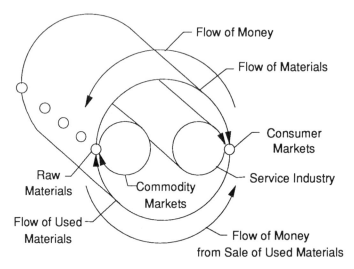

Figure 31.2. An *economic activity space* shows the flow of materials to the consumer and the flow of money from the consumer back to the supplier.

The model of the economic system shown in Part IV suggests that action can flow like products and materials in one direction, and transform signals can flow like the payments for these products in the other direction, as shown in Figure 31.2. The economic activity space shows the collection, refining, manufacturer, combining, and distribution of materials to the customer. It also shows the flow of money from the consumer to the distributor, manufacturer, and back to the supplier of raw materials. This material flow represents action behavior flowing from the source (sensors in this case) to the consumer (effectors in this case), while the money flowing in the other direction represents the feedback signal, indicating the manufactured goods were actually sold.

In an immeasurably large system such as our economy, the balance between money and material is nearly perfect most of the time. Consumer shortages or gluts and over or undersupplies of raw materials are rare in a free market system. Yet, the money for the raw material suppliers comes from one end of the system, the consumer end, and materials that the consumer needs come from the other end of the system, the supplier end. How do the kinds and amounts of materials produced exactly match the needs and wealth of the consumers? Money and materials can also flow from the consumer markets to the raw material markets in the process of *recycling*. This process closes the loop of money and materials, forming a closed system.

Economic pseudosphere

Movements in the economic activity space attempt to become lateral as materials are collected and distributed to different geographical areas. Since sources of materials may be connected to any consumer, the activity space of the model becomes a torus or donut, as shown in Figure 31.3.

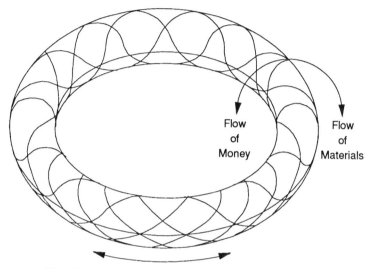

Direction of Distribution of Money and Materials

Figure 31.3. The *economic pseudosphere* is the most general representation of a complex empirical system.

Materials move from the outside of the torus, spread out along the top, where they are combined in factories, and continue to spread out through the distribution channels to the inside region of the torus representing the consumer. From there the materials are recycled back to raw material and distributed to material suppliers along the underside of the torus. Of course, money flows in the opposite direction from the consumer back to the suppliers and eventually back to the consumer as payments for recycled materials.

Movement of this type may become unstable. Some sections may speed up and others may slow down. This kind of action can be observed by rotating an elastomeric torus or O-ring in the manner described previously. It flexes, bends, resists and then springs ahead, in much the manner as worldwide economic cycles in a tortured effort to keep itself in equilibrium. These uneven motions may relate to business and economic cycles.

Conclusion

The nature of responsive behavior of an empirical control assembly is determined by the complexity of its sensor/actuator system, the variety of its control unit, the richness its environment, the dedication of its teachers and trainers, and the degree to which it is predisposed to produce successful behavior. It appears that the behavior of responsive empirical machines can be expanded to match the complexity of almost any imaginable task.

Limitations of empirical control

An empirical controller does not inherently produce the goal-setting behavior typical of men and women. This category of behavior occurs within the mind of humankind, in contrast to the responsive behavior typical of other animals and empirical controllers. The present study of empirical control concerns itself primarily with the *brain behavior* produced by most animals in contrast to the *mind behavior* attributed to humankind. However, we may need to understand brain behavior before we can deal effectively with our goal-setting mind behavior.

Real-time responsive behavior

The type of controller or brain with which we have been dealing so far may be called a *responsive brain* because it responds to sensed conditions. As conditions change, its responses change. The *real-time responsive behavior* presented in this study of empirical control is created by a machine that produces a set of responses to a set of sensed conditions. Even the decisions of a mobile robot to compete, cooperate, organize, or separate, shown in the networks in Part IV, are unmediated reactions to the environment surrounding the controller, based upon whatever mode of behavior worked best in the past. These decisions are evoked by the way outside circumstances operate upon a list of possible actions stored within the empirical controller. This contrasts the *meditative*, *thinking*, and *cognitive* mental abilities of the human mind.

The human brain

Some writers in the fields of neural networks, artificial intelligence, and other self-learning systems claim that the behavior of these systems are "humanlike." Clearly, the behavior of systems based upon these new technologies are different from predetermined or preprogrammed controlled machines, in which every aspect of their behavior is specified beforehand. However, the empirical controller and other computer-based systems do not perform the type of cognitive behavior that may be unique to humankind. The behavior of the *human brain* involves the use of words and symbols and is characterized by being thoughtful, analytical, introspective, purposeful, emotional, and logical. This type of behavior is quite different from the responsive behavior of the empirical machines discussed so far.

Goal-seeking brain

One feature of the human brain that may distinguish it from the brains of all other animals is the ability of the human brain to develop *goal-seeking behavior*. Once a person learns how to set goals, the behavior of that person is different from that of any other animal. Instead of simply reacting to the environment according to his or her modulus of behavior, a person with a set of goals actively seeks objectives, makes quests, searches for objects, and builds an environment according to his or her dream.

This particular type of behavior has not been discussed at length in this book because the empirical controllers shown in this book are not able to produce this kind of behavior. However, the empirical controllers described so far can act as if they are seeking a goal if they learn to accomplish a complex task.

Is goal seeking a learned behavior? Should we actively teach goal-seeking skills in schools? We teach other types of learned behavior, such as reasoning, creativity, artistic, and music skills to name a few. An empirical controller may provide the site for this kind of learning sometime in the future.

The animal brain

As stated, the behavior of empirical control systems, *neural networks*, and *artificial intelligence* systems may develop from experience, and they may produce unique behavior. However, none of these systems demonstrate the goal-seeking behavior typical of human beings. That is to say, none of these synthetic systems attempts to establish a sense of self, formulate goals, and pursue behavior based upon these goals. In fact, no other animal except the human animal appears to set goals and carry out behavior based upon these goals.

If goal-seeking behavior is the unique feature of humankind, then none of the human-designed systems proposed so far is truly humanlike. They may be animal-like, but not humanlike. For this reason, the study of empirical control is more of a study of the *animal brain* than the human mind. However, many aspects of the human brain are animal-like, since we also perform real-time responsive behavior as well as goal-seeking behavior. The author believes that we must first understand the animal brain before we can understand the human mind.

However, it is conceivable that an empirical system can be taught to act as if it were trying to achieve a goal. It can be encouraged to "search" for lost objects or to carry out other behavior that appears purposeful, such as stacking objects on a pallet or building block houses. This behavior can be achieved by making the searching, stacking, or building behavior the only behavior allowed by the environment and operators. Yet a responsive empirical system pursues this activity not because it has the goal to do so, but because it has developed behavior that appears to have that goal. The empirical systems discussed so far cannot be aware of what they are searching or know why they are behaving the way they do.

The animal brain is the least understood, yet the largest and the most important natural computer we have to deal with. The animal brain may be a complex empirical controller. What we learn about empirical control may contribute to our understanding of the animal brain. Teaching an empirical system to appear to have goals may become a very important activity by itself. However, the author believes that we must first build and experiment with empirical systems that carry out the wishes of human operators before we attempt to develop systems that create their own goal-oriented behavior.

Empirical and human behavior

The purpose of this book has been to describe a new world of *inorganic beings* that can create and maintain their own behavior. These nonliving organisms can create a life, free will, and existence of their own that may be completely independent of human influence. They can also develop an existence that is intimately entwined with human behavior. The study of these empirical machines suggests many parallels to human behavior.

Learning from success

An important idea suggested by the study of empirical control is that we *learn from success* rather than failure. To learn to do something, we must first do it correctly. Then it can be learned. Therefore, we should try to recognize and understand our successes when we are students, and we should try to create conditions that increase likelihood that our students will succeed rather than to fail when we are teachers. This idea is based upon the empirical algorithm, which causes empirical units to remember what they actually do rather than remember what they attempt but fail to do.

The empirical paradox

The empirical process creates the *empirical paradox*: How does an empirical machine perform the action needed for that action to be learned? If a machine can perform the action, does it not already know how to do it? The answer: A student with a good teacher can do things that he or she could not do before. Thus, an empirical unit may need a skillful teacher to learn difficult behavior.

Interdependence

Empirical units can produce successful behavior while sharing environmental input and output states with other empirical units. If some of these successful states come about because of the behavior of the other units, the success of a given unit is determined in part by the other units. If that successful behavior is to continue, part of that successful behavior must consist of staying in contact with the other cooperating units. This behavior creates the appearance of a bonding force among these units called a *behavioral bond*.

Most people are aware of the emotional bonds they have established between themselves and their parents. However, people can also establish powerful behavioral bonds with other people when they assume some responsibilities of caring for other people, or when they become dependent upon other people. In many cases, the importance of this *interdependence* goes unnoticed until the relationships are broken.

Social behavior

A great deal of our behavior consists of activities that bring us and keep us in contact with other people. Participating in social rituals, such as weddings, parties, and playing games are examples of this kind of *social behavior*. Can we expect that an empirical machine will discover that it must find other machines, offer to support them in their attempts to complete tasks, accept their support of their tasks, and produce behavior that keeps them together?

Attachment

The behavior of an empirical machine reflects the order in its environment and the behavior of the other units to which it is connected. We can assume that empirical machines will return and reflect the care and concern shown to them by their human operators. Every part of a successfully operating empirical system, including the machine and operator, must learn to act together as a single whole. If we assume that human behavior is essentially empirical, we can expect that individuals will find it difficult to separate themselves from a machine that responds to them, recognizes their commands, and reproduces the behavior that reflects their own needs and desires. This *attachment* indicates the formation of a single successfully operating empirical system.

Characteristic classes of behavior

Control systems can do more than produce negative feedback (regulation) behavior. There are other important *classes of control behavior* based upon other forms of feedback, as shown in Part I. For example, *positive inverse feedback behavior* is the basis of successful escape behavior. That behavior is based upon a unit moving away from a set point with a force that decreases as it gets away from the set point. A machine may also produce *negative direct feedback behavior,* in which it attempts to move toward a set point with a force that decreases as it gets closer to the set point. This behavior is typical of some regulators and servomechanisms that attempt to seek or chase a set point. If each unit is the set point of the other unit, they form a system with potentially never-ending *chase/escape behavior,* which evokes the utmost energy from both. The ship auto-pilot systems shown in Parts II and III are examples of chase/escape behavior, since most ships have a tendency to turn away from a straight course in an unstable manner based upon positive inverse feedback behavior, and an auto-pilot acts as a regulator that attempts to return a ship to a straight course based upon negative direct feedback behavior.

Some people tend to act like regulators with negative direct feedback behavior, by always trying to calm things down and bring situations back to known, steady conditions. The more a situation gets away from the set point, the harder they work to get things back to normal. However, people with *negative inverse feedback behavior* seek a set point, but once they have moved far enough away from their set point, they will readily let go of it. People with the *positive inverse feedback behavior* typical of escape behavior seem to want to stir things up, as if they are repelled by the status quo, but tend to calm down once they have created a big enough disturbance. People with *positive direct feedback behavior* attempt to run away from where they are and become more intense as this process continues.

Knowing the *characteristic class of behavior* of an individual may allow one to deal more effectively with that individual. For example, the best way to deal with a person with positive inverse feedback behavior is not to present them with direct threats or challenges (set points), whereas the best way to deal with a person with positive direct feedback is to keep them under close supervision and control at all times. We may identify our characteristic class of behavior and decide if it is the way we want to behave in a given situation.

Apprenticeship learning

Empirical machines learn from one another and from human operators through an *apprenticeship learning* process. That is, a student unit and a master unit work together on a task until the student unit becomes skillful enough that the master unit does not have to interfere with the behavior of the student unit and the master unit can withdraw from the task. This powerful learning process has been down played in favor of cognitive learning processes involving the exchange of ideas between teacher and student. Perhaps we should take another look at the apprenticeship system, in which student are exposed to a world more like the world in which they are expected to live. The role of the teacher in an apprenticeship system is to help students deal with the difficulties they may encounter in this world, such as learning to cook, taking care of children and older people, teaching and supervising others, interviewing prospective employers, speaking in public, balancing a checkbook, and making wise use of their money. When the students learn to deal with these activities in a more successful manner, the teacher can withdraw from the students' process of living with the expectation that the students will be more successful in their environment. Because the students will be actively engaged in this kind of behavior in the future, they may improve upon this learning through experience. In contrast, when students are taught more traditional academic skills, they may not apply these skills after graduation, and their overall skill level may actually diminish.

A philosophy of education

The study of empirical control suggests a *philosophy of education* that says that a teacher's job is to create an environment that maximizes the probability that students will discover and incorporate specific desired behavior. To do this, a teacher must create many *opportunities for success*. If a student fails to meet the expectations of the teacher, the teacher should try to find out where he or she did not create reasonable possibilities for the student to development successful behavior. The empirical paradox suggests that a good teacher can cause a student to perform better than the abilities of the student would indicate. The application of this approach toward learning may help develop better educational goals and better curricula.

Empirical controller as an educational test bed

An empirical controller can be a test bed for new ideas about learning and knowledge itself. As stated in the Epilogue: We learn when we measure. Teaching an empirical machine to accomplish specific tasks gives us a means of carrying out controlled experiments on a known learning organism. Thus, an empirical machine may serve as an *educational test bed* to develop new teaching methods.

Self-programming control

We live in the age of programmable control. To program a computer to produce some desired result, the details of a task must be identified, the logical relations of these details must be laid out in flow diagrams, and step-by-step instructions must be created telling the computer how to deal with these details. This procedure it tedious, time consuming and requires highly specialized skills. The program may have to be changed often to work out all of the unforeseen "bugs" in a program. The study of empirical control introduces us to a new *era of self-programming control*, in which an empirical machine initiates action, and the environment and its operators correct those actions that are considered inappropriate. Many of the limitations of computer programming can be overcome by this new technology, allowing a far more complex, satisfying, and effective interaction between humankind and machines and a better understanding of ourselves. Self-programming control can also create many new applications where the user does the programming, such as self-programming household controls, empirical auto-pilots, and autonomous empirical robots. In some of these application, predetermined programming techniques may not be practical because of the difficulty in determining the distribution of matrix sensitivity values beforehand.

Brain research

This volume has been prepared with the goal of presenting a method of dealing with large systems. Starting with concrete circuits, logic and mechanisms, it has presented a way of connecting systems to form more complex systems. It has dealt directly with learning and memory. It has presented principles and concepts that seem to "ring true" in the study of the brain as well as in other large systems in the fields of economics, biology and sociology. For example, the input latches in an empirical control matrix hold the values of the sensed conditions long enough so that the actual results of the sensed conditions can be associated with those sensed conditions. The empirical algorithm would not work without the input latches. These input latches create a kind of temporary or *short-term memory*, the kind of memory we need to store a phone number until we dial the number. After a transition cycle, the latches are released. This effectively erases the stored values in much the same way we forget a phone number after we dial the number.

Will the study of nonliving empirical systems help the people who are studying the structure of desire and motivation, analyzing instinct, and probing the sources of goal-seeking behavior? Is there a relation between emotion, feelings, and the concept of behavioral bonding (mutually supportive behavior)? What about love, the deepest feeling of all? Is love predetermined (innate) bonding behavior? Keep in mind that anything that can be learned by an empirical systems, such a bonding behavior, can be stored in predetermined fixed sensitivity CP cells as *innate behavior*. Thus people can have a propensity to become attached to other people through the genetic predisposition of conditional predetermined memory cells.

Concepts related to human behavior

The study of empirical control deals with many fundamental and useful principles that apply to human behavior. Similarities between empirical and human behavior are listed in Table 31.3.

Table 31.3. Similarities between empirical and human behavior

---

- Like empirical machines, our success is based upon completing what we start. This assumes that we start worthy tasks.
- Like empirical machines we learn from success. We may learn more if we pay more attention to what we do right rather than dwelling upon the many complicated ways of doing things wrong.
- Since we learn in the physical (behavioral) domain, can we devise ways of using physical behavior to improve mental ability?
- Since action is essential to learning, we may need to find more ways to test our ideas.
- Since our behavioral existence depends upon our being able to repeat what we know, we may enhance our existence by finding new ways to apply our existing skills.
- By understanding how members of a social group can expand the environment, increase the potential to act, and enlarge upon the opportunities for success of its individual members, people may be more inclined to create activities that bring and hold people together. Giving parties, living with other people instead of alone, and participating in sports, games, family, and religious activities create opportunities for successful social behavior. By understanding the mechanism of bonding, which allows us to share responsibilities with others, we may reduce the difficulty of living in a complex world.
- Similarly, the study of empirical control suggests that we are more likely to achieve independence and autonomy when we develop behavior that allows us to get along better with other people.
- Two or more individuals working together on a task may produce behavior that is more likely to be successful than the behavior produced by a single individual working alone if they establish an appropriate division of labor and efficient intercommunications. Thus, cooperation and coexistence are fundamental concerns of living beings as well as *nonliving empirical beings.*
- People may have to act in a sentient environment to sense information, and must sense information in an operant environment to act. This concept of a *bipartite network* may be useful to people working in a large corporation.
- By understanding the concept of a *tripartite organization,* people may see new ways of dealing with those individuals who work for them, their co-workers in their own and counterpart groups, and the individuals to whom they report.

- Empirical machines must change to match the organization of the system to which they are connected. Human organizations must also be flexible.
- By understanding the significance of competition and dominance, in which force, quickness, and skill play a part in the outcome of the efforts of empirical machines, we may accept competition as a natural and inherent condition necessary to solving many types of problems.
- When people attempt to solve problems, certain values of the problem variables are "latched in" and the other values are "latched out" like the input latches of a connection matrix. If the problem cannot be solved right away, it may be a good idea to "let go" of the problem for a while. This releases the previous "latched in values" and permits new values to be collected that may be more useful in solving the problem.
- By understanding how empirical machines discover and incorporate new behavior, teachers may be better able to increase the probability that their students will discover and incorporate what they are expected to learn. Can we reduce the factors that cause ourselves or other people to fail while increasing the factors that may cause us or other people to succeed?
- The ultimate test of the success of a control system is its ability to work in harmony with other control systems and the people using the control system. Thus, existence is largely a matter of coexistence. Interaction is based upon the appropriate reaction to other systems. It uses, expands, and reinforces behavioral skills, whereas interference stops behavior and reduces confidence levels. By identifying the interpersonal (interactive) skills of other people, we avoid situations that interfere with their behavior, and bring forth behavior that fosters their skills. Thus, successful interaction is shown to be a learned behavior that reflects the behavior of other systems.
- A unique quality of human behavior is setting goals and carrying them out. We are different from empirical machines that must learn whatever experiences they encounter. We can pick and choose what we learn by setting goals. Should we place more emphasis on learning how to perform goal-seeking behavior?
- We may have to choose to separate from or ignore the behavior of some people to preserve the unique qualities of our own behavior.
- We may wish to view behavioral existence as the knowledge gathered by an individual or group through experience instead of defining a living being merely through its organic presence. This enlarges the range of living entities to include wildlife habitats, migratory bird flyways, social groups, human cultures, empirical beings, etc.
- Timing of behavior is a complex problem. Usually, behavior cannot be speeded up or slowed down like a car. Changing the timing of human and machine behavior may require that new behavior be learned.
- Having a clearer view of order and information, we may attempt to organize specific values of sets of variables into a prioritized list of values of a single variable in the manner of an empirical controller.

Thus, much of our behavior is related to empirical behavior. However, empirical machines will not take over our roles as designers, builders, and leaders until they develop the ability to establish and carry out goal-seeking behavior.

Cognition and modern empiricism

What is the difference between the science of modern empiricism presented in this book and the study of cognition? *Cognition* deals with the sensing and identification of knowledge and information, while modern empiricism deals with the process of producing successful behavior.

Machine behavior

The underlying subject of modern empiricism is *machine behavior* — how machines work and fail to work properly. This is not a subject commonly discussed in schools or in the literature. So, a set of ideas has been organized in this book to deal with machine behavior. Modern empiricism is an engineering effort because its purpose is to design, build, maintain, and understand real machines. This effort to understand the behavior of machines is based upon science, mathematics, and other areas of modern technology.

Cognition

Cognition deals with the range of knowledge that is possible through perception and is concerned with how the mind abstracts, classifies, and stores these perceptions so that they can be identified in the future. For example, a measure of our cognitive ability is our ability to identify acquaintances, to read text, to be able to specify where we are, to know what day it is, etc. When the term *cognition* is used, it usually refers to the thinking or mental processes.

Modern empiricism

Modern empiricism attempts to understand how the perception of objects and conditions in our natural environment is used to influence the behavior of living and nonliving machines. This requirement puts this study within the domain of the science of cognition because both are based upon perception. However, modern empiricism is primarily interested in the process of discovering and using the natural relations in the environment to produce useful behavior rather than identifying and classifying knowledge.

While cognition has to do with perception and thought, empiricism has to do with action and experience. The primary concern of modern empiricism is useful behavior. We may never know what an empirical controller perceives, thinks, or knows. Our main concern is the appropriateness of its behavior: Is an empirical unit able to continue to act successfully within the context of its environment, the behavior of other units, and the people involved in its behavior? Does it *act* like it knows something?

Epistemology

Empirical behavior usually develops a quality that implies a knowledge of its environment or its teacher. If an empirical system discovers, incorporates, and executes doable behavior, it may seem to act as if it knows something about the features of its environment. Thus, modern empiricism involves itself with *epistemology,* in that it deals with the origin, nature, and limits of knowledge. However, modern empiricism is concerned with how this knowledge is created and stored within an organism and how it influences future actions. If thinking can be shown to be a form of behavior, then modern empiricism is also concerned with the mind.

Sensor/sensor behavior

Modern empiricism includes the study of internal superimposed behavior: the sensor/sensor, actuator/sensor, and actuator/actuator behavior of the *compound machines* described earlier. We may not be able to observe the sensor/sensor and actuator/sensor behavior that takes place within a compound machine. We may only infer the existence of this kind of behavior by observing the outward behavior of the machine. For example, the performance of a compound machine may improve by establishing a sequence of sensor/sensor states that correspond to the actual sequence of sensor states that occur in the environment as part of a line of behavior. We can assume that the sensor/sensor behavior exists if the line of behavior continues without external sensor states. A machine may seem to act like a *thinking being* if it continues to act as if it were being given a set of commands when in fact no commands are being given. This *thinking behavior* can come about because of the sensor/sensor unit in a compound machine.

Behavior of intermediate variables

The intermediate variables in a duplex and universal controller, and the intermediate variables in a direct network carry out lines of behavior that may contribute to the success of their units. What would we see and learn if we tapped into these abstract states? Could the activities of the intermediate variables create something like the unconscious mind?

Practical knowledge

An empirical machine is a mirror that reflects the laws of nature, geometries, structures, arrangements, and organization in its surroundings. What about a robot that learns to move around in rocks piled up on the beach by the ocean waves? Is there knowledge or intelligence in this rock pile? Obviously not. Has the empirical robot acquired a scientific or fundamental knowledge by learning to wend its way around a rock pile? Maybe not. But if an empirical robot is to navigate successfully within a pile of rocks, it has to discover the unique features in its environment that permit it to move without getting lost or stuck. So, modern empiricism is not the study of the knowledge we learn in school. It is the study of *practical knowledge* that leads to successful, ongoing, self-sustaining behavior of specific machines.

Private knowledge

If the environment of an empirical machine consists mainly of human operators, the behavior of the machine will reflect the knowledge, desires, and mandates of its human operators. When an empirical machine discovers its own way of dealing with its environment, it may gain some wisdom that its human operators do not have. For example, a six-legged robot may learn to walk by itself. We may find it difficult to understand how to walk on six legs since we have only two legs. How a six-legged robot moves about in difficult terrain may be its own secret. Thus, an empirical system may contain a great deal of *private knowledge* that is not available to its human operators.

The source of modern empiricism

The study of modern empiricism includes the study of living systems, such as those found in the study of biological evolution, linguistics, economics, and sociology. These subjects provide many insights into the operation of empirical systems such as:

- The study of *biological evolution* provides the concept of selection by trial and incorporation of success.
- *Sociology* provides the concept of a group of units held together by some intrinsic bonding force.
- *Business* provides the concept of profit and loss, accounting, and the survival and existence of nonliving entities such as a business enterprise.
- The study of *linguistics* provides examples of empirical systems that discover and incorporate new symbols, words, and phrases according to how successful they are in expressing the knowledge of a culture.
- *Human psychology* provides the concept of bonding through the allocation and assumption of obligations that hold people together.
- *Economics* teaches us how parallel and distributed supply and demand systems can work together to provide useful solutions to complex economic problems.

Some of these subjects do not fall within the usual domain of the cognitive sciences. Yet a concern common to all of these subjects is the spontaneous creation of new and useful entities. *Perception,* which is the cornerstone of cognition, is only a part of the whole process of empirical control. An empirical system also must act to test the validity of a perceived state. Yet, is recognition the act of perceiving?

Contributions of modern empiricism

Concepts in modern empiricism also provide insights into other fields of study, as shown in the following examples:

- The *principle of noninterference* suggests a reason for the vast differences among biological species. Species may attempt (unknowingly) to establish sources of existence that are independent from the sources of other species.
- The potential for the division of labor among empirical machines suggests existence of *behavioral bonding* in living groups.
- The *law of organization* suggests that business should make a conscious effort to determine the organization of their markets and organize their marketing efforts accordingly.
- The *law of equilibrium* suggests that a language reflects the knowledge embodied in its society.
- The concept of *behavioral existence* suggests that our sense of identity is based upon our familiarity with our surroundings.
- The *empirical algorithm* suggests that economic units, such as individuals and businesses, are empirical entities that reflect the knowledge and technology of the society in which they exist.
- The study of *machine behavior* provides many insights into animal behavior.
- *Discontinuous controlled machines* provide a simple model that shows the relationship between information and order.

Modern empiricism is not limited to the study of human and robot perception, learning, and behavior. Modern empiricism applies to the study of all natural and synthetic systems that seem to grow and act with a will of their own, and produce behavior that is based upon their own experience of what is possible and what is impossible.

The future of modern empiricism

Science has created more opportunities for us to overcome problems, reduce hardships, and understand our world. It is based upon a belief that all events have a cause that can be understood if they are measured accurately, analyzed carefully, and synthesized correctly. Empirical machines are nonliving beings similar to living beings. By designing, building, and testing empirical controllers, we may be better able to discuss human behavior and the behavior of other empirical systems. Many important features of human behavior are directly related to modern empiricism.

Accomplishments in the study of modern empiricism

What are the useful ideas generated by the study of modern empiricism?

- It explains the *spontaneous creation of self-sustaining behavior*.
- It shows that successful behavior is a sequence of *forecasts* of what is possible.
- It describes a machine that acts according to its own *volition*.
- It provides many useful examples that define *successful behavior* — completing what we set out to do.

- It suggests a *philosophy of education* based upon creating an environment in which students are more likely to discover and incorporate what we want them to learn.
- It looks deeper into the meaning of *organization* — sorting out what can be separate from what must be together.
- It discusses the concept of *hierarchy* that is inherent in all systems that consist of assemblies, subassemblies and priorities. Understanding how these systems are put together may help us deal more effectively with them.
- It proposes for the first time the concept of *bonding among machines*, which also holds people together through shared variables, division of labor, and effective intercommunication.
- It clarifies the *source of behavior* as the *modulus* of a machine.
- It demonstrates that *chase/escape behavior* is the basis of most dynamic control systems.
- It relates *social behavior* to the actions that keep members of a group together.
- It illuminates *politics* as the formation of dominant groups.
- It defines *behavioral existence* — being able to act according to what we know.
- It presents *behavioral coexistence* — shows how two or more controllers can operate on the same problem at the same time more effectively in some cases than one controller acting alone.
- It explains what it means to be human — to set goals. *Goal-setting behavior* may be unique human behavior.
- It distinguishes between *interaction* and *interference*. Interaction uses and reinforces behavioral skills. Interference stops behavior and reduces confidence levels. Successful interaction is shown to be a learned behavior.
- It compares *a priori* versus *a posteriori* control. It shows the many advantages in establishing behavior after a system is put into use.
- It contrasts *mind* and *brain behavior*. The mind manipulates abstract elements, the brain manipulates physical elements.
- It explains the difference between *information* and *order*. Information is the measure of the degree of complexity of a problem. Order is a measure of how this complexity can be expressed with a minimum number of variables.

These accomplishments extend the narrow field of machine control system design into the larger concerns of human behavior and modern empiricism.

Action needed in the study of modern empiricism

What steps need to be taken to move forward the study of modern empiricism?

- First, empirical control systems of significant size need to be built and tested. Much new knowledge will be gained by this activity.

- We need also to see how we can apply empirical principles to business, government, and international issues. Can we use the intrinsically competitive relation between local, state, and federal government to extract better *government services*? Is it possible to deal rigorously with the subject of the organization of other human groups?
- A considerable effort should be made to apply empirical control to *vision systems*. The inherent ability of an empirical controller to group variables, classify circumstances, and produce results according to different categories makes it a natural pattern recognition system. How can we apply the principles and designs of empirical control to this field? Is vision based entirely upon predetermined algorithms, or do we learn to see through an empirical process?
- An economic model based upon the empirical algorithm and the CE cell has a great deal of potential in helping us understand what is happening to our wealth/material (economic) systems. An empirical control system provides a small-scale model of an economic system. Can we identify the empirical aspects of economic behavior in contrast to the mathematical relations embodied in the *supply and demand analysis*?
- Applications for empirical control in the field of *mental health* may be far away. However, some questions should be asked: Do *brain waves* have anything to do with the complex timing relations in an empirical controller? We may benefit from looking for regulators in the brain that perform the other supply, housekeeping, or service support functions needed in an empirical controller such as latches, disconnects, time delays, ramps, and brake circuits. Is the brain a monolithic, duplex, or universal system?

These questions need to be addressed by people working within the specific disciplines that deal with these problems.

Questions raised by the study of empirical control

After building large empirical systems, we will be faced with unique training or education tasks. The following is a list of questions that come to mind at this time:

- Can we develop effective methods of teaching complex empirical systems to do what we want them to do?
- What other new and useful tasks can be handled by empirical systems?
- Can we learn more about the creation and use of *speech* by experimenting with speech communications among empirical machines?
- Can we create a *thinking machines* by using the sensor/sensor, actuator/sensor systems to generate internal states and internal lines of behavior?
- Will we find out more about the reason for *sleep*? Is sleep an attempt by the brain to fulfill the fundamental requirement of maintaining a closed line of behavior while meeting the need of the body for rest?
- Will we understand more about depression when we look at systems that attempt to hold too many variables constant?

- We need to learn more about the problem of timing in large superimposed empirical systems. Something has to keep systems with varying time delays in close synchronism. Networks that normally work smoothly together will create a cacophony of jumbled data if units get a little out of synchronism by failures in the cycle time circuits. Could this be related to *schizophrenia*?
- Will the concepts of dominance, cooperation, competition, and isolation help us deal more effectively with real *social systems*?
- Perhaps one of the most important insights developed in the study of empirical control is the mechanism of *bonding* — the creation of intertwined lines of behavior among empirical machines that act like covalent bonds in chemistry, where electrons become intertwined among two or more molecules, thus holding them together. Is there any relation between this behavioral bonding and the attachment and love, which is an important part of human relations?
- Can people develop more *successful interpersonal relationships* by understanding the concept of division of labor and bonding based upon each individual supplying some behavior that another person needs?
- There may be many cycle timing circuits that activate different sets of empirical subunits within the brain. When these subunits have an intimate connection to sensors (see Part IV), different sets are made active by certain phase relations between these timing signals, and other set of units are made active by other phase relations. Thus, changing the *phase of the cycle timing signals* by a very small amount will cause large changes in behavior. Could this effect be related to the changes in *brain waves* or changes in the phase relations of brain waves that occur with changes in behavior in living beings, such as at the onset of sleep, hypnosis, or personality changes in a person with multiple personalities?
- Is there a *single ramp cell* or set of ramp cells in the brain of living beings that is used to interrogate all of the memory cells in the brain? If this ramp cell were deactivated for some reason, the organism would appear to be in a *coma*. That is to say, the organism would not be able to produce any responses to external conditions. If the ramp system can be identified, some way may be found to reactivate it to reverse the effects of a coma in living beings.
- Are *animal muscles* sigma actuators that integrate the outputs of many separate control units? Do the sigma actuators in animals produce a *force output, compliance* (stiffness) *output*, or a *position output*?
- The different types of memory cells shown in this book appear to act somewhat like *neurons* with their ability to store information and make connects between other elements. However, could an individual neuron be a matrix of memory cell elements, giving it the ability to establish a functional rather than an absolute relationship with other neurons?
- Could *mitochondria* serve as matrix memory elements within an individual neuron? Could they provide more or less energy to specific locations on the surface of their synapse in a predetermined or empirical manner?

- Are some living brains or nervous systems made up of absolute memory cells instead of conditional memory cells? The existence of a brain that operates without brain waves may indicate that it is made up of absolute memory cells, which do not need an interrogating ramp signal.
- Are there two kinds of brain waves? One type could be ramp signals and the other type could be the cycle start signals. Could a ramp signal also be a cycle start signal?
- The logic operation used by all the conditional multivariable matrices in this study is based upon finding the output connected to the set of latched memory cells in which the memory cell with the lowest sensitivity is higher any in any other set. This is similar to the concept of a chain being as strong as its weakest link. Is this criteria used elsewhere? Is a family, club, community, or society as strong as its weakest members?

These questions may be addressed in the most effective manner by people working in the fields of neurobiology, psychology and sociology.

Conclusion

We are at the beginning of a new science of modern empiricism that combines the studies of organization, behavior, machine control, social groups, the flow and exchange of assets, and learned control. These studies are based upon the empirical algorithm, which directs a system in this way: "Continue to do what can be done, and stop doing what cannot be done." This forces empirical systems to organize themselves in an appropriate manner, to act in desirable ways, and to discover new ways of acting successfully.

These empirical systems gather information about the physical relations in their surroundings rather than operating according to predetermined relations that originate in the minds of their designers. These empirical systems keep records of their successes using tangible entities like assets in an economic system or the sensitivity values in its CE cells in an empirical controller. An empirical machine does not attempt to solve equations. Its behavior comes from its own experience about what input/output relations can and cannot happen *a posteriori* at its own sensors and actuators. This knowledge is gained only by testing possible actions for each sensed condition in the real world.

An empirical machine does not receive, classify, and store information in a passive manner. It tests and probes the world around it, like a living being, to find out what is possible and what is impossible. An empirical machine is *on trial* like every other living being. It may cease to exist if it fails to produce doable behavior.

The definition of control is to check or verify by comparison with a standard. An empirical controller does this by comparing the action it calls for to the action that is actually carried out. This is an active and potentially autonomous activity that can be carried out by an empirical machine by itself. So the main concern of modern empiricism is the spontaneous creation of self-sustaining behavior.

We are at the beginning of a new science of spontaneously learned control. It provides the opportunity to build experimental machines and to test, measure, and record the results produced by these machines in concrete engineering terms. Modern empiricism is also the study of a new kind of being: a *nonliving organism* that has the potential to create many new and useful ways of behaving. The cost of this *inorganic being* is determined primarily by the number of empirical memory cells it contains and the time and effort it takes people to teach it useful behavior. Thus, its cost is related to the complexity of the behavior it can produce.

Yet, an empirical machine can teach itself new behavior and thus acquire knowledge that may be different from anything ever created by human operators. If it is shut off or destroyed, some knowledge, understanding, or behavior that has never been uncovered before may be lost and never found again. Thus, an empirical machine has the same unique characteristic of a living being. An empirical machine has the possibility of being exceptional, of being the first of a new breed, of being a uniquely better creation, providing solutions to problems that have never been solved.

Science has helped us achieve more of our goals than any other doctrine invented by humankind. Modern empiricism is a science with the potential to help us deal with our basic need to live within our environment and with other people. It may help us understand what it means to exist, survive, and succeed as individuals in a vast network of other individuals who are also trying to do the same.

Epilogue

This epilogue is written for the reader who has finished some or all of this book and wishes to find out more about its background and preparation. Something must have excited and motivated the author to think, read, learn, and write about this subject. There must have been something important and vital in this subject that satisfied a need. A question floated around in my head as a young person that still floats around the head of many young people today: How do plants and animals come into existence? Like many people, I imagined that they are created in the same way that someone might build a house or paint a picture. Then I learned about Darwin, evolution, and natural selection. I was overwhelmed by the idea that new living plants and animals could appear without a designer and builder and continue to exist, change, and even improve simply because they could survive and reproduce. Later I asked myself an important question: Can behavior develop spontaneously and continue to improve because it can survive and repeat?

I studied engineering, physics and American literature in college. These subjects provided a good background for the effort that led to the writing of this book. My father changed the course of my life when he brought home a book by W. Ross Ashby, *A Design for the Brain* while I was on summer vacation in college. This book deals with behavior in concrete, measurable, engineering terms. It left me wondering if I also could design a machine that could learn from experience.

After reading the basic textbooks of psychology, sociology, anthropology, economics, and biological evolution, it occurred to me that all these natural systems were tied together by a common principle: Organisms find ways to reduce interferences. They choose different food supplies, ways of behaving, and different territories in which to live, to find successful ways of maintaining their existence without interferences from other organisms. The law of noninterference is based upon the hypothesis that behavior changes to reduce these interferences.

This suggested that entities such as cells, species, families, social groups and cities exist to simplify the problem of existence. This led to the development of the law of organization, which says that natural systems will attempt to separate or divide themselves into the smallest units possible to establish and maintain their existence. This is the second of several organizing principles used throughout the book used to produce the design of the natural computer.

Since my main concern is machine behavior, it was necessary to define the units of behavior, show how they can stand alone and preserve themselves, show how they can be grouped together, and show what holds them together. A machine had to be designed that could divide itself into behavioral parts that can change and disappear, while other parts are preserved because of their ability to be separate. Also, the machine had to show how the behavioral units can join with other units to form new behavioral entities.

So my first concern was to design a machine that could produce behavior at some primitive level. The first design consisted of a simple input/output (stimulus/response) matrix. This unit could respond in a unique way to unique inputs and thus produce a rudimentary form of behavior. By 1960 it was obvious that this responsive machine could learn new behavior by itself if it could produce an output under certain conditions and then measure if that output did or did not occur as called for within a given period. If the output occurs, the machine could increase the likelihood of selecting that behavior should those conditions occur again. If the output does not occur, then it is necessary for the machine to reduce the likelihood of selecting that action again and increase the likelihood of selecting the actual behavior instead. The behavior of the machine could evolve toward whatever behavior is possible and move away from whatever behavior is difficult or impossible. This became the basis of the empirical algorithm. This created a responsive empirical machine that could select and maintain a particular kind of behavior by itself through a process called transform, action, or operant conditioning.

There had to be something stronger than the machine that allowed the machine to produce certain behavior and prohibit it from producing other behavior. Otherwise, the machine would do anything it wanted to do despite teaching or environmental influences. The physical environment and machine operators can be made to restrict the behavior of a machine. This became the basis of the law of dominance.

This lead to the law of equilibrium, which says that a responsive control system (in this case an empirical controller) cannot exist by itself. It must be connected to another control system, an environment, or another responsive machine. An evaluation of the performance of the machine must be based upon how successfully the pair work together. Thus, the behavior of the empirical machine will learn to reflect the natural relations in that system.

The connected systems must provide a constant supply of new information to each other. Without these connections, a responsive controller will stop. Its behavior will cease to exist. This became the basis of the law of closure, which says that a system must have a response for each input to keep "running."

Apparently there were two ways of organizing systems having two or more input and/or output variables. In one form of organization, the variables could form a single unitary system. In the other form, the set of variables could form a set of separate diverse systems. In a unitary system, all of the variables are combined to produce a single result. In diverse systems, each unit has its own separate result. A useful system had to be able to deal with both situations and organize itself into one or the other according to the organization of its environment. This became another part of the law of organization.

In 1968 I wrote an extensive (unpublished) paper called *Empirical Control*. However, the system proposed in the paper required an astronomical number of elements if it were to be either unitary or diverse. It was obviously impractical. A breakthrough came in 1975 when I finally stumbled upon a way to reduce the number of memory elements by using two matrices, an input (sensor) matrix and an output (actuator) matrix of the duplex system. I found that the input matrix could have two or more input variables and one output variable and the output matrix could have one input variable and two or more output variables. This meant that the input matrix could be connected to the output matrix by means of a single intermediate variable. This duplex system could handle multiple input and multiple output variables in a unitary manner and could exclude some input variables at different times in an synchronous diverse manner. But this design still could not handle two or more diverse lines of behavior that could happen asynchronously. This problem was solved by connecting a monolithic input matrix to a monolithic output matrix, forming a universal duplex controller that could deal with many asynchronous diverse input and output variables.

In 1980 I realized that many duplex units could be connected (superimposed) to an assorted set of input/output variables. This allowed the units to divide themselves into separate parts or work with other units. I also realized that a behavioral bonding can occur when two or more systems share common variables in their lines of behavior. This is the basis of the law of bonding, which implies that there is a behavioral force that can hold systems together to create larger behavioral units. This allows a system to grow larger, establish a division of labor with other systems, or break apart (fragment) into smaller groupings.

In 1985 I realized how a digitized system of aggregate variables could make generalizations about the values of a variable and provide a much higher resolution of the sensor and actuator variables with far fewer empirical cells than an undigitized matrix. This technique essentially eliminated any reservations I had about the ability of an empirical controller to deal with real, high-precision control tasks.

We know that the timing of response to stimulus is often critical in producing a successful action, such as hitting a tennis ball. The timing of behavior is part of the physical domain and cannot be ignored or violated. However, systems having different timed relations between input and output can be superimposed. Then the system with the correct time relation can come to dominate the stimulus/response relation. This is the basis of the law of timing. The problem of assuring the correct timing of behavior is solved by providing a bigger control system having a greater variety of behavior.

The only problem holding back the design of a complete empirical controller was a detailed circuit that can carry out the functions required by the ten laws of behavior. It was necessary to answer the question: Exactly how is the CE cell sensitivity changing process carried out? The logarithmic subtraction mechanism and the concept of success rate provided a concrete process for carrying out the learning algorithm. Once these efforts were complete, the book was easy to write. It all just fell into place as if it were an artifact of nature. This was very encouraging, because when things fit together so nicely, one feels that one is on the right track.

A lesson I learned while working within the discipline of the scientific method is that we learn something about a problem every time we measure something about the problem. The discontinuous state ontology establishes a framework in which every concept in the study of empirical control can be tested by the measurement of a state, variable, and a value at specified moments in time. The circuits define the empirical process in great detail. They are the instruments through which the secrets of the empirical process can be revealed.

In conclusion, the book on empirical control has provided the concepts and technology needed to design, build, and understand systems that can spontaneously create self-sustaining behavior. The type of learning and empirical behavior discussed in the book may be sufficient to account for most animal behavior, and provide the sophistication and complexity of control needed for many machine and robot controllers. I hope that it also provides some understanding of empirical systems in other fields such as biology, linguistics, sociology, economics, and the animal brain.

Appendix A
Value Elements as Data

There is a more practical measure of intellectual substance than logarithmic information (the number of variables needed to represent a given variety in a given number system). It is the number of *value elements* required to record or store intellectual material. The number of value elements required to record data is roughly independent of the number system used in number systems between binary and pentary, while the logarithmic information needed to store a given amount of data varies greatly in these number systems.

Binary memory register

A value element is the same in all number systems. A computer memory register uses one gate to represent each value element regardless of the number system used by the register. For example, a binary register has the two value elements of *0* and *1*, and uses one gate for each value element, as shown in the binary memory register in Figure A.1.

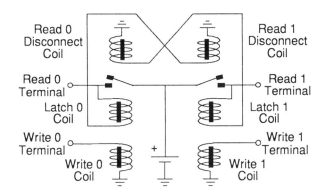

Figure A.1. A *binary computer memory register* uses two gates (latching relays) to store a one or a zero.

A *0* or a *1* is recorded according to which gate is made and held conductive by its latch coil. Like the push-button switch shown in Figure 30.2, a separate circuit is needed to shut-off each gate when another gate is energized. If one gate conducts, the other is disconnected. Thus, each gate is not a separate variable, since the value of each gate may be influenced by the value of another gate.

Trinary memory register

A trinary register has the three values elements of *0*, *1*, and *2*. A trinary memory requires three gates and three shut-off circuits, as shown in Figure A.2. A decimal register has the 10 value elements of *0* through *9*. A decade memory register needs 10 gates and 10 shut off circuits to shut off a latched gate whenever a new gate is turned on.

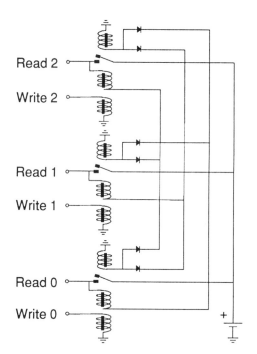

Figure A.1. A *trinary computer memory register* uses three gates (latching relays) to store a zero, one, or a two.

Absolute memory elements

Each gate in a memory register is an *absolute memory element,* though each gate can assume two positions (open and closed). Each gate is absolute, not binary, because each gate has only one output (read) terminal in which a voltage is either present or not present. The binary memory register shown has two output terminals in which a voltage can be present at either terminal, but the voltage can be present at only at only one terminal at a time. Thus, both terminals taken together represent the two binary value elements of a binary register, and each terminal is of equal purport like the output terminals of a single-pole, double-throw switch. A trinary memory register is equivalent to a single-pole, triple-throw switch.

Value elements as a measure of data

The memory size and amount of intellectual material that can be stored on a random access memory (RAM) chip is determined by the *number of value elements* (gates) etched on its surface. A typical 256 kilo "bit" chip has $n = 256{,}144$ binary registers and each register has $x = 2$ gates per register. Thus, this chip contains $x \times n = 524{,}288$ gates or absolute value elements. How these value elements are used determines how data is stored on the chip.

Each alphanumeric character in the ASCII code used in most computers has a variety of $2^8 = 256$ different possibilities. That is to say, the computer can identify 256 different kinds of characters if every combination of values of all of the eight binary variables results in a unique character. (The twenty-six letters, their capitals, and other keys on a typical keyboard use about 128 of these 256 combinations, requiring only seven binary variables. The eighth binary variables is used for a parity check in some codes.) Since most computers use 8 binary registers per alphanumeric character, called an 8-bit byte, they can store 32,768 alphanumeric characters per chip. So it may be called a 32k *byte* chip.

If these 524,288 gates were used to make 4-value (quadrary) variables, each quadrary memory register would contain 4 gates. Then 131,072 quadrary registers could be placed upon a chip with 524,288 gates. If four quadrary registers were used for each character, $4^4 = 256$ different characters could still be produced and the 131,072 quadrary registers could still store 32,768 of the 256 alphanumeric characters. Thus, the quadrary number system is just as good as the binary number system to store information in a digital computer, and no other number system is a good as these two expect for the trinary number system, which is significantly better than both (see trinary digitized systems).

The number of variables has been shown to be a measure of information. In the binary system, there would be 256,114 units (binades or bits) of information, and in the quadrary system there would be 131,072 units (quadrades) of information, although both store 32,768 eight-bit-byte alphanumeric characters. Therefore, the 524,288 gates or absolute value elements accurately defines the number of characters that can be stored on a given chip in both number systems.

All of our number, information, and control systems are made up of these *value elements,* while only binary number systems and binary computers are made up of binary registers (bits). Therefore, the value element is a much more natural and useful unit of data than the binary register (bit) for the following reasons:

1. The value element cannot be broken down into a smaller subunit.
2. A value element is independent of the kind of number system in which it is used.
3. The number of value elements corresponds exactly to the number of gates in a computer or electronic counter, regardless of the number system employed (within limits), and corresponds to the number of AP cells, CP cells, AE cells or CE cells in a responsive controller.
4. The total number of value elements is nearly the same for a given amount of variety in different number systems from base 2 to as high as base 5. (See Table 30.2. Characteristics of number systems required to produce a variety of 1000.)
5. The total number of value elements corresponds closely to the number of characters that can be stored in a memory matrix if a base 2, through 5 number system is used.

In conclusion, the binary and quadrary number systems can produce the same variety with the same number of absolute value elements, and they can produce more variety with fewer value elements than any other number system except the trinary number system. The absolute value element is the basic unit of data in all information systems. It is the unit of value of a variable. It is a gate in a computer memory and the basic memory cell in a discontinuous responsive controller. It is a flip or a flop in a binary system. It is one of the ten positions on a decade dial. It determines the complexity and therefore cost of a computer memory chip. In most cases, number of value elements represent *data* better than number of variables.

Appendix B
Problems with the Term "Bit"

There is a problem with the use of the term *bit*. It is sometimes used to mean a binary variable, and other times it is used to mean a binary value element. In some cases the terms bits, binints, and binades can be used interchangeably without creating fallacies. For example, if one refers to the transmission of serial binary data from one point to another, one refers to the number of "bits per second" or baud rate.

It is true that a sequence of binary integers (binints) appears to pass by a point along the data path. However, each binint is a value of a binary register; so one can also imagine a sequence of binary registers passing a point. One can think therefore of a binary message state in two ways: as a binary register (binade) passing a point or as a binary integer (binint) passing a point. Either way of thinking is correct.

However, there are cases where the failure to distinguish between binary variables (binades) and binary value elements (binints) can be misleading. Consider a 16-position switch. It is a 16-value (hexadecimal) variable. Though each contact can only have two states (made or not made), each contact is not a true binary variable because each is not independent of the other contacts. Only one contact can be made at a time because each contact is a value element of the device variable. So if one contact is made, no other contacts can be made. Thus, each contact is an *absolute value element*. If one were to consider each terminal as a binary register, calculations for the variety of the switch would be incorrect.

Also, if one were looking at the output terminals of an "eight-bit" parallel data bus, each terminal would still have two significant states (on/off, high/low, +/-, or 1/0). However, each combination of all eight terminals determines the meaning of the state of the bus. In this case, each terminal is an *independent* two-value variable. Each can take whatever value is given to it without forcing a change in another place in the bus. It could be called an *eight-binade bus* instead of an eight-bit bus to avoid confusion.

Notes

1 . W. Ross Ashby, *Design for a Brain*, 2nd ed. (New York: John Wiley & Sons, 1960), p. 100.

2 . W. Ross Ashby, *An Introduction to Cybernetics* (New York: John Wiley & Sons., 1963), pp. 119–126.

3 . *C.R.C. Handbook of Standard Mathematical Tables,* 12th ed. (Cleveland: Chemical Rubber Publishing Co., 1959), 12th edition, Cleveland, p. 387.

4 . *Ibid.*

5 . Robert A. Brown, "Designing Custom Robots for In-Plant Use," *Conference Proceedings, Volume 2, 13th International Symposium on Industrial Robots and Robots 7* (Dearborn, Michigan: Society of Manufacturing Engineers, 1983), pp. 20.45–20.56.

6 . Robert A. Brown, "Maximum Demand Register Using Constant Speed Drive for Periodic Subtractions Proportional to Momentary Reading of Demand Indicator," *U.S Patent # 3,325,732,* (Washington, D.C.: United States Patent Office, 1967).

7 . McClelland, Rumelhart, et al., *Parallel Distributed Processing, Vol. 1* (Cambridge: MIT Press, 1986) p. 36.

8 . Stanford Goldman, *Information Theory* (New York: Prentice-Hall, 1953) p. 4.

Index